D1672218

Biological Foundations and Origin of Syntax

Strüngmann Forum Reports

Julia Lupp, series editor

The Ernst Strüngmann Forum is made possible through
the generous support of the Ernst Strüngmann Foundation,
inaugurated by Dr. Andreas and Dr. Thomas Strüngmann.

This Forum was supported by funds from the German
Science Foundation and the European Science Foundation
EUROCORES Programme OMLL, from the EC Sixth Framework
Programme (Contract no. ERAS-CT-2003-980409).

Biological Foundations and Origin of Syntax

Edited by

Derek Bickerton and Eörs Szathmáry

Program Advisory Committee:
Derek Bickerton, Balázs Gulyás, Simon Kirby,
Luc Steels, and Eörs Szathmáry

The MIT Press

Cambridge, Massachusetts
London, England

Series Editor: J. Lupp
Assistant Editor: M. Turner
Photographs: U. Dettmar
Typeset by BerlinScienceWorks

MIT Press books may be purchased at special quantity discounts for business or sales promotional use. For information, please email special_sales@mitpress.mit.edu or write to Special Sales Department, The MIT Press, 55 Hayward Street, Cambridge, MA 02142.

The book was set in TimesNewRoman and Arial.
Printed and bound in the United States of America.

Library of Congress Cataloging-in-Publication Data

Ernst Strüngmann Forum (2008 : Frankfurt, Germany)
Biological foundations and origin of syntax / edited by Derek Bickerton and Eörs Szathmáry.
p. cm. — (Strüngmann Forum reports)
Forum held July 13–18, 2008, in Frankfurt, Germany.
Includes bibliographical references and index.
ISBN 978-0-262-01356-7 (hardcover : alk. paper)
1. Grammar, Comparative and general—Syntax—Congresses. 2. Biolinguistics—Congresses. I. Bickerton, Derek. II. Szathmáry, Eörs. III. Title.
P291.E75 2009
612.8'2336—dc22

2009023478

Contents

Brain

The Ernst Strüngmann Forum

Founded on the tenets of scientific independence and the inquisitive nature of the human mind, the Ernst Strüngmann Forum is dedicated to the continual expansion of knowledge. Through its innovative communication process, the Ernst Strüngmann Forum provides a creative environment within which experts scrutinize high-priority issues from multiple vantage points.

This process begins with the identification of themes. By nature, a theme constitutes a problem area that transcends classic disciplinary boundaries. It is of high-priority interest, requiring concentrated, multidisciplinary input to address the issues involved. Proposals are received from leading scientists active in their field and are selected by an independent Scientific Advisory Board. Once approved, a steering committee is convened to refine the scientific parameters of the proposal and select the participants. Approximately one year later, a focal meeting is held to which circa forty experts are invited.

Planning for this Forum began in 2003 and, much like a good bottle of wine, took a while to mature. In August, 2007, the steering committee was ultimately convened to identify key issues for debate and select the participants for the focal meeting, which was held in Frankfurt am Main, Germany, from July 13–18, 2008.

The activities and discourse involved in a Forum begin well before participants arrive in Frankfurt and conclude with the publication of this volume. Throughout each stage, focused dialogue is the means by which participants examine the issues anew. Often, this requires relinquishing long-established ideas and overcoming disciplinary idiosyncrasies which might otherwise inhibit joint examination. However, when this is accomplished, a unique synergism results and new insights emerge.

This volume conveys the synergy that arose from a group of diverse experts, each of whom assumed an active role, and is comprised of two types of contributions. The first provides background information on key aspects of the overall theme. These chapters have been extensively reviewed and revised to provide current understanding on these topics. The second (Chapters 7, 10, 14, and 18) summarizes the extensive discussions that transpired. These chapters should not be viewed as consensus documents nor are they proceedings; they transfer the essence of the discussions, expose the open questions that remain, and highlight areas for future research.

An endeavor of this kind creates its own unique group dynamics and puts demands on everyone who participates. Each invitee contributed not only their time and congenial personality, but a willingness to probe beyond that which is evident, and I wish to extend my sincere gratitude to all. Special thanks goes to the steering committee (Derek Bickerton, Balázs Gulyás, Simon Kirby, Luc Steels, and Eörs Szathmáry), the authors of the background papers, the reviewers of the papers, and the moderators of the individual working groups

(Frederick Newmeyer, Jim Hurford, Csaba Pléh, and Luc Steels). To draft a report during the Forum and bring it to its final form is no simple matter, and for their efforts, we are especially grateful to Maggie Tallerman, Szabolcs Számadó, Anna Fedor, and Herbert Jaeger. Most importantly, I wish to extend my appreciation to the chairpersons, Derek Bickerton and Eörs Szathmáry, whose belief in and support for this unique process overcame, at times, seemingly insurmountable hurdles.

A communication process of this nature relies on institutional stability and an environment that encourages free thought. Through the generous support of the Ernst Strüngmann Foundation, established by Dr. Andreas and Dr. Thomas Strüngmann in honor of their father, the Ernst Strüngmann Forum is able to conduct its work in the service of science. The involvement of the Scientific Advisory Board ensures the scientific independence of the Forum, and its work is gratefully acknowledged. This Forum was also supported by funds from the European Science Foundation EUROCORES Programme OMLL, from the EC Sixth Framework Programme (Contract no. ERAS-CT-2003-980409), and the German Science Foundation.

On behalf of all involved, I hope that this volume is successful in conveying a sense of this invigorating exercise and will assist in advancing the enquiry into the origins and biological foundations of syntax.

Julia Lupp, Program Director
Ernst Strüngmann Forum
Frankfurt Institute for Advanced Studies (FIAS)
Ruth-Moufang-Str. 1, 60438 Frankfurt am Main, Germany
http://fias.uni-frankfurt.de/esforum/

List of Contributors

Andrea Baronchelli Departament de Física i Enginyeria Nuclear, Universitat Politècnica de Catalunya, Campus Nord, Mòdul B4, 08034 Barcelona, Spain

Derek Bickerton Department of Linguistics, University of Hawaii at Manoa, 68–244 Crozier Loop, Waialua, Hawaii 96791, U.S.A.

Dorothy V. M. Bishop Department of Experimental Psychology, University of Oxford, South Parks Road, Oxford OX1 3UD, U.K.

Denis Bouchard Department de Linguistique, Université du Québec à Montréal, C.P. 8888, Suc. Centre-Ville, Montreal H3C 3P8, Canada

Jens Brauer Department of Neuropsychology, MPI for Human Cognitive and Brain Science, Stephanstr. 1a, 04103 Leipzig, Germany

Ted Briscoe Computer Laboratory, University of Cambridge, JJ Thomson Avenue, Cambridge CB3 OFD, U.K.

David Caplan Harvard Medical School, Neuropsychology Lab, 175 Cambridge St., Suite 340, Boston, MA 02114, U.S.A.

Nick Chater Division of Psychology and Language Sciences, University College London, London, WC1E 6BT, U.K.

Morten H. Christiansen Department of Psychology, Cornell University, 228 Uris Hall, Ithaca, NY 14853, U.S.A.

Terrence W. Deacon Department of Biological Anthropology and Neuroscience, Helen Wills Institute, University of California at Berkeley, 329 Kroeber Hall, Berkeley, CA 94720–3710, U.S.A.

Francesco d'Errico UMR 5199 PACEA, Bat. B18, Université Bordeaux 1, Avenue des Facultés, 33405 Talence, France

Anna Fedor ELTE TTK Növényrendszertani és Ökológiai Tsz, Pázmány Péter stny. 1/C, 1117 Budapest, Hungary

Julia Fischer Cognitive Ethology, German Primate Center, Kellnerweg 4, 37077 Göttingen, Germany

Angela D. Friederici Department of Neuropsychology, MPI for Human Cognitive and Brain Sciences, Stephanstr. 1a, 04103 Leipzig, Germany

T. Givón Institute of Cognitive and Decision Sciences, University of Oregon, Eugene, Oregon, U.S.A.

Thomas Griffiths Department of Psychology, University of California, 3210 Tolman Hall, Berkeley, CA 94720–1650, U.S.A.

Balázs Gulyás Department of Clinical Neuroscience, Psychiatry Section, Karolinska Institute, 17176 Stockholm, Sweden

Peter Hagoort F. C. Donders Centre for Cognitive Neuroimaging, Radboud University Nijmegen, 6500 HB Nijmegen, The Netherlands

Austin T. Hilliard Department of Physiological Science, University of California, 621 Charles E. Young Drive, South, Los Angeles, CA 90095–160606, U.S.A.

James R. Hurford Department of Linguistics and English Language, University of Edinburgh, Adam Ferguson Building, 40 George Square, Edinburgh EH8 9LL, U.K.

Péter Ittzés Collegium Budapest (Institute for Advanced Study), 2 Szentháromság utca, 1014 Budapest, Hungary

Gerhard Jäger Department of Linguistics and Literature, University of Bielefeld, 33501 Bielefeld, Germany

Herbert Jaeger School of Engineering and Science, Jacobs University, 28725 Bremen, Germany

Edith Kaan Linguistics, University of Florida, Gainesville, FL 32601, U.S.A.

Simon Kirby School of Philosophy, Psychology and Language Sciences, University of Edinburgh, Dugald Stewart Building, 3 Charles St., Edinburgh EH8 9AD, U.K.

Natalia L. Komarova Department of Mathematics, University of California, Irvine, 237 Multipurpose Science & Technology, Irvine, CA 92697–3875, U.S.A.

Tatjana Nazir Institut des Sciences Cognitives, L2C2, 69675 Bron Cedex, France

Frederick J. Newmeyer University of British Columbia, 1068 Seymour St., Vancouver, BC V6B 3M6, Canada

Kazuo Okanoya RIKEN Brain Science Institute, 2-1 Hirosawa, Wako Shi, Saitama 351-0198, Japan

Csaba Pléh Department of Cognitive Science, Budapest Technical University, St épület 3, ernelet 311, Sztoczek u. 2, 1111 Budapest, Hungary

Peter J. Richerson Department of Environmental Science and Policy, University of California, Davis, One Shields Avenue, Davis, CA 95616, U.S.A.

Luigi Rizzi Interdisciplinary Center of Cognitive Studies on Language, CISCL, University of Siena, Complesso S. Niccolò, Via Roma 56, 53100 Sienna, Italy

Wolf Singer Max Planck Institute for Brain Research, Department of Neurophysiology, Deutschordenstr. 46, 60528 Frankfurt/Main, Germany

Mark Steedman Informatics Forum 4.15, University of Edinburgh, 10 Crichton Street, Edinburgh EH8 9AB, U.K.

Luc Steels Computer Science, Free University of Brussels (VUB), AI Laboratory, Pleinlaan 2, 1050 Brussels, Belgium

Szabolcs Számadó Eotvos Lorano University, Department of Plant Taxonomy and Ecology, P′szmány Péter St. 1/c, 1017 Budapest, Hungary

Eörs Szathmáry Collegium Budapest (Institute for Advanced Study), 2 Szentháromság utca, 1014 Budapest, Hungary

Maggie Tallerman Linguistics Section, SELLL, Newcastle University, Percy Building, Newcastle upon Tyne NE1 7RU, U.K.

Jochen Triesch Frankfurt Institute for Advanced Studies, Ruth-Moufang-Str.1, 60438 Frankfurt, Germany

Stephanie A. White Department of Physiological Science, University of California, 621 Charles E. Young Drive, South, Los Angeles, CA 90095–160606, U.S.A.

Preface

If the evolution of language is "the hardest problem in science" (Christiansen and Kirby 2003), then the evolution of syntax is the hardest part of that problem. While no species other than ours has, in the wild, a communication system that in any way resembles ours, a number of species have been able to master lexical reference and engage with humans in a primitive form of protolanguage (for an overview and references, see Bickerton 2009). No member of any other species, however, has shown any capacity to acquire even a rudimentary syntax, and this capacity must therefore stand as one of the few true apomorphies of humans.

Deprived of that most useful of evolutionary tools, the comparative method, and dealing with a trait too abstract to leave any evidence in the fossil record (and only the sparsest and most ambiguous evidence in the archaeological record), how is science to proceed with the quest for the origins and biological foundations of syntax? There is one factor here that has both positive and negative consequences: the interdisciplinary nature of the inquiry.

"War is too important to be left to generals," Clemenceau said. Language is too important to be left to linguists, and even if it were not, the self-denying ordinance to which linguists subscribed at a meeting of the Linguistic Society of Paris in 1866 sufficed to keep them, with very few exceptions, out of the game prior to the appearance of three pioneering works at the beginning of the 1990s (Bickerton 1990; Pinker and Bloom 1990; Newmeyer 1991). Biologists felt stimulated by these expositions, and a few years later Maynard Smith and Szathmáry (1995) portrayed the evolution from primitive societies with protolanguage to modern societies with language as the last major transition in biological evolution.

Indeed, the evolution of language, including that of syntax, is of concern to biologists, anthropologists, and psychologists as it is a crucial part of the evolution of humans. But it is of equal concern to psychologists, neurologists, and neurobiologists as a faculty in which large areas of the brain are involved; and it is of concern as well to computer scientists and modelers, offering problems which severely test their resources but for the solution of which their discipline offers almost the only route to empirical and replicable investigations. Clearly, research in language evolution must be interdisciplinary and, equally clearly, interdisciplinary efforts offer the best prospect of arriving at a solution, even if only by using the pooled knowledge from diverse disciplines to shrink the problem space and rule out solutions which, while plausible enough within the framework of a single discipline, quickly fall victim to counterevidence from others.

Conferences on language evolution featuring contributors from diverse fields have taken place ever since the mid-1970s (Steklis et al. 1976). However, those meetings suffered from at least two drawbacks. First, the topic was simply the evolution of language; participants were free to discuss any aspect of language or language use—syntactic, semantic, lexical, phonological, pragmatic—from any perspective they chose, Second, there was no integrating structure to these conferences; people came, read papers, listened to papers by others, and left. This might give multidisciplinary exposure but there was relatively little multidisciplinary interaction. Accordingly, the negative side of interdisciplinary activities too often appeared. Each participant looked at the problem primarily or exclusively through the lens of that participant's own discipline, and each discipline came with its own baggage: its own ways of dealing with data, its own unstated assumptions and particular perspectives that came with the territory and often puzzled and mystified outsiders. Often, even the terms used in discussion had quite different meanings and connotations for members of different disciplines.

The idea for this meeting emerged in a series of conversations at the Collegium Budapest in 2002 between Derek Bickerton and Eörs Szathmáry. From those conversations came the decision to seek a different kind of venue, one that would provide true interaction between representatives of diverse fields as well as diverse approaches within each field. We originally planned to hold this meeting as a Dahlem Workshop, but due to radical changes in that institution's format, an alternative needed to be sought. It eventually took place in Frankfurt, on July 13–18 2008, under the auspices of the Ernst Strüngmann Forum, which has revived the original Dahlem tradition. At a Forum, instead of the simple presentation of a series of papers at most meetings, background papers are prepared by a substantial proportion of participants, each covering a different and substantive area of the overall field, and circulated in advance. There are no lectures or presentations; instead, the whole week is devoted to discussion, starting from the position papers but moving on to embrace every serious issue in each field. Participants were divided into four focal groups: linguistics, biology, brain sciences and computer modeling. In some sessions, groups debated by themselves, seeking to thrash out points of difference and achieve some measure of common ground; in others, members from other groups joined in and sought to merge their contributions by tackling some important issue of common interest. For many, this latter aspect of the meeting was the most valuable part of an extremely intensive week, constituting, as several said afterwards, a genuine learning experience.

This volume represents the results of our combined efforts. It includes the finalized background papers, which were modified and often extensively rewritten in light of the debate, together with reports of the discussions from each of the four specialist groups. As such, it provides a record of the first ever conference on the evolution of syntax that involved real interdisciplinary interaction. We state this, not as a boast, but as a reminder to readers that it represents the commencement rather than the completion of a task. Given the complexity of the problem, and the diversity of the methods and sources on which that problem must draw for a solution, we might have rashly hoped for, but could not honestly have expected, some stunning breakthrough that would revolutionize the field. What we got instead was steady and substantive progress in defining the overall nature of the problem, in assessing the capacities and limitations of each discipline concerned for dealing with that problem, in awareness of the missing pieces of the puzzle that will have to be assembled by future researchers, and in establishing the directions in which future inquiries must go.

The first goal in any interdisciplinary meeting is simply that of getting participants from different backgrounds to accustom themselves to each others' different cultures, methods, and tacit assumptions, and to approach positions that may sometimes appear to them bizarre or downright perverse with a genuine humility, respect, and honest desire to learn something new, rather than trying to impose their own *Zeitgeist* on others. This goal was certainly achieved in Frankfurt in full measure; all who took part were willing to listen.

There were certain areas in which members of all groups showed wide agreement. It was agreed (even by generativists who had once championed a modular "language organ") that in light of modern brain imaging techniques, neither strictly localist modules á la Fodor (1983) nor the Broca/Wernicke division-of-labor models still found in traditional neurology texts gave an accurate picture of the way in which language is organized in the brain. Nearer the probable truth is something like the "language amoeba" hypothesis of Szathmáry (2001): a wide variety of linked brain areas, perhaps involving as much as half or more of the brain, that come together to carry out language operations but which also, separately or together, discharge a variety of other functions as well. Key questions are how the microstructure of the areas involved enables the relevant neuronal networks to carry out linguistic operations in humans (as opposed to apes), and how the genes can affect the development of such neuronal networks. Needless to say, the number of cells and connections in the brain ensures that any genetic control over development must be extremely general and indirect. The unraveling of all these complex and tangled issues constitutes an awesome task, made yet more difficult by the ransom noise that the brain constantly generates; neurons may change their firing rate, but they just won't stop firing, even when they are doing nothing in particular. However, once a task has been correctly defined, one has already made a decisive step towards completing it.

Another area of agreement might seem surprising in light of many current "primate-centric" studies of language evolution (Burling 2005; Hurford 2007). Most participants felt that there were no true precursors of syntax to be found among our nearest relatives. For anything like a syntactic precursor one had to go as far afield as songbirds, whose capacities in this direction have been well known since Marler (1972). This, of course, is what is often described as "phonological syntax," the linking of sounds that do not have, either individually or collectively, any kind of referential meaning. It is also the case that the brains of songbirds are quite differently organized from those of mammals

(Avian Brain Nomenclature Consortium 2005) so any kind of homology seems intrinsically unlikely. However, the remarkable cognitive capacities of some birds, including vocal learning, tool use, logical inference, word learning from humans, etc., suggested to a number of participants that we should take much more seriously both the means by which songbirds achieve their often highly complex productions and their methods of vocal learning (another feature they share with humans). Granted, no bird has shown the crucial combination of symbolic reference and syntax, the computational capacities of species so different both ecologically and neurologically from primates surely deserves further study.

Surprisingly (or perhaps not surprisingly—civil wars are notoriously the most vicious) there was often more agreement between groups than within them. In linguistics group (Tallerman et al., Chapter 7), a considerable measure of agreement was reached over the objective properties that constituted syntax; disagreement arose as soon as ways of analyzing and, even more crucially, ways of explaining those properties came to the fore. In principle, everyone agreed that language comes from a mixture of nature and nurture, and that pure nativist or pure empiricist positions are now equally untenable. However, this does not seem to have reduced the conflict between generativists, who see the core of syntax as lying in a set of highly abstract but language-universal principles, and functionalists or constructivists who see language as mediated primarily by cultural factors and functions of general cognition. These two positions open the doors to quite different views of syntactic evolution, as Tallerman et al. make clear.

Számadó et al. (Chapter 10) considered primarily questions about the evolution and evolvability of syntax. Evolution of language, or even that of the language faculty, cannot be treated the same way as, say, the evolution of the kidney or the heart; for several reasons. One is that language itself is an adaptive system that has its own dynamics of change, even if the genetic background of language users is fixed. Second, in a social context biological fitness of the agents is crucially influenced by neighboring conspecifics who can be poorer or better language users, and this partly determines the success, hence the fitness of the focal individual. Third, in the course of the emergence of language, evolution by natural selection has probably been complemented by factors that make selection proceed more rapidly, including genetic assimilation (transformation of an initially learnt trait into a genetic one) and niche construction (where agents alter the environment to their own benefit, and that modified environment in turn becomes a selective pressure, triggering a beneficial spiral.) Consequently, much remains to be done before we can determine the relative contributions of these mechanisms to the evolution of syntax, and to find out what components of syntax can become truly innate.

In the neurological group (Fedor et al., Chapter 14), it became clear that thanks to the development of several distinct but complementary modes of brain imaging, we now have a much richer and more sophisticated view of how cognitive functions are distributed in the brain and the degree of cooperation between distinct and often distant brain areas that must accompany the simplest linguistic task. But making sense of the storm of information that imaging has unleashed is no easy task. While some propose a coherent and seemingly plausible model of how syntactic tasks are partitioned—local connections and long-distance dependencies being handled by distinct brain circuits (Friederici, Chapter 11), others find such models overly ambitious and at odds with some of the empirical data (Caplan, Chapter 12). The difficulty of resolving such disagreements prompts one to wonder if the present approach (aimed at determining what parts of the brain perform which linguistic functions, and where individual functions are located) is the only or even the best one. Ideally, one would like to see a flowchart, millisecond by millisecond, of the entire stream of relevant neural activity from the start of constructing a sentence to the completion of its utterance. It is questionable whether present imaging techniques can give the degree of spatial and temporal resolution this would require, but progress over the last two decades has been rapid enough to nurture the hope that such a goal may soon be attainable.

While neurologists have direct and immediate access to relevant data, biologists cannot run million-year experiments to see how language might evolve in some other species. This was the motivation driving the modeling group (Jaeger et al., Chapter 18). The advantage of modeling is that it forces us to think extremely clearly about our assumptions and the initial conditions that these entail. The disadvantage is that any model is only as good as its assumptions. We have the impression that modeling of syntactic evolution has thus far only set the stage for truly relevant work in the future, when really crucial questions about syntax can be asked and perhaps answered. So far we have not learned a great deal about how syntax can emerge in these models, either because they are dealing with different issues (e.g., compositionality), or because the simulated agents are equipped with faculties (such as ability to parse phrase-structure grammars), the origin of which needs to be explained in the first place. Simulated agents should have emergent symbol grounding, rather than a set of prespecified meanings, and should not be allowed to employ biologically unrealistic algorithms such as back-propagation. However, we still hope that advances made through different modeling approaches will be put together so as to test more detailed and realistic scenarios of language evolution. Without such computational models and robotic tests, a satisfactory understanding of the problem will be hard to attain.

To summarize (if it is possible to summarize so rich and varied a meeting, which will surely have repercussions throughout the field for years to come), this volume provides, for the first time, a clear view of how far we have come towards solving the problem of how syntax evolved and what infrastructure supports it. It also shows how far we still have to go and provides some vital guideposts to the routes we must follow in order to reach our goal. The importance of reaching it may not yet be fully appreciated, but it should be. Syntax lies at the very heart of what it means to be human. It is the thing, indeed the only thing that enables us to bring words together to form complex and meaningful statements. Without it, we would have no law, no science, no economics, no philosophy, no literature; all the intellectual works of which our species is so proud would collapse into meaningless word salad. That we are still unable to explain this faculty, how it evolved, what infrastructure supports it, opens up a (fortunately still unexploited) window for creationists and intelligent designers, and should therefore provide strong incentives for researchers from every relevant discipline to tackle the problem—one whose solution, a century and a half after The Origin of Species, is surely long overdue.

Background

1

Syntax for Non-syntacticians

A Brief Primer

Derek Bickerton

Abstract

Some of the most basic concepts and processes in syntax—Merge (and the hierarchical structures it creates), binding, control, movement, and empty categories (elements that are "understood" but not phonetically expressed)—are briefly and simply described and illustrated. The chapter concludes with some suggestions regarding possible avenues of approach towards a fuller understanding of how syntax is instantiated in the human mind/brain.

Introduction

For the average lay person, word order is the most significant thing about syntax. For some, it's all of syntax.

Nothing could be further from the truth. In fact you could argue that word order is an epiphenomenon, necessitated by the fact that we have only a single channel of speech, forcing words to be produced in a linear order, and that word order falls out merely from reading the terminal nodes of a hierarchical tree structure (such as that below) from left to right.

The most significant thing about syntax is its hierarchical structure. Syntactic trees are not just heuristic devices; they reflect how sentences must be constructed. Take the following example:

(1) Everyone who knows Mary says she likes Bill.

In linear order, *Mary* and *says* are adjacent and seem to show evidence of subject–verb agreement; indeed, *Mary says she likes Bill* is in itself a complete and fully grammatical sentence. In hierarchical structure, however, *Mary* and *says* are far apart from one another and not in a subject–verb relationship.

The proper relationship between *Mary* and *says* is shown when we create a simplified hierarchical tree structure for the sentence (many details omitted):

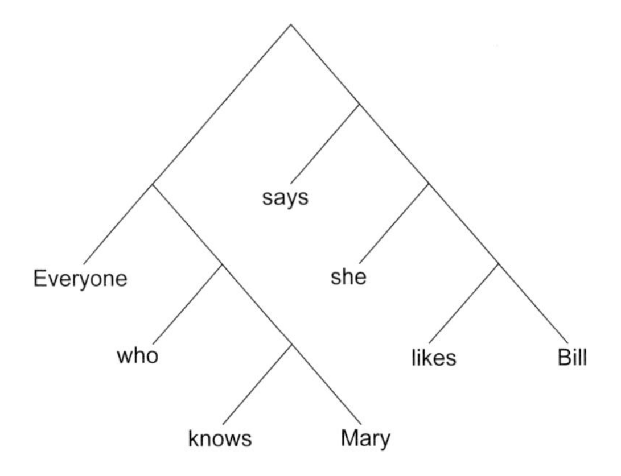

It will be noted that, as stated above, a reading of the node terminals from left to right yields the sentence described by, and analyzed in, the syntactic tree.

Merge as the Central Process in Syntax

As the above suggests, what one might intuitively describe as the "closeness" of words is illustrated by their positions in the tree. The closest relationship is that known as "sisterhood"—one between two words exclusively dominated by a single node, such as *knows* and *Mary*, or *likes* and *Bill*. Although *Mary* and *says* have a linear adjacency similar to that between *knows* and *Mary*, the relationship between them is remote, since the only node that dominates both is the node that dominates the entire sentence. The degree of closeness is further shown by the fact that words cannot normally be inserted between sisters: we can say, *She obviously likes Bill* or *She likes Bill, obviously*, but not **She likes obviously Bill.* (An asterisk in front of a sentence indicates that the sentence is ungrammatical.)[1]

The basic syntax-creating process proposed by Chomsky's Minimalist Program (Chomsky 1995) is Merge: a process that takes any two units (words, phrases, clauses…) and forms them into a single unit, subject to "feature-matching." For instance, *likes* has the feature [+transitive] that must be satisfied by a Theme argument in object position, while a noun like *Bill* must seek a verb that still requires an argument. The unit [*likes Bill*] requires a subject that is third-person singular; *she* supplies this deficiency and is therefore merged with [*likes Bill*] to yield [*she* [*likes Bill*]]. (Note that bracketing is simply a

[1] It should be noted that while in English a verb and its direct object are generally sisters, in many European languages an adverb or other verbal modifier, if present, may be a sister of the verb (at least in surface structure; see Rizzi, this volume).

notational alternative exactly mirroring the structure of a syntactic tree.) The process is repeated until the sentence is complete, and naturally it cannot be completed until all the requirements of the lexical items it contains have been satisfied. Here are a few simple examples of sentence failures:

(2) (a) *broke the bottle (lacks an agent argument in subject position)
(b) *Mary see John daily (third-person agreement required)
(c) *Reciprocity invited Bill (verb needs [+human] agent)
(d) *There seems someone to be asleep (*someone* inappropriately merged; dummy subject *there* appropriate only if *someone* is first merged with *asleep*)

The order in which the various operations of Merge are performed will, of course, determine the linear order of constituents. In English, the general rule is that verbs will first be merged with their direct objects (if present); then the resultant verb–object combination will be merged with indirect objects (if there are any) and adverbial phrases, subjects being merged last into the structure. Other languages may adopt different orders. The overriding consideration is that the argument structure of the sentence be recoverable—in other words, that it is possible to determine who did what to whom, when, and where (if the latter are specified). Languages may adopt strategies other than sequential ordering of Merge (e.g, case marking) in order to achieve a greater degree of freedom in ordering constituents. (In general, earlier in the sentence equals old information, later means newer, so some degree of flexibility in ordering may be communicatively advantageous.)

Although this was not the purpose for which it was originally designed, sentence construction via Merge forms a plausible model of how the brain may actually operate in creating sentences. Much evidence suggests that, rather than sending individual words directly to the motor organs of speech, the brain combines neural signals representing several words (at least up to the level of phrases or short clauses, perhaps demarcated by intonational phenomena) before it dispatches them for utterance. Such segments may well correspond to Chomsky's proposed "phases" (Chomsky 2001) which are dispatched individually to Phonetic Form for Spell-out, after which such phases are inaccessible for further syntactic computation. It should be apparent that Merge (as distinct from the purely linear bead-stringing process underlying protolanguage, where it is assumed, on the basis of phonological behavior by speakers of pidgin [Bickerton 2008, chap. 8; Bickerton and Odo 1976], that single words are dispatched separately to the speech organs) automatically creates hierarchical structures.

A crucial question for anyone examining the evolutionary origins of syntax is whether Merge involves recursion. It has been claimed in one widely cited paper (Hauser et al. 2002) that it does, and that recursion is unique to humans, requiring us to assume either a special mutation or the exaptation of some task-specific mechanism that predated—hence originally had nothing to do with—language. In linguistics, recursion is generally defined as the ability to insert one structure inside another of the same kind. However, this concept seems to have originally arisen from early (and long-abandoned) generative analyses (Bickerton 2009, chap. 12). According to a reviewer of an earlier version of this paper, it is only if one adopts an unlabeled version of Bare Phrase Structure (as in, e.g., Collins 2002) that the usual definition of recursion fails to apply. This view, however, seems to confuse notational with operational criteria. The issue is not whether, in some descriptive notation, the label for one structural type does or does not fall within the boundaries indicated by the label of a similar structural type (CP within CP, NP within NP...). It is whether (assuming Merge to be an actual operation performed by the brain) the brain, in the course of constructing sentences, actually inserts a member of one structural type within another member of the same type. Were this the case, the claim by Hauser et al. (2002) that some uniquely human adaptation is required for recursion might stand a chance of being true. However, if the brain obeys Merge, it does not insert anything within anything, but merely merges ever-larger segments of lexical material with one another until a complete sentence is achieved.

It is, of course, possible to adopt a looser definition of recursion, one that makes Merge itself a recursive process, by defining the latter as any process that uses the output of one stage as the input to the next. This solution is adopted by Rizzi (this volume). However, recursion defined in these terms could apply to almost any process, including processes routinely executed by other species—a bird building its nest, for example:

Step 1: Weave two twigs together.
Step 2: Interweave a third twig with the interwoven pair.
Step 3: Interweave a fourth twig with the interwoven three, etc.

With this second definition, any justification for claiming recursion as a uniquely human component of the narrow language faculty simply disappears. In short, Merge itself can be treated as an iterative rather than a recursive operation, requiring no specialized development in the human brain.[2]

---

[2] It is useful to relate these two approaches to linguistic recursion to concepts of recursion in mathematics and computer science, especially since the modeling approaches are gaining momentum. With some simplification, Rizzi's second definition is like any primitive recursive function that can repeatedly be applied to its own output; most well-known functions are like this (such as addition, multiplication, exponentiation, etc.). In computer science, solving a problem using recursion means the solution depends on solutions to smaller instances of the same problem. A loop in a computer program, when executed, calls itself again and again until a certain condition is met, when the program jumps out of the loop. The very definition of a factorial is recursive in this richer sense, since $fact(n) = 1$ if $n = 0$, and $fact(n) = n\ fact(n - 1)$ if $n > 0$. Merge could be regarded as an operation that calls itself in the course of sentence construction: the accomplishment of Merge at the lowest levels of the tree requires some linguistic features that trigger a further application of Merge, until the sentence is completely assembled. However, it should be noted that the recursivity of the process depends crucially on the materials merged, i.e., words. It is the fact that words (and combinations of words) have

However, while Merge is central to syntax, it far from exhausts the phenomena that an evolutionary account of syntax must explain. Among the most salient of these are binding, control, movement, and the reference of empty categories. All of these processes (with minor variations specific to individual languages) are found universally in language, and have in common the fact that they can only operate over limited domains. In other words, all seem to involve some principle or principles of "locality" that demarcate the sections of trees over which processes can operate.

Binding

Binding is a relationship that exists between anaphors (e.g., reflexives, reciprocals, pronouns) and their antecedents. Typically, pronouns are free in reference. Take the sentence we already examined:

(3) Everyone who knows Mary thinks she likes Bill.

On pragmatic grounds, we prefer an interpretation that identifies *she* as *Mary*. However, this is not necessarily the case; *she* could, in principle, refer to any individual female. However, in the sentence *Mary likes her*, *Mary* and *her* cannot refer to the same person, whereas in *Mary likes herself*, *herself* can only refer to Mary.

In general, anaphors cannot occur outside the clause that contains their antecedent. For instance, in the sentence *Mary asked Susan to wash herself*, the meaning cannot be that *Mary asked Susan to wash Mary*. However, there are exceptions in both directions. The following sentence, for example, is grammatical, with *himself* co-referring outside its clause with *John*.

(4) John believes that stories about himself are exaggerated.

But then so is:

(5) Mary saw a snake near her.

Here, a pronoun within the same clause may refer to *Mary*. Definition of a binding domain has therefore not proven easy, even for English (without taking the complications introduced by other languages into account). Consequently, the fact that native speakers of any language can acquire, without explicit training, distinctions that have eluded trained linguists suggests that there must be a limited set of possible algorithms for binding, innately established, and that

dependencies that must be filled which drives repeated applications of Merge until the problem is solved—that is, until the complete grammatical sentence is generated. In other words, it is not simply Merge, but rather Merge + lexical material that constitute the recursive process, as well as force it to result in a hierarchical structure. This suggests that Pinker and Jackendoff (2005) were correct when, contra Hauser et al. (2002), they listed words as a uniquely human part of the language faculty.

children can select among these on the basis of very limited data (what kind of data, how much, and much more remain topics for future inquiry).

Control

The subjects of certain nonfinite verbs in subordinate clauses are "controlled" (given obligatory reference) by constituents of higher clauses. Typical are sentences of the following kind:

(6) Mary expects to arrive shortly.

Although there is no overt subject for *arrive*, that verb is understood to be coreferential with the subject of the higher verb, *Mary*. However, if the higher verb also has an object, this usually supplies the (understood) subject of the lower verb:

(7) Mary told Bill to leave immediately.

Here, *Bill* is taken as the subject of *leave*. There are some exceptions to this rule, for example:

(8) Bill promised Mary to leave immediately.

Here, *Bill* is understood as the subject of *leave*.

Control, as compared to binding, seems to involve a relatively straightforward algorithm involving locality: select the "closest" (in terms of tree structure) possible antecedent for the verb in question, subject to lexical exceptions (like *promise*) that are, in turn, determined by purely semantic considerations.

Movement

Do constituents of sentences "really" move around, or can the empirical data that suggest they do be explained some other way? This question, in one form or another, has concerned generativists since the dawn of generative grammar, even though Chomsky himself once suggested that movement and alternative explanations might merely be "notational variants" of one another. (The question's most recent incarnation involves attempts to derive Move from—or incorporate it somehow into—Merge.)

Phenomena that suggest the existence of movement involve cases where some constituent seems to have been "displaced" from its "normal" position:

(9) Who did Mary see___?

Since *who* clearly functions as the object of *see*, movement theorists hypothesize that *who*, originally in the object position, has been moved to the left periphery of the sentence. Seemingly confirming this is the parallel evidence of "express surprise/request confirmation" pseudo-questions such as:

(10) Mary saw who?

What is referred to as the copy theory of movement removes some, but not all, of the problematic features of movement by assuming an intermediate stage in the derivation consisting of *Who did Mary see who*? followed by deletion of the lower occurrence. (Other possible solutions—the leaving of "traces" at movement sites, or the assumption that those sites contain "empty categories" subject to general rules of interpretation for such categories—need not concern us here, although we return to the second issue below.) What concerns us here is that the relationships between "extraction sites" and overt positions—often referred to as long-distance dependencies (LDDs)—fall under locality restrictions somewhat different from those that apply in cases of control and binding.

Some pre-minimalist versions of generative grammar sought to capture those restrictions with a notion known as subjacency (originally "lying nearby but lower"), which defined the domain within which LDDs could hold as (roughly) the clause that had the moved constituent at its left periphery plus any chain of subcategorized clauses below that clause in the tree. Thus a *wh*-word could be moved to the left periphery of any sentence along the lines of:

(11) Who do you think Bill believes that Mary saw ___?

A subcategorized clause is one that is required by a higher verb; *think* and *believe* both take sentential complements. However, if a subordinate clause is not subcategorized, movement results in ungrammaticality:

(12) (a) *Who did Bill have a bad cold ever since he met ___?
(b) *What did John sing ___ and Mary accompanied him?
(c) *Who did the fact that Mary likes ___ irritate Bill?
(d) *What did Mary see the man who played ___?

Note that

(13) (a) Bill has had a bad cold ever since he met John.
(b) John sang German folksongs and Mary accompanied him.
(c) The fact that Mary likes John irritates Bill.
(d) Mary saw the man who played the tuba.

are all perfectly grammatical sentences that would be appropriate answers to the questions in (12), if those questions could be asked. Items in the underlined positions, however, cannot be questioned, at least not in the ways shown.

The ungrammaticality of sentences like in (12) cannot be dismissed as a result of semantic or pragmatic factors. Sentences are assembled automatically, at very high rates of speed—too fast for any conscious monitoring (or even conscious awareness) of the process. There must surely be some kind of algorithm the brain executes to ensure that (except through very occasional error) sentences like the starred examples do not occur.

Reference of Empty Categories

As we saw above, hypothesized movement creates empty categories—sites where you would expect to find an overt constituent but do not, where you have to search among overtly expressed items to determine what these empty categories refer to. But empty categories also occur in sentences where no movement—or at least, no overt movement of the kind discussed in the previous section—has occurred. In many of these latter cases, in contrast to overt movement cases, empty categories may alternate with pronouns, yielding pairs of sentences both of which are fully grammatical but which differ sharply in meaning. Consider examples like the following:

(14) (a) Mary is too angry to talk to.
(b) Mary is too angry to talk to her.

In the first example, there are two empty categories, respectively subject and object of *talk to*:

(15) Mary is too angry ___ to talk to ___.

At first sight, the first empty category might seem to fall under "control," as described above; however, the algorithm that determines empty-category reference has to start with the most deeply embedded empty category (the one nearest the bottom of the syntactic tree). This must be identified with an overt referential constituent in the sentence, and there is only one, *Mary*, which must accordingly be identified as the object. However, since no two items in the same clause may co-refer (unless that fact is marked by a reflexive pronoun or some other explicit marker of co-reference), *Mary* cannot be interpreted as the subject in this case. Thus the first empty category must be assigned arbitrary reference, i.e., interpreted as *people* (unspecified) or *anyone*, hence the meaning of the sentence is *Mary is too angry for anyone to talk to Mary*.

The presence of a pronoun in the second member of the pair changes the dynamics of the sentence. Since the first empty category (subject of *talk to*) is now the only one in the sentence, it becomes free to co-refer with *Mary*, and since clause-mates cannot co-refer, *her* can then no longer be interpreted as *Mary* but must be read as *some female person other than Mary*.

Note that, in a superficially similar pair of sentences, the references of pronoun and empty category are reversed; the pronoun can be (and will be, with a very high degree of probability) interpreted as *Mary*, while the empty category cannot be thus interpreted.

(16) (a) Mary needs somebody to talk to.
(b) Mary needs somebody to talk to her.

As before, the first sentence contains two empty categories, in the same positions and with the same functions as before:

(17) Mary needs somebody ___ to talk to ___.

In this case, in contrast to *Mary is too angry to talk to*, there are two referential constituents in the sentence: *Mary* and *somebody*. As before, the deepest-embedded empty category is identified first, and since its nearest possible antecedent is *somebody*, it is thus interpreted, leaving the first empty category to select the next-closest referential unit, *Mary*, as its antecedent. In the second member of the pair, however, the presence of the pronoun leaves the subject of *talk to* as the only empty category requiring a co-referent. Since *somebody* is the closest referential item, it is interpreted as the subject, leaving *Mary* as a possible (and for pragmatic reasons, by far the likeliest) antecedent.

It would be satisfying if we could take all cases of empty-category reference assignment, including those cases involving movement and control, and subsume them under a single algorithm. However, this does not seem to be possible, since the item co-referent with an extraction site may be found indefinitely far from that site, while the site itself must obligatorily remain empty (i.e., cannot alternate with a pronoun, unlike the cases just discussed). For our present purposes, it is unnecessary to attempt to find a common algorithm; we need merely to understand that in order to handle syntactic processes, the brain must be sensitive both to the nature and extent of specific domains (e.g., subcategorized and non-subcategorized clauses and phrases) and to relative distances (in terms of position in a tree, not serial order) between referential constituents.

For those who would claim that the aspects of syntax discussed above could constitute learned behaviors, rather than the results of automatic, autonomic neural processes, the reference of empty categories represents the most difficult counterevidence to settle. How do you learn a nothing? Here the burden of proof clearly lies with those who would argue for learning as an adequate explanation of syntactic processes.

Beyond Analysis

To summarize, syntax consists of a process for progressively merging words into larger units, upon which are superimposed algorithms that determine the reference of items (in various types of structural configuration) that might otherwise be ambiguous or misleading. Many other factors—far more than can be mentioned here—are implicated in syntax, but the processes described above are central to it, and if these can be accounted for in evolutionary and neurobiological terms, we will have taken a massive step forward.

Although strictly speaking this lies outside the scope of the present paper, it may be worthwhile to briefly glance at some further considerations that affect the problems facing us, and suggest some lines of inquiry that might prove fruitful, or at worst, provide a null hypothesis that may serve as a basis for further investigation.

If we are looking at biological foundations and origins, we would do well to forget about the distinction between competence and performance. This distinction proved useful, in the early days of generative grammar, to determine what was valid empirical evidence and what was mere accident and happenstance. The distinction between ungrammatical and grammatical sentences was crucial to the enterprise—no one previously had thought about what *couldn't* be said, merely about what could—and the distinction seemed best made on the basis of a presumed body of invariant knowledge, either Universal Grammar or the grammar of a specific language. So was born the distinction between competence (knowledge of language) and performance (how that knowledge was expressed in the creation of actual sentences, subject to all the potential confounds of memory limitations, slips of the tongue, and so forth), with competence as the primary focus of research and performance generally downgraded, ignored, or postponed for future inquiry.

There is good reason, however, to believe that the distinction between competence and performance has outlived its usefulness. It is surely not by accident that a standard reference work on the cognitive sciences (Wilson and Keil 1999) contains no entry for "competence–performance distinction" (even though most other issues in generative grammar are extensively documented) but merely refers the reader to three other entries, none of which mentions the competence–performance distinction. Any evolutionary account surely demands that we treat language as an acquired behavior rather than a static body of knowledge, a behavior with deep roots in our biology but expressed through real-time actions of neurons, axons, synapses, and speech organs. Our focus should be firmly fixed on what humans, and their brains, had to be able to do in order to rapidly, automatically, and unconsciously produce sentences that would fall within the quite narrow bounds that delimit human language. What was required for these tasks might in principle represent (or be derived from) either capacities shared with other animals but selected for novel purposes, or novel, purpose-built capacities unique to the human species. But the null hypothesis clearly is that given symbolic units to combine, the processes used to combine them were ones of a fairly general nature, already present in the genome and shared with other mammalian (and perhaps even avian) species. While some rewiring of the brain may well have been required, it was most likely a rewiring of existing areas (areas that had previously had a variety of functions), rather than one that required the superimposition of some task-specific mechanism unique to (and uniquely used for) language.

Aside from what was required to produce Merge (in all probability a domain-general, iterative mechanism already used to integrate sensory inputs within single modalities and perhaps also cross-modally), it is worth considering the likeliest mechanism for use in dealing with the other syntactic processes outlined above. These, as we have seen, require sensitivity to distance between individual constituents and sensitivity to boundaries between syntactic units. The most parsimonious solution would, as with Merge, employ one (or more than one) already-existing domain-general mechanism, and a plausible candidate would be the brain's ability to establish sequence by means of the fading of neural signals.

All neural signals degrade over time, so that *ceteris paribus* the equations stronger = later and weaker = earlier should hold over all brain operations. Since Merge combines constituents in a temporal series of sequential operations, it should be possible for the brain to keep track of sequence by exploiting this aspect of neural signaling.

Note that focusing on processes and how the brain might execute them, rather than on hypothesized linguistic knowledge, might help to restrain what has been one of the most recurrent (as well as the most frustrating) developments in generative grammar. That theory has passed, since its inception, through at least five avatars: Chomsky's original (1957) formulation, the Standard Theory, the Extended Standard Theory, the Principles and Parameters framework, and the Minimalist Program. Each new avatar has undertaken a radical revision of its predecessor, and each revision has been prompted by the fact that the preceding version had proliferated and complicated itself to an inordinate extent. However, each of these versions eventually succumbed to the very failings it was created to correct. The latest, the Minimalist Program, is no exception. In its origins the leanest and most elegant expression of generative grammar (see Rizzi's admirably clear and concise description in this volume), it has become a jungle of arcane technicalities, daunting for any neurobiologist who might hope to reconcile it with the workings of the human brain.

The need is obvious: to nail down the biological foundations of syntax demands scholars who combine a thorough training in both linguistics and neurobiology, enough so that they give neither precedence over the other in their thinking. Given the traditional Balkanization of science, this may be a lot to ask. But until such scholars are available, researchers in both communities should remain as open to one another as possible. There is a great temptation to say things like, "That linguistic model could never be realized by the brain," or "That neurological model is far too simplistic to account for language." Such remarks, however plausible they may seem, are no more than assumptions. The issues themselves are empirical. The brain creates sentences somehow, and the way it actually does this must constitute the real and only true grammar of language.

2

The Biological Background of Syntax Evolution

Anna Fedor, Péter Ittzés, and Eörs Szathmáry

Abstract

It is difficult to gain an understanding of language since we do not know how it is processed in the brain. Many areas of the human brain are involved in language-related activities, including syntactic operations. Aspects of the language faculty have significant heritability. There seems to have been positive selection for enhanced linguistic ability in our evolutionary past, even if most implied genes are unlikely to affect only the language faculty. Complex theory of mind, teaching, understanding of cause and effect, tool making, imitation, complex cooperation, accurate motor control, shared intentionality, and language form together a synergistic adaptive suite in the human race. Some crucial intermediate phenotypes, such as analogical inference, could have played an important role in several of these capacities. Pleiotropic effects may have accelerated, rather than retarded, evolution. In particular, it is plausible that genes changed during evolution so as to render the human brain more proficient in linguistic processing.

Introduction

Natural language is a fascinating phenomenon, and it is undoubtedly partly biological. Apes, dolphins, and parrots are unable to acquire language by learning, no matter how hard they try. There must be something in our genetic endowment that makes humans "ready" for language. Some people think that this readiness is simply due to higher intelligence. Although this may be true in a very broad sense, such claims do not explain how this intelligence differs from that of, say, apes. Humans seem to have gained an insight into the cause and effect in the physical domain, and we are able to produce and use tools by the so-called subassembly strategy. To us it seems that humans possess a few neural procedural capacities that are shared by several important faculties, and that these exist only in very rudimentary forms in other animals. One such procedural element is the ability to handle hierarchical structures efficiently: in the language domain, this is the recursive element of syntax; in the tool-making domain, this is the subassembly strategy; in the theory of mind domain, this is second-, third-, and fourth-order intentionality. We propose that over the past ca. 5 million years, there has been selection on several different capacities in the hominine lineage; these capacities are partly overlapping, and the intensity of selection most likely shifted (perhaps several times) over these domains. There may have been a period during which tool use and tool making were primarily favored, but then the positively selected genetic variants could well have turned out to be favorable in some of the other critical domains.

Fisher and Marcus were right when they stated:

> In short, language is a rich computational system that simultaneously coordinates syntactic, semantic, phonological and pragmatic representations with each other, motor and sensory systems, and both the speaker's and listener's knowledge of the world. As such, tracing the genetic origins of language will require an understanding of a great number of sensory, motor and cognitive systems, of how they have changed individually, and of how the interactions between them have evolved (Fisher and Marcus 2006, p. 10).

The study of language origins has, however, been hampered by the fact that there is a critical lack of detailed understanding at all levels, including the linguistic one. There is no general agreement among linguists as to how language should be described: widely different approaches exist, and their proponents often have very tense scientific and other inter-relationships. As biologists, we maintain that symbolic reference combined with complicated syntax (including the capacity of recursion) constitutes the least common denominator in this debate. Within this broad characterization, we would like to draw attention to two approaches which have, perhaps surprisingly, a strongly chemical character. The first is the Minimalist Program of Noam Chomsky (1995), where the crucial operator is Merge, the action of which triggers certain rearrangements of the representation of a sentence. There is a broad similarity between this proposal and chemical reactions (Maynard Smith and Szathmáry 1999). An even closer analogy between chemistry and linguistics can be detected in the second approach: Luc Steel's Fluid Construction Grammar (Steels 2004; Steels and de Beule 2006). Here, semantic and syntactical "valences" have to be filled for correct sentence construction and parsing. We should note that the roots of genetic inheritance are, of course, in chemistry, and that even at the phenomenological level Mendelian genetics was a stoichiometric paradigm, influenced by contemporary chemical understanding (elementary units that can be combined in certain fixed proportions give rise to new qualities). Chemical reactions can be also characterized by rewrite rules. In-depth study is required to consider the ramifications of this analogy: the deeper it goes, the more benefit one can hope from taking the analogy seriously.

There are now attempts to rethink Fluid Construction Grammar in terms of replicator dynamics *within* the brain. This may sound surprising, but one should not forget that the question of whether thinking or language is performed in the brain by processes analogous to evolution by natural selection is wide open. Ever since the ideas of William James (1890), some suspect that brain dynamics will be proven to include a crucial evolutionary element, just as it is now known to apply for the immune system. The immune system is particularly interesting because it is both an evolutionary (D'Eustachio et al. 1977) and generative system (Jerne 1985). A few selectionist approaches to brain epigenesis and function exist (Dawkins 1971; Changeux 1983; Finkel and Edelman 1985) but they lack one crucial component: multiplication. The only existing attempt was carried out by Calvin (1996), and this triggered further research on how language may be processed in the brain (Calvin and Bickerton 2000). The main problem with Calvin's mechanism is the lack of *connectivity copying* from one brain site to another. Such a mechanism is crucial if we think that connectivity encodes information about alternative hypotheses, among which some reward mechanism selects the better ones (Fernando et al. 2008), and it is also needed to close the gap between neuroscience and the evolutionary epistemology of, say, Campbell (1974).

Language needs certain prerequisites. Some, however, are not especially relevant to the main problems addressed in this volume. For example, apes do not possess a descended larynx nor do they have cortical control over their vocalizations. Undoubtedly, these traits must have evolved in the human lineage, but we do not think that they are indispensable for language as such. One could have a functional language with a smaller number of phonemes, and sign language does not require either vocalization or auditory analysis (Senghas et al. 2004; MacSweeney et al. 2008). Thus, our focus is primarily on the *neuronal implementation of linguistic operations*, irrespective of the modality. It is difficult to imagine the origin of language without capacities for teaching (which differs from learning), imitation, and a complex theory of mind (Premack 2004). Apes are limited in all these capacities. It is fair to assume that these traits have undergone significant evolution because they were evolving together with language in the hominine lineage. To this one should add—not as a prerequisite, but as a significant human adaptation—the ability to cooperate in large non-kin groups (Maynard Smith and Szathmáry 1995). Together, these traits form an *adaptive suite* that is specific to humans. We suggest that in any selective scenario, capacities for teaching, imitation, some theory of mind, and complex cooperation must have been rewarded, because an innate capacity for these traits renders language evolution more likely (Szathmáry and Számadó 2008b).

From the neurobiological perspective, we call attention to the fact that some textbooks (e.g., Kandel et al. 2000) still present a distorted image of the neurobiological basis of language. It is very simplistic to assign the Wernicke and Broca areas of the left hemisphere to semantics and syntax, respectively. The localization of language components in the brain is extremely plastic, both between and within individuals (Neville and Bavelier 1998; Müller et al. 1999). Surprisingly, if removal of the left hemisphere happens in the first few months after birth, the patient can nearly completely retain his/her capacity to acquire language. This stands, of course, in sharp contrast to the idea of anatomical modularity. It also severely limits the idea that only the afferent channels changed during the evolution of the human brain: modality independence and the enormous brain plasticity in the localization of language favor the idea that *whatever has changed in the brain to render it capable of linguistic processing must be a very widespread property of the neuronal networks* (Szathmáry 2001). Components of language (as well as those of other capacities) get localized *during development* somewhere in any particular brain in the most functionally "convenient" parts available (cf. Karmiloff-Smith 2006). Language is just a certain activity pattern of the brain that finds its habitat, like an amoeba in a medium. The metaphor "language amoeba" expresses the plasticity of language, but it also calls attention to the fact that a large part of the human brain is apparently a potential habitat for it; no such habitat appears to exist in nonhuman ape brains (Szathmáry 2001).

For a long time, there has been a dogma concerning the histological uniformity of homologous brain areas in different primate species. Recent investigations, however, do not support this claim (DeFelipe et al. 2002). In fact, the primary visual cortex shows marked cytoarchitectonic variation (Preuss 2000), even between chimps and man. Therefore, one cannot at all exclude the possibility that some of the species-specific differences in brain networks are genetically determined, and that some of these are crucial for human language capacity. As discussed above, these language-critical features must be a rather widespread network property. Genes affect language through the development of the brain. Thus the origin of language must be to a large extent an exercise in the linguistically relevant developmental genetics of the human brain (Szathmáry 2001).

Consider the existing data on genetic changes that are more directly relevant to language. The FOXP2 gene was discovered to have mutated in an English-speaking family (Gopnik 1990, 1999). It has a pleiotropic effect: it causes orofacial dyspraxia, but it also affects the morphology of language. Affected patients must learn or form the past tense of verbs or the plurals of nouns case by case, and even after practice they do so differently from unaffected humans (for a review, see Marcus and Fisher 2003). FOXP2 underwent positive selection (Enard et al. 2002) in the past, which demonstrates that there are genetically influenced important traits of language other than recursion (Pinker and Jackendoff 2005), contrary to other opinions (e.g., Hauser et al. 2002). In addition, there is a known human language that apparently has no recursion: the Pirahã language in the Amazon (Everett 2005). It would be good to know how these particular people manage recursion in other domains, such as object manipulation or "action grammar" (cf. Greenfield 1991).

The capacity to handle recursion appears to differ from species to species. Tamarin monkeys, for example, have demonstrated insensitivity to auditory patterns defined by more general phrase structure grammar, but they discover violations of input conforming to finite state grammar (FSG; Fitch and Hauser 2004). Human adults are sensitive to both violations (but see discussion below). Needless to say, it would be very interesting to know the relevant sensitivities in apes and human children (preferably just before they master grammar of natural language). One should design an experiment capable of producing consistent patterns of such a capacity in evolving neuronal networks, and then reverse engineer proficient networks to discover the evolved mechanisms for this capacity (see below).

We share the view that language is a complex, genetically influenced system for communication that has been under positive selection in the human lineage (Pinker and Jackendoff 2005). The task of the modeler, then, is to try to model intermediate stages of a hypothetical scenario and, ultimately, to reenact critical steps of the transition from protolanguage (Bickerton 1990) to language. It cannot be denied that language is also a means for representation. This is probably most obvious for abstract concepts, for which the generative properties of language may lead to the emergence of a clear concept itself. This is well demonstrated for arithmetics. For instance, an Amazonian indigenous group lacks words for numbers greater than 5; hence they are unable to perform exact calculations in the range of larger numbers, but they do have approximate arithmetics (Pica et al. 2004).

Language changes while the genetic background also changes (this must have been true especially for the initial phases of language evolution), and the processes and timescales involved are interwoven. This opens up the possibility for genetic assimilation: some changes that each individual must learn at first can later become hardwired in the brain. Some have endorsed the importance of this mechanism in language evolution (Pinker and Bloom 1990), whereas others have raised doubt (Deacon 1997). Deacon's argument against it is that linguistic structures change so fast that there is no chance for the genetic system to assimilate any grammatical rule. This is true, but not very important. There are linguistic operations—performed by neuronal computations and related to compositionality and recursion among others—that must have appeared sometime in evolution. Whatever the explicit grammatical rules are, such operations must be executed.

Hence, a much more likely scenario for the importance of genetic assimilation proposes that many operations must have first been learned, and those individuals whose brain was genetically preconditioned to a better (faster, more accurate) performance of these operations had a selective advantage (Szathmáry 2001). Learning was important in rendering the fitness landscape more climbable (Hinton and Nowlan 1987). This view is consonant with Rapoport's (1990) view of brain evolution. This thesis is also open for experimental test.

The origin of language is an unsolved problem; some refer to it as the "hardest problem of science" (Christiansen and Kirby 2003). What makes it difficult is the fact that physiological and genetic experimentation on humans and apes is very limited. The uniqueness of language prohibits, strictly speaking, application of the comparative method, so infinitely useful in other branches of biology. Fortunately, some *components* of language lend themselves to a comparative approach, as we shall see in relation to birdsong. Nevertheless, limitation of the approaches calls for other types of investigation.

We believe that simulations of various kinds are indispensable elements of a successful research program. Yet a vast range of computational approaches have brought less than spectacular success (cf. Elman et al. 1996). In our opinion, this is attributable to the utterly artificial nature of many of the systems involved, such as connectionist networks using back-propagation, for example (for a detailed criticism, see Marcus 1998).

In this chapter, we review some key findings and ideas concerning the genetic, neurobiological, and evolutionary background of the "language problem." In addition, we provide an update on some of the previous suggestions, present our model for a minimalist neural network parsing context-free grammar, and discuss arguments in favor of a human-specific adaptive suite.

Genetic Background of Language

Information about the human and the chimp genome (Chimpanzee Sequencing and Analysis Consortium 2005) is now "complete," and one can ask how far previous optimism seems justified in light of comparative studies based on this information. It is clear that much work lies ahead. Knowing all the genes of chimps and humans is not everything: we need to know how the genotype is mapped to the phenotype, and this is a formidable problem. Genes are expressed in specific ways, under the influence of other genes and the environment. Interaction between genes is not the exception but the rule. One gene can affect several traits (pleiotropy) while actions of different genes do not affect traits (including fitness) independently (epistasis). It is the *network of interactions* that is of importance, and one must not forget that there are networks at different levels: from genetic regulatory networks through protein interaction networks and signal transduction pathways to the immune system or neuronal networks. The question is how the effect of genes percolates upwards. Genes act on expressed molecules (proteins and RNA) that do their job in their context. There is something amazing about the fact that hereditary action on such primitive molecules percolates upwards, resulting in heritability of complex cognitive processes, including language.

The chimp and human genomes are indeed similar, but one should understand clearly what this means (Fisher and Marcus 2006). Substitutions make up for 1.23% of difference between the two genomes; this translates into 35 million altered sites in the single-copy regions of the genomes! Insertions and deletions yield a further 3% genomic difference. It is convenient to distinguish between altered structural and regulatory genes. The first codes for altered enzymes or structural proteins; the latter codes for altered transcription factors, for example. Both kinds of changes happened since humans diverged from chimps, and both affected language in critical ways.

There seems to have been an acceleration in the changes of neural gene expression patterns in human evolution, although this should be evaluated against the background that liver and heart expression patterns have diverged a lot more between chimps and humans. The usual interpretation is that neural tissue is under stronger stabilizing selection. Another observed tendency is the up-regulation of human neural gene expression relative to the chimp, but the functional significance of this finding is unclear (it may be a more or less direct result of recent genomic region duplications).

It is not yet clear what gene expression differences exist behind the cytoarchitectonic differences among the Brodmann areas: the most known differences between chimps and humans are common to all cortical regions. Recently, this view has been refined. Oldham et al. (2006) analyzed gene co-expression patterns in humans and chimps and were able to identify network modules that correspond to gross anatomical structures including the cerebellum, caudate nucleus, anterior cingulated cortex, and cortex. The similarity of network connectivity between the respective human and chimp areas decreased in that order, consistent with the radical evolutionary expansion of the cortex in humans. It is intriguing to note that in the cortical module there is a strong co-expressive link between genes of energy metabolism, cytoskeletal remodeling, and synaptic plasticity.

There are genetic changes that probably did boost language evolution but in a general, aspecific way. Genes influencing brain size are likely to have been important in this sense. Note, however, that genes involved in primary microcephaly seem to have been under positive selection in the past, but children with this syndrome can have rather normal neuroanatomical structures despite the fact that their overall brain size may be reduced to a mere one third of the normal. They show mild to moderate mental retardation but pass several developmental stages. Fisher and Marcus conclude:

> In our view the honing of traits such as language probably depended not just on increased "raw materials" in the form of a more ample cortex, but also on more specific modification of particular neural pathways (Fisher and Marcus 2006, p. 13).

Perhaps the most revealing recent finding concerning genetic brain evolution is the identification of an RNA gene that underwent rapid change in the human lineage (Pollard et al. 2006). It is expressed in the Cajal-Retzius cells of the developing cortex from 7 to 19 gestational weeks. It is co-expressed with reelin, a product of the same cells, which is important in specifying the six-layer structure of the human cortex.

Even if some of our linguistic endowment is innate, there may not be much genetic variation for the trait in normal people, just as most people have ten fingers. In contrast, our linguistic capacity may be like height: whereas all people have height, there are quantitative differences in normal people. To be sure, children as well as adults differ in their linguistic skills. However, to what extent do genes account for this variation?

Surveying many studies, Stromswold (2001) concluded that twin concordance rates are significantly higher for monozygotic than for dizygotic twins. Twins are concordant for a trait if both express the trait or if neither expresses it. Twins are discordant for a trait if one exhibits the trait and the other does not. If the concordance rate for language disorders is significantly greater for monozygotic than dizygotic twins, this suggests that genetic factors play a role in language disorders such as dyslexia and specific language impairment (SLI). The concordance rates for written and spoken language disorders are similar. For both written- and spoken-language disorders, the mean and overall concordance rates were approximately 30% higher for monozygotic than for dizygotic twins, and genetic factors accounted for between one-half and two-thirds of the written and spoken language abilities of language-impaired people. In studies of twins with no language impairment, between one-quarter and one-half of the variance in linguistic performance was attributable to genetic factors depending on the aspect of language being tested. People have been tested on phonological short-term memory, articulation, vocabulary, and morpho-syntactic tasks. It seems that different genes may be responsible for the variance in different components of language and that some genetic effects may be language-specific.

The sum of all genetic effects is usually not much greater than 50% for various aspects of cognition (Stromswold 2001). Most individual genes are expected to have small effects. Candidate genes affect functions such as the cholinergic receptor, episodic memory, dopamine degradation, forebrain development, axonal growth cone guidance, and the serotonin receptor. It is a great problem that cognitive skills are likely to have been, at least in part, inadequately parsed; thus so-called intermediate phenotypes with a clearer genetic background should be sought. By this token, schizophrenia as such does not exist; rather, different genes may go wrong and the symptoms such as hallucinations are emergent outcomes (Golbderg and Weinberger 2004). The situation may be similar to that of geotaxis in *Drosophila,* where the individual involvement of different genes that collectively determine this capacity is counterintuitive (Toma et al. 2002).

It is worth calling attention to the fact that the genetics of human cognitive skills is a notoriously difficult problem. One common reason is that usually the clinical characterizations are not sufficient as descriptions of phenotypes (Flint 1999). A consensus seems to emerge that the genes involved are so-called "liability genes" which, when present in the right allelic form, significantly enhance the probability of developing the respective cognitive skills.

Perhaps the most important neurodevelopmental syndrome for our topic is SLI, where there is significant difference between verbal and nonverbal skills. Several candidate chromosomal regions have been identified (SLI consortium 2002).

A by now famous gene is FOXP2, first identified by Gopnik (1990). In a certain English-speaking family, a dominant allele was found to cause the syndrome developmental verbal dyspraxia (DVD), formerly grouped under SLI. No one disputes the fact that SLI is real. What is contested is how closely it is limited to, or rooted in, a specific grammatical impairment. The Gopnik (1990, 1999) case has been very stimulating because of its characterization as "feature-blind" dysphasia and its obvious genetic background (a single dominant allele); however, cognitive skills are affected as well (Vargha-Kadem et al. 1998). More evidence with other linguistic groups is accumulating (Dalalakis 1999; Rose and Royle 1999; Tomblin and Pandich 1999). One study (Van der Lely et al. 1998), sadly without genetics, claims to demonstrate that grammatically limited SLI does exist in "children" (although only one child is analyzed in the paper).

The FOXP2 protein is an ancient transcription factor present in vertebrates, and there is evidence that it has been under positive selection in the human lineage. It seems to affect development of distributed neural networks across the cortex, striatum, thalamus, and cerebellum. DVD differs from SLI, but speech and language deficits are *always* present, even in otherwise normal children. In other affected individuals, general intelligence is impaired and grammar deficits (difficulty with morphological features such as the suffix *–s* for plural or *–ed* for past tense) occur in written language as well. The selective sweep that affected this gene in the human lineage occurred within the past 200,000 years (Enard et al. 2002; Zhang et al. 2002).

Analysis of the expression patterns of FOXP2 in other species suggests that this gene has been involved in the development of neural circuitry processing sensorimotor integration and coordinated movements, lending support to the notion that language has its roots in motor control (e.g., Lieberman 2007). This makes the involvement of basal ganglia in speech and language less than surprising.

Recent studies (reported by White et al. 2006) demonstrate that FoxP2, although without accelerated evolution, plays a crucial role in the development and seasonal activation of relevant brain areas in songbirds. Interestingly, although the avian and human forms are very similar, neither of the human-specific mutations has been found in the *FoxP2*. Also of interest is the fact that the ganglia involved in birdsong learning seem to be analogous to the basal ganglia involved in human vocal learning (Scharff and Haesler 2005).

Researchers have called attention to the fact that in songbirds and humans both FoxP2 and FoxP1 are expressed in functionally similar brain regions that are involved in sensorimotor integration and skilled motor control (Teramitsu et al. 2004). Moreover, differential expression of FoxP2 in avian vocal learners is correlated with vocal plasticity (Haesler et al. 2004). Using songbirds as analogs to human learning of speech, Haesler et al. (2007) proved that birds with *FoxP2* knockout suffer from incomplete and inaccurate vocal imitation.

Mice, like humans, have two copies of the Foxp2 gene as well. If only one of them is affected, pups are severely affected in the ultrasonic vocalization upon separation from their mother. This suggests a role for this gene in social communication across different species. The Purkinje cells in the cerebellum are affected in the pups (Shu et al. 2005). Determination of the expression pattern in the developing mouse and human brain is consistent with these investigations: regions include the cortical plate, basal ganglia, thalamus, inferior olives, and cerebellum. Impairments in the sequencing of movement and procedural learning may thus underlie the linguistic symptoms in humans (Lai et al. 2003). According to Vernes et al. (2007), the targets of this regulatory gene in mice include loci involved in modulating synaptic plasticity, neuronal development, axon guidance, and neurotransmission. Spiteri et al. (2007) identified transcriptional targets of *FOXP2* in human basal ganglia and the inferior frontal cortex. Many target genes play roles in neurite outgrowth and plasticity. Fujita et al. (2008) inserted a human-specific *FOXP2* gene into mice, which in homozygous condition die early, have (among other impairments) abnormal Purkinje cells, and show severe ultrasonic vocalization and motor impairment.

Recently, Krause et al. (2007) attempted to date the fixation of the two human-specific amino acid substitutions in FOXP2 by claiming that the gene was shared by Neanderthals and hence the substitutions had been fixed more than 300 thousand years ago. This conclusion was challenged by Coop et al. (2008) on methodological grounds, so the jury is still out on this issue.

It is important to emphasize that the link between genes and mental capacities is extremely indirect (e.g., Karmiloff-Smith 2006): genes encode for RNA and protein molecules, and every effect on behavior must penetrate "upwards" through a large molecular and cellular interaction network. Williams syndrome provides a good case in point. These patients are regarded as handicapped in spatial orientation and yet are good at language: 28 contiguous genes are involved in the phenomenon. However, a closer analysis of their language reveals that it is not "normal" either, and it also develops late in life. Even a mouse "model" exists with a mutant LIMK1 (a protein kinase gene expressed in the developing brain) for the spatial problem. This gene is expressed not only in regions responsible for spatial orientation, and some human patients have an impaired LIMK1 gene yet they *do not* exhibit Williams syndrome. By the same token, many children with SLI have *no* problem in the *FOXP2* gene. Here it is instructive to quote from Karmiloff-Smith (2006, p.15):

> WS [Williams syndrome] is caused by a deletion of some 28 genes on one copy of chromosome 7; DS (Down syndrome) is caused by an additional whole chromosome 21; Fragile X is caused by a mutation of a single gene on the X chromosome; velocardiofacial syndrome (or di George syndrome) is caused by a large deletion on chromosome 22. Yet all four syndromes display both delay and

deviance, mental retardation, gross and/or fine motor deficits, impaired sleep patterns, memory deficits, number impairments, and often hyperactivity. Three of them show better language skills than spatial skills.

We believe that the biologically motivated dissection of the language faculty is of primary importance. Put differently: *What are the intermediate phenotypes that make up language?* This question cannot be answered, we believe, without an appropriate formulation of aspects of language. Thus linguistic theories must ultimately be biologically constrained. A good start in this direction may be Fluid Construction Grammar (Steels and de Beule 2006). To date, though, there is not much coupling of details of linguistic theories to those of brain mechanisms.

Brain and Language

The analysis of neural activity during the performance of cognitive tasks has become a burgeoning industry. Sensitivity has increased over the years, and these methods have been increasingly applied to the recording of brain activity during linguistic performance. The crucial observation to bear in mind is that localization of certain functions to particular brain areas during normal development does not necessarily mean that the particular region is a "hard-wired region in the modular sense" for that particular function. Brain development, especially in the first few months in life, is very plastic and many cognitive skills can be (nearly) spared due to plastic recovery after early injury. The same applies to components of language, even syntax. One can surely learn about the neurobiological foundations of syntax by studying Broca's area in normal people or patients with late lesion of that area, but at the same time one should also ask how the relevant tasks are performed in patients who do not have Broca's area at all!

Where Is Language in the Brain?

The recognition that neural localization of language can be plastic is widely known (Nobre and Plunkett 1997; Neville and Bavelier 1998; Musso et al. 1999). Studies of brain injury have shown that when damage to the left hemisphere is sustained before a critical period, the right hemisphere is able to take over the necessary functions (Müller et al. 1999). This does not contradict the finding that in normal people Broca's area seems specialized for syntax (Embick et al. 2000). It appears that the common left-hemisphere localization of language is just the most likely outcome in the absence of genetic or epigenetic disturbance. What is more, both the cortical and subcortical areas contribute to language processing; reward systems and motor control provided by basal ganglia and the cerebellum seem to be critical components of our language faculty (Lieberman 2002, 2007).

The conclusions that we can draw from brain studies are as follows:

- Localization of language is not fully genetically determined: even large injuries can be tolerated before a critical period.
- Language localization to certain brain areas is a highly plastic process, both in its development and end result.
- A surprisingly large part of the brain may sustain language: there are (traditionally recognized) areas that seem to be most commonly associated with language, but they are by no means exclusive, either at the individual or the population level, during either normal or impaired ontogenesis.
- Whereas a large part of the human brain can sustain language, no such region exists in apes.
- Language processing has a distributed character.

It is instructive to look at the evolutionary patterns of the sensory neocortex in mammals (Krubitzer and Kaas 2005). Auditory, somatosensory, and visual fields (contiguous brain tissue regions) have changed in location and size in different species. Fields can change in absolute and relative size, as well as in number. Connections of cortical fields can also change. Such alterations can be elicited by manipulation of either the peripheral morphology or activity, or that of the expression level of certain genes. Phenotypic within-species variation can be extremely broad; however, little is known about the relative magnitude of the genetic part of this variation. A good example of genetic influence is the variation in the cortical area map of inbred mice, reflecting strain identity (Airey et al. 2005).

Evolution of the vertebrate brain has produced an increase in cortical size and elaboration of the cortical circuit diagram (Hill and Walsh 2005). Most importantly, cortical layers II and IIIb, IIIc of the chimp differ from layers IIa, IIb and IIIa, IIIb and IIIc, respectively, in humans. A tentative conclusion, based on "rewired" ferrets and three-eyed frogs, is that layers form independently of patterned input, and instructive electrical signals play a crucial role in fine network development, which also affects intracortical connections (Sur and Learney 2001).

Genetically determined patterning of parts of the brain follows mechanisms well-known from conventional developmental studies. For example, during the formation of the retinotopic map, axons from the retinal ganglion cells find their targets in the tectum as a result of matching between two receptor/ligand pairs (Schmitt et al. 2006), both expressed according to four gradients: two in the eye and two in the tectum.

Several researchers, including Greenfield (1991), have suggested the involvement of tool making in the evolution of language, for example, in the form of "action grammar" which can be recursive when agents use the "subassembly" strategy in the "nesting cups" experiment. The idea is that selection for efficient tool use could have aided language evolution and vice versa. Stout and Chaminade (2007) investigated the neurobiological bases of this via brain imaging in modern naïve (i.e., untrained) humans by requiring them to implement a 2.5 million year-old Oldowan practice of tool making. (Incidentally, in that material culture we see evidence for the uniquely human practice of using a tool to make another tool.) Premotor cortex was activated in the task but not the prefrontal executive cortex (involved in planning) nor the inferior parietal cortex. The activation of caudal Broca's area in this task underlines the possible link between language and tool making, and is consistent with views of the importance of "mirror neurons" in language evolution (Rizzolatti and Arbib 1998)—the latter, however, are by no means sufficient for language, as many animals possess them. As we learn words by imitation, and tool-making requires an "action grammar," it is unlikely to be accidental that human Broca's area evolved from structures that are involved in these capacities beyond (and prior to) language. Caplan (2006b) suggests that Broca's area is involved in syntactic processing, not merely because it is evolutionarily related to the dorsolateral prefrontal cortex or its original involvement in sensorimotor functions, but because of its intrinsic neural organization. However, this leaves the very essence of the suggested neural organization obscure.

Grodzinsky and Santi (2008) present evidence in favor of the view that Broca's area is specifically involved in syntactic movement (for a discussion of what Movement means, see Bickerton and Rizzi, both this volume) rather than syntactic complexity per se, although they also accept that it is involved in language production and working memory (Friederici, this volume). Note that this conclusion is drawn from analysis of either normal people or patients in which the right hemisphere has not taken over language processing. Thus, once again, Broca's area is exciting but by no means exclusively so. The lesson, however, is to figure out in which way could Movement require specific neuronal mechanisms relative to other linguistic operations.

A Minimalist Neural Network Parsing Context-free Grammar

A crucial element of syntax is center-embedded recursion (Hauser et al. 2002), which has been regarded as specific to humans. This view was recently challenged by Gentner et al. (2006), who believe to have demonstrated that European starlings recognize context-free grammar (CFG; Figure 2.1) with center-embedding. This experimental design was influenced by the former experiment of Fitch and Hauser (2004), who interpreted their results as showing that, whereas tamarin monkeys as well as human students recognize FSG, only the latter recognize CFG.

The methodological problem with these studies is that because there is no need for real center-embedding (bracketing), the task can be solved by counting (Corballis 2007a, b). Consonant with this approach is the experimental finding that humans also perform poorly on learning center-embedded structures when other strategies (such as counting) are not allowed in *artificial* grammar learning (de Vries et al. 2008).

Sun et al. (1998) implemented a hybrid system in which a recurrent neural network was coupled to an external nonneural stack memory. After training with back-propagation, the system was able to infer a CFG from input. In another study (Bodén and Wiles 2000), continuous time recurrent networks without a stack can learn both context-free (A^nB^n) and context-sensitive ($A^nB^nC^n$) languages in a prediction task, using back-propagation through time. Since there were no long-range dependencies connecting words within the sentences, performance of these systems boiled down to counting (Rodriguez at al. 1999). Chen and Honavar (1999) proposed an artificial neural network architecture for syntax analysis through the systematic composition of a suitable set of component symbolic functions realized using neural associative processor modules. The neural associative processor is a 2-layer perceptron that can store and retrieve arbitrary binary pattern associations. Their model is a fairly complex system which can avoid the problem mentioned above.

Recently, we performed a study to examine these issues further (Fedor et al., submitted). Our aim was to handcraft a minimalist neural model that can parse real center-embedded structures with established associations between AB word pairs as mentioned above (Figure 2.1, right panel). Although the proposed network is not directly biologically realistic, we believe it can be smoothly transformed into such an architecture. We rely on the observation that CFG requires *some* implementation of a stack, with the necessary *pop* and *push* operations (Hopcroft and Ullmann 1979). The task is then to come up with a near-realistic and minimalist neural architecture.

Our proposed network is simpler than the above solutions and avoids back-propagation. It rests on the assumption that gating of synaptic connections is critical for complex cognitive processes (e.g., Gisiger et al. 2005; O'Reilly 2006). There are four main components of the neural network: the input layer, the clocked stack, the pairing module, and the end-of-sentence neuron (Figure 2.2). The input layer receives one word at a time from the sentence. The stack

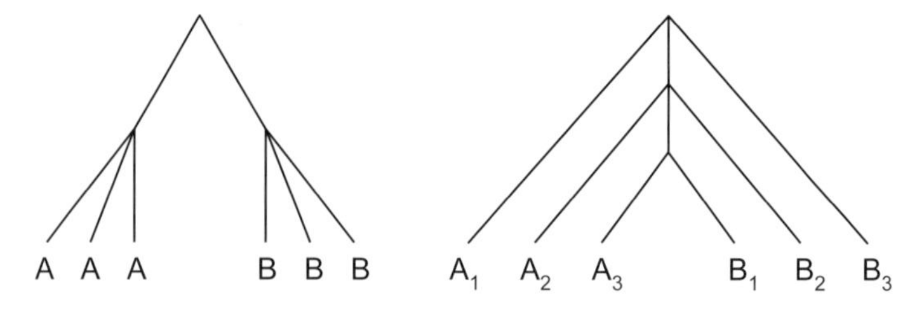

Figure 2.1 Apparent (left) and real (right) context-free grammar. The left structure can be parsed by simple counting; the tree on the right needs some knowledge of the context-free grammar because of the long-range dependencies (word pairs). After Corballis (2007a, b).

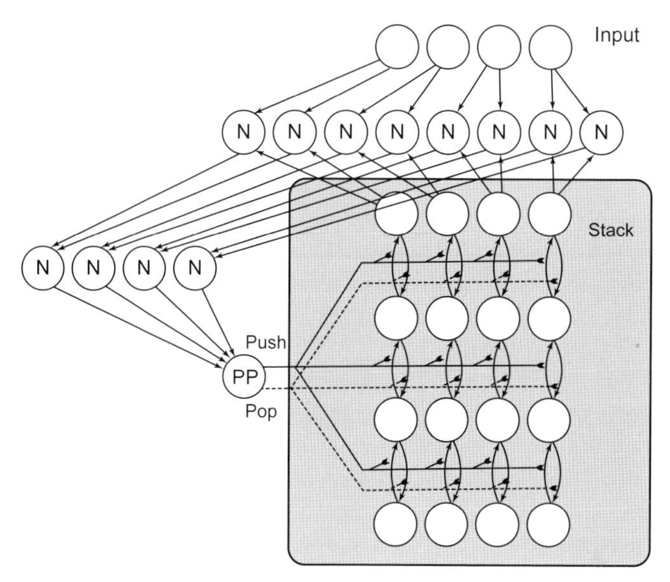

Figure 2.2 Architecture of the grammar parsing network. The stack is depicted by the gray background. Neurons (N) are part of the pairing module that compares the input with the top of the stack. The PP (push/pop) neuron blocks or activates the different synapses of the stack and in this way either pushes down the next input or pops up the word from the top of the stack. The end-of-sentence neuron is not depicted in the figure. Only one word from the input sentence is represented here.

consists of a number of layers, each capable of storing one word at a time. The task of the pairing module is to compare two words: one on the input layer and the other on the top of the stack. The result is given to the push-pop neuron (PP), which signals 1, if the words are pairs, and signals –1, if the words are not pairs. The PP neuron performs gating on the inter-layer synapses of the stack. It is connected to every upward synapse with weights of 1 and to every downward synapse with weights of –1. As a result, if the words are pairs, it inhibits the downward connections in the stack; hence upward connections will predominate, and each layer will take the value of the layer below it (a pop action). In this case, the bottom layer will take the value of the empty sign. If, however, the words are not pairs, the PP neuron inhibits the upward connections in the stack, and downward synapses will predominate such that every layer will take the value of the layer above it (a push action). In this case, the top layer takes its next value from the input. Finally, there is an end-of-sentence neuron, which signals 1 only if there is the end-of-sentence sign on the input. The stack should be empty at the same time, if the sentence is grammatically correct; if the sentence is grammatically incorrect, some words get stuck in the stack.

This architecture was tested with input conforming to FSG and CFG. Each synaptic weight was trained independently with the simple perceptron learning rule during several learning sessions until weights converged. After learning, testing was performed with a mix of novel CFG, FSG, and grammatically incorrect sentences, and the performance of the network was scored. Two types of performance were measured: the percentage of word pairs recognized and the percentage of sentences correctly categorized to grammatical and nongrammatical. The recognition of word pairs is irrespective of stack depth, and the network can learn word pairs correctly provided that every possible word pair was present in the learning set. Sentence categorization, however, is dependent on the depth of the stack: to parse an n-word-long CFG sentence successfully, an n-layer-deep stack is required, whereas to parse an FSG sentence, a 1-layer-deep stack is always enough.

Apart from the FSG and CFG grammars mentioned above, the network was exposed to another type of CFG, called the palindrome. These types of sentences are similar to the center-embedded structure, but bounded words are identical. The network can learn palindrome grammar with equal ease, provided that the depth of the stack is at least half the number of the words in the palindrome sentence. Of course, if a network is trained on palindrome sentences, it will not recognize the original FSG and CFG grammars as correct, and vice versa.

The stack implemented here follows a design embedded in the chemical literature (Hjelmfelt et al. 1992) that rests on gating. We strongly believe that gating will be found to be crucial for hierarchical tasks, just as for complex cognition in general (Gisiger et al. 2005; O'Reilly 2006). The fact that it has readily evolved in a reinforcement-learning task in a simulated honeybee neural network (Soltoggio et al. 2007) supports this idea. The design of our minimalist network is very pragmatic in that it includes a perceptron (the pairing module) which can be substituted by a more realistic neuronal network if necessary. In contrast, we suggest that the introduction of the neural stack memory (pushdown automaton) will turn out to be substantial for any biological "hierarchical processor." The performance of our network naturally depends on the depth of the stack, and as such it can be replaced by a finite state automaton (Hopcroft and Ullman 1979). However, in this sense, human parsing ability is also limited: no person can parse sentences with arbitrarily many hierarchical layers (Pinker 1994). The likely hierarchical processor (maybe even supramodal) in humans with normal development is Broca's area (Friederici 2006; Tettamanti and Weniger 2006). Sadly, we know next to nothing about the relevant "internal wiring" of this area: we propose that it is likely to contain a neuronal stack, wherein gating will be found important.

Brain Epigenesis and Gene–Language Coevolution

It has to be admitted that on the whole we do not understand how the brain works. Nevertheless, some crucial elements seem to emerge. One is that development of the normal brain is enormously plastic, even though the power of genetic factors is obvious. One classic example is that in the same brain areas of identical twins, the two hemispheres of the same individual resemble each other more closely than the same hemispheres in the two people (Changeux 1983).

Another insight is that a tremendous amount of variation and selection transpires during brain ontogenesis. This is a Darwinian-type process, no doubt. As William James recognized a long time ago, natural selection of heritable variation is the only known force that can lead to adaptations, so let's apply it to brain ontogenesis and problem-solving as well (James thought that even learning is the result of selection of variation within the brain). There are several expositions that all regard the brain, one way or the other, as a "Darwin machine" (Calvin 1987). Here, for simplicity, we stick to the formulation by Changeux (1983), who stated that the functional microanatomy of the adult cortex is the result of the vast surplus in initial stock of synapses and their selective elimination according to functional criteria (performance).

In the previous section we learned that a very large part of the human brain can process linguistic information, including syntactical operations. This means that there is no fixed macro-anatomical structure that is exclusively dedicated to language, but some functional micro-anatomical structure *must* be appropriate, otherwise it could not sustain language. This further suggests that there is some *statistical connectivity feature* of a large part of the human brain that renders it suitable for linguistic processing (Szathmáry 2001). From the selectionist perspective there are three options: the initial variation in synaptic connectivity is novel; the means of selection on functional criteria is novel; or both. Maybe both component processes are different in the relevant human brain areas, and we do not dare to speculate about their relative importance.

This idea must be seen in close connection to the one presented by Rapoport (1990) concerning the coevolution of brain and cognition (within a population of humans). The traditional view is the so-called bottom-up mechanism: that is, a genetic change of some neural structure is subjected to selection and, based on its performance, it either spreads or it does not. There is, however, a so-called top-down mechanism, which could have contributed more significantly to the evolution of human cognitive skills, including language. The crucial idea is as follows:

- Due to the plasticity in brain development, enhanced demands on a certain brain region lead to less synaptic pruning (a known mechanism).
- Less synaptic pruning is assumed to lead to more elaborate (and more adaptive) performance.
- Any genetic change contributing to the growth of the brain area thus affected will be favored by natural selection.

Two important connections deserve to be highlighted. First (observed by Rapoport himself), the top-down mechanism is a more detailed exposition of the late Allan Wilson's idea (Wyles et al. 1983). Thus an increased brain, due to its more complex performance, alters the selective environment (in social animals composed of conspecifics to a great extent), which selects for an even larger brain, and so on. Second, and perhaps more important, this mechanism is also a neat example of a Baldwin effect (or genetic assimilation), when "learning guides evolution." As Deacon (1997) pointed out, it is trickier to apply the idea of genetic assimilation to language than usually thought. The reason for this is that the performed behavior must be sufficiently long lasting and uniform in the population. It is thus hard to imagine how specific grammatical rules, for example, could have been genetically assimilated. This point is well taken, but here we speak of a different thing: the genetic assimilation of a general processing mechanism that is performed by virtue of the connectivity of the underlying neural structures.

Our claim is that the most important, and largely novel, faculty selected for was the ability of the networks to process syntactical operations on symbols that are part of a semantically interwoven network. The specific hypothesis is that linguistically competent areas of the human brain have a statistical connectivity pattern that renders them especially suitable for syntactical operations. In conclusion, we think that:

- The origin of human language required genetic changes in the mechanism of the epigenesis in large parts of the brain.
- This change affected statistical connectivity patterns and dynamical development of the neural networks involved.
- Due to the selectionist plasticity of brain epigenesis, coevolution of language and the brain resulted in the genetic assimilation of syntactical processing ability as such.

An intriguing possible example of gene–culture coevolution has recently been raised by Bufill and Carbonell (2004), who call attention to a number of facts. First, human brain size has not increased over the past 150,000 years; in fact, it has decreased somewhat in the last 35,000 years. Second, a new allele of the gene for apolipoprotein E (ApoE4) originated sometime between 220,000 and 150,000 years ago. This allele improves synaptic repair (Teter et al. 2002). The original form entails a greater risk of Alzheimer disease and a more rapid, age-related decline in general (Raber et al. 2000). More importantly, ApoE4 impairs hippocampal plasticity and interferes with environmental stimulation of synaptogenesis and memory in transgenic mice (Levi et al. 2003). Interestingly, the ancestral allele decreases fertility in men (Gerdes et al. 1996). Taken together, these facts indicate, but do not prove, a role in enhanced synaptogenesis during a period when syntactically complex language is thought to have originated. More evidence like this would be welcome in the future, since one such case can at best be suggestive.

Selective Scenarios for the Origin of Language

On the Human-specific Adaptive Suite

Various people (e.g., Premack 2004) have called attention to the fact that besides language, efficient teaching (which differs from learning), imitation, and a developed theory of mind are also uniquely human resources. We would add to this the trait of human cooperation (Maynard Smith and Szathmáry 1995), which is remarkable in that humans are able to cooperate even in large nonkin groups. We propose that these traits did not appear by accident together. They form an adaptive suite, and presumably they have coevolved in the last five million years in a synergistic fashion (Szathmáry 2008; Szathmáry and Számadó 2008b). A relevant image is a coevolutionary wheel (Figure 2.3): evolution along any of the radial spokes presumably benefited all the other capacities, even if the focus of selection may have changed spokes several times. This hypothesis is testable; and there is indeed already evidence in its favor. Take the case of autism, for example. Affected people have a problem with theory of mind and communication, and they can be seriously challenged in the strictly linguistic domain as well (Fisher and Marcus 2006). The prediction is that there will be several to many genes found that will have pleiotropic effects on more than one spoke of the wheel in Figure 2.3.

Penn and Povinelli (2007) review evidence in favor of the idea that apes have no understanding of cause and effect in the physical domain. Wolpert (2003) points out that this seriously limits apes' capacity to make and use tools;

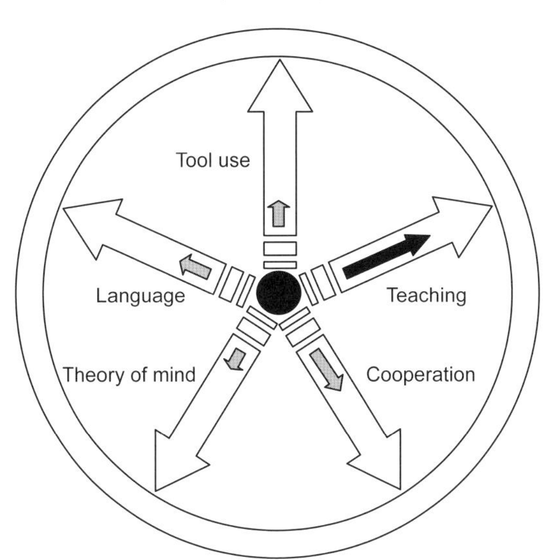

Figure 2.3 The coevolutionary wheel and the human adaptive suite. In this example, direct selection on genetic variation is on teaching/docility (black arrow). This leads to some improvement, to varying degrees, in other dimensions (gray arrows).

he even suggests that understanding causality may have given us language. But why are we so good at causal inference? Penn and Povinelli suggest that it is our capacity of analogical reasoning which helps us to figure out more causes:

> It is well known that human subjects often learn about novel and unobservable causal relations by analogy to known and/or observable ones: The structure of the atom, for example, is often described by analogy to the solar system; electricity is conceived of as analogous to a flowing liquid; gravity is like a physical force (Penn and Povinelli 2007, p. 111).

It is here where we see a possible close connection to language. First, many grammatical constructions function such that if one element appears, it causes the mandatory appearance of another. Second, a sentence like *Mary loves John* is analogous to *Susan loves Jim*: language rests on an unlimited variety of such analogical constructions. So, besides recursion, analogy seems to be a strong connection between tool making and language; incidentally, it links strongly to causal inference as well. Regarding the sequential order of these adaptations in evolution we stress again that it may be an ill-formulated question: as stated at the onset of this chapter, it is more likely that the target of selection shifted among components of the human adaptive suite, and that improvement in one component may have given advantage in some others as well. In fact, any genetic change providing pleiotropic advantage in more than one domain is more likely to become fixed than one with the same positive effect in only one domain.

Analogical reasoning is likely to be an intermediate phenotype that is necessary for causal reasoning, tool making, and language. The same may hold for shared intentionality (Tomasello and Carpenter 2007): analysis of child development suggests that gaze following develops into joint attention, social manipulation transforms into cooperative communication, group activity develops into collaboration, and social learning develops into instructed learning. A third such intermediate skill is recursive processing in intentionality, tool making, and language. Fine motor control (Calvin 1991) is an executive skill that is required for the efficient implementation of human adaptations.

It is apparent that components of the human-specific adaptive suite can be tentatively grouped into two categories: (a) indispensable procedural components (handling hierarchies, analogical reasoning, imitation, shared attention and intentionality, and fine motor control) and (b) complex adaptive faculties (docility, complex cooperation and theory of mind, language, and tool making).

This proposal raises a question about the evolution of such a complex network of procedures and interactions. As stated, pleiotropy is likely to be the rule here. If so, a concern can be raised about the evolutionary plausibility of such a suite, given the fact that many think that pleiotropy retards evolution (Orr 2000). Griswold (2006) has approached this question from a different perspective. First, he demonstrated that if a mutation with pleiotropic effects is overall beneficial, then it is likely to be beneficial for more than one trait. Second, such a mutation, although rare, will spread faster and have a higher rate of fixation in the population than others. Third, the rate of evolution of a phenotypic character may not decline when that character is pleiotropically associated to an increasing number of other characters, provided that the characters are under pure directional selection such that they are far from their optima relative to the average magnitude of a mutation. Griswold notes that such a situation is typical for adaptive radiations. We think that in the past 5 million years, such a radiation has happened in the hominine lineage.

Selective Scenarios for the Origin of Language

The origin of human language has provided fertile grounds for speculation and alternative theories have been proposed (Table 2.1).

Most of the theories that suggest a given context for the evolution of human language attempt to account for its functional role. Given that all of these theories are functionally more or less plausible, it is almost impossible to decide on their usefulness based only on this criterion. However, recent game theoretical research can help us evaluate various contexts. These criteria concern the interest of communicating parties and the cost of equilibrium signals.

The central issue is whether early linguistic communication was honest. If signal cost is the same for all signallers, then honest cost-free signalling can be evolutionarily stable only if there is no conflict of interest between the participants (Maynard Smith 1991). If the cost of signals varies with the quality of the signaller, then the situation is more complicated. In this case, it is possible to construct cost functions that give an arbitrarily low cost at equilibrium even if there is a conflict of interest (Hurd 1995; Számadó 1999; Lachmann et al. 2001) (Table 2.2). In the case of human language, the most obvious way to construct such a cost function is to punish dishonest signallers (Lachmann et al. 2001). This solution assumes, however, that dishonest signallers can, on average, be detected (i.e., signals can be cross-checked); it also assumes that dishonest signallers are punished (which is a nontrivial assumption). Thus, one can conclude that "conventional" signals will be used when communicating about (a) coincident interest or (b) verifiable aspects of conflicting interest; "costly" signals will be used otherwise (Lachmann et al. 2001). Although theory thus far says nothing about the evolution of such systems of communication, there are a few computer simulations which suggest that honest, cost-free communication evolves only if there is shared interest between the participants (Bullock 1998; Noble 2000; Harris and Bullock 2002).

What does this tell us about the emergence of human language? The production cost of speech or gesturing appears to be low; thus human language consists of cost-free or low-cost signals at equilibrium (not counting time constraints). Based on the above criteria, one should favor either those theories which propose a context with no conflict of interest (e.g., hunting, tool making,

Table 2.1 Alternative theories to explain language evolution.

Theory (Source)	Description
Gossip (Power 1998)	Menstrual ritual can be a costly signal of commitment; hence participating in such rituals can create female groups of shared interest in which sharing information about the social life of others (i.e., gossiping) can be beneficial
Grooming (Dunbar 1998)	Language evolved as a substitution for physical grooming. The need for this substitution derived from the increasing size of the early hominid groups.
Group bonding and/or ritual (Knight 1998)	Language evolved in the context of intergroup rituals, which first occurred as a kind of "strike action" against non-provisioning males. Once such rituals were established, a "safe" environment was created for further language evolution
Hunting (Washburn and Lancaster 1968); (Hewes 1973)	Our intellect, interests, emotions, and basic social life are evolutionary products of the success of the hunting adaptation. Later, Hewes (1973) argued that the probable first use of language was to coordinate the hunting effort of the group.
Language as a mental tool (Burling 1993)	Language evolved primarily for the function of thinking and was only later co-opted for the purpose of communication
Mating contract and/or pair bonding (Deacon 1997)	Increasing size of the early hominid groups and the need for male provisioning also necessitated "social contract" between males and females
Motherese (Falk 2004)	Language evolved in the context of mother–child communication: Mothers had to set their babies down to collect food efficiently, and their only option to calm them was to use some form of vocal communication
Sexual selection (Miller 2001)	Language is a costly ornament that enables females to assess the fitness of a male. According to this theory, language is more elaborate than a pure survival function would require.
Song (Vaneechoutte and Skoyles 1998)	Language evolved rapidly and only recently through cultural evolution, assuming two important sets of preadaptations: the ability to sing and better representation abilities (i.e., thinking and mental syntax).
Status for information (Desalles 1998)	Language evolved in the context of a so-called "asymmetric cooperation," where information beneficial to the group was traded for status.
Tool making (Greenfield 1991)	Assumes a double homology: a homologous neural substrate for early ontogeny of the hierarchical organizations shared by two domains—language and manual object combination—and a homologous neural substrate and behavioral organization shared by human and nonhuman primates in phylogeny.

Table 2.2 To evaluate the properties and the explanatory power of the various theories, the following questions were asked: (1) Can the theory account for the *honesty* of early language, i.e., is there a shared interest between the proposed communicating parties? (2) Are the concepts proposed by the theory *grounded* in reality? (3) Can the theory account for the *power of generalization* unique to human language? (4) Can the theory account for the *uniqueness* of human language? T: thought; V: vocalization; G: gestures.

Theory (Source)	Modality	First words	Topic	(1)	(2)	(3)	(4)
Gossip (Power 1998)	V	"Faithful," "philander"	Social life	No	No	Yes	No
Grooming (Dunbar 1998)	V	?	?	Yes	No	No	No
Group bonding and/or ritual (Knight 1998)	?/V	?	?	Yes	No	No	No
Hunting (Washburn and Lancaster 1968); (Hewes 1973)	G/V	Prey animals	Coordination of the hunt	Yes	Yes	Yes	No
Language as a mental tool (Burling 1993)	T	?	?	Yes	No	Yes	No
Mating contract and/or pair bonding (Deacon 1997)	?	?	Social contract	No	No	No	No
Motherese (Falk 2004)	V	"Mama"	Contact call	Yes	Yes	No	No
Sexual selection (Miller 2001)	?	?	Anything	No	No	No	No
Song (Vaneechoutte and Skoyles 1998)	V	?	?	No	No	No	No
Status for information (Desalles 1998)	?	?	Valuable information	No	No	Yes	No
Tool making (Greenfield 1991)	?	?	?	Yes	Yes	Yes	No

motherese, grooming or group bonding, and/or ritual theory) or a context in which there might be a conflict of interest but where signals can be easily cross-checked. None of the theories fit the second context. For example, both mating contract and gossiping assume a context in which conflict of interest exists and signals cannot be easily cross-checked.

Explaining the evolution of human language is likely to remain a challenge for the coming decade. Presently, no single theory is able to answer sufficiently all the questions about honesty and groundedness, power of generalization, and uniqueness. Table 2.2 gives a summary of these criteria (Számadó and Szathmáry 2006). As one can see, most of the theories fail to answer the majority of the questions. Perhaps the easiest criterion to fulfil is shared interest, as there are a number of social situations which assume shared interest between communicating parties (e.g., hunting or contact calls). Only two theories—tool making (Greenfield 1991) and hunting (Washburn and Lancaster 1968)—do significantly better than the others, as they can answer three out of the four questions asked (Table 2.2). Thus, it might be tempting to say that some combination of the two could provide a series of selective scenarios that would fit all of our criteria. The most notable conclusion, however, is that all theories fail to explain the uniqueness of human language. Thus, even though indirect evidence strongly suggests that the evolution of human language was limited by selection, it remains difficult to envisage a scenario that would explain why.

Although the different scenarios suggest all kinds of selective forces, none of these scenarios has been consistently implemented in a family of models. Given the limitations of experimentation on humans and chimps, researchers should consider implementing the different scenarios in various model-based settings. Ultimately, researchers should be able to reenact the emergence of language in artificial worlds, many of which will probably involve robots.[1] The use of robots offers a unique and probably indispensable way of symbol grounding (basic words, via concepts, should be linked to physical reality; Steels 2003) and somatosensory feedback (actions, or results of actions, on behalf of the agent feed back into its own cognitive system via sensory channels; Nolfi and Floreano 2002).

Some major transitions in evolution (such as the origin of multicellular organisms or that of social animals) happened a number of times, whereas others (the origin of the genetic code, or language) seem to have been unique events (Maynard Smith and Szathmáry 1995). One must, however, be cautious with the word "unique." Due to a lack of the "true" phylogeny of all extinct and extant organisms, one can give it only an operational definition (Szathmáry 2003). If all the extant and fossil species, which possess traits due to a particular transition, share a last common ancestor after that transition, then the transition is said to be unique. Obviously, it is quite possible that there have been independent "trials," as it were, but we do not have comparative or fossil evidence for them. What factors, then, can lead to "true" uniqueness of a transition? (a) The transition is variation-limited. This means that the set of requisite genetic alterations has a very low probability. "Constraints" operate here in a broad sense. (b) The transition is selection-limited. This means that there is something special in the selective environment that can favor the fixation of otherwise not really rare variants. Abiotic and biotic factors can both contribute to this limitation. For example (Maynard Smith 1998), a single mutation in the hemoglobin gene can confer on the coded protein a greater affinity for oxygen: such a mutation got fixed in some animals which live at high altitudes only (such as the lama or the barred goose, the latter migrating over the Himalayas at an altitude of 9000 m).

[1] One example is the ECAgents, a project sponsored by the Future and Emerging Technologies program of the European Community; see http://ecagents.istc.cnr.it/.

There are interesting subcases for both types of limitation. For (a), one can always enquire about the time-scale. "Not enough time" means that given a short evolutionary time horizon, the requisite variations have a very low probability indeed, but this could change with a widened horizon. An interesting subcase of (b) is "pre-emption," meaning that the traits resulting from the transitions act via a selective overkill and sweep through the biota so quickly that they competitively suppress further evolutionary trials. The genetic code could be a case in point.

Acknowledgments

The authors thank Mauro Santos, Luc Steels, Michael Corballis, Peter Gärdenfors, and Chrisantha Fernando for useful discussions. Partial support of this work has been generously provided by the National Office for Research and Technology (NAP 2005/KCKHA005) and the Hungarian Scientific Research Fund (OTKA NK73047).

3

Functional Neuroimaging and the Logic of Brain Operations

Methodologies, Caveats, and Fundamental Examples from Language Research

Balázs Gulyás

Abstract

With the advent of positron emission tomography (PET), followed by functional magnetic resonance imaging (fMRI), magnetoencephalography (MEG), and other complementary methods, cognitive neuroscience is now equipped with a unique methodological array of functional neuroimaging tools. Through these methods, active neuronal populations can be mapped during brain activation, thereby allowing the internal logic of the brain to be deciphered. Functional imaging techniques are ideally suited to explore the neuronal correlates of language. Exploration, however, is subject to existing methodological limitations as well as the spatial and temporal constraints of the technique used.

This chapter reviews the methodological background of functional neuroimaging and explores the pros and cons of its general use and application in cognitive research, with special emphasis on language research. Methodological limitations are discussed and caveats provided, and the history of neuroimaging in language research is briefly presented. The objective is not to give a detailed, comprehensive analysis of advanced neuroimaging studies relative to the neurobiological underpinnings of language, but rather to provide an introduction to the field.

Introduction

Functional neuroimaging techniques entered the research battery of neurosciences over two decades ago. By using positron emission tomography (PET), fMRI, or magnetoencephalography (MEG), it is generally accepted that the active neuronal populations responsible for sensory, motor, cognitive, or emotional processes can be identified, localized, and visualized in the living human brain. The seemingly uncontested theory of the physiological foundations of brain imaging (i.e., the origin of the imaging signal and its correlation with neuronal events) claims that the functional architecture of the human brain can be mapped during various sensory, motor, or cognitive tasks using these techniques.

Can we indeed explore, interpret, and understand the neurobiological basis of the human brain's mental processes with imaging? Are these techniques helping us to reveal the neurobiological underpinnings of cognitive processes? Can we use them to uncover the neuronal underpinnings of language understanding, language generation, and various biological aspects of syntax and semantics?

The essence of the conundrum arises from the fact that brain activities are multidimensional and can be approached from various points of view, using different methodologies. For example, to measure brain activation directly, blood flow and metabolic changes, changes in the electrical activity of cells and cell populations, neurotransmitter dynamics, as well as other consequent biochemical, physiological, and/or physical parameters (e.g., neuromagnetic changes) can be utilized (Figure 3.1).

When we use imaging techniques that are based primarily on blood flow and metabolic measures, we must remember that even during the simplest task, all of these processes operate in a closely interacting manner. Therefore, before final conclusions about brain function are drawn from pure imaging data, the exact relationship between the levels of neuronal organization and function must be clarified (Figure 3.2).

PET and fMRI are the most widely used techniques because of their usefulness in localizing the active neuronal processes through at least two approaches.

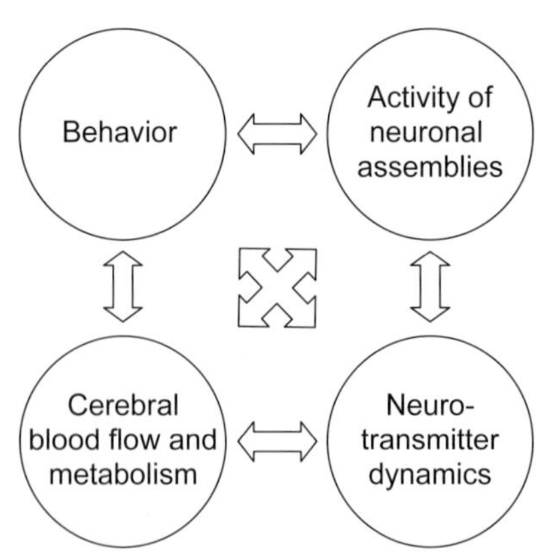

Figure 3.1 Four facets of the functioning human brain: neuronal activity of cortical micro-networks (action potentials, field potentials); neurotransmitter dynamics (e.g., transmitter release, binding, uptake, re-uptake, modulatory effects); regional cerebral blood flow and metabolism; and behavior (Kéri and Gulyás 2003).

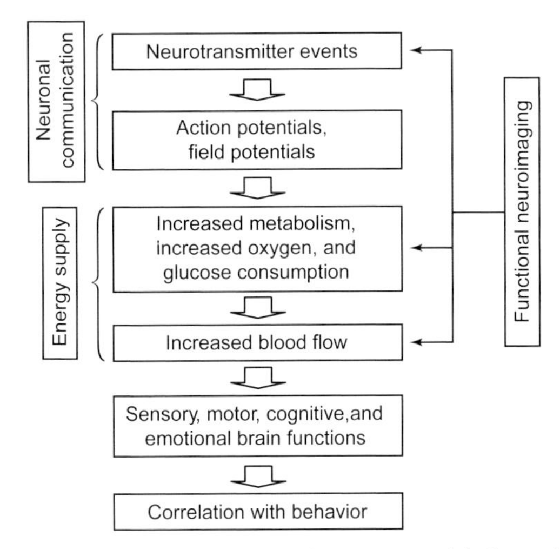

Figure 3.2 Classical logic of brain activation: Neuronal information transmission processes are initiated through the release of neurotransmitters, resulting in neuronal communication and action potentials. These processes all require energy. The outcome is brain activation in the sensory, motor, cognitive, or emotional systems of the brain. The measurable output is a behavioral event. Functional neuroimaging can uncover neurotransmitter events as well as metabolism and blood flow-related changes (Kéri and Gulyás 2003).

First, since active neuronal populations that participate in information processing in the brain require more energy than those not involved directly in such processes at any given time, their glucose and oxygen consumption will be increased. This entails amplification of the regional cerebral blood flow as well as increased oxygen extraction in the active brain regions, which, in turn, can be localized using PET or fMRI (Roland 1993). Second, increased neuronal activity results in more regional electromagnetic activities in the activated neuronal populations as well as in their projection pathways, which can be quantitatively detected and localized with MEG or electroencephalography (EEG).

A Caveat

Despite more than ten thousand publications to date on the localization of cortical functions with functional neuroimaging techniques, a few caveats should be noted. The origin of the imaging signal is not fully clarified in either the PET or the fMRI technique.

First, it is accepted without question that neuronal activation (i.e., increased firing at the neuronal level, which is the essence of all brain processes) is an energy-requiring process that entails increased brain metabolism, oxygen and glucose consumption, and blood flow. The increase in neuronal activation and spiking frequency during information processing in the brain needs to be compared, however, to the spontaneous firing rate and/or baseline activities of the neurons. Does ordered neuronal firing necessarily require more energy, and consequently higher blood flow and metabolism, than spontaneous firing (Singer 1999)?

Second, the detailed contribution of and the balance between excitatory and inhibitory neuronal activities in pattern generation are not fully understood. Inhibitory interneurons significantly outnumber excitatory neurons; however, the body size and axon length (alongside which the action potential should proceed) of the excitatory neurons (mainly pyramidal cells) significantly exceed those of the interneurons. Consequently, the energy requirement of an excitatory neuron is higher than that of an inhibitory interneuron. What happens, though, in neuronal pattern generation? What are the realistic proportions of activated interneurons and excitatory neurons, and what is the price of activation and inhibition? What is the real price of neuronal pattern generation during information processing in the brain? Despite several explanatory theories and models (e.g., Lennie 2003; Buzsáki et al. 2007), these questions remain unanswered. Nevertheless, despite the shortcomings of our knowledge, we accept that functional neuroimaging is a useful way toward a better understanding of brain functions, including language-related brain processes.

Neuroimaging Techniques

The spatiotemporal domain of the human brain can be addressed almost entirely using a wide variety of functional neuroimaging techniques (Figure 3.3), each of which has different strengths and weaknesses. The spatial resolution of the techniques lags behind that of the best morphological imaging technique (MRI), but PET and fMRI still have better spatial resolutions than either MEG or EEG, which in turn have better temporal resolution than PET and fMRI. Thus, complementary use of various anatomical and functional imaging techniques can enhance imaging conditions to produce optimal results with respect to covering both the spatial and temporal domains. This fact is indeed of paramount importance in the functional imaging of language-related brain processes, which spatially may span fractions of a cortical area (i.e., the submillimeter range) to large and extended cortical macro-networks (i.e., the 10 cm range), whereas temporally it covers the subsecond range (from some 10 ms, the reaction time of recognizing familiar syllables or names, to several hundred milliseconds for the processing of larger semantic units).

Functional Neuroimaging with Positron Emission Tomography

The principle of PET can be traced back to Georg de Hevesy's classic discovery that radioisotopes can be used as tracers of biological processes. In PET,

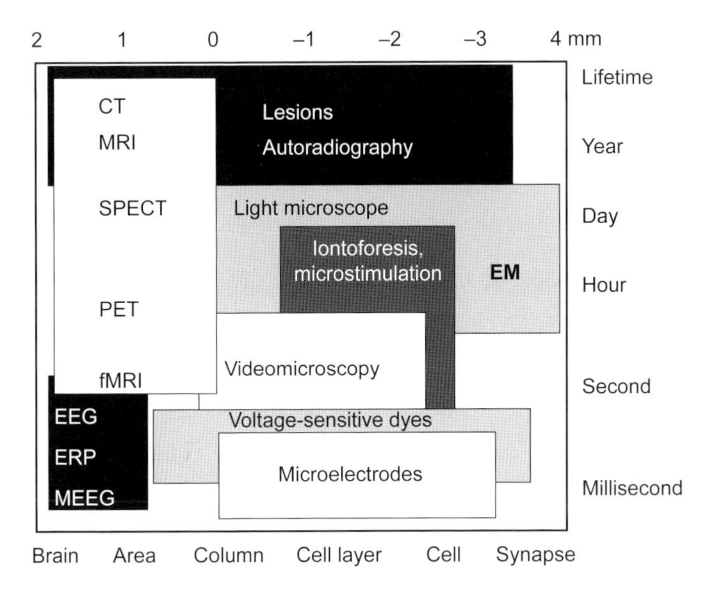

Figure 3.3 The place of functional imaging techniques in the battery of neuroscience methodologies, covering the spatiotemporal domain of the human brain.

positron-emitting radionuclides, such as 11C, 13N, 15O, or 18F, are incorporated into a biologically inert or biologically active molecule and administered to a living human or animal as "radiopharmaceuticals" or "radiotracers." Due to the decay of the radionuclides, positrons are continuously generated. When they encounter an electron, positrons are annihilated and along one side of the axis, two 511 keV gamma photons are emitted from the annihilation site in two directions (Gulyás and Sjöholm 2007). With an appropriate detector system, these generated gamma photons can be detected, and the original "annihilation maps," representing the radioactivity distribution inside the detector ring of a PET scanner, can be faithfully recreated.

To explore the higher functions of the human brain, two radiotracer categories are of considerable importance: blood flow tracers and radioligands for neuroreceptors. Blood flow tracers (e.g., 15O-water or 15O-butanol) are inert chemicals that enter the bloodstream and circulate with the blood. The global cerebral blood flow in primates, with special regard to humans, is relatively high: approximately 20% of the total circulation. Regional cerebral blood flow is directly proportional to neuronal activity. Cortical neuronal populations with increased neuronal activity require more oxygen, more glucose and, consequently, more blood. Increased regional cerebral blood flow is therefore a direct indicator of increased neuronal activity in the brain. Thus, regional cerebral blood flow measurements with PET can be used to localize task-specific neuronal activity in the human brain. This technique is widely used and, until the introduction of the fMRI BOLD technique, was the primary methodological approach of functional localization in neuroimaging. The other option is for radionuclides to be incorporated into molecules that bind to neuroreceptors or transporter molecules. The resulting radioligands can be used in various paradigms to explore the distribution of central neuroreceptors in the human brain as well as their functional state in rest or under physiologically or pharmacologically challenging conditions.

Functional Magnetic Resonance Imaging

Functional magnetic resonance imaging, or fMRI, can quantitatively map various types of physiological information about the working brain, including cerebral blood volume and perfusion (and the corresponding changes), blood oxygenation, oxygen extraction, and oxygen consumption. The common basis of all fMRI techniques is that of magnetic resonance imaging (for an overview, see Bandettini 2007).

A large part of the atomic nuclei has a magnetic moment due to the fact that some of their components (e.g., protons) have a quantum mechanical property called spin. Certain nuclei (e.g., 1H, 7Li, 13Na, or 31P) possess two spin states: an "up" and a "down" state. When these atomic nuclei are placed in a strong external magnetic field, the spins precess around an axis parallel to the direction of the field, and the protons align in two energy "eigenstates" separated by a quantum of energy.

In the MRI scanner, a strong static magnetic field causes the spin of certain atomic nuclei within the body to be oriented parallel or anti-parallel to the magnetic field and the nuclei process about the magnetic field with a frequency. In the static magnetic fields commonly used in MRI (0.5–9 Tesla), the energy difference between the nuclear spin states corresponds to a photon at radio frequency wavelengths, and a tiny excess of protons are in the higher energy state. This gives a net polarization parallel to the external field. For instance, at 1.5 Tesla, 1,000,000 protons of the H atoms in the body that have a higher energy state correspond to 1,000,010 protons that have a lower energy state. In addition to the magnet that provides the static and homogeneous magnetic field in the scanner, a number of gradient coils are also used to encode the positions of protons spatially, by varying the magnetic field linearly across the imaging volume.

Magnetic resonance occurs when a radio frequency pulse excites the nuclear spins and raises them from the lower energy state to the higher energy state. In other words, the nuclei absorb energy resulting in a shift between lower and higher energy states. In the MRI scanner, a radio frequency coil is used to cause excitation of the nuclear spins. When the radio frequency pulse is turned off, the excited nuclei dissipate their excess energy to their neighborhood, the lattice, and return from excited state to ground state. This process produces an oscillating magnetic field, which can be detected as an induced current in a receiver coil built into an MRI scanner.

A number of different schemes have been created for combining field gradients and radio frequency excitation to create an image, using the tomographic approach. A revolutionary development of the application of MRI imaging appeared when the BOLD technique was developed. This technique is based on the fact that blood oxygenation is lower in capillaries and veins than in arteries, due to the extraction of oxygen from the blood needed for cellular metabolism. Compared to the rest of brain tissue, water, or oxyhemoglobin, deoxyhemoglobin is paramagnetic. Thus, it creates a distortion in a magnetic field that can be quantitatively measured and localized using MRI.

Increased regional neuronal activity results in greater oxygen extraction as well as in increased regional blood flow. The net sum of the two physiological phenomena is a decreased level of deoxyhemoglobin and, consequently, a decrease in magnetic field distortions in the loci of the activated neuronal populations. In other words, neuronal activation will appear as signal increase in the MRI image.

The advantage of fMRI BOLD imaging over PET blood flow imaging is that it is noninvasive: no radiation exposure is involved, sampling frequency is much higher, and thus measurements can be repeated several times in the same subjects.

Electroencephalography and Magnetoencephalography

Despite advances in functional neuroimaging techniques over the past two decades (in particular, PET and fMRI), EEG and MEG continue to be indispensable tools in the hands of cognitive neuroscientists. Due to their methodological advantages, their importance has steadily increased (for an overview, see Bagic and Sato 2007).

The source of the EEG signal is an extracellular field potential, which is basically the cumulative postsynaptic activity of the cortical pyramidal cells. EEG signals are detected on the scalp. Because pyramidal cells are arranged radially to the surface of the cortex, the source of the EEG signal is radial to the skull, where the cortex is on the surface of the brain (ca. 30%); it is, however, to a varying extent, tangential to the surface of the cranium in the cortical sulci (70%). An EEG measures the sources from both radial and tangential orientations, although the predominant source is radial. The best temporal resolution of the technique is a few milliseconds. Because EEG sensors are usually relatively far from the signal source, the spatial resolution of the technique is not well defined, compared to other functional imaging techniques (e.g., fMRI). Using up to 128 sensors, the area of the active cortex "seen" by the technique occupies only a few square centimeters.

In contrast to EEG, MEG measures the magnetic field changes that are generated by the magnetic dipole sources on the cortical surface. The magnetic fields produced by electrical activity in the brain are a result of the intracellular dendritic currents. Compared to the Earth's magnetic field (30–60 micro-Tesla), they are orders of magnitude weaker (a few femto-Tesla). For this reason, extremely sensitive devices, such as superconducting quantum interference devices (SQUIDs), must be used, and this requires magnetic shielding as well as the maintenance of extremely low temperatures around the device.

The temporal resolution of the technique is in the millisecond range, whereas the spatial resolution of the more advanced devices is in the millimeter range. Although EEG and MEG signals have a common origin, compared to the EEG signal, the MEG signal is less prone to disturbances due to the resistive properties of the skull and scalp. Whereas EEG is capable of detecting both radial and tangential signal sources and can thus measure neuronal activities on the surface of the brain as well as in the depths of the sulci, MEG is used only to detect tangential signals (i.e., those generated on the cranial surface of the cortex).

Functional Neuroimaging in Cognitive Neuroscience Research

The classic paradigm used during the early phase of neuroimaging was the subtraction paradigm. The idea can be traced back to the Dutch physiologist and ophthalmologist, Franciscus Cornelius Donders, in the 19th century, when Donders applied "sequential" stimulation paradigms to "decompose" the neurophysiological basis of complex brain functions with superimposed components. In his paradigm designs, Donders used realistic behavioral reaction times, which, in addition to his revolutionary "functional decomposition technique," greatly advanced the then maiden field of experimental psychology. In modern subtraction paradigms, two experimental situations (say, A and B) are presented and the underlying neuronal activities are measured with a neuroimaging technique. The two paradigms differ from each other in only one stimulus dimension, one stimulus feature. The hypothesis behind the subtraction paradigm states that the difference in brain activation measurements, displayed on the PET or fMRI images, highlights those neuronal populations that are responsible for the processing and analysis of the given stimulus feature. For instance, in the case of color vision, presentation of the same image in full color and in monochromatic gray results in different regional cerebral blood patterns, the difference of which may indicate those neuronal populations that are highly dedicated to the analysis of color information (Gulyás and Roland 1991, 1994b) (Figure 3.4).

The human brain is never in a real resting state; each sensory, voluntary, or cognitive process causes perturbation on an already existing activation state. This has been included in other, more complex, multifactorial paradigm

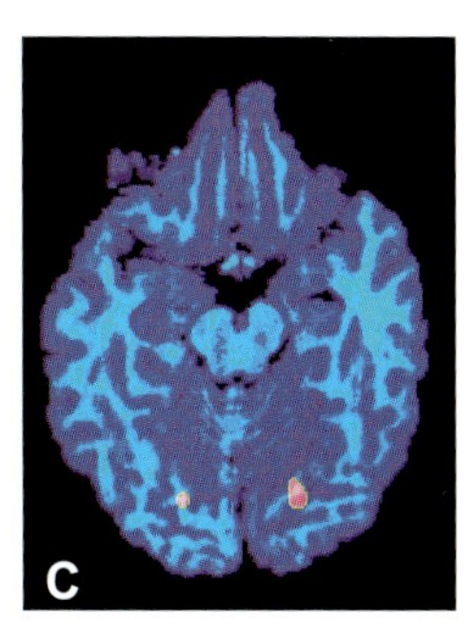

Figure 3.4 The subtraction paradigm: visual scenes (A and B), differing from each other in only one stimulus dimension (color), result in PET images of regional cerebral blood flow distribution. The subtraction of this results in an image displaying the neuronal populations engaged by the processing and analysis of color information (C).

designs currently in use in functional neuroimaging studies. The analysis of imaging data can be based upon different approaches. Using a data-driven approach, fundamental statistical tests (e.g., t-test, ANOVA, ANCOVA) are applied to the image data sets, usually comprising a large number of volume elements or voxels. This approach is not biased with respect to a preliminary working hypothesis or observer expectations. Hypothesis-driven approaches can also have legitimacy under various conditions, especially when expectations from certain anatomical regions are evident. The most commonly used analysis of PET images, similarly to that of other neuroimaging modalities, is based upon the general linear model, one of the most important and widely used statistical models applicable to biological and social data sets. The most commonly used version of the general linear model is statistical parametric mapping (Friston et al. 2007).

The Logic of Brain Operations

Convergence and Divergence

From the onset, a leading question concerned the involvement of one or more cortical neuronal populations in the processing and analysis of simple brain tasks (e.g., a perceptual task). This issue can be traced back to the age-old localization problem in neurology: one cortical region, one brain function. The extreme version of the hypothesis suggests that one brain cell in higher cortical regions is specialized to one highly defined function: the recognition of the observer's grandmother by "the grandmother cell" (Gross 1992; Barlow 1995). Of course, as complex brain functions cannot be bound to one single neuron, the moderate version of the hypothesis is inclined to state that certain well-defined brain functions and the function of well-circumscribed, anatomically coherent neuronal populations correlate. A classic example is the "color area of man": a well-defined region in the human fusiform gyrus which, according to the original protagonists of the hypothesis, specifically underlies color perception (Lueck et al. 1989; Zeki et al. 1991).

Detailed neuroimaging studies on simple perceptual or motor functions indicate, however, that even the simplest perceptual or motor tasks engage a number of cortical neuronal populations (Gulyás and Roland 1991, 1994a, b; Claeys et al. 2004). For instance, in the case of visual perception, incoming visual information is distributed in the visual cortex, and a number of cortical regions are engaged in the processing and analysis of the various visual submodalities of the visual image before a unified, integrated visual percept is generated. A body of literature indicates clearly that for the processing of even the simplest perceptual, motor, or cognitive tasks, a *network* of neuronal populations is activated (i.e., functional networks)—not single cortical areas. This is even true for the processing and analysis of the simplest visual submodalities, such as color, form, or disparity (Figure 3.5). And, indeed, this is valid for the whole range of brain operations, from simple ones to complex highest level cognitive operations (Mesulam 1998).

Analysis of a large number of imaging (and other) studies in the sensory fields indicates that the processing of incoming information diverges: the information reaches, either by parallel or serial channels, a number of cortical neuronal populations which participate in the processing and analysis of the given sensory information. This is referred to as the *divergence principle* (Gulyás 2001).

The question then is: Can the same neuronal populations, which form a part of a functional network X, be used to form a part of another functional network Y? Can and do the same cortical regions participate in different functional networks, underlying different perceptual, motor, or cognitive processes in the brain? Stated differently, are the very same neuronal populations multifunctional such that they can participate in the processing and analysis of various tasks?

Again, comparative analysis of several functional neuroimaging studies with sensory, motor, or cognitive functions indicate that different functional macro-networks may include the very same cortical regions. For example, we have shown that cortical areas involved in color and disparity processing, or the analysis of spatial frequency and orientation information, may be congruent at the macro-network level. This does not necessarily mean, however, that at the cellular level the very same neuronal populations are active. In blobs and/or interblobs of the primary visual cortex (V1 or Brodmann 17), for example, neurons specialized for various tasks (e.g., orientation, disparity, color) may overlap with each other, and it is possible that during the processing of one or another visual submodality, only a fraction of the neurons in a cortical region is predominantly active (Gulyás and Roland 1994a, 1995) (Figure 3.6).

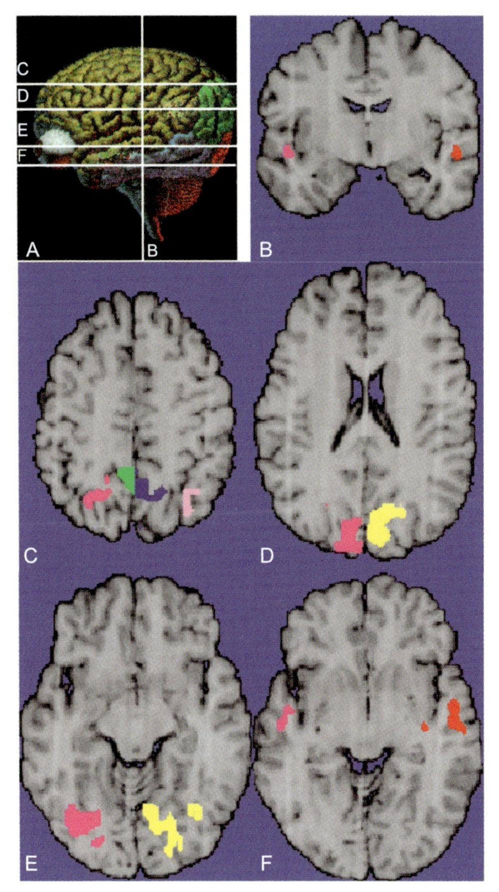

Figure 3.5 Divergence: Cortical fields activated by the processing and analysis of color information. Meta-analysis of four color detection related PET experiments. The horizontal image slices in panels B–F intersect the image plans indicated in panel A.

During the processing and analysis of sensory information, the same "hubs" in various cortical macro-networks can be involved; this indicates that convergence is also a key phenomenon in cortical information processing. The very same cortical fields may participate in the processing and analysis of various information (e.g., sensory, motor, or cognitive). This can be referred to as the *convergence principle.*

In short, the basic logic of information processing in the human brain can be traced back to two elementary principles: divergence and convergence. Divergence means that the very same information reaches various cortical neuronal populations and is processed and analyzed by divergent cortical neuronal populations. Convergence means that the very same cortical neuronal population may participate in the processing and analysis of various cortical information processes, both bottom up and top down.

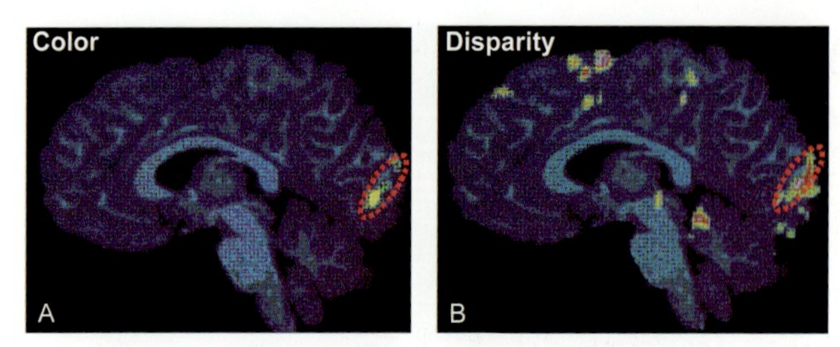

Figure 3.6 Convergence: Cortical fields activated by the processing and analysis of color (A) and disparity (B) information. The area of overlap of the neuronal populations participating in both tasks in the occipital cortex is indicated by a dotted ellipse in each panel (meta-analysis, cf. Gulyás et al. 1994).

Core Networks and Recruited Fields

Cognitive tasks vary according to their complexities. Some tasks are simple, requiring relatively little neuronal activity. Others are more complex and may extensively use the human brain's processing capacity.

During the processing of, for example, a simple visual task, it appears quite natural for the lateral geniculate nucleus, the primary visual cortex, and a few other visual cortical areas to be involved, independently of the submodalities of the visual information. Indeed, during the processing of various visual tasks, a number of visual cortical fields are present in the various cortical macronetworks underlying the different tasks. These networks contain a number of cortical neuronal populations as a central core (a "core network") which is, in the case of sensory information processing, sensory modality dependent. The "core network" is absolutely necessary for the given task, but it may be not sufficient. Additional cortical neuronal populations, which are essential for task performance under various stimulus or task conditions, may be recruited to join this "core network." These fields can be termed "recruited fields" or "recruited neuronal populations." For instance, in the case of a visual task on feature uncertainty analysis, using gratings with varying orientations and spatial frequencies, we can identify a core network of cortical neuronal populations present in all task performances, whereas certain cortical fields are only present when the decision is made along one specific stimulus modality (e.g., orientation or spatial frequency; Gulyás and Roland 1995) (Figure 3.7).

My colleagues and I have attempted to explore the task, stimulus, and input modality of the cortical networks, with special regard to the core and the recruited fields. In one experiment, we used the same stimuli under different task conditions (Vidnyánszky et al. 2000). Subjects visually inspected identical stimuli: gratings in the center or placed out of center in a rectangle. The task was either to make a discrimination regarding the form (regular or irregular grating) or the position (centered or not) of the stimulus; that is, the stimuli were identical, but the tasks were different. The resulting cortical networks contained a core network and recruited fields that were task dependent,

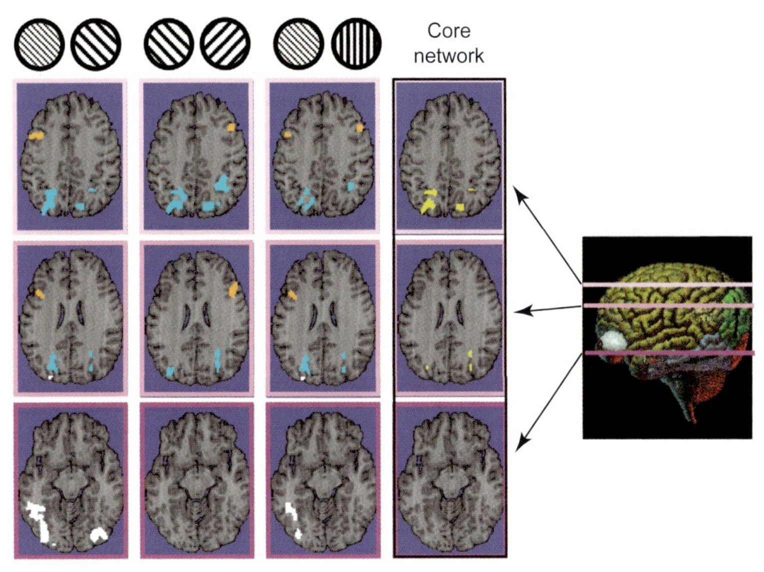

Figure 3.7 Core networks and recruited fields. Cortical fields activated by spatial frequency (first column), orientation (second column), or conjoint spatial frequency–orientation (third column) tasks, using identical stimuli (gratings with varying spatial frequencies and orientations). The cortical fields activated by all tasks (core network) are shown in the fourth column. The position of the horizontal slices is shown in the right hand panel (Gulyás and Roland 1995).

indicating that recruited cortical neuronal populations may indeed be task dependent (Figure 3.8).

In addition, investigations were conducted to examine to what extent the recruited cortical fields are stimulus dependent when the tasks are identical. Subjects were asked to inspect small objects (so-called parallopipeda) that were presented in three-dimensional (i.e., as real-life objects) or two-dimensional (e.g., photographs) form. The objects were presented as consecutive pairs (first object followed by the second), and the subjects had to say whether they were identical or not. Regardless of whether the object was 2D or 3D, a core network was evident from which stimulus-dependent cortical neuronal populations were recruited (Kovács et al. 1998).

How does this work in higher-level sensory tasks (e.g., the recognition of the same visual form on the basis of various visual input cues)? It is a common fact that visual contours can be obtained on the basis of various visual information: luminance, color, disparity, texture, and motion can each create visual contours. Consequently, identical form stimuli can be generated by each of these visual cues. In an experimental series, which used an "odd one out" paradigm, we explored whether the generation of identical form stimuli (rectangles) is dependent on input cues or not. Our findings indicated clearly that although the resulting form percepts were identical, the cortical networks

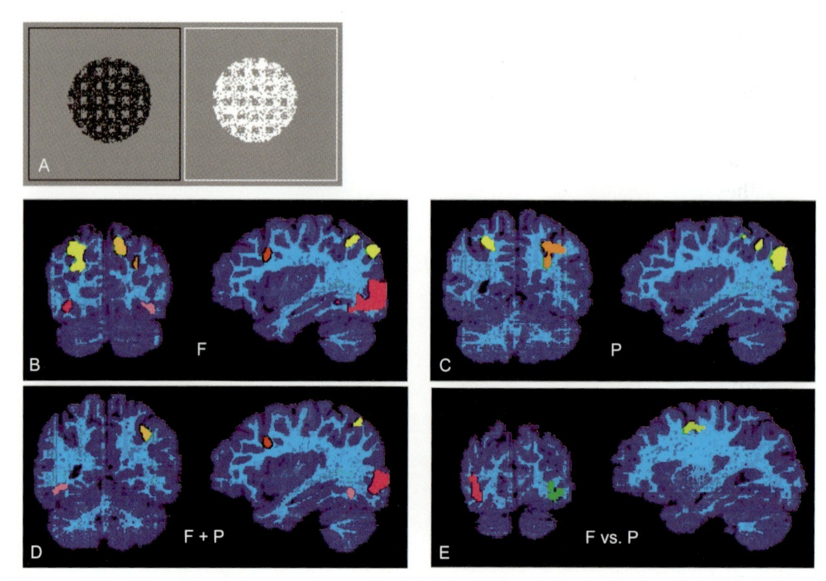

Figure 3.8 Task dependence of recruited fields: Using identical stimuli (A) in a form or position discrimination task, the form (B) and position (C) tasks activate partially overlapping networks of cortical fields. Some constituents of the congruent part of these networks (core network) is shown in (D), whereas recruited fields present in the form discrimination task alone are shown in (E) (Vidnyánszky et al. 2000).

contributing to the generation of the percept were input cue-dependent, thus demonstrating the importance of recruiting various cortical neuronal populations for the different submodalities (Gulyás and Roland 1994a, b; Gulyás et al. 1998) (Figure 3.9).

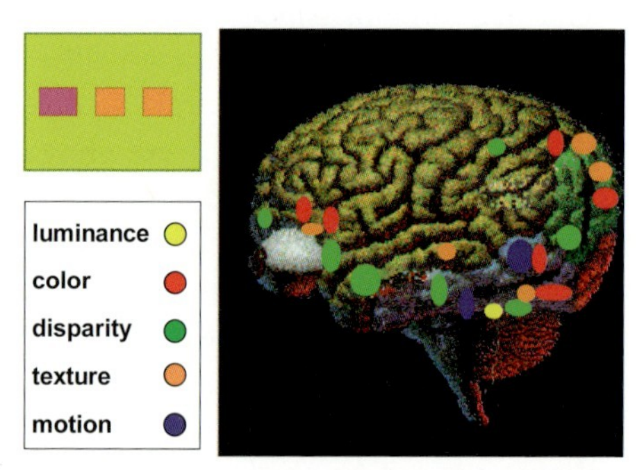

Figure 3.9 Input cue dependence of cortical networks generating identical form percepts on the basis of different input cues. The "odd one out" task is shown in the left upper panel, whereby, the middle rectangle served as the reference figure and; the one left or right to the reference figure represents the "odd one out," which the subjects had to identify. The contours of the shape of the rectangles were made up by luminance, color, disparity, texture, or motion cues. In the color-coded image, the input cue-dependent cortical fields are shown in a schematic brain image. They represent, in fact, the recruited fields that were dependent on the input cues.

Core Networks and Recruited Fields in Language Generation

How do core networks and recruited cortical fields impact higher cognitive functions? Let us consider, for instance, speech or word generation on the basis of various input cues. We can repeat words or generate a given text by listening to another speaker (input cue: hearing), by reading a text (input cue: reading), or by using tactile stimulation to analyze the meaningful surface texture, as in Braille letters (input cue: somatosensory). Does the aforementioned logic hold for such a higher process, such that the input cue (vision, audition, somatosensation) would complement the very same core network of word understanding and word generation in the human brain? This question has been explored by different groups, not as a unified experiment but by focusing on the various input modalities. What would we expect? The best established model on speech generation can be traced back to Paul Broca's discovery of the prefrontal motor speech area (the "Broca area"), followed by Karl Wernicke's discovery of the "sensory speech area." A series of observations based predominantly on lesion studies led to the Wernicke–Geschwind model of speech generation. In this model, the three key "players" involved in the cortical basis of speech understanding and speech generation are the motor speech area or Broca area, the sensory speech area or Wernicke area, and the fiber tract that connects these two cortical regions, the fasciculus arcuatus. Depending on the input modality, other sensory modality-specific regions (visual, auditory, somatosensory) may be recruited to this core network.

Indeed, imaging studies on speech generation that are based upon different sensory inputs indicate that the core network is the same in each case. The other cortical fields recruited to this core network are sensory modality dependent and include, respectively, visual, auditory or somatosensory cortical areas. These observations lend broad support to the validity of the model; namely, certain cortical core networks are the basis of cortical activation patterns, including those underlying syntax, and these networks are complemented by additional "recruited" fields dedicated to the various aspects of the complex operation.

Core Networks, Necessary Networks, and Sufficient Networks

Historically, the localization of functions in the human brain has been based on lesion studies. Neurological lesions teach us about the fundamentals of cortical organization and reveal specific brain areas that participate in a neuronal operation at macro-network levels. Further to the original and seminal observations of Paul Broca (1861, 1865) and Paul Wernicke (1874), *in vivo* and postmortem studies of brain lesions have contributed for over a century to the fundamental models of brain structures and, consequently, neuronal operations that underlie language (Geschwind 1985; Damasio and Geschwind 1984).

Recently, studies utilizing transcranial magnetic stimulation (TMS) have complemented observations on lesion patients, as TMS is capable of producing well-planned transient local brain dysfunctions, thus mimicking local lesions (e.g. Devlin et al. 2003).

In the literature of cortical language networks, two terms used often are "necessary networks" and "sufficient networks." Definitions of these terms were originally based on lesion studies; however, recent neuroimaging studies have contributed to a more accurate interpretation. In a cortical macro-network performing an operation in the brain, a number of cortical neuronal populations or areas are necessary for the successful performance of the operation. For example, in conducting a semantic similarity judgment task on normal subjects, functional neuroimaging studies demonstrate that the left temporal, parietal, and inferior frontal cortices are engaged when the task is successfully performed (Koenig et al. 2003). In patients with local brain damage, temporal and parietal region lesions resulted in semantic deficits, indicating that these cortical regions are necessary for task performance. In contrast, damage to the inferior frontal cortex did not impair task performance, indicating that the inferior frontal cortex might not be necessary.

Further neuroimaging and neuropsychological studies on lesion patients show that despite extensive cortical damage in some patients, adequate task performance is possible through the activation of a network of cortical neuronal populations sufficient for the task (Price et al. 1999). In other words, what is absolutely necessary for a successful task performance may not necessarily be enough. By definition, sufficient cortical networks include necessary cortical networks and more.

Core networks may correspond to necessary networks, since without the cortical neuronal areas of a core network, a successful behavioral operation is not possible. Recruited fields, in turn, may correspond to elements of the sufficient brain network that complement the activity of the core networks elements under the given stimulus or task conditions.

Do Cortical Macro-networks in Language Include Only the Broca and Wernicke Areas?

Many earlier studies on other dedicated brain systems, including the primate visual system, have demonstrated that not only complex cognitive but even simpler brain functions (e.g., visual perception) engage large parts of the human brain, and not merely those subsystems that belong to the core networks of modality-specific information processing. In the case of language, the Broca and Wernicke areas have been identified as essential parts of a "language core network." However, similar to other observations related to cortical sensory or motor macro-networks that engage widespread cortical neuronal populations, language-related cortical macro-networks also occupy extensively the cerebral cortex and include the cerebellum and subcortical structures, as well.

The cortical localization of syntax is of particular relevance to the discussion in this volume. Only one central issue finds full agreement in the literature: Broca's area is absolutely necessary as a component of the cortical macro-network underlying syntax. However, many have questioned how many and which other cortical neuronal populations are also involved in syntactic processing. An overview of 14 imaging studies of syntax (Table 3.1) shows that there is growing consensus among imagers regarding the complexity of the cortical macro-network that underlies syntactic operations.

Devlin (2008) surveyed several language-related neuroimaging studies and mapped the results onto the brain. His results indicate clearly that the cortical neuronal populations engaged by language are not limited to the "classical" Broca and Wernicke areas, but rather they extend over most of the cortex (Figure 3.10).

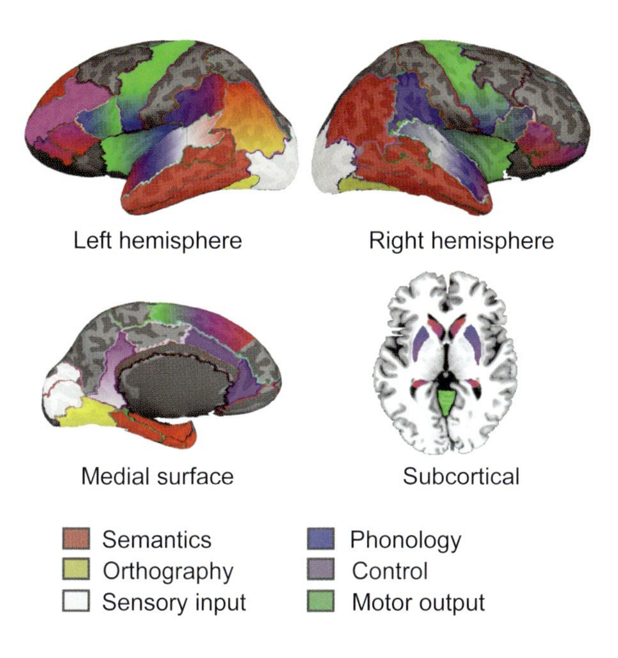

Figure 3.10 Using the Freesurfer (Desikan et al. 2006) inflated cortical model, a summary of various language-related neuroimaging studies is projected onto the cortex. The regions are color coded according to the type of information attributed to them. Composite colors indicate more than one function. For instance, the angular gyrus' red-yellow label indicates the area's involvement in both semantic and orthographic functions. Figure used with permission from J. Devlin.

Table 3.1 Overview of imaging studies demonstrating the complexity of the cortical macro-network underlying syntax.

Inferior frontal gyrus		Mid frontal gyrus		Superior temporal gyrus		Mid temporal gyrus		Other cortical structures	Reference
L	R	L	R	L	R	L	R		
X									Stromswold et al. 1996
X				X				transverse temporal gyrus L, temporal pole L+R, supramarginal gyrus L, interior parietal lobe R	Dapretto and Bookheimer 1999
X		X						superior parietal lobule L	Caplan et al. 1999
X	X								Moro et al. 2001
X		X						cingulate gyrus L	Caplan et al. 2000a
X	X								Tettamanti et al. 2001
X	X	(X)	(X)	X	(X)	X	(X)		Kang-Kwong et al. 2002
X						X			Roeder et al. 2002
X	X								Heim et al. 2003
X	X			X				cingulated R, hippocampus R	Waters et al. 2003
X				X		X		parietal-occipital junction L+R, mid occipital gyrus L, precuneus R, cuneus L+R, parahippocampal gyrus L, angular gyrus L, superior parietal region R	Peck et al. 2004
X									Haller et al. 2005
X		X						several areas in occipital and parietal lobes	Golestani et al. 2006
X		X	X			X	X	superior frontal gyrus L, orbital gyrus L	Grodzinsky and Friderici 2006
X									Santi and Grodzinsky 2007b
X									Momo et al. 2008
X									Uddén et al. 2008

Personal Closing Remarks

I have written this overview from the perspective of a "brain imager"—one without direct and extensive experience in imaging language-related cortical areas in the human brain. My background nonetheless enables me to view language-related cortical operations objectively as one possible form of higher brain functions alongside other complex sensory, motor, or cognitive functions. Surveying the recent neuroimaging literature of language has convinced me that there must be some basic principles behind how the human brain organizes its operations. Language processing, similar to other cognitive processes in the brain, is anchored to cortical macro-networks that are anatomically defined and which comprise core constituents, such as the Broca or Wernicke area complexes. These basic components of language-related activities include language acquisition, understanding, and production. Core networks, in turn, are complemented by recruited cortical fields (i.e., cortical neuronal populations required for performing a specific language task). Thus, a large part of the human brain may be directly or indirectly involved in language processing. Indeed, many neuroimaging studies support the idea that language faculty widely exploits cortical neuronal populations as well as processing pathways, making language one of the most "invasive" cognitive faculties of the human brain.

Syntactics

4

Some Elements of Syntactic Computations

Luigi Rizzi

Abstract

This chapter focuses on some current directions of research on the nature of syntactic computations. I illustrate how certain issues have taken shape in the course of the last half century and discuss how they are addressed in current syntactic models, with special reference to "Principles and Parameters" and Minimalist models. The first issue concerns the expression of the open-ended character of natural language syntax. Minimalism has introduced an extremely simple structure building rule, Merge, which is able to reapply indefinitely to its own output, thus expressing syntactic recursion, a property permitting the generation of an unlimited number of sentences. Merge builds hierarchical structures, which may be modified by the other fundamental syntactic operation: movement. I discuss the typology of movement processes and illustrate how Merge and Move interact to determine basic word-order properties of natural languages. As is to be expected given the parsimonious, economy-based design of natural langue syntax, movement takes place to satisfy requirements of other components: phonology–phonetics, requiring full linearization of the hierarchical structures of syntax; morphology, requiring the formation of well-formed words; and semantics–pragmatics, dealing with the meaning and use of linguistic expressions. Focusing on verb movement in the functional structure of the sentence, certain interplays between morphology and syntax are illustrated which generate significant diachronic and comparative predictions. The final part of the chapter is devoted to the issue of invariance and variation, and to how the universality and variability of human language is expressed by parametric models.

Introduction

In this chapter I present current directions of research on the nature of syntactic computations. After illustrating the roots of these directions in classical work in generative grammar, I discuss current understanding of the issues. Coverage of topics is not exhaustive, the selection being based on my competence and taste. I will assume the basic conceptual structure of the Principles and Parameters framework, and will shape much of the presentation in terms of the formalism and tools of the Minimalist Program. According to its general guidelines, Minimalism puts a strong emphasis on sticking to the bare minimum of the formal apparatus required to express the fundamental generalizations of syntax, and this effort of conciseness and simplicity seems to be particularly appropriate when posing questions as to the biological foundations and evolutionary origins of mechanisms.

Creativity and Recursion

One salient property of the human knowledge of language is the so-called creativity manifested in normal language use. We constantly produce and understand new linguistic objects, sentences that we had never encountered in our previous linguistic experience, and our linguistic capacities give us the possibility of expressing an indefinitely large number of messages. No other species possesses a communication system with such characteristics.

The importance of this creative aspect is not a new observation: it was clear, in essence, as early as in the seventeenth century. René Descartes pointed out that the capacity to organize words into an unlimited number of contextually appropriate sentences distinguishes the dumbest man from the most intelligent ape, and from the most sophisticated machine. What is the essence of this "familiarity of the newness," and of the unbounded character of the human linguistic capacities?

What was missing until the middle of the twentieth century was a technical device to address these questions in a precise manner. In the first half of the century, structural linguistics conceived of language (saussurean "langue") as a systematic inventory of linguistic signs, each of which consisted of a sound–meaning pairing, basically a theory of the lexicon. An inventory, however, is limited by definition; therefore, this approach was intrinsically unable to address the fundamental question of creativity, except through some vague notion of analogy: the infinite possible combinations of linguistic signs were, at least in part, relegated by Saussure to "parole" the actualization of the system of "langue" in individual linguistic acts, and "langue" basically contained a repertoire of frozen idioms, not productive syntax. Saussure was probably dissatisfied with this conclusion, as certain oscillations in the *Cours de linguistique générale* suggest (Saussure 1916/1985).[1] Natural language syntax is

---

1 For example, "*...des phrases et des groupes de mots* [*sont*] *établis sur des patrons réguliers... répondent à des types généraux...*" but "*... il faut reconnaître que dans le domaine du syntagme il n'y a pas de limite tranchée entre le fait de langue, marque de l'usage collectif, et le fait de parole, qui dépend de la liberté individuelle*" (sentences and groups of words are established on regular patterns...but... it is necessary to recognize that in the domain of the phrase there is no sharp limit between the fact of langue, characterized by the collective usage, and the fact of parole, which depends on individual freedom). The interpretation of Saussure's ideas about the open-ended character of syntax may be controversial, as an anonymous reviewer points out, but the fact that linguistic approaches of the early twentieth century did not work

clearly regular, a rule-governed process, but linguistics at the beginning of the twentieth century did not possess a formal device to express this regularity.

Introducing such a device was Chomsky's first critical contribution to the study of language. Chomsky showed that the core notions of the theory of recursive functions, developed in the study of the foundations of mathematics, could be adapted to language. A recursive procedure is one that can indefinitely reapply to its own output, giving rise to a hierarchical structure. The following examples illustrate some simple cases of recursion in natural language, in which a phrase of a given type can be embedded into a phrase of the same type (phrases are delimited by brackets); this can go on indefinitely:

(1) The preface [of the first book [on the discovery [of the …]]]

(2) I believe [that people wonder [whether Mary thinks [that someone said ...]]]

(3) I met [the boy [who bought [the book [which pleased [the critics [who wrote [the review ...]]]]]]].

Notice that it is not the mere iterability of the procedure which makes it recursive, but its capacity to create a hierarchical structure, as happens at different levels of organization of linguistic structures. Thus, for example, the iteration of the motor program activated in walking, one step after the other, can go on indefinitely, but it does not give rise to any hierarchical structure. By contrast, the stringing together of words in a sentence does, determining the bracketed representations as in (1)–(3) or, more perspicuously, tree-like representations such as the ones we will consider below. Such representations, far from being mere artifacts of the adopted formalism, contribute crucially to determining properties of form and meaning of linguistic expressions.

It is sometimes said that recursion is not as critical to natural language syntax as the approach just introduced assumes, because the normal use of language, as emerging from corpora of ordinary conversation, typically consists of rather short sentences. This objection does not, however, take into account the fact that a system capable of producing very simple phrases, like *John's book* or *the picture of the girl*, is already recursive, as it allows a nominal expression to be embedded within a larger nominal expression. Thus, a system capable of generating such simple structures already yields automatically the unbounded character of language. Imposing a limitation to sentences of a fixed length would complicate the system in an arbitrary and unwarranted manner: it would be about as arbitrary as defining the number system as ending with a particular number *N* on the basis of the observation that people use only fairly small numbers in everyday life.

Let us now try to express the role of syntax in a more comprehensive model of the human linguistic capacities. Phrasing things at a very general level, we can say that to know a language means to possess certain inventories of elements, somehow stored in memory, and the computational procedures to

out an operative formal device to capture this property is clear (Saussure 1916/1985, p. 173).

combine the elements of the inventories to form entities of a higher order. This fundamental inventory is the lexicon and consists of two major systems of lexical items: (a) the *contentive lexicon*, consisting of nouns, verbs, adjectives, etc. (i.e., elements endowed with descriptive content characterizing events, arguments, qualities, etc.) and (b) the *functional lexicon*, consisting of grammatical words and morphemes such as determiners, complementizers, auxiliaries and copulas, expressions of tense and aspect (i.e., elements that have a more abstract semantic content and somehow define the configurational structure in which the contentive elements are inserted).[2]

There are other lists of elements which must be stored in memory, ready to be used in linguistic computations (e.g., features, phonemes, syllable structures, morphemes, idiomatic expressions) and which define the structure of the lexicon. When this cascade of levels leaves the lexicon and enters productive syntax, and we start putting words together, the computational procedures become recursive and give rise to higher-order entities, phrases, and sentences, which are indefinitely extendable.

More generally, language is sound with meaning (abstracting away from languages that use different modalities, such as sign language; the generalization to these cases is straightforward). Thus, a model of language must be able to connect representations of sounds with representations of meanings over an unbounded domain. The following structure is quite generally assumed in the tradition presented here:

(4)

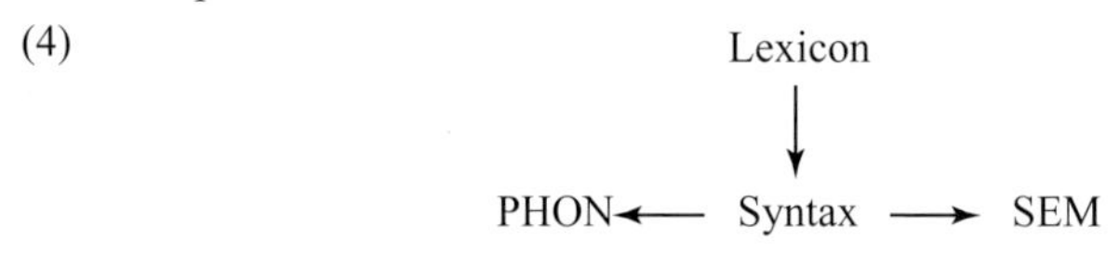

Items are selected from the functional and contentive lexicon and strung together through the recursive procedures of syntax. Hence, interface representations of sound (PHON) and meaning (SEM) are computed (also called Phonetic Form and Logical Form, respectively) and accessed by other systems: the auditory–articulatory systems and the conceptual–intentional systems. In this conception, syntax is the generative heart of the system, the device that generates an unbounded number of linguistic representations; it is also, in a sense, ancillary to the external systems that deal with sounds and meanings as it subserves the needs of such systems. This is clearly expressed by certain concepts of economy which are assumed to apply to syntactic computations within Minimalism: the assumption is that there is no true syntactic optionality; a syntactic device is used only when it is needed to obtain a certain interface effect (Fox 2000; Reinhart 2006). This conception led to a reanalysis of many apparently optional syntactic processes and yielded significant empirical results: very often an apparent optionality reveals detectable interpretive

2 I include in the functional lexicon the system of functional heads which structure the clause and trigger important operations, such as movement.

differences (in phenomena like scrambling, "free" inversion; e.g., Belletti 2004) upon careful analysis. The fundamentally ancillary character of syntax makes intuitive sense. What really matters, for the expression of thought and communication, is the articulation of sound–meaning pairings. Syntax is the powerful mechanical device which makes the generation of the pairing possible over an unbounded domain.

A model like (4) raises the question of the "timing" of the transfer to the interface systems. How is it done? Traditional models of the Extended Standard Theory assumed that complex syntactic structures (with embeddings, etc.) are computed entirely by the syntactic component; thereafter, the entire configuration is transferred to PHON and SEM (or the equivalent interface levels). Among other more technical drawbacks, this radical "syntax first" assumption had the effect of divorcing the model of linguistic competence from the functioning of the processor: clearly, when we parse and interpret a complex utterance, we do not complete the syntactic analysis before starting to build the interpretation, even though it is plausible that the syntactic analysis is the necessary initial step (e.g., Frazier 1987a; Friederici 2000). In current minimalist models, the syntactic computation is assumed to proceed by *phase* (Chomsky 2001, 2007; Nissenbaum 2000): relatively small chunks of syntactic structures, the phases, are computed (roughly corresponding to simple clauses, but assumptions vary on the exact size of the phase) and sent to the interface, and then the syntactic component computes another phase and sends it to the interface, and so on.[3]

Merge and Structure Building

Various recursive techniques have been adopted in the different linguistic models which have been proposed since the 1950s, ever since *Syntactic Structures* (e.g., generalized transformation, rewriting rules, X-bar theory; Chomsky 1957). Jumping ahead almost fifty years of syntactic research, the ultimate distilled format of syntactic recursion is the operation Merge, the fundamental structure-building procedure assumed by the Minimalist Program (see Chomsky 1995, 2000 and related work), which takes two elements, A and B, to form a composed expression C:

(5)

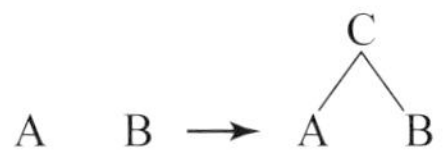

```
                 C
                / \
A   B  -->     A   B
```

[3] Standard minimalist models involve "bottom-up" derivations of the kind illustrated in what follows. Other models, motivated by both linguistic and psycholinguistic considerations, involve variants of Merge and phase theory consistent with "top-down" derivations (Phillips 2003; Chesi 2005). I do not address this issue here and will adhere to standard assumptions.

The operation is recursive in that it can reapply indefinitely to its own output, generating a hierarchical structure. Thus, A and B can be two elements taken from the lexicon or complex expressions already formed by previous applications of Merge.

Merge strings words together and, at the same time, expresses the hierarchical structure of the sentence giving rise to tree representations. Merge, as in (5), creates a minimal subtree with two sister nodes, A and B, and a mother node, C. In the very impoverished computational system assumed by minimalist syntax, the computational component cannot introduce new labels (i.e., labels not already present in the lexical items involved: the inclusiveness principle). Thus the label C of the AB constituent in (5) must be inherited from one of the merged elements, the one which "projects": either C = A, or C = B. The element that projects is the "head" of the construction.

Let us consider a concrete case. For Merge to apply there must be some kind of "affinity": some selectional relation between A and B. If A is a transitive verb and B is a noun, Merge can apply, forming a transitive verb phrase (say, *meet Bill*). In the obtained configuration, the selector is the head, the element possessing the label which projects, and gives a name to the whole structure; thus, in the case of a verb–object construction, the obtained C constituent would be a verbal projection, a verb phrase in more traditional terminology:

(6)

V

V N

Successive applications of Merge can give rise to complex structures like the following, expressed here in terms of the approach known as Bare Phrase Structure, a component of Minimalism: The verb *meet* is merged with the noun *Bill* to give rise to the verbal constituent *meet Bill*, which is then merged with the tense-bearing element, here the modal *will*. After merger of the subject (cf. later discussion for an important refinement), the sentence thus created, *Mary will meet Bill*, is merged with the complementizer *that*, an element which transforms a sentence into a complement, available to be selected by, and merged with, a higher verb, *said*, and so on.

(7)

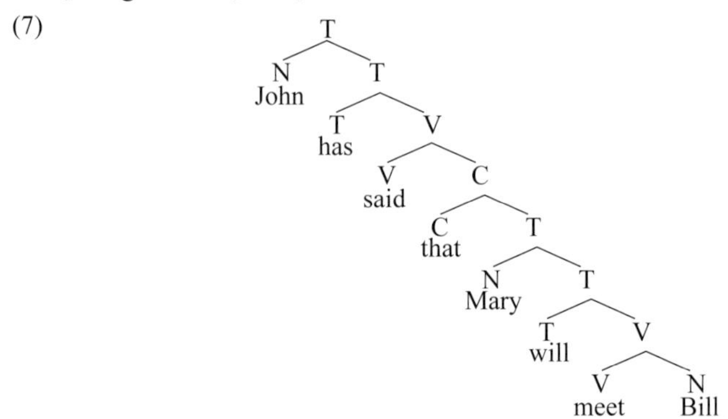

These representations express key properties of the structure of sentences (e.g., what words are related to other words, what units are formed) and enter directly into the determination of form and meaning of the sentence. These representations are transferred to the interface systems at the end of a phase, in a phase-based model, and determine, on the PHON side, intonation and other prosodic patterns; on the SEM side, they determine properties of argumental semantics (who does what to whom), of referential dependencies (the interpretation of pronouns, anaphors, and the like), and of the scope-discourse semantics (e.g., scope of operators, informationally related properties such as topicality and focus). We will return to this shortly.

Going back for a moment to the properties of Merge, we can observe that it is an extremely general formal operation. Minimalism accepted the difficult challenge of showing that all the fine details of syntactic structures uncovered in half a century of formal syntax could be traced back to this operation, in interaction with reasonable assumptions on lexical specifications and interface requirements. The challenge is remarkably successful, even though many problems remain.

The hypothesis that Merge represents the core of syntax has opened new perspectives for the study of language evolution. Hauser et al. (2002) speculate that the availability of recursion in the right "spot" of the human cognitive map, perhaps in the form of the Merge operation, may have been a sudden and recent event in evolutionary history, perhaps the major computational consequence of a minor reorganization of the brain, and that this single evolutionary event may be at the root of the emergence of the language faculty and possibly, even more broadly, of what paleoanthropologist Ian Tattersall calls "the human capacity," the collection of cognitive capacities that distinguishes our species from the others (Tattersall and Schwartz 2001). The following are examples of the many questions raised by this fascinating hypothesis:

- Is there an identifiable neural substrate which implements the recursive property for language, and, if so, at what granularity of analysis could it emerge?
- If so, how does it relate to the other major human capacity—namely, the capacity to count (Dehaene 1997), which deals with discrete infinities?
- Do the mechanisms underlying linguistic and numeric capacities relate in a nontrivial way to the mechanisms responsible for other kinds of hierarchical structures in other cognitive domains (e.g., vision, motor control, the theory of mind and the other cognitive capacities that govern social interactions)?
- How did the mastery of recursion for communication and other human cognitive systems evolve in the natural history of the species?[4]

---

[4] Derek Bickerton (pers. comm.) points out that Merge, so broadly construed, can hardly be seen

These and many related questions define a broad and ambitious long-term program, but it is imaginable that partial answers to some of these questions may be within reach, through the conjoined efforts of formal modeling of cognitive capacities, the study of pathology, and brain imaging techniques.

Movement

A pervasive property of natural language syntax is movement. We use this term to refer to the fact that linguistic expressions are very often pronounced in positions different from the positions in which they are interpreted (or, more exactly, in which they receive crucial elements for their interpretation). Consider:

(8) (a) [Which book] did Mary want to buy ___?
(b) Mary wanted to buy [this book].
(c) Which book did you say…John believed…Mary wanted…. to buy ___?

To understand a sentence like (8a), it is necessary to interpret the expression *which book* as the object of the verb *buy*, much as *this book* in (8b); however, *which book* has been displaced to the initial position and, in fact, it may end up being, in the surface configuration, indefinitely far away from the immediate structural context of *buy*, as in (8c). Still, it must be interpreted as belonging to the argument structure of *buy*.

To illustrate the pervasiveness of movement, consider the following French example:

(9) Qui rencontreras-tu ?
"Whom will-meet you-NOM? = Whom will you meet?"

---

as a domain-specific operation. For instance, the stone-chipping technique, which consists of successive applications of the same operation to build an organized object and was presumably available to hominids long before the advent of modern humans, is reminiscent of sentence building through successive applications of Merge. The point is well-taken. In fact, it is precisely the great generality of Merge which invites comparisons across cognitive capacities in search for cognitive and neural invariants involved in different types of computations. Such comparisons would have been hopeless with more complex models of Universal Grammar (UG), whose constructs (e.g., "the specified subject condition," "the complex NP constraint") sounded deeply rooted in language-specific notions and categories, but are perfectly sensible in a Minimalist, Merge-based conception of UG. Granting that Merge-like operations may well exist in other cognitive domains and in other species, perhaps supported by analogous cognitive and neural mechanisms, the question of the recent evolution of syntax clearly cannot be phrased in terms of the appearance of Merge *tout court*: what remains specific to modern humans is the availability of a Merge-like operation in the system of signals for communication and the expression of thought. If we phrase things in this way, a central question for the evolution of syntax then becomes: How and when did a Merge-like operation, previously available to other cognitive systems, penetrate the system for the expression and communication of thought?

French is an SVO (subject–verb–object) language like English, but in this kind of structure the order is reversed to OVS by various applications of movement, as indicated by the arrows:

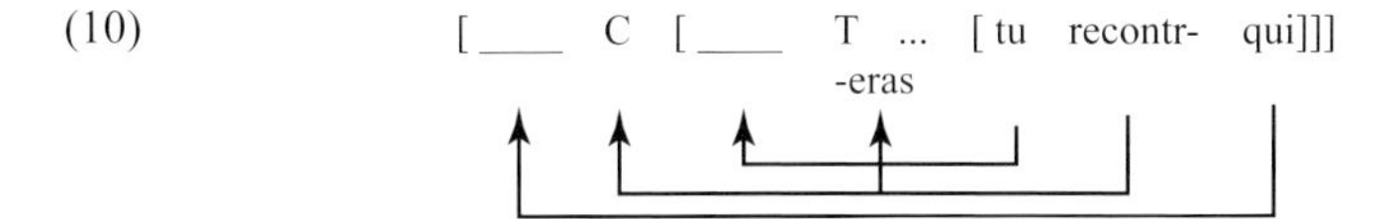

This gives rise to a surface configuration like the following:

(11) [Qui rencontr+eras C [tu ___ … [___ ___ ___]]].

where the blanks indicate the positions vacated by movement, or the "traces" of movement. In (11) the verb phrase, the initial nucleus in which thematic roles like agent and patient are assigned, is completely vacated by movement, and this is by no means an exceptional situation. There are at least three kinds of movement involved in (10):

- Head movement, forming the complex word *rencontrera*, hence associating the lexical verb to the tense affix,
- A-movement, which moves the subject from its thematic position to its canonical clause-initial position (in languages like French),
- A′-movement (read "A-bar"), which moves the interrogative pronoun *qui* to clause-initial position.

I will now illustrate these in the appropriate structural context.

Clausal Structure, Head Movement, and A-movement

Let us follow a derivation step by step, looking first at an example in English. The verb *meet* takes two arguments: an agent and a patient. The verbal root is thus initially merged with two nominals, which receive the two "theta roles," or argument roles, of agent and patient; this operation satisfies the argument structure of the verb, giving rise to the thematic nucleus of the clause: the verb phrase (VP, in traditional informal notation):

(12) [you [meet Mary]].

Then, the functional structure is added, which forms the structural backbone of the clause; in particular, it includes the tense specification. Tense has a number of functions, which affect both the form and the interpretation of the expression.

In the system presented here, tense has at least a dual function of relevance for the interpretive systems. It locates the described event in time with respect to the speech time (present–past–future), and it (or some element close to it in the functional structure of the clause) creates the subject–predicate articulation.

Suppose that the tense (T) element merged to (12) is a future marker, expressed in English by the modal *will*:

(13) will [you [meet Mary]].

In languages like English, T attracts the closest nominal to its left-adjacent position (its specifier), thus creating the subject–predicate articulation, or "aboutness" structure ("about argument *X*, I'm presenting event *Y* concerning *X*"). This is an instance of A-movement, or movement of an argument to a subject position:

(14) You will [___ [meet Mary]].

In English, future T is expressed by an autonomous word, the modal *will*. In some languages, T is always expressed by an autonomous particle (e.g., typically in Creole languages), but this is by no means the general case.

In French, as in many other languages, future T is expressed by an affix, an element which does not form an independent word:

(15) Tu rencontr-eras Marie
"You will-meet Marie."

In this, as in many other cases, linguists have profitably followed an intuition of uniformity and assumed that the clausal structure of French is exactly the same as in English. Thus, future T, expressed by the affix *-eras*, is merged to the verb phrase nucleus, much as in (13):

(16) -eras [tu rencontr- Marie].

Example (15) is derived by a double movement: *tu* A-moves to the canonical subject position. Moreover, as *-eras* is not a morphologically well-formed word in French, but an affix, something must happen to associate it to the verbal root. This is also done through movement. So, the verb moves to T, giving rise to the complex inflected verb *rencontr-eras*:

(17) tu rencontr+eras [___ ___ Marie].

This is a case of head movement: a head, the verb, moves to the next higher head and combines with its content. The reason for this process is to align syntax and morphology: syntactic units (the heads) and morphological units (the words) do not match perfectly; in particular, there are heads which are not complete words. Head movement aligns the two systems by forming complex inflected words: a lexical root moves to higher heads "picking up" the morphological specifications expressed in them. The system is illustrated in (17) with a single functional head for T, but it readily generalizes to more complex cases involving a richer functional structure expressing mood, tense, aspect, voice, the markers of agreement with the subject, and other arguments (Cinque 1999).

Important evidence supports this syntactic conception of how inflectional morphology works, and offers straightforward explanations for complex cross-linguistic patterns of adverb distribution (Emonds 1978; Pollock 1989).

Consider the different position of a frequency adverb like *often/souvent* in English and French:

(18)	(a)	You	often	meet	Mary.
		You will	often	meet	Mary.
	(b)	Tu rencontres	souvent		Marie.
		Tu rencontreras	souvent		Marie.

The adverb precedes the verb phrase (VP) in English, whereas it interpolates between the inflected verb and the object in French. Following again a fundamental intuition of uniformity, Emonds (1978) and Pollock (1989) have proposed that the adverbial position is the same in the two languages: the adverb is merged with the VP it modifies. Moreover, in both languages the subject moves to the initial position, the specifier of T. What varies is the independent difference we have just seen in how the morphology–syntax interface is addressed: French involves verb movement to T, which raises the verbal root past the adverbial position, whereas in English the lexical verb does not move. Thus, here we have a single movement in English and a double movement in French:

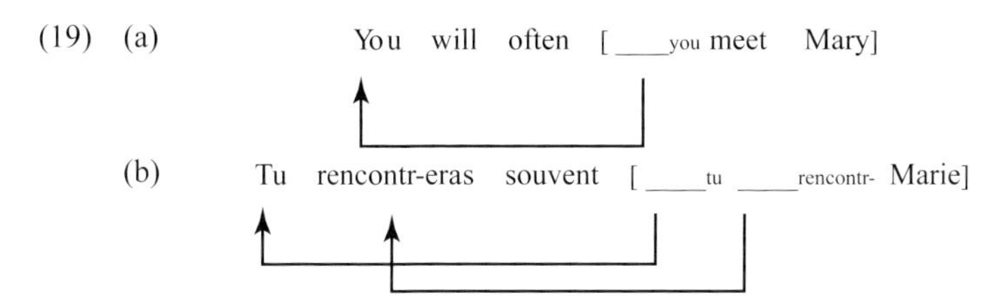

This mode of explanation connecting morphology and syntax has proven extremely fruitful. First, it has prompted an in-depth exploration of syntactic structures, giving rise to the so-called cartographic projects, which look at the fine details of the structural articulation of clauses and phrases using, as guidelines, the morphological properties of the expression of tense (but also mood, aspect and voice) and the distributional properties of adverbials and other kinds of elements. Second, it has produced an effective technique to address major cases of variation in word order within and across languages. Third, it has favored the exploration of diachronic patterns of change in the morphosyntax of languages.

Comparative and Diachronic Implications

It does not seem to be a pure syntactic accident that French requires head movement of the lexical verb to the inflectional system, while Modern English does not. There seems to be a relation between the richness of the paradigm of verbal morphology (in particular, the well-differentiated expression of agreement) and the syntactic movement of the verb: roughly speaking, a rich inflectional system is able to attract the verb out of the VP, a more impoverished system does not. This is immediately suggested by certain comparative facts. Consider, for instance, Icelandic in comparison with the mainland Scandinavian languages Swedish, Norwegian, and Danish.

Icelandic has a well-differentiated agreement paradigm both in the present and preterite tenses; mainland Scandinavian varieties, illustrated here by Danish, lost the agreement specification completely and have a single verbal form that co-occurs with subjects with all persons and numbers:

(20) Icelandic (heyra "hear")
present: heyr-i, heyr-ir, heyr-ir, heyr-um, heir-ið, heyr-a
preterite: heyr-ði, heyr-ði-r, heyr-ði, heyr-ðu-m, heyr-ðu-ð, heyr-ðu

(21) Danish (høre "hear")
present: hør-er
preterite: hør-te.

Not surprisingly, in Icelandic the lexical verb raises to T past the negative adverb *ekki*, much as in French, whereas in Danish the lexical verb does not leave the VP and appears lower than the negative adverb *ikke* (we have to use embedded clauses to see these phenomena in order to control for Verb Second, a major phenomenon radically modifying word order in main clauses in most Germanic languages):

(22)	…að	hann	keypti	ekki	bokina (Icelandic)
	"that	he	bought	not	the book"
(23)	…at	han		ikke kobte	bogen (Danish)
	"that	he		not bought	the book."

There are immediate diachronic implications. Platzack (1987) showed that until the seventeenth century Swedish had rich inflection and verb movement, much as modern Icelandic; both properties were lost, however, in the following history of the language. Faroese, a Scandinavian variety spoken in the Far Oer islands, roughly halfway in between Iceland and the continent, seems to be in an unstable transitional state, with dialectally variable morphological richness and verb movement.

A clear diachronic effect is also straightforwardly observable from the history of English, as highlighted by Roberts (1993). He observes that sixteenth-century English showed clear signs of verb movement, as illustrated by the interpolation of negation and various kinds of adverbials between the lexical verb and the direct object:

(24) (a) If I gave *not* this accompt to you (1557)
(b) In doleful way they ended *both* their days (1589).

In parallel, Roberts observes, verbal morphology expressed a richer paradigm of agreement, with some variation between different varieties:

(25) Is: cast Ip: cast(-e)
IIS: cast-est IIp: cast(-e)
IIIS: cast-eth IIIp: cast(-e).

Again, loss of verb movement and loss of a rich morphological expression of agreement seem to have gone hand in hand. Much research has been devoted to the exact characterization of the notion of "morphological richness" (Roberts 1993; Vikner 1997; Rohrbacher 1999) as well as to the important question concerning the direction of the causal effect, from morphology to syntax, or vice versa. According to the first view, the properties of inflectional morphology cause the syntactic behavior directly: if the morphology reaches a certain threshold of "richness," syntactic movement is automatically triggered. According to the alternative view, morphology merely registers what syntax does, and "rich agreement" is simply the way in which morphology typically underscores the fact that syntactic movement took place. The debate revolves around certain exceptions to the generalization "Rich agreement if and only if syntactic movement" in Scandinavian dialects and other regional varieties, and represents a very lively chapter of current research.

VSO Languages

The same ideas that have been proposed to address the syntax–morphology interface and the position of adverbials have shown a clear explanatory power with respect to other major questions of word order across languages. One concerns the analysis of VSO (verb–subject–object) languages, traditionally a serious puzzle for syntactic theory.

(26) (a) Cheannaigh siad teach anuraidh (Irish)
"Bought they a house last year"
(b) Chuala Roise go minic an t-amharan sin
"Heard Roise often this song."

In the other major language types, SVO and SOV, the verb and object are adjacent and seem to always form a constituent, the verb phrase, merged with the subject. In fact there are very good reasons to assume structures like S [VO] and S [OV]. But how can such a binary subject–predicate articulation be expressed in a VSO language, in which the subject apparently interpolates between the two constituents of the verb phrase? A traditional approach was to assume that VSO languages are different in that they instantiate a flat ternary structure, with subject and object generated as sisters of the verb. This approach is suspect, both on theoretical and empirical grounds. Theoretically, if the structure-building operation is Merge, structures can only involve binary branching; the point that tree branching is binary was forcefully asserted by

Kayne (1984). Empirically, all of the standard evidence showing that the subject is structurally higher than the object applies to VSO languages as well. For instance, the subject typically can bind a reflexive in object position (*John washed himself*), but not vice versa (**Himself washed John*). This property can be shown to be sensitive to structural prominence, not just linear order (e.g., in constructions with a clause-final subject in Italian, the subject can bind a preceding object, and not vice versa: *ha votato per se stessa anche la sorella di Gianni*, "voted for herself also Gianni's sister"). The asymmetry clearly holds in VSO languages as well, with subjects binding objects but not vice versa. Thus, subject–object asymmetries emerge which would not be expected under a "flat structure" analysis. Moreover, phrasal idioms typically involve the VO sequence, with the subject remaining a freely referential position (*John kicked the bucket*) but virtually no idiom involves the SV sequence leaving the object position referential, a fact which follows from the fact that V and O form a constituent (to which a special idiomatic interpretation can be attached), while S and V do not. Idioms pattern exactly like that in VSO languages.

The natural solution is that the VSO order is derived via verb movement to an inflectional head X from an underlying SVO (or SOV) order (Emonds 1980):

(27) X [N [V N]]

Perhaps X is T, and VSO languages have V to T movement as in French. However, they differ from French in that the subject does not move to the specifier of T (i.e., only one of the two movements involved in (19b) takes place here). More plausibly, X is a functional head higher than T, and V-movement takes this extra step to X with respect to a language like French, where it stops in T (that the subject moves in VSO languages is suggested by the possibility of adverb interpolation between the subject and the object in examples like (26b).

When an auxiliary verb is raised, as in the Welsh example (28b), the SVO order with the lexical verb resurfaces:

(28)	(a)	Cana	i	yfory (Welsh)	
		"Will-sing	I	tomorrow"	
	(b)	Bydda	i	'n canu	yfory
		"Will-be	I	singing	tomorrow."

Thus, recalcitrant VSO languages can be traced back to familiar ingredients: a basic order permitting the subject–predicate articulation and head movement of the verb to the functional system.

A′-Movement and the Interface with Semantics and Pragmatics

Let us now consider the class of cases that most straightforwardly illustrates the phenomenon of movement, the displacement of an operator-like element to the beginning of the clause. This is what linguists call A′-movement and is illustrated by the English constructions in (29), which give rise, respectively, to a question (a), a topic–comment structure (b), a focus–presupposition structure (c), a relative (d), and an exclamative construction (e):

(29) (a) Which book should you read ___?
 (b) This book, you should read ___.
 (c) (It is) THIS BOOK (that) you should read ___ (rather than something else).
 (d) The book which you should read ___ is here.
 (e) What a nice book I read ___!

A′-movement, in the clear cases, has a very straightforward "teleological" motivation. It is a device that associates to a linguistic expression two types of semantic properties:

- properties of *argumental semantics* (e.g., thematic roles, who does what to whom), and
- properties which Chomsky refers to as expressing *scope-discourse semantics*: the scope of various kinds of operators (e.g., interrogative, relative, exclamative) and such discourse-related and informationally related properties as topicality and focus.

Natural language expressions may be assigned both properties. How is this done? Among the many solutions that would be a priori possible, natural languages opt for movement: the expression is inserted in a position dedicated to argumental semantics and "picks up" the scope-discourse property through movement, by moving to a position dedicated to this kind of interpretive property. So, for instance the nominal expression *which book* in (29a) is to be interpreted both as the patient of the verb *read* and as an interrogative operator with main clause scope in a direct question, to yield a semantic representation which is something like "for what x, x a book, you read x." This is achieved through Merge and Move: the phrase is merged with *read*, and in the local configuration with the verb it receives the thematic role of patient; then it is moved to the initial periphery of the clause, where it "picks up" its scope property.

How is movement triggered? In the other cases of movement (head movement and A-movement to a subject position), we have seen that it is always a functional head which attracts another element, a head or a phrase, so it is natural to assume that in movement to a scope-discourse position, this would also hold. Here I will make the somewhat controversial assumption that in fact all kinds of movement work in this manner. In other words, I will assume that the representations of (30) all involve a functional head in the left periphery of the clause which attracts the question operator in (30a), the topic in (30b), the focus in (30c), the head of the relative clause in (30d), and the exclamative operator in (30e).

(30) (a) Which book Q should you read ___?
 (b) This book TOP you should read ___.

(c) THIS BOOK FOC you should read ___.
(d) The book which R you should read ___.
(e) What a nice book EXCL I read ___!

This system of heads governing A′-movement is silent in English. None of them is pronounced. However, often the corresponding constructions show overt counterparts in other languages, italicized in the following examples:

(31) (a) Ik weet niet [wie *of* [Jan ___ gezien heeft]]
(Dutch varieties; Haegeman 1996)
"I know not who Q Jan seen has"
(b) Un sè [do [dan lo *yà* [Kofi hu ì]]]
(Gungbe; Aboh 2001)
"I heard that snake the TOP Kofi killed it"
(c) Un sè [do [dan lo *wè* [Kofi hu ___]]]
(Gungbe; Aboh 2001)
"I heard that snake the Foc Kofi killed"
(d) Der Mantl [den *wo* [dea Hons ___ gfundn hot]]
(Bavarian; Bayer 1984)
"The coat which R the Hans found has"
(e) Che bel libro che [ho letto ___]! (Italian)
"What a nice book Excl I read!"

Many languages overtly manifest a Q marker (e.g., the dialectal Dutch *of*), topic and focus markers (*yà and wè* in Gungbe), a relative marker (the dialectal German *wo*), and an exclamative marker (*che* in Italian). It is tempting to assume that the difference between English and these languages is very superficial and has mainly to do with the phonetic realization of a system of heads that is always present and syntactically active across languages, but pronounced only in some. These constructions have analogous (sometimes identical) syntactic properties across languages, are interpreted uniformly at the SEM interface (apart from a few possible parametrizations), and the fundamental variation only involves the superficial property of being pronounced or not at the PHON interface.

In conclusion, according to this approach the functional lexicon of every language specifies a number of heads, creating positions dedicated to scope-discourse semantics. These heads attract phrases from their thematic positions, and the created configurations are handed over to the interpretive systems where specific interpretive routines are triggered (e.g., scope, topic-comment, focus-presupposition). Thus, a typical A′-construction connects a thematic position and a scope-discourse position. A phrase moved from one to the other is interpreted as carrying both kinds of semantic properties; and both kinds of interpretive properties are assigned in analogous configurational structures: dedicated head of the substantive lexicon (for thematic roles) or of the functional lexicon (for scope-discourse properties) assign such properties to their immediate dependents, specifiers and complements:

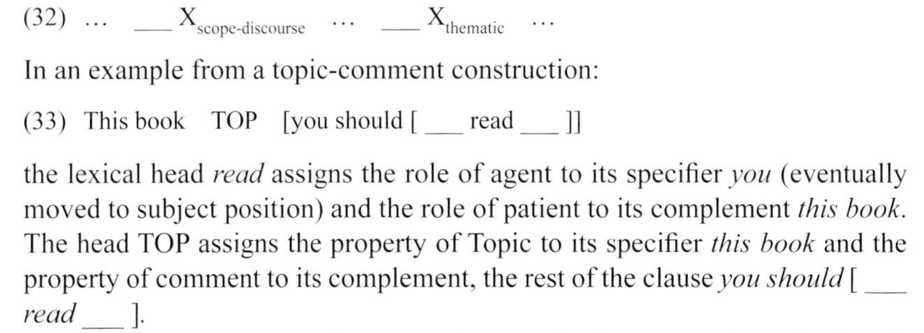

(32) … ___ $X_{\text{scope-discourse}}$ … ___ X_{thematic} …

In an example from a topic-comment construction:

(33) This book TOP [you should [___ read ___]]

the lexical head *read* assigns the role of agent to its specifier *you* (eventually moved to subject position) and the role of patient to its complement *this book*. The head TOP assigns the property of Topic to its specifier *this book* and the property of comment to its complement, the rest of the clause *you should* [___ *read* ___].

The next observation is that A′-positions typically tend to pile up in the left periphery of the clause, often in a fixed order, partly universal and partly subject to parametric variation. For example, the left periphery of Italian permits the co-occurrence of a topic, a focus and a preposed adverbial modifier in the space in between the declarative complementizer and the rest of the clause:

(34) Credo [che [a Gianni TOP [*QUESTO* FOC [oggi Mod [gli dovreste dire]]]]], non qualcos' altro.
"I believe that to Gianni *THIS* today you should say, not something else."

There are then "cartographic" issues that arise in this domain as well. The attempts to draw maps as precise as possible of the left periphery of the clause have given rise to a very lively area of research, with the identification of different positions and ordering constraints, ruled by specific principles and parameters (e.g., Rizzi 1997; Cinque 2002; Belletti 2004; Rizzi 2004b).

The following example illustrates a reasonable approximation to a map of the left periphery that is valid quite generally across the Romance languages:

(35) [Force … [TOP … [INT … [FOC … [Q … [MOD … [FIN [Clause]]]]]]]].

This zone of structural layers is then assumed to include the clause, specifying a space of positions dedicated to scope-discourse semantics. A′-movement is movement of an element to one of these external layers, triggering the expression of the relevant interpretive property.

Movement: A Subcase of Merge

What kind of formal operation is movement? In traditional generative approaches, movement was performed by a class of formal rules, the transformations, which were completely different from the formal rules building the structural representations, the phrase structure rules, or X-bar theory. This sharp distinction raised a serious conceptual problem. Emonds (1970) observed that the core cases of transformations are "structure-preserving" in that they create configurations that are independently generated by phrase structure rules. Why, however, would two completely different types of rules converge to create exactly the same kinds of structures? The problem of structure preservation led a number of researchers to explore the possibility that it was a mistake to postulate two distinct rule systems for creating and modifying structures, and that phrase structure and transformational rules could be unified.

A full unification is achieved in the Minimalist Program, which contained a single structure building operation, Merge, repeated here for convenience:

(36)

A B → [A B]

What varies is the origin of A and B, the elements undergoing Merge. If they come directly from the lexicon, or one or both are independent complex entities already created by previous applications of Merge, we have *external merge*; if one (let's say A) is taken from within the other (B in our case) we have *internal merge*, which amounts, in traditional terms, to moving A from within B to the position sister of B. Using the format of (36), we could depict the global operation of internal merge as follows:

(37)

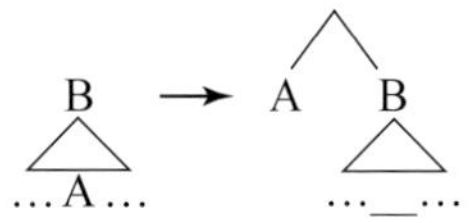

B [… A …] → [A B [… ___ …]]

This way of expressing things is, however, misleading: (37) is not an operation distinct from (36). What differs in the two cases is simply the way in which the two candidates of Merge, A and B, are selected through a search in the available computational space: they are separate objects in (36) whereas one is contained within the other in (37). Once the two candidates are selected through some kind of external or internal search, the formal operation that strings A and B together is the same.

Consider the (simplified) derivation of an interrogative: *What will you say?* This will illustrate how a series of applications of external and internal merge are interspersed in the computation of a clausal structure from a selection from the lexicon like the one given in (38):

(38) Selection from the lexicon: {say, what, you, will, Q}

(39) Derivation:

(a) [say what]	Ext Merge
(b) [you [say what]]	Ext Merge
(c) [will [you [say what]]]	Ext Merge
(d) [you [will [___ [say what]]]]	Int Merge
(e) [Q [you [will [___ [say what]]]]]	Ext Merge
(f) [will+Q [you [___ [___ [say what]]]]]	Int Merge
(g) [what [will+Q [you [___ [___ [say ___]]]]]]	Int Merge.

In this view, the structure-preserving property of movement is immediately explained: movement is structure preserving because it is a particular case of the fundamental structure-building operation, Merge.

Locality

Let us now focus on the procedure through which the candidates of Merge are selected. The significant case is internal merge. So, consider again the derivational stage in which the Q head, indicating an interrogative, is merged with the rest of the structure. The Q head starts a search for an interrogative operator, also endowed with the question feature.

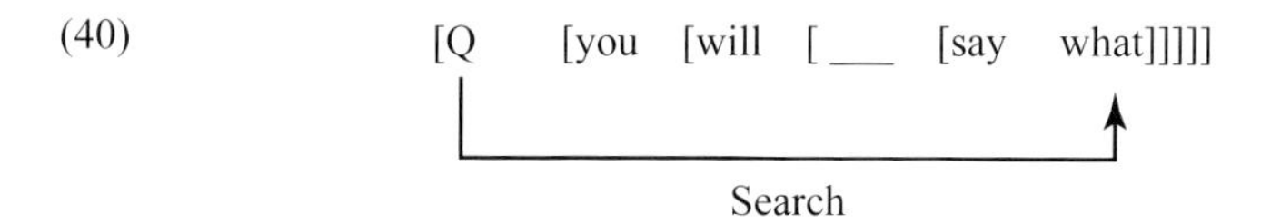

This is, in essence, the operation that Chomsky calls "agree," but I opt here for the more general term "search." Once an element is selected through the search operation (the *wh*-word *what* in our case), the search is terminated and the selected element becomes a candidate for internal merge with the whole structure. Through other operations, a sentence like (39g) is derived.

Search is a "local" operation involving a "probe" (i.e., the head activating the search) and a "goal" (i.e., the element which is reached); respectively, Q and *what* in (40). The operation is local in that it is blocked when an element intervenes between the probe and the goal, and the intervener bears some kind of structural similarity, to be precisely defined, to the elements involved in the relation. For instance, a *wh*-operator can be freely extracted from an embedded declarative (42a), but not from an indirect question, as in (42b). The structures from which extraction is attempted are given in (41):

(41) (a) I think Bill behaved like that.
   (b) I wonder who behaved like that.

(42) (a) How do you think Bill behaved ___?
   (b) *How do you wonder who behaved ___?

This is, in essence, the effect of Relativized Minimality: the locality principle barring local relations when an element intervenes which bears some structural similarity to the elements that should be connected.

(43) Relativized Minimality: in a configuration like X … Z … Y a local relation connecting X and Y is blocked if Z has the same feature specification as *X* and *Y* (Rizzi 1990, 2004a).

In the case under consideration, the derivation of (42b) is barred because the search connecting the main Q element and *how* in the embedded clause is blocked by the intervention of *who*, which also is an interrogative pronoun:

(44)

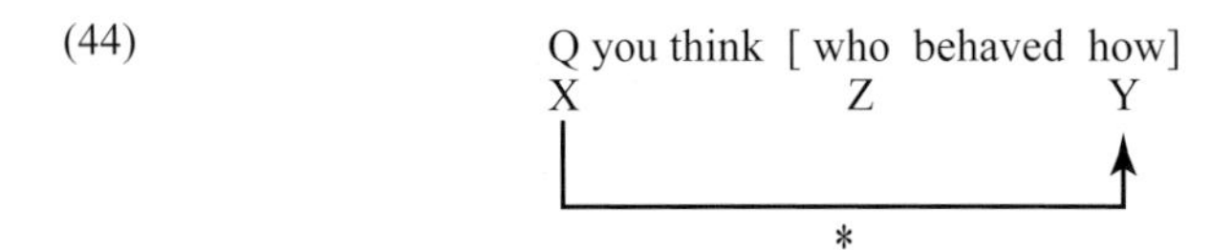

Search is often discussed in the context of analyses of movement, as a prerequisite for internal merge. There are reasons to believe, however, that it is a much more general operation, perhaps encompassing all the cases in which a local relation is established between two positions, independently from movement. For instance, the binding of a reflexive element by an antecedent, or the control of the null pronominal subject (PRO) of an infinitival clause in so-called "obligatory control" constructions—two relations, expressed by co-indexation in (45), constrained by a kind of intervention locality similar to the principle operative in (44)—may plausibly involve variants of the search operation:

(45) (a) John_i saw himself_i in the mirror.

(b) John_i decided [PRO_i to leave].

Thus, the search operation that identifies the candidate for internal merge may well be a particular case of a more general operation which can connect two positions in the tree if intervention locality is respected.

Invariance and Variation

How can this kind of approach address the problem of invariance and variation in natural language? This is a fundamental question for virtually every aspect of the study of language, clearly of central relevance for the question of language evolution. The theoretical entities that have been referred to within the generative tradition to address this issue are the concepts of Universal Grammar (UG) and particular grammars. The traditional conception of the 1960s and 1970s, which was based on the idea that particular grammars are systems of rules specific to a particular language, did not provide adequate tools to factor out the invariant properties across languages (among many other drawbacks).

Things changed radically around the late 1970s with the Principles and Parameters (P&P) approach, based on very different ideas (e.g., Chomsky 1981; Rizzi 1982; Kayne 1984). The key notion became UG, which was construed as an integral component of particular grammars. UG was conceived of as a system of principles containing certain parameters, binary choice points expressing the possible cross-linguistic variation. Particular grammars could be seen as UG with parameters fixed or set in particular ways. This conception went with a particular model of language acquisition. Acquiring a language meant essentially setting the parameters on the basis of experience. This is not a trivial task, as several researchers observed: in a number of cases, the evidence available to a child may be ambiguous between different parametric values, and there are complex interactions between parameters (Gibson and Wexler 1994). Still, despite such problems, parameter setting is a much more workable concept than the obscure notion of rule induction, which was assumed by previous models. More generally, the P&P approach introduced a very effective technical language to express, in a concise and precise manner, what languages have in common and where languages differ. Modern comparative syntax flourished once the P&P approach was introduced, and language acquisition studies took a new start.

Let me just mention here one basic example of parametrization. In some languages (e.g., VO languages), the verb precedes the object: *love Mary* (English) or *aime Marie* (French). Other languages (e.g., Japanese) have object–verb (OV) order. To address these properties, we need some kind of parameter operating on Merge and having to do with linear order. In some languages, the head (the verb) precedes the complement, whereas in other languages the head follows the complement. For a different approach to this kind of parametrization and deriving certain orders via movement, see Kayne (1994).

This simple ordering parameter has pervasive consequences in languages that consistently order heads and complements one way or the other. (Other languages, a minority according to typological studies starting with Greenberg (1963), are "incoherent" in this respect, as they opt for distinct ordering options for different types of heads.) Thus, two examples like the English sentence (46a) and its Japanese counterpart (46b) differ dramatically in order and structure, as illustrated by the two trees (47a, b):

(46) (a) John has said that Mary can meet Bill.

(b) John-wa [Mary-ga Bill-ni a - eru- to] itte-aru
"John-Top [Mary-Nom Bill-Dat meet-can- that] said-has"

(47) (a)

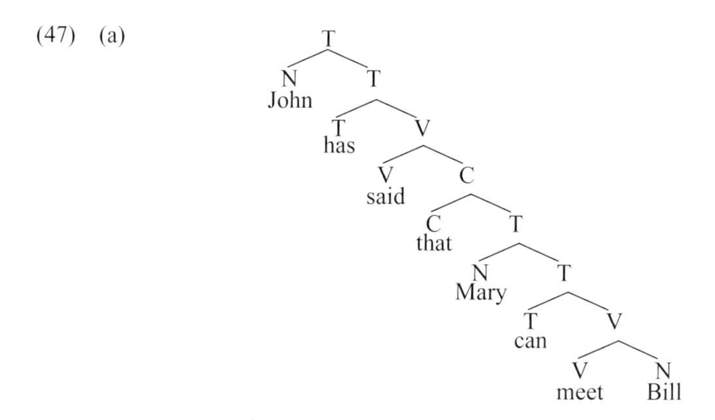

(b)

```
                    T
                 /     \
               N         T
          John-wa      /   \
                      V      T
                    /   \   -aru
                  C       V
                /   \    itte-
              T       C
            /   \     to
          N       T
     Mary-ga    /   \
               V      T
             /   \   -eru-
           N       V
        Bill-ni    a-
```

English expressions have a fundamentally right-branching structure, whereas Japanese expressions follow a fundamentally left-branching structure. The two are not perfect mirror images because certain ordering properties (such as the subject–predicate order) remain constant; however they are almost a mirror image of each other.

We have parameters on external merge, as the one just illustrated, and parameters on internal merge or movement. For example, we have seen that the lexical verb moves to T in French and Welsh, but not in Modern English. Also, the subject moves to the clause-initial position in English and French, but not in Welsh, yielding the VSO order. In addition, all Germanic languages except English have a Verb Second constraint in main clauses which involves a double movement of the inflected verb and another phrase to the left periphery.

A third parameter concerns Spell-out: the phonetic realization of the various expressions. There are certain elements that can or must be left unpronounced in particular configurations in some languages. One classical case is the Null Subject parameter: subject pronouns can be left unpronounced in languages like Italian and Spanish (e.g., in sentences like *parlo italiano*; I speak Italian), and this property relates in a nontrivial manner to other properties of the language (e.g., Rizzi 1982).

After a few years of developing these ideas, a crucial question arose concerning how to express the format of the parameters. Is it the case that anything can be parameterized in UG, or is there a specific locus for parameters? The first idea on the locus for parameters was that parameters were expressed directly in the structure of principles. This was probably suggested by the fact that the first parameter discussed in the late 1970s had to do with a particular locality principle, Subjacency, the proposed parametrization involving the choice of the nodes that would count as bounding nodes or barriers for locality (the S/S′ parameter; see Rizzi 1982, chap. 2). On the basis of this case, it was assumed for some time that parameters were generally expressed directly in the structure of the principles, and that could be the general format. Among other things, this assumption raised certain expectations on the question of how many parameters one should expect in UG: as the UG principles were assumed to be relatively few, if parameters were expressed in the structure of principles one could expect an equally (relatively) small number of parameters.

This view was abandoned fairly quickly, for a number of good reasons. One reason was that some principles turned out not to be parameterized at all. In no language, as far as we know, does a structure like *He thinks that John is crazy* allow for co-reference between *He* and *John* (principle C of the Binding Theory). That pronouns cannot be referentially dependent on some expression in their c-command domain is a general, invariable property of referential dependencies; many other principles are categorical in a similar way.

The second reason was that some macro-parameters (i.e., big parameters initially assumed to characterize basic cross-linguistic differences) turned out to require reanalysis into clusters of smaller parameters. One case in point was the so-called "configurationality parameter." Some languages have a much freer word order than other languages. Originally it was thought that there was a major parameter that divided languages between those with and without free word order. It quickly turned out, however, that there are different degrees of free word order: some languages are freer in the positioning of the subjects, others are freer in the reordering of the complements (scrambling), etc. There is a sort of continuum: not in a technical sense, but in the informal sense that there are different degrees of freedom, so that the big "non-configurationality" parameters really needed to be divided into smaller parameters.

The third reason was that some parametric values turned out to be intimately related to specific lexical items. For instance, consider the long-distance anaphor parameter (i.e., that certain reflexives roughly corresponding to English *himself* in some languages allow for an antecedent that is not in the same local clause, e.g., in Icelandic). This turned out to be the specific property of certain lexical items: if the language has such special lexical items (i.e., anaphors of a certain kind), then these anaphors work long-distance. Thus, we are not looking at a global property of the grammatical system, but simply at the presence or absence of a certain kind of item in the lexicon. These considerations led to the general view that parameters are not specified in the structure of principles, but rather are properties specified in the lexicon of the language. In fact, assuming the fundamental distinction between the contentive and the functional lexicon, parameters could be seen as specifications in the functional lexicon.

To summarize, a reasonable format for parameters would be: H has F, where H is a functional head and F is a feature triggering one of the major syntactic operations. This view implies important differences with respect to the view expressing parameters in principles. For instance, the order of magnitude of parameters is now related not to the number of principles but to the size of the functional lexicon.

If one combines this view on parameters with the cartographic approach (Belletti 2004; Cinque 1999, 2002; Rizzi 1997, 2004b), assuming very rich functional structures, the implication is that there can be a very rich system of parameters. Putting together the theory of parameters as specifications in the functional lexicon, some minimalist assumptions on linguistic computations and cartography, we end up with something like the following typology of parameters.

(48) For H a functional head, H has F, where F is a feature determining H's properties with respect to the major computational processes of Merge, Move (internal merge), and Spell-out.

Merge parameters:	What category does H select as a complement?
	To the left or to the right?
Move parameters:	Does H attract a lower head?
	Does H attract a lower phrase to its specifier?
Spell-out parameters:	Is H overt or null?
	Does H license a null specifier or complement?

We have parameters determining the capacity of a functional head to undergo Merge: What categories does it select, and does it take complements which precede or follow it? Perhaps even more fundamental properties include: Does the language use that particular functional head? Thus it may be the case that (certain) heads of the cartographic hierarchy may be "turned on" or "turned off" in particular languages. In terms of the move parameters, heads function as attractors: they may attract a lower head which incorporates into the attractor, or a phrase which moves to the attractor's specifier. Does the tense marker attract the lexical verb, as it does in the Romance languages but not in English or most varieties of continental Scandinavian? For spell-out parameters, we ask: Is a particular head overt or not? For instance, the topic head is realized in some languages (e.g., Gungbe, possibly Japanese *wa*) but not in others (e.g., in Romance clitic left dislocation). Does a head permit null dependents? For instance, does the verbal inflection license a null subject? That is one of a number of possible ways of looking at the null subject parameter in current terms.

This general picture is rather widely accepted at present, though many points remain controversial. As there are many more parameters than was originally assumed in the early days of the P&P approach, it turns out that the different parametric choices will enter into various complex kinds of interactions, generating many possible configurations of properties, so that the superficial diversity to be expected is very significant. For this reason, the parametric approach turned out to be particularly well-suited for the comparison of historically and structurally close languages, in which it is easier to control for very complex interactions; the likelihood of pinning down truly primitive points of differentiation is high, as Kayne (2000) indicated. This also explains the remarkable success of the approach to account for dialectal varieties (e.g., Italian and Scandinavian dialects) within language families (e.g., Romance, Germanic) and the renewal of dialectological studies based on this theory-driven approach.

Despite the apparent vastness of the parametric space, the abstract structure of parametric options still reduces to very few schemata, perhaps along the lines of (48). The deductive interactions between P&P remain very tight, so that there are many logical possibilities that are excluded in principle. A profitable research strategy remains the search for attested and unattested clusters of properties under large-scale surveys. Consider, for example, Cinque's (2005) discussion of Greenberg's (1963) Universal 20, which showed how certain systematic gaps in the ordering of various kinds of nominal modifiers can receive a principled explanation. Empirical claims about universals can now be checked against a data base that is enormously richer than it was 25 years ago. In addition, the search for nonaccidental gaps in the clustering of properties preserves all of its heuristic value and naturally complements the study of the micro-parametrization within very close systems, in the attempt to understand different facets of invariance and variation across languages.

Acknowledgments

I would like to thank Derek Bickerton, Noam Chomsky, and an anonymous reviewer for helpful comments on an earlier draft.

5

The Adaptive Approach to Grammar

T. Givón

Abstract

This chapter describes the adaptive function of grammar (morpho-syntax) as, primarily, an instrument of communication rather than mental representation. Morpho-syntactic constructions map onto specific communicative intents. These communicative intents pertain to the mental representation, in the mind of the speaker–hearer, of the interlocutor's constantly shifting epistemic and deontic states during ongoing communication. Grammar is thus a streamlined, highly automated theory-of-mind processor. The three developmental trends that shape human language—diachrony, ontogeny, and evolution—are responsible for the rise and instantiation of universals of grammar. The evolution of grammar is discussed in the context of the evolving cultural and communicative ecology of *Homo sapiens*. The neurology of grammar and its putative evolution are surveyed.

Preamble

My charge in writing this chapter was to explore basic issues in the neurobiology and evolution of grammar. The questions I am bound to raise are constrained by an approach to grammar whose basic premise can be captured in the following quote from a standard anatomy text:

> Anatomy is the science that deals with the structure of the body...physiology is defined as the science of function. Anatomy and physiology have more meaning when studied together (Crouch 1978, pp. 9–10).

In pursuing a functional-adaptive research program, I defer to what has been a basic tenet of evolutionary biology not only since Darwin, but ever since Aristotle: biological structures are shaped and selected by their adaptive functions. I will thus take it for granted that grammar, like any other biologically based structure, evolved through adaptive selection. To understand the adaptive niche of grammar, however, one must view it in its wider context; namely, that of human cognition and communication.

Representation and Communication

The two core adaptive functions of human language are the *representation* and *communication* of information (knowledge, experience). Given the overwhelming evidence from animal communication, child language development, and neuropsychology (Geary 2005; Cheney and Seyfarth 2007; Ungerleider and Mishkin 1982; Schmahmann et al. 2007; Givón 2002), I will take it for granted that cognitive representation preceded communication in human evolution, is present in prehuman species, and is a developmental prerequisite to child language acquisition. What the evolution of human communication added to the preexisting primate cognitive-representation platform are two specific communicative codes: phonology and grammar.

Cognitive psychologists have long recognized three major systems of mental representation in the human mind/brain (Atkinson and Shiffrin 1968). The linguistic equivalents of these systems are transparent.

(1)

Cognitive level	Linguistic equivalent
Semantic memory	The mental lexicon
Episodic memory	The current text
Working memory and attention	The current speech situation

Semantic Memory

The functional-adaptive properties and structural organization of the mental lexicon have been reviewed extensively elsewhere (Geary 2005; Givón 2005). Its interaction with episodic memory is well documented. The core of the lexical semantic network has been tentatively ascribed to two brain locations: one in the left inferior frontal gyrus (IFG) of the prefrontal cortex (BA 47); the other in the medial temporal lobe (Posner and Pavese 1998; Martin and Chao 2001; Bookheimer 2002; Badre and Wagner 2007). But the network is more extensive. First, older subcortical limbic areas have also been implicated (Tucker 1991), and, second, relevant "experiential" brain loci—sensory cortical and affective subcortical—are co-activated as part lexical semantic activation (Caramazza and Mahon 2006; Pulvermüller 2003). Finally, the neurocognitive system of conceptual representation is cross modal (e.g., verbal, visual, tactile) rather than modality specific (Humphreys and Riddoch 1987).

Episodic Memory

Propositional information about unique events, states, situations, or individuals, or about their concatenations in longer chunks of coherent discourse, is represented in episodic memory (Gernsbacher 1990; Kintsch 1994; Ericsson and Kintsch 1995). Both visual and linguistic input are represented in the same system, first in the subcortical mediotemporal system (hippocampus and amygdala; Squire 1987; Petri and Mishkin 1994; Goodale 2000; Mesulam 2000). The limbic-based early episodic memory is the one most relevant to on-going human communication (Ericsson and Kintsch 1995).

Working Memory and Attention

Working memory represents what is available for immediate attentional activation. It thus overlaps partially with the attentional network (Schneider and Chein 2003; Posner and Fan 2008). It is a storage-and-processing buffer of small capacity and short duration, where material is kept temporarily pending further processing choices. It has a cross-modal conscious component that interacts with *executive attention* (Schneider and Chein 2003; Ericsson and Kintsch 1995), as well as several modality-specific nonconscious components (Gathercole and Baddeley 1993).

Information in the working memory buffer must receive some type of attentional activation in order to reach longer-term episodic representation. Retrieval of information from episodic memory requires attentional reactivation and thus repeated representation in working memory.

While these three cognitive representation systems are neurologically distinct, they display multiple interactions (Gathercole and Baddely 1993; Ericsson and Kintsch 1995; Givón 2005), and their brain representation involves distributive networks.

Communicative Codes

Two human-specific communicative codes, *phonology* and *grammar*, map onto two language-coded representational levels.

(2) Mapping between representation levels and communicative codes

Cognitive	Linguistic	Communicative code
Semantic memory	Lexicon	Phonology
Episodic memory-I	Propositional semantics	Simple clause grammar
Episodic memory-II	Discourse pragmatics	Complex clause grammar

Propositional semantics involves the structure of states and events ("argument structure"; "who did what to whom where and how"). Multi-propositional pragmatics involves the *communicative use* of propositional information (discourse context, speaker–hearer relations, communicative intent). Much of the confusion about the adaptive function of grammar arises from its dual mapping, onto both propositional semantics and discourse pragmatics.

Grammar

Grammatical Coding Devices

Grammar is a much more complex and abstract code than the phonological–lexical code. At its most concrete level, the primary grammatical signal combines four major coding devices.[1]

(3) Primary grammar-coding devices
 (a) Morphology
 (b) Intonation: clause-level contours; word-level stress or tone
 (c) Rhythmics: pace or length; pauses
 (d) Sequential order

Of these, morphology and intonation are more concrete, relying on the sensorimotor lexical codes. But these concrete devices are integrated into a complex system with the more abstract devices (rhythmics, sequential order) that are no doubt second- or third-order constructs. The most concrete element of the grammatical code, grammatical morphology, is a diachronic derivative of lexical words (Givón 1971, 1979; Traugott and Heine 1991).

The primary coding devices that make up grammatical structure (3) are, in turn, used to signal more abstract levels of organization.

(4) (a) Hierarchic constituency
 (b) Grammatical relations
 (c) Syntactic categories (noun, verb, adjective; noun phrase, verb phrase)
 (d) Scope and relevance relations (operator-operand, noun-modifier, subject-predicate)
 (e) Government and control relations (agreement, co-reference).

The structural elements in (3) and (4) combine to create various grammatical *constructions* (clause types). These constructions, with their attendant morphology, map most directly onto various communicative functions. Constructions are thus the concrete building blocks of syntax.

Constructions

Chomsky's (1965a) distinction between simple ("deep structure") and complex ("transformed") clauses remains fundamental to our understanding of syntax.

[1] The first-order formal properties cited here are relatively concrete and perceptually accessible. More abstract approaches to syntax may reject some of those, including the entire notion of syntactic construction (Chomsky 1992), and may count other abstract properties not mentioned here.

It underlies the approach to syntactic structures as concrete *constructions*, directly perceived and processed by the hearer in language comprehension, then translated into both propositional information and communicative intent; and directly produced by the speaker to code his/her communicative intent. This approach was later rejected by Chomsky (1992) as he veered toward his minimalist account of syntax.

One may classify syntactic constructions or clause types as follows (Givón 2001):

(5)

Simple	Complex	Complex clause types
Main	Subordinate	REL-clause, v-comp., ADV-clause
Declarative	Nondeclarative	Imperative, interrogative
Affirmative	Negative	Negative
Active-transitive	De-transitive	Passive, antipassive, inverse
Default topic/focus	Marked topic/focus	Left dislocation, cleft-focus

To illustrate how the same simple clause (propositional information, deep structure) may be transformed into a multiple complex one (surface structures), consider:

(6) (a) *Simple*: Marla saw John
(b) *REL-clause (obj.)*: The man [Marla saw]...
(c) *V-complement (modality)*: Marla wanted [to see John]
(d) *V-complement (manipulation)*: Betty told Marla [to see John]
(e) *V-complement (cognition)*: Betty knew [that Marla saw John]
(f) *ADV-clause (temporal)*: When Marla saw John,...
(g) *Imperative*: Go see John, Marla!
(h) *Interrogative (y/n)*: Did Marla see John?
(i) *Interrogative (wh-obj.)*: Who did Marla see?
(j) *Negative*: Marla didn't see John.
(k) *Passive*: John was seen (by Marla).
(l) *Left dislocation (obj.)*: As for John, Marla saw him (later).
(m) *Cleft-focus (obj.)*: It was *John* that Marla saw.

The best evidence speakers–hearers have about morphosyntactic structures and how they differ from each other are the regular form-function correlations between specific constructions and specific semantic-pragmatic functions during communication. To the speaker–hearer, thus, the syntactic variations (6a–m) on the same "deep structure" make a contrastive difference because they are manifestly paired with specific communicative functions.

Likewise, the best evidence the linguist has that the syntax of complex clauses—the bulk of syntax—could not possibly be about propositional representation, but is rather about communicative function, hinges on the fact that constructions (6a–m) all share the very same event-structure information. However, the communicative function of syntactic constructions can only be discovered by studying their use in their natural communicative context.

Many of the propositional-semantic and discourse-pragmatic functions of syntax are coded primarily by the *grammatical morphology* associated with constructions. While morphemes always cliticize on particular lexical words, their functional scope may be narrow (word scope), broader (phrase scope), or broader yet (clause scope). Thus consider:

(7) Functional scope of grammatical morphemes:
   (a) Word scope (gender): They hated the Empr-ess
   (b) Noun-phrase scope (case, determiner, number): She talked to-the-[new phone operator]-s.
   (c) Verb-phrase scope (negation): They didn't-[give the award to John].
   (d) Simple-clause scope (tense-aspect-modality): [She will sleep in the basement].
   (e) Complex clause scope (subordinators): The woman that-[he gave the book to].

Syntactic constructions are not flat and linear, but rather they exhibit a *hierarchic constituent structure*. This hierarchic structure is not observed directly in the speech signal, but is inferred from semantic, pragmatic and prosodic information. As an illustration of a fairly complex hierarchic structure, consider the phrase structure of example (7e) above:

(8)

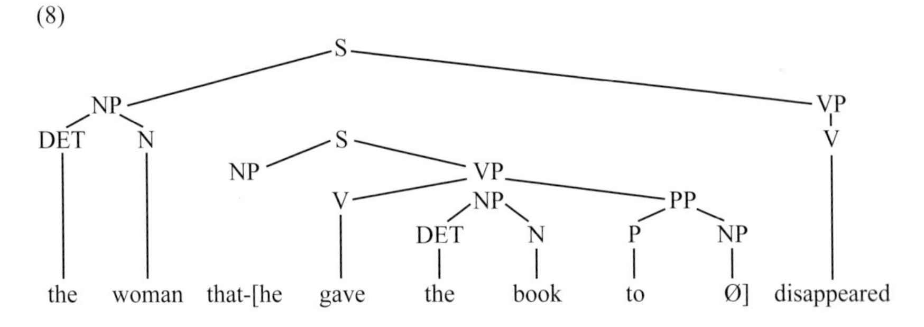

Universals

The balance between the great diversity of human languages and the universal properties of human language has been the subject of much debate in philosophy and linguistics going back to antiquity. On this issue, which mirrors the balance between species diversity and universals of biology, two reductionist schools have emerged in linguistics, represented most conspicuously by Leonard Bloomfield and Noam Chomsky. Bloomfield (1933) felt that language diversity was so immense and unconstrained that there was no point in looking for universals, except "purely inductive" summaries of the facts. This position is the precursor to the *surface universals* approach taken by some typologists (e.g., Comrie 2008; Haspelmath 1990). Adherents of this approach consider *concrete traits* of language(s), such as constructions, morphemes or phonemes, to be potential universals. Since some traits are more widespread cross-language than others, concrete universals may be a matter of degree. Indeed, more extreme practitioners of this approach may on occasion deny the existence of any morphosyntactic universals (Dryer 1997).

Chomsky's position is the extreme opposite of Bloomfield's: the nuts-and-bolts of surface structure are figments of the linguist's descriptive method, and underneath them lie *deep universals* that are rather remote from concrete features (morphemes, constructions). They are the highly abstract Principles and Parameters of Universal Grammar (Chomsky 1992).

Bloomfield and Chomsky each exaggerate a different aspect of the complex relationship between diversity and universals. Bloomfield did so by looking for universals of the type "all languages have surface feature X," while ignoring general principles that may control the diversity of attested surface features. What is more, he discounted the possibility that universal principles control extant surface diversity indirectly, through the three developmental trends that directly shape the surface features of language(s): diachrony, ontogeny, and phylogeny.

Chomsky, on the other hand, exaggerated universality and trivialized diversity by developing a descriptive framework (Minimalism) that misrepresents surface morphosyntax, positing extremely abstract underlying structures and few universal operations (Merge, Move). Like Bloomfield, Chomsky also discounted the possibility that universal principles do not control synchronic structure directly, but rather indirectly through the three developmental trends.

A middle-ground alternative to the two extreme approaches to language universals may be ascribed to Joseph Greenberg (1974, 1976, 1978, 1979) but is, in fact, rooted in the work of nineteenth-century diachronic-cognitive typologists such as Hermann Paul (1890). This middle-ground alternative echoes, explicitly or implicitly, evolutionary biology's traditional balance between diversity and universality (Givón 1979, 2002, 2005, 2008):

- Universals are not simply nuts-and-bolts surface features, though some of those may trivially exist ("all living organisms have cells").
- Universals reveal themselves in the control of development (i.e., ontogeny and phylogeny).
- Universals are not exception-less but rather flexible, due to the interaction among conflicting adaptive pressures. Such interaction often yields surface features that are an *adaptive compromise*.[2]

A closely related polarization is found in neurolinguistics, concerning whether processing modules/circuits are *dedicated* to syntax rather than to general cognitive functions. An extreme Chomskian position tends to opt for the language or grammar specificity of brain modules (e.g., Ben-Shachar et al. 2003, 2004).

---

[2] Ernst Mayr (1969, 1976) discusses this in the context of the difference between the laws of physics and generalizations in biology.

The opposite extreme denies the language or grammar specificity of *any* neurological module, noting that all of them still perform their pre-linguistic cognitive functions. Their involvement in syntactic processing is then viewed as an instance of "emergence" (Hagoort, this volume; Kaan, this volume), human specific only in the adaptive-communicative context of acquisition or usage.

A more modulated middle-ground position may be found in the work of Friederici (this volume), where lower-level modules may still function in either their linguistic or pre-linguistic capacity, but higher-level circuits or networks may already be dedicated to linguistic functions. This approach is in line with current views on attentional networks (Schneider and Chein 2003; Posner and Fan 2008). It is also compatible with Dehaene and Cohen's (2007) work on the more recent recruitment, during learning and development, of older cognitive modules to perform novel cultural tasks (reading, arithmetic).

The modulated middle-ground position in linguistics and neurology takes for granted that many, but perhaps not all, syntactic features have clear pre-linguistic cognitive foundations in memory, attention, perception, motility, etc. Pre-linguistic cognition deals only with mental representation, but human language involves both representation and communication. Syntax, therefore, does not simply "fall out of" or "emerge from" mental representation. Human language is certainly old enough to have, in principle, evolved dedicated, genetically specified neurological mechanisms, especially distributed multi-module networks/circuits.

Developmental Trends

Universal principles and/or mechanisms exert their influence on grammatical structure through the three developmental trends: diachrony (historical change), ontogeny (first- and second-language acquisition), and evolution. While we have no direct evidence about the latter, we have considerable evidence about the first two. And while in some purist circles language diachrony and ontogeny are deemed irrelevant to language evolution (Slobin 2002; Botha 2006a), they are still the best evidence of a step-wise, gradual development of language and grammar. Relying on such data to develop evolutionary hypotheses requires caution and justification (Givón 2008), but rejecting them outright is not a viable position in science, where any hypothesis is preferable over none at all. Hypotheses can be falsified; nil hypotheses are untestable.

The following developmental trends can be extracted from the diachrony of grammar (Heine and Kuteva 2007; Givón 2008):

- words > morphemes,
- concrete > abstract,
- concatenations (parataxis) > subordination (syntaxis),
- syntactic complexity > morphological complexity.

The following developmental trends can be extracted from first-language ontogeny of grammar (Givón 2008):

- lexicon before grammar,
- one-word clauses (nouns) before multi-word clauses (nouns and verbs),
- mono-propositional before multi-propositional coherence,
- pre-grammar (pidgin) before grammar,
- concatenation (parataxis) before subordination (syntaxis).

The following developmental trends can be extracted from second-language acquisition and pidginization (Givón 2008):

- lexicon before grammar,
- pre-grammar (pidgin) before grammar,
- concatenation (parataxis) before subordination (syntaxis).

Whether a putative resemblance between language diachrony/ontogeny and language evolution are merely analogical is debatable; evolutionary biology certainly has not closed the door on the relevance of both adaptive lifetime behavior and ontogeny to phylogeny (Gould 1977; Fernald and White 2000).

Grammar as an Adaptive Function

Representation vs. Communication

The greatest misunderstanding about the function of grammar, long licensed by formal linguists and adopted uncritically by others, is that grammar is a set of strict rules that governs the combination of words and morphemes into propositions (clauses), or, at best, that grammar codes primarily event structure (propositional semantics). The following quotation is in this respect typical:

> Reading or hearing a sentence such as *The little old man knocked out the giant wrestler* demonstrates the crucial role of syntax in normal language understanding. Identifying **who did what to whom** enables humans to understand the unlikely scenario that is described here. Thus, syntactic information helps us combine words we hear or read in a particular way such that we can extract the **meaning of sentences** (Kaan and Swaab 2002, p. 350; boldface added).

This misunderstanding about grammar's adaptive niche in human communication is natural, given two entrenched habits of formal linguistics: (a) a methodology that inspects isolated clauses ("propositions") apart from their natural communicative context; and (b) a theoretical perspective that emphasizes event frames ("argument structure") at the expense of communicative intent. The most cogent articulation of these habits may be found in Chomsky's *Aspects* (1965a), where chapter 1 licenses the methodology of de-contextualized isolated clauses ("competence"). Chapter 2 then concedes simple clauses ("deep structures"; event frames) a coherent adaptive-functional correlate—propositional semantics. This, again, is natural, given that the simple clause (main, declarative, affirmative, active) is the most frequent clause type in natural communication, the most semantically transparent, and cognitively the easiest to process (Givón 1995, chap. 2). Chapter 3 goes on to discount the adaptive function of complex ("transformed) clauses as "stylistics." As it turns out, it is in the study of this much larger set of syntactic structures (6b–m), and of their use in natural communication, that one finds the best clues to the communicative function of grammar.

Pre-grammar

The adaptive function of grammar comes into sharp relief when one notes that humans can, in some developmental, social, or neurological contexts, communicate without grammar. In such contexts, we use a well-coded lexicon together with some rudimentary combinatorial rules; that is, we use *pre-grammatical pidgin* (Bloom 1973; Bowerman 1973; Scollon 1976; Bickerton 1981; Schumann 1978; Andersen 1979; Givón 1979, 1990, 2008).

Pre-grammatical pidgin speakers seem to deploy readily the cognitive representation systems of human language:

- Lexical semantics
- Propositional semantics
- Multi-propositional coherent discourse.

They also seem to have acquired some use, however rudimentary, of the *lexical phonology* of their target language. Where they seem to be deficient in their target language is in the more abstract, more complex and phylogenetically younger communicative code of human language: *grammar*. That is, they lack grammatical morphology and well-marked syntactic constructions.

To illustrate coherent, multi-propositional second-language pidgin discourse, consider the following narrative from a Japanese–English Hawaiian Pidgin speaker (Bickerton and Odo 1976):

(9) Japanese–English pidgin: [F 59 yrs. old; 40-year resident of Hawaii; telling her life story; husband present]
   (a) …me born three three month, then Japan go home, see?
   (b) and then, ey, Japan go school see?
   (c) me all together Japan go school ten year…ten years-*no*, *uchi* ["we"]…
   (d) me two year old Japanese school go, you know?
   (e) and then Hawaii come…nowadays me…ah, not across, so see?
   (f) yeah, and then my fa-…my husband ho, ah, marry,
   (g) ah nineteen thirty-six, August thirty, little more whole forty year…
   (h) Papa, what's come talk to me, now, now, you by me, by…
   (i) how many place no more light place go?
   (j) Kauai Lia go, but Honolulu man Lilia, she don't know, no?…

(k) yeah, anyway Kanturi go, see? Kanturi go, Hilo go…
(l) before over here plantation sugar-mill get,
(m) old-town, now only *make*-["dead"] man place, yeah,
(n) only one man, Hawaiian man *hapai* ["carry"] he stay over there,
(o) now *make*-["dead"] man place…

Early childhood communication, between the one-word (Bloom 1973) and two-word (Bowerman 1973) stages (ca. 1–2 yrs), is another instance of pre-grammatical pidgin. As an illustration, consider a conversation between Naomi (age 2;2) and her mother (CHILDES transcripts, MacWhinney and Snow 1985; Givón 2008):

(10) [Context: discussing objects in the immediate environment]
Naomi: Baby **ball**.
Mother: Baby has a ball.
Naomi: Got [???].Got [???].
Mother: What?
Naomi: Got **shoe**.
Mother: Got show, yeah. Yes. The baby has a dress on.
Naomi: **Jacket** on.
Mother: And a jacket on, right.
Naomi: **Shoes** on.
Mother: Yes, Daddy has shoes on.
Naomi: **Knee**.
Mother: Yeah. Daddy has knees. Where is the baby's **elbow**?
Naomi: Elbow.
Mother: Do you know where the elbow is?
Naomi: Elbow [pointing to Daddy's **head**].
Mother: No, that's Daddy's head.

Lastly, the communication of Broca's aphasia patients is also a recognizable instance of pre-grammatical pidgin, as the following narrative illustrates (Menn and Obler 1990, p. 165):

(11) …I had stroke…blood pressure…low pressure…period…Ah…pass out…
Uh…Rosa and I, and…friends…of mine…uh…uh…shore…uh drink, talk, pass out…
…Hahnemann Hospital…uh, uh I…uh uh wife, Rosa…uh…take…uh… love…ladies…uh Ocean uh Hospital and transfer Hahnemann Hospital ambulance…uh…half'n hour…uh…uh it's…uh…motion, motion…uh… bad…patient…I uh…flat on the back…um…it's…uh…shaved, shaved… nurse, shaved me…uh…shaved me, nurse…[sigh]…wheel chair…uh...

The structural and functional differences between pre-grammatical pidgin and grammatical communication may be summarized as follows (Givón 1989):

(12)

Properties	Grammatical	Pre-grammatical
Structural:		
(a) Morphology:	abundant	absent
(b) Constructions:	complex, embedded, hierarchic	simple, conjoined, nonhierarchic
(c) Word order:	grammatical (S/O)	pragmatic (topic/comment)
(d) Pauses:	fewer, shorter	copious, longer
Functional:		
(e) Processing speed:	fast	slow
(f) Mental effort:	effortless	laborious
(g) Error rate:	lower	higher
(h) Context dependence:	lower	higher
(i) Processing mode:	automated	attended
(j) Development:	later	earlier
(k) Consciousness:	subconscious	more conscious

The heavy dependency of pidgin communication on the lexicon tallies with the fact that the lexicon is acquired before grammar in both first- and second-language acquisition, as well as with the fact that abstract vocabulary is the diachronic precursor of grammatical morphology in grammaticalization. Pre-grammatical children, adult pidgin speakers, and agrammatic aphasics comprehend and produce coherent multi-propositional discourse, albeit at slower speeds and higher error rates than those characteristic of grammatical communication. That grammar is a highly automated, subconscious, high-speed processing system has been suggested by Givón (1979, 1989), Blumstein and Milberg (1983), Neville (1995), Neville et al. (1992), Kintsch (1992), Pulvermüller and Shtyrov (2003), and Pulvermüller et al. (2008).

Grammar and Other Minds

> Mind reading pervades language (Cheney and Seyfarth 2007, p. 244).

Earlier it was noted that the primary adaptive function of grammar was to code the communicative (discourse) context of the proposition/clause, rather than its propositional semantics (event structure). This notion of context-as-text is, however, only a methodological heuristic. In cognition, and even more so in communication, context is not an objective entity, but rather a mental construct. What the use of grammar is sensitive to, what grammar is adapted to do, is to represent—systematically, in the mind of the speaker–hearer—the constantly shifting epistemic and deontic states of the interlocutor during ongoing communication. In other words, grammar is an adapted code for the mental representation of *other minds*, or what is currently known as *theory of mind* (ToM).

Communicating without a ToM is either implausible or inordinately slow, cumbersome, and error prone. This was the implicit message of Grice's (1975) paper on the pragmatics of communication. For the sake of brevity, I will give only two illustrative examples.[3]

Mental Models of Epistemic States

The first example is taken from the grammar of referential coherence ("referent tracking"), a domain that comprises a huge number of constructions and grammatical morphology (Givón 2005, chap. 5). Consider the following mid-discourse narrative:

(13) (a) There was *a* man standing near the bar,
(b) but we ignored *him* and went on across the room,
(c) where *another man* was playing the pinball machine.
(d) *We* sat down and ordered a beer.
(e) *The bartender* took his time,
(f) I guess *he* was busy.
(g) So *we* just sat there waiting,
(h) when all of a sudden *the man standing next to the bar* started screaming…

In marking *man*, introduced for the first time in (13a), with the indefinite article *a*, the speaker cued the hearer that he does not expect him/her to have an *episodic memory* trace of the referent. In coding the same referent with the anaphoric pronoun *him* in (13b), the speaker assumes that the referent is not only accessible, but is still *currently activated*; that is, it is still under *focal attention*. An additional other referent is introduced for the first time in (13c), this time with the indefinite marker *another*. In using of the first-person pronoun *we* in (13d), the speaker assumes that his/her own referential identity is accessible to the hearer from the immediate speech situation, available in *working memory*. *The bartender* is introduced for the first time in (13e), but marked as *definite*. This is so because the prior discourse had activated *bar*, which then remained activated by the persistence of the narrated situation. *Bar tender* is an automatically activated connected node of the lexical frame *bar*, and thus a consequence of the cultural specificity of *semantic memory*. In continuing with the anaphoric pronoun *he* in (13f), the speaker assumes that the referent is both accessible and currently activated (i.e., still under focal attention). In using the first-person pronoun *we* in (13g), the speaker assumes that his own identity is accessible to the hearer in the *speech situation*, still held in *working memory*.

Finally, the *man* introduced in (13a, b), and then absent for five intervening clauses, is re-introduced in (13h). The use of a definite article suggests that the speaker assumes that this referent is still accessible in the hearer's

[3] The literature on ToM is mind boggling, multidisciplinary, and still growing, going back to Premack and Woodruff's (1978) seminal contribution. For an extensive review, see Givón (2005).

episodic memory, but that the hearer's memory search is not going to be simple. *Another man* was mentioned in the intervening (13c) as *playing the pinball machine*. Both referents are assumed to still be accessible in *episodic memory* and would thus compete for the simple definite description *the man*. To differentiate between the two, a restrictive relative clause is used, matching *standing next to the bar* in (19h) with the proposition *a man standing near the bar* in (13a). In using this grammatical cue, the speaker reveals his/her assumption that the hearer still has an episodic trace of both the referent and the proposition in (13a).

Mental Models of Deontic States

Example (13) reveals another important feature of our presumption of access to other minds: Our mental models of the mind of the interlocutor shift constantly, from one clause to the next, during ongoing communication. As speakers release more information, they constantly update what they assume that the hearer knows; that is, the hearer's constantly shifting *epistemic* (knowledge) states. Here we will see that speakers also possess running mental models of the hearer's constantly shifting *deontic* (intentional) states.

The deontic (and epistemic) states which we will consider are coded by the cluster of grammatical subsystems that mark *propositional modalities* (Givón 2005, chap. 6). The most conspicuous of these subsystems, and the easiest to illustrate, is the grammar of *speech acts*.

The study of speech acts has traditionally centered on a set of *felicity conditions* ("use conventions") which are associated with declarative, imperative, interrogative, and other speech acts. These conventions have enjoyed an illustrious history in post-Wittgensteinean philosophy and linguistics (Austin 1962; Searle 1970; Grice 1975; Levinson 2000).

As an illustration, consider the following, somewhat schematic but still plausible, dialogue between speakers A and B:

(14) A-i: So she got up and left.
B-i: You didn't stop her?
A-ii: Would you?
B-ii: I don't know. Where was she sitting?
A-iii: Why?
B-iii: Never mind, just tell me.

In the first conversational turn (14A-i), speaker A executes a declarative speech act, which involves, roughly, the following presuppositions about hearer B's current mental states (in addition to the speaker's own mental states):

(15) (a) Speaker's belief about hearer's epistemic state:
- Speaker believes hearer doesn't know proposition (14A-i).
- Speaker believes hearer believes that speaker speaks with authority about proposition (14A-i).

(b) Speaker's belief about hearer's deontic state: Speaker believes hearer is well-disposed toward the speaker communicating to him/her proposition (14A-i).
(c) Speaker's own epistemic state: Speaker believes he/she knows in proposition (14A-i).
(d) Speaker's own deontic state: Speaker intends to inform hearer of proposition (14A-i).

In the next turn (14B-i), B, the speaker now executes an *interrogative* speech act (yes/no question), which involves, roughly, the following presuppositions about hearer A's current mental states (as well as the speaker's own):

(16) (a) Speaker's belief about hearer's epistemic state:
- Speaker believes hearer knows the declarative proposition underlying question (14B-i).
- Speaker believes hearer knows speaker does not know that proposition.

(b) Speaker's belief about hearer's deontic state: Speaker believes hearer is willing to share their knowledge of that proposition.
(c) Speaker's own epistemic state: Speaker is not certain of the epistemic status of the proposition underlying (14B-i).
(d) Speaker's own deontic state: would like hearer to share his/her knowledge.

In turn (14B-iii), lastly, speaker B executes a *manipulative* speech act, which involves, roughly, the following presuppositions about hearer A's current mental states (as well as the speaker's own):

(17) (a) Speaker's belief about hearer's epistemic state: The hearer believes the hearer knows that the desired event (*You tell me*) is yet unrealized.
(b) Speaker's belief about hearer's deontic state:
- Speaker believes hearer is capable of acting so as to bring about the desired event.
- Speaker believes hearer is well-disposed toward acting to bring about the desired event.

(c) Speaker's own epistemic state: Speaker believes the desired event (*You tell me*) is yet unrealized.
(d) Speaker's own deontic state: Speaker would like the event (*You tell me*) to come about.

At every new conversational turn in (14), not only do the speaker's own belief-and-intention states change, but also his/her mental representation of the hearer's belief-and-intention states. One would assume that a similar fast-paced adjustment also occurs in the hearer's mental model of the speaker's belief-and-intention states.

Evolutionary Considerations

Whether or not one is happy with a razor-sharp Cartesian boundary between human and prehuman communication, one must acknowledge that the rise of well-coded human communication had a profound impact on our species' ability to extend mental framing operations beyond the bounds of prehuman cognition. Whether this extension turns out to be a matter of kind or degree is for the moment unclear, and perhaps even irrelevant.[4]

Phonology and Semantic Memory

Both communicating social animals and pre-linguistic children possess a rich cognitive representation system which is fundamentally not all that different from that of human adults (Cheney and Seyfarth 2007), and yet this does not necessarily enable them to communicate the bulk of their cognition. Animal and pre-linguistic child communication is hopelessly context dependent. In the absence of well-coded vocabulary, one must scan the speech situation constantly for clues to the interlocutor's referential intent. The less explicitly coded is the communication, the more it depends on the vagaries and ambiguities of the shared cultural and situational context. This restricts the generic cultural domains that are lexicalized and communicated about to a few adaptively urgent topics: mating, aggression, dominance, territorial defense, feeding, predator warnings, and social grooming. Likewise, the specific spatiotemporal domain of reference is restricted to the immediate speech situation: here-and-now, you-and-I, this-and-that visible. The addition of the phonological code and the rise of the well-coded lexicon is an order-of-magnitude boon to the speed, fidelity, and referential range of communication. It no doubt also contributes to a richer, more explicit network of semantic categories, and thus a more efficient access to and activation in semantic memory.

Grammar and Episodic Memory

It is fairly clear that communicating social animals, much like pre-linguistic children, have some rudimentary version of ToM, however implicit, unsystematic, and subconscious. As Cheney and Seyfarth (2007) note however, evidence supports the existence of prehuman ToMs in the domain of intention (deontics) more than in the domain of belief (epistemics). This, again, is a matter of degree and explicitness, since deontics always implies some epistemics, contrasting an existing state with an intended one.[5]

Communicating social species grope for the state-of-mind of their conspecifics haltingly, inefficiently, and often inaccurately—often to their adaptive

4 Cross-species differences, here as elsewhere, may be a matter of degree, as conceded by Tomasello et al. (2005); see also de Waal (2001) and Cheney and Seyfarth (2007).

5 Premack and Woodruff (1978) suggested that the mental representation of deontics evolved before that of epistemics. But the intentional *I'd like to eat the apple* presupposes the epistemic states, the factual *I haven't yet eaten the apple* and the irrealis *I will eat the apple*. In both language diachrony and ontogeny, epistemic senses are a later development out of deontic ones (Diessel 2005; Givón 2008).

pain. The rise of the grammatical code may be viewed as the second crucial evolutionary step of liberating communication from the tyranny of context. In this case, the burden involved the laborious, unsystematic, inefficient, and attention-demanding guessing game of reading the intentions and beliefs of one's interlocutor. What grammar does is streamline, systematize, conventionalize, and automate our access to other minds.

As in the case of phonology, grammatical code does not leave the preexisting representation system the same. Rather, grammar acts as an efficient in-and-out access channel to episodic memory, making for a much more efficient storage and retrieval system, perhaps also enriching the hierarchic organization of episodic information (Kintsch 1992; Ericsson and Kintsch 1995).

Early Hominid Communicative Ecology

The cultural ecology of social primates is eerily reminiscent of that of hunting-and-gathering hominids prior to the advent of agriculture. This social adaptation may be called the *society of intimates*. Its most salient characteristics are (Givón 2002, chap. 9):

- Small social group size: The size of foraging hominid social groups ranged between 50 and 150 individuals (Dunbar 1992). Baboon societies tend toward the upper limit of the hominid range, with the group comprised of several female-headed matrilineal families (Cheney and Seyfarth 2007). Bonobos and chimpanzees have a more flexible split-and-merge social organization, ranging from 25 individuals in the extended matrilineal family (plus associated males) to 120 members of multi-family "tribes" (deWaal and Lanting 1997).
- Kin-based social organization and cooperation: Primate social organization is kin-based, and within it cooperation is organized most prominently along kin-based lines, particularly in female-headed matrilineal families (Cheney and Seyfarth 1990, 2007).
- Restricted territorial range: The widest range recorded for chimpanzees in the wild is ca. 20 miles foraging radius.
- Genetic homogeneity: The social unit is kin-based, though provisions for exogamy are made, through either male or female migration.
- Cultural homogeneity: Social differentiation within the group follows primarily gender and age lines, augmented by personal charisma and foraging and social skills. Cultural and foraging skills are distributed relatively evenly across the group, with relatively low occupational specialization.
- Flat, nonhierarchic social organization: While leadership of the female-headed families is rigid and determined by seniority, larger-group leadership is neither hereditary nor permanent, but depends on character, dominance, and physical and social skills.

These features of primate cultural ecology inevitably predict the last characteristic—the one most directly relevant to the evolution of communication:

- Informational stability and homogeneity: Taken together, the territorial stability, genetic homogeneity, cultural homogeneity, and great cultural stability of prehuman primate societies indicate that this is the most important parameter of prehuman and early hominid communicative ecology. When all members of the social group know each other intimately, when the terrain is stable and well known to everybody, and when the culture is time-stable and cultural diversity is minimal, then the bulk of relevant generic knowledge (i.e., the conceptual semantic map of the physical, social and mental universe) is equally shared by all group members and requires no elaboration. In the intimate social unit, day-to-day specific episodic information is also largely shared, by virtue of the shared here-and-now situation. The communication system that springs out of such social ecology is indeed predictable: (a) there is minimal explicit coding and (b) most information is inferred from the context.

Human Communicative Ecology

The transformation between prehuman and human communicative ecology is striking and involves three core features.

Rise of Displaced Spatiotemporal Reference

Both early childhood and prehuman communication are heavily weighted toward here-and-now, you-and-I, and this-or-that referents which are perceptually accessible in the immediate speech situation. When all referents are equally accessible to all participants in the shared speech situation, the lexical coding of the type of referent is superfluous. Mere pointing (deixis)—orientating the interlocutor to achieve joint attention—will suffice.

Mature human communication is, in contrast, heavily tilted toward spatiotemporally displaced referents, be they individuals, objects, states, or events. This is reflected first in the lopsided use frequencies of displaced reference. It is also reflected, in turn, in the fact that much of our grammatical machinery is dedicated to communicating about displaced referents, states, and events (Givón 2001).

Referents in the shared immediate speech situation are mentally represented in the working memory/attention system. Such representation shifts—with motion and attention—from one moment to the next and is thus temporally unstable. In contrast, displaced referents are more likely to be represented in episodic memory, either as memories of past experience or future projections, plans, or imaginations. Compared to working memory, episodic memory is a much more stable mental representation, and this temporal stability may have contributed to the objectivization of verbally coded referents, including mental predicates.

The rise of the human phonological–lexical code may now be understood as an adaptation designed to accommodate the shift to displaced reference in human communication. When the adaptively relevant topics of communication became, increasingly, the spatially displaced past experiences or future plans of some individual rather than of everybody present on the scene, deixis, or pointing, ceased to be a viable tool of referent identification.

From Manipulative to Informative Speech Acts

Spontaneous prehuman communication is confined almost exclusively to manipulative speech acts (Tomasello and Call 1997; Savage-Rumbaugh et al. 1993; Pepperberg 1999; Cheney and Seyfarth 2007), a tendency also observed in early childhood communication (Carter 1974; Bates et al. 1975; Givón 2008). In striking contrast, mature human discourse is tilted heavily, at the use-frequency level, toward declarative speech acts (Givón 1995, chap. 2), and the bulk of the grammatical machinery of human language is invested in coding declarative speech acts (Givón 2001).

The emergence of declarative speech acts may have enhanced the liberation of epistemic mental predicates from their erstwhile subordination to deontic predicates. In turn, the separate and more explicit representation of epistemic mental states (e.g., think, know, see) may have contributed to a heightened consciousness of *mental framing* operators, first those referring to one's own mental states and then, by extension, those referring to the mental states of others.

The emergence of declarative communication also points toward the increasing adaptive relevance of *displaced reference*. Manipulative speech acts are confined to here-and-now, you-and-I—the immediate speech situation; that is, primarily to what is represented in working memory and current focal attention. Declarative and interrogative speech acts, on the other hand, are utterly superfluous when the referents are equally available to both interlocutors here and now. Why bother to tell the other guy if he already knows what you know? Why bother to ask them if you already know what they know?

It is the emergence of displaced reference as the more prevalent topic of communication that endows declarative (and interrogatives) speech acts with their adaptive motivation: They are designed to carry the load of reporting (and querying) about inaccessible referents and past or future events that are not available to all interlocutors. Displaced reference creates an *informational imbalance* in the erstwhile intimate social unit, and declarative/interrogative speech acts are the adaptive response to such an imbalance.

From Mono-propositional to Multi-propositional Discourse

Both early childhood and primate communication are overwhelmingly mono-propositional (Tomasello and Call 1997; Savage-Rumbaugh et al. 1993; Cheney and Seyfarth 2007; Bloom 1973; Givón 2008). In contrast, mature human communication is, at the use-frequency level, overwhelmingly multi-propositional. This is also reflected in the fact that the bulk of the machinery of grammar is invested in coding multi-propositional, cross-clausal coherence (Givón 2001).

As noted above, grammar primarily codes the mental representation of the interlocutor's putative ever-shifting epistemic and deontic states during communication. The high automaticity of grammar may mean, among other things, that the evolution of grammatical communication was motivated, at least in part, by the strong adaptive pressure of having to deal with a high frequency of *perspective shifting* (MacWhinney 2002); perhaps an order-of-magnitude higher than what prehuman social species encountered.

One may view the rise of multi-propositional discourse as the next step in the rise of declarative communication. As the volume of adaptively relevant information about displaced reference became greater, the faster, more streamlined processing of such voluminous information became more adaptively pressing, especially in terms of the constant perspective shifting involved in the processing of larger stretches of coherent discourse. The rise of grammar may thus be viewed as an adaptive response to the need to process this explosion of declarative multi-propositional information.

Neurology and Grammar

Overview

The Geschwind (1970) model divided the load for language processing in the cortical left hemisphere between two main loci: Broca's area in the IFG was said to be responsible for grammar. Wernicke's area in the posterior superior temporal gyrus (STG), or thereabout, was said to be responsible for meaning. The two areas were presumed to connect through a dense cross-cortical channel, the arcuate fasciculus.

The two-module view of language processing in the brain has been largely superceded and refined by the immense amount of work conducted after the pioneering lesion-based insights of Broca and Wernicke. The accumulation of new studies has taken advantage of an array of brain-imaging techniques, which allow much finer spatial (PET, fMRI) and temporal (ERP) resolution of brain loci and their specific processing activity. In this connection, Bookheimer observed:

> It is apparent that large-module theories are clearly incorrect; rather, the language system [in the brain] is organized into a large number of relatively small but tightly clustered and interconnected modules [each] with unique contribution to language processing. There is increasing evidence that language regions in the brain—even classic Broca's area—are not specific to language, but rather involve more reductionist processes that give to language as well as nonlinguistic functions (Bookheimer 2002, p. 153).

Similar conclusions, more specifically addressing the representation and processing of syntax, have been voiced by Kaan (this volume, p. 132). Both the frontal and temporal sites, as well as the connecting channels, turn out to have many anatomically and functionally distinct subcomponents. Those sites, in both the prefrontal (old Broca's) and temporal (old Wernicke's) areas, appear to be joined into a number of distributive networks or circuits, each with its own pattern of spatial connectivity and temporal activation. Below I will survey some of the relevant literature pertaining to the distribution of these function-specific networks related to grammar, and when possible the pattern of their interactions.

Linguistic Representation Systems

Lexical Semantics and Words

Earlier work by Petersen et al. (1988) and Raichle et al. (1994), using PET scans, probed the localization of word-meaning representation, implicating one inferior prefrontal site and one medial temporal site. Subsequent ERP imaging studies (Posner and Pavese 1998) suggested that these sites were functionally and temporally distinct. The prefrontal site was activated by purely lexical tasks, and its activation peaked at ca. 200 ms, the well-known time range for the resolution of both word meaning (Swinney 1979) and visual object meaning (Treisman and DeSchepper 1996; Barker and Givón 2002). The medial temporal site was activated by noun–verb combinatorial tasks, and peaked at ca. 800 ms. This is within the time range of the resolution of clausal-propositional meaning (Swinney 1979) and visual event meaning (Barker and Givón 2002).

A finer yet spatial resolution of the core lexical network was reported by Badre and Wagner (2007), identifying the ventrolateral prefrontal cortex (VLPFC; pars orbitalis; BA 47/12) as the prefrontal lexical representation locus. Additional brain areas were, however, implicated in more specific processing tasks related to lexical semantic representation.

The work of Badre and Wagner (2007) suggests that lexical semantic representation-*cum*-processing implicates a network of brain submodules. In her general review of the localization of lexical semantics in the brain, Bookheimer (2002) suggests just that, implicating (a) the left VLPFC site (pars orbitalis; BA 47/12); (b) a left temporal site responsible for object and concept categorization, perhaps the posterior STG or MTG; and (c) right hemisphere sites responsible for the comprehension of contextual and figurative meaning (pragmatics). The frontal temporal connection for lexical semantic processing is now said to go via the extreme capsule (Schmahmann et al. 2007). In addition, various lexically related phonological functions, including visual, auditory and articulatory coding, also seem to implicate some left inferior prefrontal (IFG) sites (Fietz 1997; Price et al. 1997). Support for this distributive view of lexical semantics comes from work by Caramazza (2000), Pulvermüller (2003), and Hauk et al. (2004). The combined thrust of this research is that concrete, sensorimotor, or affective-visceral words prompt the activation of relevant peripheral sensorimotor or visceral-affective regions, endowing meaning with its "wet" experiential contents.

Given such a wide distribution of sites, Bookheimer (2002) posed the crucial question about the overall brain organization of lexical semantics: Is there a controlling core component? Both her question and answer are worth noting:

> Assuming that different aspects of sensory, conceptual and associative semantic information have separate and diffuse organization in the brain, how do we then integrate such knowledge in the service of language? Martin and Chao (2001) argue that a good candidate model for this integration could be observed by the left anterior IFG [BA 47/12] as discussed in detail above. This region is likely involved in the executive control of semantic information processing, including retrieving, integrating, comparing, and possibly selecting the diverse pieces of semantic information in the brain (Bookheimer 2002, p. 174).

The representation of lexical semantic meaning thus seems to be a network of nodes and connections, within which lexical items are specific clusters of coactivated nodes (Spitzer 1999; Givón 2005, chap. 3). For each word, coactivated peripheral nodes endow it with category-specific senses (Martin et al. 1996). But a more abstract, central *core network*, possibly in the left anterior IFG (BA 47/12; Martin and Chao 2001), is responsible for integrating concepts and relating them to other concepts in the global network.

Propositional Semantics and Simple Clauses

The left temporal site identified by Posner and Pavese (1998) was activated by noun–verb *combinatorial* tasks. One may thus identify it with clause-level event representation. The peak activation time (ca. 800 ms from stimulus presentation) is consonant with the clause-processing time frame of Swinney (1979) and the visual-event processing time frame (Barker and Givón 2002). This activation contrasts with that of the left inferior prefrontal semantic site (at ca. 200 ms).

Recent work has clarified the locations of these centers more precisely (Badre and Wagner 2007; Friederici and Frisch 2000; Friederici et al. 2006a, b; Grodzinsky and Friederici 2006; Bahlmann et al. 2008; Friederici, this volume).

By combining ERP and fMRI imaging, Friederici and her colleagues identified two combinatorial syntactic prefrontal-to-temporal circuits. The first, relevant to combinatorial/propositional semantics, connects the frontal operculum (fOP; posterior VLPFC; pars opercularis; BA 44) with an anterior STG site. The connecting venue here is the fasciculus uncinatus. This "simple" ("local," "phrase structure") circuit is responsible for processing clause-level combinatorial clusters. It is distinguished, both spatially and temporally, from another prefrontal temporal circuit, the "complex" ("global") one, which is responsible for processing complex clauses and/or longer-distance dependencies.

The simple/local circuit of Friederici et al. may correspond to Posner and colleagues' combinatorial temporal lobe site, although the peak activation times do not quite match.[6] It also corresponds, at least conceptually, to Bickerton's (2008) clause-level Merge operation.[7] It probably also corresponds to Pulvermüller and associates' serial-order module (Pulvermüller 2002; Pulvermüller and Assadollahi 2007); as well as to Hagoort's (this volume) unification function.

One may suggest, lastly, that the simple/local frontal temporal circuit (Friederici, this volume) may have a prehuman precursor, given the rhesus macaque work of Perret et al. (1989). Using single-cell recordings, Perret and his colleagues found that a site in the anterior superior temporal sulcus (STS), the lower part of the STG, is activated by pictures of a hand-held object being moved to the monkey's mouth—in essence a complex three-argument event/verb. Friederici (this volume) notes that the simple/local combinatorial circuit matures earlier in children, has a clear homologue in prehuman primates, and is older, phylogenetically, than the complex/global circuit (see below).

Complex Clauses and Long-distance Dependencies

Earlier work implicated Broca's area (BA 44/45) in grammar as a single mega-module (Geschwind 1970; Neville et al. 1992; Neville 1995), as well as in processing sequential-hierarchic structures (Greenfield 1991). More recently, Friederici and her colleagues have identified a second syntax-related prefrontal-to-temporal circuit, implicated in processing complex clauses or longer-

[6] Posner and associates suggested peak activation for the combinatorial temporal site at ca. 800 ms, in the clause-processing time range identified by Swinney (1979). Friederici et al. suggest a much earlier activation for their simple/local circuit (early left anterior negativity, ELAN, at ca. 150 ms), as opposed to a later activation for the complex circuit (late centro-parietal positivity, ca. 600 ms; P600). The latter timing for complex syntax corresponds to Posner and Pavese's (1998) timing range for the combinatorial task, presumably the clause-level simple circuit (BA 44 to anterior STG). The timing data of Posner et al. is consonant with Swinney's (1979) clause activation timing.

[7] Bickerton (2008) suggests no neurological identification of his Merge operation, but contends that the same operation accounts for both local (clause-level phrase structure) and global (complex clause) processing. The work of Friederici et al., if I am not misjudging, pretty much scuttles this parsimony-driven suggestion.

distance dependencies (Friederici and Frisch 2000; Friederici et al. 2006a, b; Grodzinsky and Friederici 2006; Bahlmann et al. 2008; Friederici, this volume). This circuit connects Broca's area proper (BA 45a/45b; mid-VLPFC; pars triangularis) with a site in the posterior STG. The connecting venue here is the fasciculus longitudinalis superior (FLS).

Other studies purport to further differentiate sites responsible for specific aspects of syntactic processing. Thus, for example, Ben-Shachar et al. (2003) report that a site in Broca's region (BA 45) responds selectively to "transformed" (as opposed to simple) clauses. In another study, Ben-Shachar et al. (2004) identified a prefrontal-to-parietal circuit responsible for processing "movement" transformations; that is, constructions such as *wh*-questions, cleft, dative shift, and object-relative clauses where the surface order differs from the canonical order of simple clauses. The implicated brain sites here are the left IFG and a bilateral activation in the posterior STS. This circuit is not clearly distinguishable from Friederici's (this volume) complex/global circuit. Finally, Bornkessel et al. (2005) have reported different circuits for word order, morphology, and verb-frame semantics.

Bookheimer (2002), Kaan (this volume), and Hagoort (this volume) caution against ascribing dedicated linguistic functions to brain modules that may be general and not language-dedicated, and that still perform older pre-linguistic tasks. Such modules partake in (i.e., are co-opted by) various language-processing circuits. Therefore, it is not yet clear that the work on various syntax-specific brain sites and circuits has identified those sites as being solely dedicated to language, rather than shared by both linguistic and pre-linguistic tasks. The task-sharing view of such modules is consonant with the idea that complex syntax is probably the last evolutionary addition to language processing. As such, the brain sites that process complex syntax have the highest probability of not yet being dedicated exclusively to grammar or language.

Unresolved Issues

Grammar as a Distributed Network

The three neural microcircuits identified above—lexical, clausal, and complex clausal—are but a small part of the neural mechanism that partakes in grammatical language processing. With many other circuits, they form a higher-level distributive network. The most likely members of this macro-network are (a) working memory (modality-specific and executive; Gathercole and Baddeley 1993; Ericsson and Kintsch 1995; Fernández-Duque 2009); (b) episodic memory (Gernsbacher 1990; Kintsch 1994; Ericsson and Kintsch 1995); (c) attention (implicit and executive; Schneider and Chein 2003; Givón 2005); (d) the cerebellum as temporal coordinator (Agyropoulos 2008); and no doubt others. This hierarchic distributive network organization is characteristic of higher cognitive functions (Mesulam 2000; Schneider and Chein 2003; Posner and Fan 2008; Dehaene and Cohen 2007).

Morphology

Classical Broca's aphasia communication shows impaired use of *both* grammatical morphology and syntactic constructions. The same coupling of these two components of grammar is seen in early childhood pidgin (Givón 2008) and in adult second-language pidgin (Givón 2008). What is more, the diachronic rise of morphology and syntactic constructions are two aspects of the very same process (Givón 1971, 1979). All these considerations suggest a single locus—or distributive network—for processing morphology and syntactic constructions.

On the other hand, grammatical morphology invariably arises from the reanalysis—both functional and structural—of lexical words, and this morphogenesis takes place during ongoing communicative behavior. The neurological site that codes and/or processes grammatical morphology may thus be, at least in principle, distinct from the site(s) that process hierarchic syntax, both simple and complex. This is especially intriguing given the adjacency of the lexical semantic site (BA 47/12) and the complex syntactic sites (BA 45) in Broca's area. Is there a submodule for grammatical morphology at the upper fringe of BA 47/12? Or at the lower fringe of BA 45? Or spanning the borderline of both in a connected circuit?[8]

A suggestion that grammatical morphology may have its own neurocognitive niche comes from the facts of grammaticalization. During morphogenesis, the lexical precursors of grammatical morphemes are *de-semanticized*, gaining more abstract meanings. That is, they cease to partake in the lexical semantic network of nodes and connections, and lose the typical lexical capacity for *semantic priming* and *spreading activation*. This militates against an unmodified representation of grammatical morphology within the lexical network (BA 47/12).

An added complication, partly methodological, arises from the fact that grammatical morphology—both nominal (articles, demonstratives, pronouns, adpositions/case markers, plural markers) and verbal (tense-aspect-modal markers, pronouns, transitivity markers, speech act markers) —appears copiously in simple clauses. The stimuli used in most of the investigation of the neurology of syntactic constructions seldom compare the presence vs. absence of grammatical morphology. A study by Bornkessel et al. (2005) has attempted to address this issue, reporting differential brain activation for word order

---

8 There is a neurological precedent for interactive cross-modal sites located between two modality-specific sites. In the optic tectum (superior culliculus) of the barn owl, a cross-modal representation area is located between two modality-specific areas: the visual and auditory sites. In the first 90 days of the neonate owl's life, the visual system trains the auditory system in 3D spatial representation (Takahashi 1989).

(object fronting) and morphology (case marking) in German: an IFG site (BA 44) for word order, and a posterior STG site for case marking. However, both sites partake in the two syntactic circuit reported by Friederici (this volume). It is thus not clear whether the temporal site is specific to morphology. In addition, the experiment did not address the much richer, more complex and cross-language component of morphology; namely, verbal morphology.[9]

Conjoined vs. Embedded Clauses

Experimental studies by Just et al. (1996), Booth et al. (2000), and Caplan et al. (2006) compared stimuli of two conjoined clauses vs. a complex two-clause construction—all outside their natural communicative context. The more specific studies by Friederici and her colleagues compared simple clauses with complex/embedded clauses, again outside their natural communicative context. However, as noted above, in both diachrony and ontogeny, complex-embedded *syntactic* constructions arise from *paratactic*—chained/conjoined—precursors. Such paratactic precursors to either relative clauses or verb complements perform the very same communicative functions as their embedded counterparts. That is, in Friederici's terms, they exhibit the same long-distance (global) dependencies.

In both diachrony and ontogeny, the rise of complexity progresses from one-word clauses to simple chained clauses (parataxis) to complex-embedded clauses (syntaxis). It would thus be of great interest to investigate a bit more carefully the neurocognitive status of the intermediate step in the rise of syntactic complexity: clausal conjunction or chaining. However, this can only be done using language stimuli that take into account the natural communicative context.

Representation of Other Minds

The facts of grammar use during communication strongly suggest that it is a streamlined, high-speed processor of the presumed, rapidly shifting deontic and epistemic mental states of the interlocutor during ongoing communication. The neurological location of this capacity, and its relation to other aspects of ToM, remains to be explored by experimental neurolinguistics.

[9] In most language families, but most conspicuously in Iroquois, Algonquian, Athabaskan, Southern Arawak, Uralic, Bodic (Tibetan), Turkic, Nilotic, Semitic or Bantu, verbal morphology is rich and complex, signaling multiple communicative functions. In such languages, the processing load is tipped heavily from constructional syntax to morphology. Even elsewhere, most syntactic constructions find their morphological coding primarily on the verb. What is more, in connected natural discourse in all languages, subject and object NPs tend to be anaphoric, thus either zero-marked or coded as pronominal affixes on the verb. To assess more realistically the processing role and brain localization of morphology, cognitive neurologists must design stimuli that take account of these facts.

Module Sharing, Circuit Sharing, and Evolution

In a recent thought-provoking article, Dehaene and Cohen (2007) discuss brain circuits, or cortical maps in their terminology, that are shared between recently acquired cultural-symbolic functions, such as reading and arithmetic, and older precursor functions in the primate brain. The recency of these cultural symbolic pursuits virtually guarantees that they have not yet acquired any specifically evolved dedicated brain modules. Rather, the brain circuits that process them are assembled during development and learning from available, functionally amenable precursors. Dehaene and Cohen (2007) suggest that *module sharing* of this type may involve all levels of brain hierarchic organizations: micro-maps (millimeter-size columns), meso-maps (centimeter-size circuits), and macro-maps (larger-size networks). There are two questions I would like to raise about this proposal, which may turn out to be applicable to the evolution of language-related neurology.

First, reading and arithmetic are recent cultural innovations with no evolved, dedicated brain-processing mechanism. The opportunistic recruitments, during development and learning, and subsequent sharing of modules that evolved for other functions, are thus plausible at all hierarchic levels of brain organization, micro to macro. Syntax, on the other hand, is a much older function, within the time range of, perhaps, 50,000–500,000 years or more. Some evolutionary changes specific to syntax, however subtle, may have already had time to occur, at least at the higher levels of *distributive networks* organization. One could thus raise the possibility that while lower-level micro-modules (e.g., the prefrontal BA 47, BA 45, BA 44, or the temporal and parietal sites) may still perform both their pre-linguistic and linguistic/syntactic functions, the more global circuits and networks (i.e., those that group together multiple low-level modules in specific spatiotemporal interaction patterns) may have already become dedicated to syntax, and thus in some sense have an evolved neural basis.

Second, if modules are shared between distributive networks (circuits, macro-maps) that perform different functions, what are the control mechanisms that instruct a module, or a whole circuit, to perform one function rather than another? Is there a default mechanism that instructs the module to partake in its old pre-linguistic circuit rather than its new linguistic/syntactic circuit, absent counter-default switching instructions? Is attention involved in the control of switching a module from one functional circuit to another (Mike Posner, pers. comm.)? Or could the context by itself—say, motor activity vs. visual memory vs. communication—trigger the automatic switching of a module from one circuit to another?

I have no concrete answer to these questions, which apply to many other brain networks that support the processing of complex functions. However, they must be addressed eventually, if we are to develop a more comprehensive evolutionary model of the neurology of grammar.

Acknowledgments

I am indebted to Derek Bickerton and Eörs Szathmáry for including me in this Ernst Strüngmann Forum and to Julia Lupp for her hospitality, perseverance, and grace. I have been blessed with an incredible group of patient tutors. Many participants at this Forum have shared with me, generously and enthusiastically, their knowledge about related and less related but just as exciting topics. Most specially: Angela Friederici, David Caplan and Edith Kaan for the neurology; Eörs Szathmáry, Terry Deacon and Szabolcs Szamadó for evolutionary biology; Julia Fischer, James Hurford, Stephanie White and Kazuo Okanoya for animal communication and ethology; Luigi Rizzi for formal syntactic theory; Francesco d'Errico for paleo-archaeology; and Balázs Gulyás for an incredibly cheerful spirit and infallible good vibes. Several members of the Symposium on the Genesis of Syntactic Complexity, Rice University, March 2008, made my learning more rewarding than it could have been: Don Tucker, Diego Fernández-Duque, Holger Diessel and Brian MacWhinney. Last but not least, I am indebted to Mike Posner, who started me thinking about the neurology of language.

6

Fundamental Syntactic Phenomena and Their Putative Relation to the Brain

Edith Kaan

Abstract

One fundamental question related to the biological foundation of syntax and its origin is to what extent syntax is "hardwired"; that is, to what extent specific neural areas or operations are dedicated to syntax. The increasing availability of human electrophysiology and brain imaging techniques has allowed researchers to assess these issues. Electrophysiological studies investigating various syntactic phenomena typically yield a P600 component for syntactically difficult continuations of a sentence. The P600 component may be preceded by a left anterior negativity (LAN) response in case of local violations. These responses are elicited regardless of the specific syntactic phenomenon tested. Similarly, studies investigating changes in cerebral blood flow report a network of areas active for syntactic processing, which does not vary systematically with syntactic distinctions. Furthermore, responses obtained in both electrophysiological and brain imaging studies are not unique to syntax. This suggests that syntax is not hardwired in the sense that there is a one-to-one correspondence between syntactic phenomena and brain areas. Instead, a network of areas and processes is involved, which are largely shared with other cognitive functions. This is compatible with the idea that syntax evolved from existing cognitive functions. The lack of detail obtained in cognitive neuroscience experiments may be partly due to the limited sensitivity of the techniques and analysis methods available, but also results from the lack of detailed processing models of syntactic phenomena, their interrelation with other cognitive processes, and how these processes can be instantiated in the brain.

Introduction

One fundamental question related to the biological foundation and origin of syntax is to what extent syntax is "hardwired," or to what extent specific neural areas or neural operations are dedicated to syntax. One way to assess this question is by investigating what cognitive operations and which brain areas are involved in syntactic processing, and to what extent these are unique to syntactic processing. Until about three decades ago, the only way to investigate language in the brain was by studying patients with brain lesions. On the basis of selective impairment of specific aspects of language, and by mapping the corresponding lesion sites, models were constructed of where in the brain certain aspects of language processing take place (Geschwind 1979). The drawback of this approach, however, is that it is based on brain damage. The findings, therefore, do not necessarily generalize to normal brains. In addition, lesions often cover large areas, making it difficult to pinpoint which area is responsible for what type of language processing. Furthermore, with some exceptions, these models are rather coarse and often do not distinguish between various aspects of syntax. With the advent of techniques such as functional magnetic resonance imaging (fMRI) and event-related potentials (ERPs), researchers have been able to investigate detailed aspects of syntactic processing in healthy brains. In this chapter, I introduce these techniques and the methodological issues involved. I discuss results from studies that investigate various syntactic phenomena and what these may tell us about the putative relation between syntactic phenomena and the brain.

Methods for Studying Syntax in the Brain

Currently, various techniques are available to study brain activation in healthy volunteers. These techniques are either based on differences in electrical activity or magnetic fields in the brain (electrophysiology and magnetoencephalography, respectively), or differences in blood flow. The latter are hemodynamic techniques and include fMRI, positron emission tomography (PET), and optical imaging. Alternatively, transcranial magnetic stimulation (TMS) enables the researcher to temporarily "lesion" cortical areas in healthy participants and to study the effect of this disruption on performance. I will restrict the discussion below to techniques that are currently used most often in language processing research, namely fMRI, PET, and ERPs. For more details on these and other neuroimaging techniques, see Gulyás (this volume).

Hemodynamic Techniques

When brain areas become more active in response to a certain stimulus or task, blood flow to these areas will increase. Hemodynamic techniques are techniques that track such changes in blood flow. The most popular technique of this kind is fMRI. When blood flow to a brain area increases in response to a certain task, the composition of the blood changes as relatively more oxygenated blood versus deoxygenated blood flows into these areas. This change in blood composition can be traced by placing participants in a strong magnetic field and quickly switching radio frequency waves on and off. This causes molecules to rapidly turn away from and reorient back to the magnetic field. Energy released in this way can be recorded and processed to reconstruct an image of the brain. Importantly, the magnetic resonance properties of blood cells with oxygen differ from those without, and this enables researchers to see which areas show a change in blood composition and are differently active in response to the task in which the participant is engaged.

A somewhat older technique to study changes in blood flow is PET. Using PET, blood flow is traced by injecting a radioactively labeled substance (usually water) into the bloodstream and recording the gamma rays emitted. Brain areas where a lot of gamma rays originate are the areas that receive the most blood. The advantage of PET, as opposed to fMRI, is that PET scanners do not produce much acoustic noise. This makes it easier to investigate brain activity to subtle auditory distinctions. In addition, PET allows for easy investigation in areas around sinuses, such as the anterior temporal lobe and medial frontal cortex, where fMRI often does not give a clear signal due to susceptibility artifacts. One of the disadvantages of PET is that it is more intrusive since it involves inserting a radioactively labeled substance into the body, limiting the number of scans that can be taken of a participant. Another disadvantage is that it takes several minutes for an image to build up, as opposed to several seconds in fMRI. This implies that, when using PET, only one type of stimulus can be presented to research participants during the multi-minute scan time. This may induce changes in participants' arousal, processing strategies, and anticipation of the stimuli. By contrast, fMRI allows the researcher to alternate the presentation of various types of stimuli rapidly, which somewhat reduces changes in the way the participants process the stimuli over the course of the experiment.

The advantage of hemodynamic techniques, in general, is that one can localize the activated areas with an accuracy of a few millimeters. A disadvantage of these techniques is that changes in blood flow occur rather slowly. Typically, the increase in blood flow reaches its peak only after about 6 seconds after the onset of the stimulus. Given the fact that humans can understand sentences at a rate of three words per second or even faster, these techniques do not give us insight into the fast processes involved in language processing. Another problem concerns the interpretation of data. Typically, a multitude of areas will be activated in response to the task, many of which may not necessarily be involved in the process under investigation. For instance, if a brain area shows increased activation in a syntactic task, this does not mean that this area processes syntax; it may be involved, for example, in attentional processes. In addition, an increase in activation can mean two things: excitation or inhibition. If an area of the brain shows increased blood flow, this does not necessarily mean that this area is working harder to execute the process under investigation. It may show increased activation because it is inhibiting some other area.[1]

Event-related Potentials

In contrast to blood flow, electrical activity in the brain changes very quickly in response to a stimulus, typically within a few milliseconds. By placing electrodes in the scalp, changes in the ongoing electrical activity in the brain can be recorded while the participant is presented with stimuli or engaged in a particular task. One way to analyze such data is to average the recorded activity over forty or more trials of a particular type. Averaging will enhance the brain signal that is time-locked to the stimulus onset (the ERP) and reduce activation that is not systematically related. The resulting ERP signal is a series of positive- and negative-going deflections. When deflections are found to vary with a particular cognitive task or stimulus type, they are called components. As discussed below, several components have been found to be sensitive to language processing. These components reflect simultaneous (mainly postsynaptic) activation of large groups of neurons (Kaan 2007).

The advantage of using ERPs for studying language processing is that processing can be tracked continuously with a temporal accuracy of 3 milliseconds, or even faster. A disadvantage of this technique is that the localization of activity is problematic. Given a certain pattern of electrical activity recorded at the scalp, a potentially infinite number of solutions is available concerning the location and number of the neural sources that may have generated this pattern. One can only *estimate* the source of the activation, typically by recording from many electrodes and restricting the solution space with data from hemodynamic techniques. Also, the pattern of current flow is easily distorted by the various layers of tissues and fluids between the activated neurons and the scalp. An alternative technique is to record magnetic fields rather than electrical potentials (magnetoencephalography, MEG). Because magnetic fields are less susceptible to irregularities in the intervening layers, this technique is preferred by investigators interested in source estimation.

Methodological Considerations

When studying the cognitive and neural underpinnings of syntax, research participants must be studied while they are *processing* syntactic constructions, either by producing, listening to, or reading phrases or sentences. This brings up a number of methodological issues. First, a sentence is not read or heard at

[1] Inhibition may be a very important aspect of cognitive functioning. For instance, according to Crosson et al. (2003), the right basal ganglia actively inhibit right frontal activation during language production, thus prohibiting interference from the right onto the left hemisphere. Such inhibition is therefore rather important for language processing.

once, but unfolds over time and is necessarily processed in a piecemeal fashion from beginning to end. This often entails temporary ambiguity. Consider:

(1) I wonder what you bought…

The phrase *what* can potentially be the direct object of *bought*, as indicated by the blank in (2).

(2) I wonder what you bought ___ at the store.

However, it can also be the object of a following preposition:

(3) I wonder what you bought that gadget for ___.

To address such ambiguities, the human sentence processor has developed a number of strategies. These strategies are not necessarily syntactic in nature. In fact, psycholinguistic research has shown that parsing decisions or preferences are driven by factors such as frequency, plausibility, and the discourse context in which the sentence is uttered.

In addition, processing can be modulated by task demands. Readers will process syntactic aspects of a sentence differently when asked to give grammaticality judgments than when asked to answer comprehension questions about the sentence or to spot a change in font type. Moreover, in most experiments on syntactic processing, sentences are presented in isolation, which may not reflect natural language processing where sentences are typically part of an ongoing discourse. The context, task used, and instructions given, as well as individual differences among participants, may therefore affect the kind and depth of processing. According to some models of sentence processing, readers and listeners do not construct a full syntactic representation of the sentence, but use lexical information and word-order heuristics to construct a "good enough" representation of the sentence. Evidence for this is that healthy young adults sometimes do not even notice anomalies such as *The cat was chased by the mouse* (Ferreira and Patson 2007).

When trying to isolate syntactic processing, researchers must be certain that participants in their experiments do not bypass syntax and simply rely on lexical semantics and stored templates, but are actively constructing a detailed syntactic representation and using this syntactic representation to guide the interpretation of the sentence. Even if one could guarantee that participants are actively constructing a syntactic representation, isolating syntactic processes from other processes is quite challenging. To do so, one needs to compare a "syntactic" condition with a "control" condition. Ideally, this control condition is identical to the syntactic condition, except for (aspects of) syntactic processing. This means that the stimuli used in the control condition need to be exactly equal to those in the syntactic condition in terms of the words used, their length, semantics, prosody, attentional load, memory load, etc.; stimuli can differ from the "syntactic" condition only with respect to the aspect of syntax one is investigating. Needless to say, such close minimal pairs are nearly impossible to construct. Over the years, researchers have used several types of comparisons in attempting to isolate syntactic processes (Kaan and Swaab 2002):

- comparing full sentences with lists of unrelated words,
- comparing sentences containing pseudo-words (i.e., meaningless strings such as *grop*) with normal sentences,
- comparing syntactically complex sentences (e.g., with noncanonical order) with syntactically simpler sentences (canonical word order),
- comparing sentences containing syntactic violations with sentences that are grammatically correct.

The assumption underlying all these comparisons is that the mechanisms involved in syntactic processing are more engaged in the first-mentioned condition than in the controls. The brain areas found active, or the differences in the ERP components observed, can then be said to at least partially reflect syntactic processes. However, none of the above comparisons is perfect. For instance, processing complete sentences, compared with a list of isolated words, not only involves syntactic operations, but also semantic operations and differences in prosody, working memory, and attention. Comparing syntactically incorrect with correct sentences does not fare much better. The assumption underlying the use of violations is that the mechanisms involved in syntactic processing are interrupted and get additionally taxed. However, syntactic errors may also trigger completely different processes, including error detection, disruption of semantic processing, attempts to repair the error or resolve the conflict between, for example, the semantic and syntactic representation as well as working memory and attentional processes. In addition, since the grammatically correct control sentence has also a syntactic structure, comparison of the ungrammatical with the correct sentences will obscure processes involved in regular syntactic processing, which are shared between the two conditions. It is thus extremely difficult to isolate syntactic processes from other processes.

In addition to the above issues, important questions that must be considered are: What qualifies as syntax, and how does syntax map onto processing? Most syntacticians would agree that syntax includes (a) the hierarchical combination of elements which are stored in the lexicon and (b) the establishment of relations among elements that are not necessarily adjacent in the sentence (Tallerman et al., this volume). However, this does not tell us anything about how structures are built during language production and comprehension. According to some models of processing (Hagoort 2005 and this volume), words have their syntactic projection (phrase template) stored in their lexical entry. For instance, a particular noun, such as *cat*, is stored together with a noun phrase frame, including a slot for a determiner and a modifier; a verb, such as *devour*, will have a verb phrase template stored with it, which includes a slot for the direct object, and the semantic and syntactic features that this direct object is likely to have. In such models of language processing, reading a phrase, such as *the cat*, hardly involves any combinatory processing, but simply associates phrasal templates in the lexicon. Does this process of activating and associating stored information qualify as syntax, or is syntax only active computation? If so, when does such active computation occur? Without an answer to these questions, it is hard to know what we should look for when investigating the hardwiring of syntax in the brain.

Studies on Syntactic Phenomena in the Brain

Syntax and Semantics

Despite the fact that it is difficult to isolate syntactic from semantic and other processes, electrophysiological and brain imaging studies have found rather systematic differences in brain responses for semantic and syntactic processing. One should note that the term "semantic" in the brain imaging literature is often taken to denote conceptual meaning and world knowledge, rather than sentential semantics and thematic role assignment. An example of a semantic violation in this sense is the last word in *She spread the warm bread with socks* (Kutas and Hillyard 1980). In ERP studies, words that constitute semantic violations, as in this example, generally yield a larger N400 response compared with a plausible control. The N400 is a negative deflection that peaks at about 400 ms after onset of the word. This component is typically largest over the middle of the scalp. In contrast, syntactic violations, such as agreement violations (see below), typically elicit a P600 response (a positive deflection for the violating word versus a correct control, peaking roughly around 600 ms, but more typically at 500–900 ms). This component is largest at posterior scalp positions. The N400 and P600 represent clearly two different brain responses that are distinct in time, polarity, and scalp distribution. In addition, hemodynamic studies often report a slight difference in the location of activation for semantic versus syntactic tasks, with the semantic loci being slightly more posterior/inferior than the syntactic loci of activation (Ben Shalom and Poeppel 2008; Vigneau et al. 2006).

These differences in brain responses suggest that semantic and syntactic information are processed by at least partially different mechanisms in the brain. This, of course, does not tell us whether the processes involved in syntax are unique to syntax, what these processes are, or whether they are even reflecting syntactic computation rather than, for example, error detection or working memory. To determine the relation between syntax and the brain, it may be more informative to examine to what extent different types of syntactic violations or syntactic processes elicit different types of brain responses. If indeed different brain responses are obtained for different syntactic phenomena, we can assume at least a coarse relation between syntactic theory and brain processes. Below I will discuss results from studies looking at various kinds of syntactic dependencies. For a more exhaustive overview of ERP studies on various syntactic phenomena, see Kutas et al. (2006).

Local Dependencies

Syntactic dependencies are relations between elements in a syntactic representation (e.g., between a subject and a verb). An important distinction is that between local and nonlocal syntactic relations (Tallerman et al., this volume). Local dependencies are relations between elements in a syntactic representation that are close to each other in the hierarchical syntactic structure. For example, in English, a determiner or possessor needs to be followed by a noun rather than a preposition, as in the example from Neville et al. (1991):

(4) We admired John's sketch of the landscape/John's *of sketch the landscape.

Violations such as in (4) have been labeled phrase structure or word-order violations, since they are a violation of the rule that determines the order of elements in a phrase. This type of violation typically elicits two kinds of ERP component. The first is an (early) LAN, a negativity that is largest at left frontal electrode sites. This effect typically occurs between 100–200 ms for phrase structure violations (Friederici et al. 2002; Neville et al. 1991). The second is a P600, a positivity which is largest at the back of the head and peaks between 500–900 ms. These results suggest that violations of this kind are perceived quickly, sometimes as fast as 100 ms, and involve at least two different processes, reflected by the LAN and P600, respectively.

Another type of a rather local dependency is agreement, for example, between a subject and the finite verb (5) (example from Hagoort et al. 1993), or auxiliary and form of the lexical verb (6) (example from Osterhout and Nicol 1999).

(5) The spoiled child throws/*throw the toys on the floor.
(6) The cats won't eat/*eating the food that Mary gives them.

One could argue that agreement violations are different from the phrase structure violations above, since the agreement operates upon an established syntactic structure and/or involves morphology, whereas phrase structure violations occur when the phrase structure representation is built. However, the ERP responses obtained for agreement violations are very similiar to those for phrase structure violations: a P600 component (Hagoort et al. 1993; Osterhout and Nicol 1999), which is sometimes preceded by a LAN component (300–500 ms) (Coulson et al. 1998b). Although some researchers claim that the early LAN found for phrase-structure violation is different from the later LAN for agreement violations (Friederici 2002), some studies have shown that the timing of the components depends on the position of the inflection that defines the syntactic category or agreement on the critical word (Deutsch and Bentin

2001; Hagoort et al. 2003; Neville et al. 1991). Brain responses are therefore not much different for agreement and phrase structure violations.

Studies using hemodynamic techniques have investigated local dependencies by either comparing sentence or phrases with local violations versus their correct counterparts, or by comparing the processing of syntactically simple sentences with that of lists of unrelated words. Typically, activation differences are found in the left, and sometimes also right, temporal areas, in particular, the anterior temporal areas. In general, Broca's area (left inferior frontal gyrus) shows more activation for local dependencies only when the stimuli involve violations (see overviews by Grodzinsky and Friederici 2006; Kaan and Swaab 2002; Stowe et al. 2005), or when the linear distance between the locally dependent elements (e.g., subject and verb) increases (Makuuchi et al. 2009; Friederici, this volume). Parietal and subcortical areas have been shown to be involved in local dependencies as well (Moro et al. 2001; Vigneau et al. 2006). One study comparing agreement versus word-order violations (Moro et al. 2001) finds largely overlapping activation for both types of violations, with the word-order violations showing higher activation in a subcortical area (caudate nucleus) and the insula.

The above results suggest that the processing of local dependencies is not localized in one particular area, but involves a network of frontal, temporal, parietal, and subcortical areas. Results from the ERP studies suggest that parsing local dependencies involves multiple processes, which do not differ much depending on the kind of relation tested (phrase structure or agreement).

However, one could object that local dependencies do not yield much insight into how the brain processes syntax, since local dependencies do not necessarily involve active computation. Most of these local relations may be precomputed and stored, especially frequently used ones. More insightful data about syntactic computation may be obtained by studying nonlocal dependencies, which are less likely to be stored and more likely to involve active combinatory procedures.

Anaphora

One syntactic phenomenon that involves a nonlocal dependency is the interpretation and distribution of pronouns and reflexives. A well-known linguistic observation is that in English, a reflexive, such as *himself* and *themselves*, needs to refer to the subject of the predicate if it is a direct argument of the predicate. For instance, *himself* in (7) is ungrammatical as opposed to *themselves* (example from Harris et al. 2000):

(7) The pilot's mechanics browbeat themselves/*himself.

Pronouns, on the other hand, cannot refer to the subject of the local predicate. For instance, in (8), the pronoun (*them*) cannot refer to the local subject (*mechanics*):

(8) The pilot's mechanics browbeat them.

Note that the sentence in (8) is strictly speaking grammatical, since the pronoun can refer to a group of people not mentioned. As long as the pronoun does not refer to the subject of its clause, the grammar does not further restrict the interpretation of pronouns. For instance, in (9), the pronoun *she* can refer either to the subject or to another person in the discourse. Only the latter reading is possible for *he* in (10) (example from Osterhout and Mobley 1995).

(9) The aunt heard that she won the lottery.
(10) The aunt heard that he won the lottery.

Even though the distribution of reflexives and pronouns is determined by different grammatical principles, ERP studies have not found much difference between the two. Reflexives that did not refer to the local subject, as in (7), elicited a P600 component (Harris et al. 2000; Osterhout and Mobley 1995). A similar effect was obtained at the pronoun in cases such as (10) in readers who judged this sentence to be unacceptable (Osterhout and Mobley 1995). Recall that a P600 effect was also found for the phrase structure and agreement violations mentioned above. This ERP response therefore does not distinguish between violations of the conditions on anaphora (reflexives and pronouns), versus other violations.

Polarity

A P600 effect has also been found for the violation of another type of nonlocal dependency, namely, the dependency between a negative polarity item (such as *ever* and *anymore*) and a negation, see (11). Negative polarity items are ungrammatical when the clause does not contain a negation (12), or when the negation occurs in an incorrect position (13) (example from Drenhaus et al. 2006):

(11) No man who had a beard was ever happy.
(12) *A man was ever happy.
(13) *A man who had no beard was ever happy.

Generally, P600 effects have been reported for polarity items when a negation or other licensing element is missing. This effect is preceded by an N400 effect, which is typically associated with conceptual semantic and world knowledge violations (Drenhaus et al. 2006). Again, even though negative polarity licensing is a phenomenon that is syntactically different from agreement, word order, or the interpretation of pronouns, the "syntactic" ERP response does not seem to reflect this.

Wh-movement

The above manipulations all involved violations. It can thus not be excluded that the LAN and P600 responses reflect error detection or meta-linguistic repair processes rather than syntactic processes themselves. One dependency that can be studied without using violations, and which may therefore be more informative, is *wh*-movement. An example is the formation of dependent questions, as illustrated below (example from Kaan et al. 2000):

(14) Emily wondered *who* the performer had imitated ___ for the audience's amusement.

(15) Emily wondered whether the performer had imitated *a pop star* for the audience's amusement.

In (14), the *wh*-phrase *who* is the direct object of *imitated*, but does not appear immediately to the right of the verb as is usually the case in English, see (15). Instead it is "moved" to the front of the clause. Psycholinguists distinguish among various processes that occur when sentences with *wh*-dependencies, such as (14), are read or listened to as they unfold. First, at the word *the* following *who* in (14), it is clear that *who* cannot be the subject and cannot be directly integrated into the syntactic structure. In ERP research, this process, and subsequent storage of the unintegrated *wh*-phrase into working memory, is associated with a LAN component (Kluender and Kutas 1993). Note, however, that this could also be viewed as a temporary local violation: if there is a strong preference to interpret *who* as the subject, one would expect the next word to be a verb rather than a determiner. In this view, the same processes may be involved as in processing phrase structure or, in languages in which the *wh*-phrase is case marked, agreement violations.

Second, the *wh*-phrase, or, at least some of its features or a place holder, must be kept in memory until it can be integrated into the syntactic structure and assigned a thematic role. Some ERP studies have reported a slow negative wave, starting from the point at which it is clear that the *wh*-phrase cannot be integrated, and spanning multiple word positions (King and Kutas 1995). Although not all studies have reported such an effect (Kaan et al. 2000; McKinnon and Osterhout 1996), this slow negative wave may be associated with maintenance of the *wh*-phrase in working memory.

Third, at the verb or other subcategorizing head, the *wh*-phrase is retrieved from memory and is integrated into the syntactic and thematic structure. This integration of the dislocated element with the verb or its base position is associated with a P600 effect (Kaan et al. 2000). A LAN component has been reported for word positions following the gap (Kluender and Kutas 1993).

Finally, *wh*-movement is syntactically constrained (Tallerman et al., this volume). For instance, in (16) the *wh*-phrase *which candidate* has been extracted out of an adjunct clause, yielding an ungrammatical sentence (example from McKinnon and Osterhout 1996). This violation becomes clear at *when*, when the reader has good reasons to assume that the extraction is out of the adjunct clause and the sentence is ungrammatical. ERPs studies comparing *when* in (16) with the same word in a grammatical control sentence without movement have reported a P600 effect (McKinnon and Osterhout 1996; Neville et al. 1991).

(16) *I wonder which of his staff members the candidate was annoyed when his son was questioned by ___.

To summarize the ERP findings on nonlocal dependencies, different ERP components are modulated by different aspects of processing dependencies: a LAN for (apparent) local violations, a P600 for general syntactic difficulty or ungrammaticality, and a slow negative wave for maintenance in working memory. Except for slight and nonsystematic differences in timing and scalp distribution, the ERP components are not sensitive to the specific type of syntactic construction, or even to whether the sentence can or cannot be eventually grammatically correct.

Moving now to studies using hemodynamic techniques, quite a number of studies have investigated the processing of sentences containing *wh*-movement. These studies have systematically found an increase in activation in Broca's area for sentences containing movement (complex sentences) versus those without movement (Kaan and Swaab 2002). However, given the poor temporal resolution of these techniques, it is hard to distinguish among the first three aspects of processing *wh*-dependencies mentioned above: recognition that the *wh*-phrase is in a noncanonical position, maintenance of the *wh*-phrase in memory, and integration of the *wh*-phrase. In a clever experiment, Fiebach et al. (2005) successfully distinguished maintenance from the two other processes. This experiment was conducted with embedded *wh*-questions in German, in which the *wh*-phrase was either the object of the clause (OSV order) or the subject (SOV order). In addition, the *wh*-phrase was either directly followed by the second noun phrase (subject or object: *wh*–S/O–V) or was separated from the second noun phrase by a prepositional phrase (*wh*–prepositional phrase–S/O–V). Note that separating a *wh*-phrase from the following noun phrase will increase the memory load when the *wh*-phrase is the object and the following noun phrase is the subject (OSV order). In this case, the object *wh*-phrase must be maintained in memory at least until the second noun phrase (the subject) is encountered for it to be integrated in the structure. The comparison between the long and short versions of the OSV sentences, in which the w*h*-phrase was the object and the second noun phrase was the subject (*wh*O–prepositional phrase–S–V versus *wh*O–S–V), showed an increase in activation primarily in Broca's area. The comparison between object initial (OSV) and subject initial (SOV) clauses showed only a weak activation in the left anterior temporal area. This suggests that Broca's area is involved in the retention of nonintegrated material, whereas temporal areas are involved in processing noncanonical word order and syntactic integration.

Recent work (Makuuchi et al. 2009) showed that the activation in and around Broca's area differed depending on what kind of relations needed to be stored in working memory: The inferior part of the pars opercularis (part of

Broca's area) was more active the more *wh*-relations needed to be stored; the left inferior frontal sulcus, an area more anterior and superior to Broca's area, was increasingly active when more words intervened between the subject of a clause and its finite verb. These two areas connect with different parts of the superior temporal gyrus and may be involved in different aspects of syntactic processing (Friederici, this volume).

Summary

ERPs do not appear to be very sensitive to distinctions made by linguistic theories. Violations of all kinds elicit a P600 response, which may be preceded by a LAN. A LAN is found only for genuine or apparent violations of local dependencies, although its occurrence is not systematic and its scalp distribution varies. A P600 is quite systematically found for any type of actual violation, apparent violation, or syntactic difficulty. Some researchers distinguish subcomponents of the P600, each having a different timing and/or scalp distribution. These subcomponents have been proposed to reflect various aspects of handling real or apparent violations, regardless of the syntactic phenomenon investigated (Friederici et al. 2002; Hagoort et al. 1999a; Kaan and Swaab 2003).

Studies using hemodynamic techniques typically report a network of areas active for syntactic processing, involving subcortical areas, the temporal, parietal, and frontal lobe. The foci of activation may differ depending on the task and materials used. The left frontal lobe (Broca's area), for instance, shows increased activation in the case of violations or complex sentences, with different parts being active depending on the kind of elements that need to be stored in working memory (*wh*-dependencies, SV dependencies; see Makuuchi et al. 2009). Therefore, syntactic processing is not confined to one process and one location in the brain, but involves an entire network of areas and various processes. This is not surprising, given that other cognitive functions, such as attention, also involve a multitude of processes and networks of brain areas (Posner and Fan 2008; Schneider and Chien 2003).

Syntax, Its Putative Relation to the Brain, and Putative Origin

Based on the above, there is some evidence that the brain distinguishes semantic from syntactic processing. Although, thus far, the brain does not seem to be sensitive to all distinctions made by syntacticians, it is capable of distinguishing some general aspects of syntactic processing, which suggests that at least some aspects of syntactic processing are "hardwired." When discussing the origin of syntax, it is especially important to ask to what extent the brain areas and processes found to be involved in syntactic processing are unique to syntax or are shared with other cognitive processes.

The ERP components observed in response to syntactic manipulations (P600, LAN, and slow negative wave) are not unique to syntactic processing, or even to language processing in general. The P600 component, initially labeled "the syntactic positive shift" by some researchers (Hagoort et al. 1993), has also been observed for difficulties related to discourse processing (Kaan et al. 2006), violations of musical structure (Besson and Macar 1987; Patel et al. 1998), sequencing (Núñez-Peña and Honrubia-Serrano 2004), and mathematical rules (Lelekov et al. 2000). This suggests that the P600 is either an index of structural integration, in general, or reflects attempts to resolve conflicts between representations (Kuperberg 2007).[2]

As for LAN, there is a strong similarity between this response and a mismatch response found for infrequently presented auditory stimuli that deviate in some respect from more frequently presented stimuli (Pulvermüller and Shtyrov 2003). In addition, violations of musical chord sequences elicit an anterior negativity with a right lateralized distribution, which has been shown to share some generators with the LAN elicited in linguistic tasks (Koelsch et al. 2005). Moreover, LAN can be elicited in nonlinguistic symbol manipulation tasks (Hoen and Dominey 2000).

The slow negative component found between the *wh*-phrase and the verb is also not unique to syntax. Memory tasks that involve retention of letters, colors, or location for several seconds have found such a slow wave, with the scalp distribution varying slightly depending on the type of materials that need to be maintained (e.g., Ruchkin et al. 1990).

Similarly, the brain areas found activated in syntactic processing tasks are not dedicated to this particular cognitive domain. Broca's region is involved in various non-syntactic and even nonlinguistic functions, such as working memory, inhibition, or resolving conflict among representations (Grodzinsky and Amunts 2006; Novick et al. 2005). The (anterior) temporal lobe found active for syntactic processing is also involved in semantic priming and discourse processing (Van Petten and Luka 2006). Parietal areas are involved in attention (Raichle 1998), reading, semantics (Price 2000), and working memory (Hickok et al. 2003), among other functions, whereas subcortical areas found active in syntactic studies are involved in a great variety of tasks. Therefore, none of the brain areas activated and ERP components elicited in syntactic tasks are unique to syntactic processing.

The general idea from current findings is that syntactic processing involves the combination of various, more general cognitive processes. Lexical information (including syntactic information) is stored in the temporal lobes. The left inferior frontal gyrus (Broca's area) serves a synthetic or integrative function in combining the information activated in the temporal lobe and operating

[2] The issue of whether P600 is specific to syntax is related to the controversy of whether P600 is actually a P300 component observed for unexpected stimuli in general (Coulson et al. 1998a, b; Osterhout and Hagoort 1999).

upon it (Hagoort, this volume), or in resolving conflicts between various levels of representation (Novick et al. 2005). The parietal area has a more analytic function, separating elements in the input (Ben Shalom and Poeppel 2008). As for ERPs, LAN may be an index of a violation of a strong expectation based on the structural aspects of the preceding input, regardless of whether "structure" refers to musical, syntactic, or a more abstract structure (see also Lau et al. 2006). P600 may be an index of general integration difficulty or an attempt to resolve conflict between various representations (semantic, syntactic) (Kuperberg 2007); finally, the slow negative wave may be a general index of memory load and anticipation of a target (response probe or base position of the moved element).

The observation that the brain areas found active and ERP components observed during syntactic processing are not unique to this task supports the idea that syntax originated from, or coevolved with, other cognitive functions, such as working memory, conflict resolution, sequencing, and conceptual combination (see also Hagoort, this volume). These functions needed to be orchestrated in a particular way to produce and understand sentences, or precursors thereof. Eventually, this combination of processes may have resulted in a network (or set of networks) specialized for syntactic processing, even though each part constituting the circuit(s) may be involved in other functions.

Open Questions

As far as the "hardwiring" of syntax is concerned, data from current research in cognitive neuroscience suggest that syntax is not localized in one area, and that the areas and processes involved are not unique to syntactic processing, or even language processing. For syntacticians, it is often disappointing to see that ERP responses and brain areas activated are largely similar for syntactically different phenomena. However, we must bear in mind that whenever cognitive neuroscience techniques are used to study syntax, research participants in those studies are *processing* phrases or sentences. It may thus be the case that the brain does not distinguish among the various types of syntactic phenomena proposed by grammarians as reflected by differences in the time, location, and manner of processing or storage of information.

It may also be that within the language network, different subareas or subnetworks can be distinguished that address different aspects of syntactic knowledge (Santi and Grodzinsky 2007a). Many researchers agree that slightly different subnetworks are involved in syntactic versus semantic versus phonological processing (Ben Shalom and Poeppel 2008; Vigneau et al. 2006), with phonological areas located more superior/anterior to areas involved in syntax, and semantic areas located more inferior/posterior. Therefore, separate subnetworks may address, for example, binding, structure building, and *wh*-movement. To date, however, very few studies have closely compared various syntactic phenomena using the same participants and techniques, and controlling for factors such as differences in task difficulty and working memory. It may therefore be that these subnetworks have not been uncovered.

Another problem may be that the techniques available thus far, or the analysis techniques used, do not have the resolution to segregate specific syntactic processing networks. fMRI data are typically spatially smoothed and merged into a standard brain format, which impoverishes the spatial resolution. Alternatively, we may have not been looking at the brain in the appropriate way. Until recently, cognitive neuroscience of language was primarily aimed at localizing the "syntax area," the "semantics area," etc. However, it has become increasingly clear that functions are not localized in a few particular areas, but that a whole network is involved. It may be that diversity in syntactic operations is reflected in subtle differences in the interaction and functional connectivity between the parts of the network. Future research should be directed at uncovering the nature of the connections between brain areas in the network.

Obviously, many other open questions remain. For example, as discussed above, detailed models are needed of what kind of syntactic information is stored, and what processes take place when people process syntactic constructions, and what the effect is of the task and participant instructions. This pertains to operations on linguistic representations, as well as more general cognitive operations such as retrieval, inhibition, working memory, and attention. Such a model is crucial to specify how various syntactic phenomena are processed, how processing will differ depending on the task and participant characteristics, and what to expect in terms of differences in brain activation. Related to this, appropriate techniques and analysis methods need to be developed to be able to uncover such differences.

Even if such a detailed model were available and systematic differences in brain activation were found between various syntactic phenomena, the question remains of how this all relates to the actual neural mechanisms: the workings of neurons, neurotransmitters, and, eventually, genes.

The ultimate goal for the cognitive neuroscience of syntax is to answer the question: Why is syntax the way it is? Why are certain syntactic operations allowed and others not? Why can a *wh*-phrase not be moved out of an adjunct clause or coordination? The answer lies necessarily within the workings of the human mind/brain. However, it is unlikely that answers will be found by looking at the workings of neurotransmitters. As more information becomes available about the cognitive neuroscience of syntax, we need to envision what kind of answers would be satisfactory and at what level they can be framed. Obviously, research in the cognitive neuroscience of syntax still has a long and challenging way to go.

Acknowledgments

The author would like to thank Milla Chappell, Brent Henderson, Tom Gívon, Ratree Wayland, and an anonymous reviewer for reading and commenting on an earlier draft of this paper. The author is currently sponsored by NIDCD-R03 DC06160-01.

Overleaf (left to right, top to bottom):
Fritz Newmeyer, Edith Kaan, Luigi Rizzi
Derek Bickerton, Mark Steedman, Marina Turner
Frankfurt Institute for Advanced Studies, Denis Bouchard
Julia Lupp, Tom Givón, Maggie Tallerman

7

What Kinds of Syntactic Phenomena Must Biologists, Neurobiologists, and Computer Scientists Try to Explain and Replicate?

Maggie Tallerman, Frederick Newmeyer, Derek Bickerton, Denis Bouchard, Edith Kaan, and Luigi Rizzi

Abstract

This chapter summarizes the extensive discussions of the working group whose mandate was to review the major syntactic phenomena that require an explanation as part of the evolved biological capacities of human beings. The main building blocks of syntax are outlined, starting with lexical categories and functional categories. Hierarchical structure and recursion are examined, particularly in light of proposals that recursion is the only uniquely human characteristic in the language faculty. A typology of dependencies between syntactic elements is discussed and the relationship between such dependencies and the needs of the human parser is considered. Syntactic universals of various kinds are outlined, their treatment within different grammatical traditions is examined, and possible responses to exceptional constructions are considered. Some evidence is presented in favor of abstract, high-level grammatical principles. Finally, using examples from the development of creoles from pidgins, from the diachronic processes involved in language change, and from the ontogenetic development of language in infants, consideration is given as to whether evidence of the (unobservable) evolution of syntax itself can be gathered from an examination of different kinds of observable syntactic phenomena.

Introduction

Two major linguistic stances regarding the investigation of syntactic phenomena emerged during the course of the twentieth century: broadly, these can be characterized as the *functionalist* approach and the *formalist* approach. In this report we represent both viewpoints, where possible, with an indication of the strengths and weaknesses of each. Functionalist syntacticians seek explanations for the observed phenomena in terms of the functions which language evolved to fulfil; this approach draws, for instance, on evidence from language processing, on pragmatic properties and principles, and on principles governing information flow in discourse. Linguists working within this approach focus generally on why languages exhibit the properties that they do. Formalist syntacticians, on the other hand, seek explanations in terms of simple and maximally general formal principles, such as Merge, which are proposed to be an *innate* part of the human language faculty. Often, such principles appear quite abstract and connect to the actual linguistic data through sometimes complex deductive chains, which may make their structure, motivation, and function hard to grasp to those outside the field. However, the underlying motivation is the desire to abstract away from the superficial linguistic diversity found in the world's 6000 or so languages and thus identify the essential properties of human language. Of course, postulating some principle to be innate does not explain its evolution, but merely pushes the burden of explanation back a stage.

The roots of modern functionalism lie in the approach to syntactic description taken by many European linguists (especially in the Prague School) in the 1920s and 1930s. During the 1960s, functionalism received new impetus with the advent of the typological approach to syntactic universals, represented particularly by the work of Joseph Greenberg (1963, 1978). This involves comparing the surface properties of large numbers of languages, ideally using a database which is representative of existing language stocks in terms of areal and genealogical diversity. The roots of modern formalism lie with the program initiated by Noam Chomsky from the late 1950s onwards (1957, 1965a). Traditionally, this has tended to investigate in depth the less obvious syntactic properties, crucially seeking explanations for *ungrammatical* (i.e., impossible) utterances as well as grammatical utterances in a language, and largely drawing on native speaker intuitions as evidence. This approach attempts to uncover the properties of Universal Grammar, otherwise characterized as the initial state of the language faculty in human infants before any exposure to linguistic data. Examples of work undertaken within each approach will be presented in the sections that follow. It should be noted, however, that comparing these two approaches can be difficult, because they have typically not investigated the same syntactic phenomena. Even when the same phenomena are investigated, assumptions are determined in large part by the way the data are interpreted under the particular theory adopted: facts are observational propositions, and these propositions are part of a theory, not external to it.

We begin by introducing some of the major building blocks of syntax, focusing specifically on the lexical/functional syntactic category distinction, and on syntactic constructions. One obvious (but open) question concerns the phylogenetic origins of the lexical/functional split, especially interesting since nothing remotely analogous appears in any other animal communication system. Thereafter we will look at the properties of hierarchical structure and investigate the phenomenon of recursion. This is followed by a catalogue of the main kinds of dependencies that occur between syntactic elements. Here an important question concerns the nature of the relationship between elements which appear to occur simultaneously in two distinct places within a sentence; for example, a fronted *wh*-phrase and the "gap" or trace associated with it, occurring in the canonical position in the clause. Critically, such dependencies are not necessarily local, but can be formed across many clause boundaries. Such phenomena, pervasive in natural language and again apparently uniquely human, are much in need of a biological explanation. We then consider syntactic universals from the perspective of each of the two major theoretical standpoints noted above, asking also whether the two approaches can be reconciled and briefly discussing some recent work which suggests possible avenues for future research.

Perhaps the main problem facing enquiries into the biological foundations and origin of syntax is that there is no direct evidence for the evolution of syntax in *Homo sapiens* (or in some earlier species). It has become commonplace to note that language does not fossilize. However, there may perhaps be "fossils" or traces of evolutionary processes in syntactic phenomena currently observable. We consider whether clues as to the origins of syntactic phenomena can be found in the development of creoles from pidgins, in diachronic changes occurring in languages in historical time, and in the ontogentic development of language in infants. All of these possible "windows" on language evolution are controversial, as we can have no idea whether the same conditions existed in the phylogeny of our species. However, informed speculation along these lines may supplement work done within comparative biology, archaeology, and computer modeling.

Building Blocks of Syntax

Despite the many differences among theoretical models of syntax, there is a broad consensus that grammatical theory needs to recognize two major categories in the human lexicon: *lexical* categories and *functional* categories (which correspond roughly to the distinction between open- and closed-class elements).[1] In fully modern (i.e., all attested) languages, the lexical class of items contains such syntactic categories as nouns, verbs, and adjectives, which are probably universal, as well as certain contentful adpositions (i.e., prepositions and postpositions) and adverbs. Functional elements are typically stripped down semantically, especially compared to the "content" words above, and are generally shorter than the lexical elements. The functional class comprises such grammatical categories as articles, quantifiers, auxiliaries, tense, aspect and mood markers, numerals, complementizers, polarity markers (affirmative and negative), question markers, voice markers, and potentially many others. Not all members of these categories are free morphemes, in the sense of being stand-alone words, but may well be represented by bound affixes or clitics, with much cross-linguistic variation existing. Languages also seem to vary enormously in terms of which of the many possible functional categories they actually instantiate, at least superficially, but it is typically reckoned that within a sentence in any given language, members of functional categories occur in a roughly 50/50 proportion to members of lexical categories.

In language we see a clear division of labor between functional and lexical items. For instance, a verb specifies the kind of action or event and selects the semantic roles of the participants in that action, while tense and aspect markers, for instance, locate events in time and indicate whether an action is completed or ongoing. In the noun system, similarly, functional elements determine such properties as definiteness, specificity, and so on. In current generative theories of syntax, much syntactic action is determined by functional elements, whether they are independent words or just affixes. These are, in a sense, the engines of syntactic action, driving processes such as syntactic movement (or displacement) by attracting the lexical verb or noun to raise to a higher head position. Within the Minimalist Program, for instance (Chomsky 1995; Boeckx 2006), all major kinds of syntactic movements are regarded as attraction by a functional head, including focus movement, question formation, and so on (Rizzi 1977). In such theories, too, functional categories form the locus of cross-linguistic variation, a fact which entails syntax being sensitive to morphology. In other theories, the same dependencies (discussed below) are captured by different mechanisms, some directly mimicking movement and some not.

It is worth noting that not all syntactic constructions are correlated in any obvious way with an actual morpheme. For instance, the process known as scrambling, whereby an object can move optionally to the left of the subject (as in Japanese), is not morphologically driven. It is also interesting to note that syntactic processes of the typical kind found cross-linguistically, such as passivization, *wh*-movement and extraposition, do not operate solely in terms of

[1] The term "functional" is used in this section in a very different sense than in the distinction between functionalist and formal approaches referred to in the Introduction. To avoid ambiguity, some linguists refer to the distinction between word classes as "contentive" vs. "grammatical," but here we continue to use the more widespread terms "lexical" vs. "functional" categories.

words or morphemes, which we might consider to be obvious building blocks of language, but may also involve entire phrasal categories, which are themselves made up of words and morphemes.

Despite these caveats, there is a broad measure of agreement between linguists that functional and lexical elements form the two central building blocks of the human lexicon.

Given the centrality of the distinction between functional and lexical elements, one important question concerns the evolutionary origins of these two main classes of elements. It is well known that in attested languages, a bundle of processes known as *grammaticalization* is responsible for creating functional elements from lexical elements. Loss of semantic content occurs in the course of the reanalysis of lexical items as functional elements, and in the process, phonetic reduction typically occurs, thereby producing the characteristic short and relatively content-free functional elements found cross-linguistically. Some examples from the history of English are the development of modal verbs (*will*, *might*) from lexical verbs and auxiliary *have* (*She has left*) from the possessive *have* (*She has a cat*).

If lexical and functional categories are core building blocks, what of constructions themselves? In the generative transformational tradition, constructions are regarded as epiphenomenal, whereas in some syntactic models, including construction grammar, cognitive linguistics, and most varieties of functional linguistics, they are central. Construction-oriented approaches to syntax suggest that the building blocks of syntactic description, and, by extension, of syntactic processing in the mind/brain, are morphosyntactic constructions of various size and complexity (Givón 1979, 1995, 2001). Syntactic constructions may be word size, phrase size, or clause size. Is there any evidence that we need to invoke constructions as building blocks? Possibly: It has been argued that children learn constructions as whole units; that in conversation, greater-than-word-level collocations are employed by participants; and, furthermore, that there is evidence from psycholinguistic studies that units larger than the word level are stored in memory.

An alternative viewpoint is that constructions are always decomposable into their component features. One argument for such a position comes from the fact that constructions always display features which occur elsewhere, in other constructions. For instance, consider passivization. The central property of the passive construction is that the logical object becomes the grammatical subject (*John was murdered* ___). But far from being unique to the passive, this property is also found in constructions with very different semantic and pragmatic effects, such as the "unaccusative" construction in (1b), in which the syntactic subject of a verb is semantically and thematically its object, as the comparison with (1a) indicates:

(1) (a) The sun melted the ice.
    (b) The ice melted.

Another property of the passive, namely the appearance of a *by*-phrase, also occurs in nominal expressions in many languages (e.g., *a book by Chomsky*) or in the causative construction, for instance in French:

(2)	Il	fait	laver	la voiture	*par Pierre.*
	he	make.PRES.3S	wash.INF	the car	by Pierre

"He makes Pierre wash the car."

Taken to its logical conclusion, facts such as these suggest that all constructions might be decomposable into more fundamental elements.

There is also the question of how the grammatical primitives posited by linguists pair with their meanings. Many morphemes have no fixed or clearly identifiable meaning, despite the textbook definition of the morpheme as the smallest unit of meaning in languages. Some obvious examples are "cranberry" morphs (*cranberry*, *inept*) and also fossilized Latinate morphemes such as *-ceive*, as in *receive*, *perceive*, *deceive*, etc., where there is no consistent meaning across lexical items. We assume that meanings are evolutionarily prior to the existence of forms to express them, but meanings and forms have clearly not evolved in any simple one-to-one relationship.

Recursion: What Is the Nature of Hierarchical Structure?

Syntactic structures are hierarchical in that words, or more generally the elements of syntactic computations, are combined to form larger units, namely phrases, and phrases are in turn combined with other phrases to form larger phrases. This successive phrase formation gives rise to a hierarchical structure which can be represented by using tree diagrams.

The hierarchical nature of syntactic representations is related to the recursive character of natural language syntax, but the two notions are not coextensive, in that it is clearly possible to have hierarchy without recursion. Recursion has been defined in two distinct ways within generative grammar: (a) as the capacity to insert a structural unit of a particular type (e.g., noun phrase, sentence) within another unit of the same type, or (b) as the property whereby certain formal rules or rule systems reapply to their own output (see Rizzi, this volume). For (b), recursion may produce structures of the form $[_X \ldots [_X \ldots]]$, as the first definition does, but is not obliged to do so. The operation known as Merge (see below) does not in fact require the insertion of any unit within any other unit, and indeed does not require anything beyond iterated applications of the Merge procedure. However, it is clear that both iteration and the creation of a hierarchical structure are necessary components of recursion. Simple iteration is not sufficient, since we can indefinitely iterate a given motor program (e.g., walking) without producing a hierarchical structure (3–4–5 steps do not combine to form a larger step!). Conversely, there are hierarchical structures which cannot be indefinitely extended. For example, phonemes are organized in hierarchical syllable structures (the nucleus and the coda form a rhyme which combines with the onset), but then the system stops. A combination of syllables does not form a larger syllable. Phonology thus involves a kind of "syntax," but the procedure is limited by an inflexible structural boundary—in this case the boundary of the syllable. Morphology offers a similar example. Here, the merging of morphemes to create a complex word is constrained by a more flexible yet still unbreakable barrier: only a finite number of semantically appropriate affixes can be attached to any word. However, when words themselves are merged into phrases and sentences, there are no boundaries, and since any sentence contains points where potential dependencies can be satisfied, thus extending the structure, the syntactic system acquires its open-ended scope. Thus, we can only truly speak of recursion if we adhere to the original definition.

It is sometimes said that recursion is not as crucial to natural language syntax as linguists have often assumed, because if one looks at corpora of natural conversation one typically finds relatively short and simple sentences, with few embeddings. This objection, however, misses the point: a system capable of generating very simple structures like *the teacher's book* is already recursive, because it has the ability to produce a nominal expression within a larger nominal expression. Once this property is available, the system is capable of generating an unbounded number of expressions, unless it is stopped by introducing an ad hoc limitation,

A number of technical proposals exist to formalize the recursive property of natural language syntax, among which are phrase structure rules and X-bar theory. The Minimalist Program (starting with Chomsky 1995) proposes that the generation of an unlimited set of syntactic structures is performed by the simplest combinatorial operation one can imagine, namely Merge, which takes two elements A and B, and forms the unit [AB]. The elements undergoing Merge can be simple lexical items or complex expressions already formed by Merge. Hence the combinatorial procedure can indefinitely extend a syntactic structure.

An important question concerns the evolution of the Merge process, if we are to believe that it reflects what actually happens in the brain: Are Merge and other processes postulated by syntacticians literally replicated in the brain, and, if so, how did these evolve? If we follow the line taken by Hauser et al. (2002), Merge is the only uniquely human element in language, which entails that its evolution must represent a very significant step in hominin evolution.

Hierarchy itself is widespread in cognitive spheres outside language (e.g., in planned action) and allows exponential savings in the search for plans. On the other hand, true recursion may be quite rare. In particular, the mere ability in animals to discriminate context-free "languages" (such as the set of strings consisting of any number of some symbol *a* followed by the same number of another symbol *b*) should not be confused with the possibility that they possess a truly recursive context-free *grammar*. Crucially, there are other mechanisms, such as counting, which can be used to recognize such "languages" (see Friederici, this volume).

Linguists believe human language to be truly recursive; some would suggest that this rests on the recursive nature of semantics itself. Despite the possibilities of finite state grammars covering unbounded tail- and head-recursion of the kind exhibited in possessives like *John's mother's brother's lover's chauffeur*, and the acute performance limitations on depth of center embedding in examples like *The cat the dog the rat bit chased ran away*, the compositional semantics itself is agreed to be truly recursive. In the first case here, the semantics is a recursive term such as *chauffeur* (...*lover* (*brother* ((*mother* (*John*)))...), nested to arbitrary depth. The incomprehensibility of center embeddings follows from some memory limitation on the processor: our recursive capacity is used by a system with finite and limited memory, and there are other external constraints which limit the length and complexity of the structures actually produced by humans, but these limitations are determined by such external factors, rather than by some inherent limitation in the generative mechanism.

To conclude, if we seek an evolutionary origin for recursion in human language, it is possible that we are really asking about the origin of recursive *semantics*, which may in turn rest upon intrinsically recursive concepts, such as kinship relations or concepts of other minds (Tomasello 1999).

Dependencies: What Kinds of Dependencies Exist among Syntactic Elements?

Dependencies that exist in natural languages pose a challenge to neurologists, biologists, and computer scientists, as they seek to account for such phenomena in evolution. Below we outline, on a purely descriptive level, the kinds of dependency phenomena that are observed:

(3) *Selectional and thematic restrictions*: These are the requirements of verbs to occur with particular arguments. For instance, **He devours* is ill-formed, since the verb is missing an argument, and **He devoured the grammatical rule* is ill-formed, because the predicate *devour* not only requires both an agent and a patient, but moreover requires the patient to have the lexical property of edibility. The dependencies here are both local, existing between a head and its arguments. Other dependencies are potentially much less local.

(4) *Agreement* refers to the copying of certain features that are inherent to one lexical item onto another item, or the copying of features from one constituent to another. Inherent features are typically such properties as person, number, and gender, and the kinds of agreement dependencies found in natural languages include SV agreement, OV agreement, and agreement within a noun phrase (NP), whereby the properties of the head are reflected on dependents such as articles and adjectives. Agreement may also be local (e.g., articles and adjectives agreeing with the head noun in a NP) but is not necessarily (at least not in the surface order), as shown by examples like *Which girls do you think are/*is leaving today?* Agreement is a

purely syntactic phenomenon, appearing to have no bearing on meaning whatever. It may potentially aid parsing, however, by indicating the relationships between elements in a phrase or a clause.

(5) *Dependencies involving displacement* (also known as movement) refers to the appearance of noncanonical ordering within phrases and clauses. The basic hallmark of displacement phenomena is that elements which are grouped together semantically are not linearly adjacent in the syntax. Displacement gives rise to dependencies between an element and its base position. Various kinds of displacement phenomena are observed in natural languages, including:

(a) Passivization and similar grammatical function-changing operations: in the passive construction, a logical object (the thematic patient) becomes the grammatical subject, creating a dependency between the canonical position of the moved element and its surface position (*John was murdered* ___).

(b) Scrambling, the term for the free constituent order phenomenon found, for example, in Japanese, where the object is freely moved across the subject position, deriving a noncanonical OSV order from the canonical SOV order.

(c) *Wh*-dependencies of all kinds, including *wh*-questions and relative clauses, where the filler and the associated "gap" may be separated by numerous clause boundaries. These are known as "unbounded" dependencies, since the dependency covers a potentially infinite syntactic space (see below for examples).

(d) Verb movement, for instance of the kind that derives the SVO order found in Dutch main clauses from a hypothetical underlying SOV order (i.e., the order found in subordinate clauses).

(e) Extraposition, the term for the rule which creates discontinuous constituents involving rightwards movement. For instance, from *A student with really terrible tooth decay came into my office* we can derive *A student came into my office with really terrible tooth decay*. The displaced phrase forms a dependency with the remainder of the phrase to which it is canonically attached.

(f) Scope-related displacement: in *Fred drinks only martinis*, *only* is adjacent to the item over which it takes scope, *martinis*. However, we can interpret the scope in the same way if the scopal element is nonadjacent, as in *Fred only drinks martinis.* Hence, a nonlocal dependency exists between the two elements. The idea that scope involves displacement is controversial. Some analyses express the dependency purely as a relationship between two positions.

(6) *Dependencies not involving displacement*, of which there are three main types:

(a) Anaphora, exemplified by the antecedent–anaphor relations of the type illustrated in reflexives: *Jill really adores herself/*her*. Here the reflexive object *herself* must be in a dependency relation with the subject *Jill*, but the pronominal object *her* cannot be, as it is within the same clause and must therefore refer to a female other than Jill. Dependencies involving ordinary pronouns can be established across clauses over an infinite distance, however (*The girl thought my father said...the policeman would arrest her*), and a pronoun can even take its antecedent across utterances and speakers: A: *Do you know that woman? B: I used to work with her.*

(b) Cataphora, as in *Before she left, Jill paid the waiter*, where a dependency is (optionally) established between the pronoun and the lexical noun *Jill*.

(c) Polarity: Certain negative polarity items, such as *ever*, establish a dependency with a negative antecedent, as in *No teacher ever says that*, but not **A teacher* / **Everybody ever says that.*

Various ways of classifying these dependencies have been suggested in the literature. Many syntactic models posit no movement operations. Nonetheless, displacement effects must still be accounted for in these models. This is typically accomplished by indexing operations which establish a relation between the dependent elements, or by using semantically compositional syntactic operations (see Steedman 2000).

The set of dependencies (3–6) varies greatly in terms of the property of *locality*, with those at the top of the list involving extremely local dependencies, while those lower down potentially involving dependencies that are much less local. Locality constraints can be expressed in terms of tree architecture, with, for instance, the dependency between a verb and its direct object involving a sisterhood relation, the closest possible relationship between two nodes in a tree.

Another typology of dependencies classifies them according to whether they involve *bounded* or *unbounded* processes. The former are local, typically, and are bounded by some domain, such as the word or the subcategorization domain of the verb (e.g., a verb syntactically and thematically specifies its arguments and their thematic properties, as outlined in Pt. 3 above). Unbounded dependencies, on the other hand, cross clause boundaries and penetrate into embedded domains. Central examples of bounded dependencies are the following:

(7) Binding of reflexives: Harry thinks Sally likes herself/*himself
(8) Raising and control: *John seems that Fred assumes [___ to be a fool].
(Cf. It seems that Fred assumes John to be a fool.)

In (7), the antecedent of the reflexive pronoun must be within the same clause. In (8), *John* cannot "raise" over the *assume* clause. Thus, such operations are bounded. Compare the typical unbounded dependencies below, where any number of clause boundaries can separate the "gap" from its antecedent:

(9) Relativization: (This is) a man who I think Fred said that Jill believes that Tom thought that Bill likes ___.
(10) Topicalization: Harry, I think Fred said that Jill believes that Tom thought that Bill likes ___.

Another central issue in syntactic theory concerns the various *constraints* on forming dependencies that exist in natural languages. A massive amount of literature has focused on such topics as "island" constraints, which are described in the syntactic literature with varying degrees of success, though not thereby explained. Essentially these are prohibitions against forming dependencies across certain types of structure. The metaphor suggests that a constituent on an island is stranded there, unable to form a dependency with elements outside that island. To illustrate:

(11) *Who did she wonder whether Jim liked __ ?
(Cf. She wondered whether Jim liked Jill.)
(12) *What did you believe the claim that the biologist proposed __?
(Cf. I believed the claim that the biologist proposed the correct answer.)
(13) *What were they eating beans and ___ ?
(Cf. They were eating beans and chips.)

Note in (11), however, that the constraint holds for dependencies created by movement, but not by pronominalization: *That woman, I wonder whether John likes her*.

An interesting question concerns whether or not the constraints on dependencies can be grounded in meaning. Many syntacticians argue that there are no semantic reasons for the observed phenomena. For instance, it is impossible in the majority of languages to extract an element from within a complex NP, such as a relative clause or noun complement clause, as in (12). Evidence that this constraint is not semantically driven comes from the fact that the element can be easily parsed when *in situ*, as the grammatical sentence included for comparison in (12) illustrates clearly.

The needs of processing appear to provide a more promising explanation for the origins of constraints than do semantic causes. Intuitively speaking, the constraint-violating sentences appear difficult to process, as the relation between the filler and the gap is rather indirect in (11) and (12). Even the less obviously processing-related dependencies, such as agreement, are useful in resolving ambiguity, by identifying antecedents. Complicating matters, however, is the fact that constraints on forming dependencies seem subject to extension. That is, they appear to be extended far beyond what is useful for parsing purposes. For instance, take the Coordinate Structure Constraint, illustrated by (13). In parsing terms, there is no obvious reason for the ungrammaticality of the construction, yet extracting an element from a conjunct is believed to be impossible in natural languages generally. However, it may be a mistake to assume that all constraints arise in the service of the parser: some clearly do, but in other cases there is no support for the idea that some particular constraint reduces ambiguity.

Moreover, any theory that suggests basing the origins of some aspect of grammar (such as constraints) on processing needs runs up against a problem: it is often still unclear how the processes postulated by syntacticians translate into parsing operations or other cognitive processes. Consider one well-known analysis (Emonds 1978; Pollock 1989) of the distinction between French and English in terms of the position of an adverb in relation to the finite verb:

(14) Je lis **toujours** le même journal.
I read.PRES.1S always the same newspaper
"I always read the same newspaper."
(15) I **always** read the same newspaper.

This analysis presupposes a single canonical underlying order for both languages. Under these proposals, the adverb is consistently adjoined to the verb phrase (VP) over which it has scope; to derive the different surface orders, the finite verb raises over the adverb in French, but not in English. Therefore, under this analysis, the order V–Adv–O in French involves more movements than the order Adv–V–O in English. Can we expect to find any correlation with this increased syntactic complexity, for example, in processing or in acquisition? At present this remains an open question. The depth of the problem of ascertaining the relationship between the grammar and the parser can be illustrated by the central Merge operation. Merge applies from the bottom up (i.e., right to left in languages like English), yet sentences are produced and processed linearly in real time from left to right. The basic problem is that (correctly or incorrectly) syntactic theories have, in general, not been concerned with the question of what happens when one produces or comprehends a sentence (for moves in this direction, cf. Phillips 1997; Gibson 1998).

Related to this, it is as yet unclear to what extent posited movement processes reflect neurological operations performed in the brain (Kaan, this volume). Clearly, the brain is in a real sense "aware" of dependencies between items in a structure, yet the relationship between grammatical operations and neurological "operations" is still largely unknown (for a discussion on current research, see Friederici, this volume).

What Syntactic Phenomena Appear to be Universal and How Much Variation Is Possible among Nonuniversal Phenomena?

Two distinct approaches to syntactic universals have been taken in the literature over a period of many decades. The first is concerned with what we might call *surface universals*. This approach investigates easily identifiable and relatively superficial features of language and, in particular, the surface morphosyntactic features. Within this broad approach, encompassing the field known as *linguistic typology*, the major concerns are to identify what properties may or must occur in every language, and the implicational relationships among the properties. As noted, such an approach stems from the work of Joseph Greenberg, in the early 1960s, and involves the postulation of dozens of mainly implicational universals, many of which involve word order. For instance, consider correlates of basic word order along the following lines: if a language is OV (i.e., if the object precedes the verb) as in Japanese, we can predict that it will probably have postpositions rather than prepositions, complementizers following the clause, genitive–noun order, and so on; if the language is VO (i.e., if the object follows the verb) as in Welsh, we can predict that it will probably have prepositions rather than postpositions, complementizers preceding the clause, noun–genitive order, and so on.

While many statistical universals of this nature have been proposed, subsequent research shows that there are few examples of absolute universals of this superficial kind and that counterexamples are commonplace. Thus, for instance, although the vast majority of VO languages are indeed prepositional, some are postpositional (Dryer 1991, 1992), and although it was long believed that no language exhibits syntactic ergativity in the absence of morphological ergativity, at least one exception to this has now been discovered (Donohue and Brown 1999). The hierarchical implicational universals are also less secure than was previously thought: one generally quite robust example has been shown not to be inviolable, namely the Keenan and Comrie (1977) Relativization Hierarchy. This is a hierarachy of relative clause formation on the various grammatical function positions, with the order Subject < Object < Indirect Object < Oblique < Object of comparison. The generalization is that if a language can form a relative clause on, say, indirect object position, it will always be able to form a relative clause on all of the higher positions on the hierarchy. Even here, there are systematic exceptions, so that languages are indeed found which can form relative clauses on lower but not higher positions (e.g., Larsen and Norman 1979).

Typologists have traditionally offered functional explanations for the universals observed. For instance, object initial languages are rare, and the explanation offered for this fact involves the pragmatic principles of ordering old and new information. According to this explanation, the "natural" order presents old information before new information: subjects are topics and express old information; they are thus ordered before predicates (including objects), which express new information. However, object-initial languages have indeed been found, so the correlation is far from perfect. As noted, this kind of imperfect correlation is characteristic of typological universals.

The second approach to syntactic universals explores what we might call *deep universals*. In the generative approach, proposed universals are typically highly abstract properties of grammars, which may be related to surface observation only in indirect ways. Universals within this tradition are characterized as "substantive" or "formal." Substantive universals are available to all languages and include syntactic categories, such as V, N, Adj, and P, their phrasal projections (VP, NP, AdjP, PP) and syntactic features such as +V, +N. One problem here is that not all of the posited universals are found in every language. For instance, Vietnamese has no tense morphology. Do we say that this language has a tense phrase node but that nothing fills it? Or do we say that Universal Grammar is more like a toolkit from which languages choose the parts they want (Jackendoff 2002), thereby allowing Vietnamese to "abstain" from having a tense phrase? These are currently unanswered questions within this paradigm.

Formal universals (more abstract in nature) are at the heart of the generative program and include the principles and constraints that are said to govern the grammars of all languages. Under some current conceptions, these include such basic operations as Move and Merge (see Rizzi, this volume), as well as locality and economy constraints. Under other current conceptions, all such principles reduce to a single unified notion of syntactic composition, and many of these constraints are eliminated or relegated to the processor.

Various issues arise as well. First, why do these universals exist? In the generative model, the standard answer is that they are innate, provided by Universal Grammar. Evidence for innateness is often drawn from poverty of the stimulus arguments. The central idea here is that certain properties are observed that cannot be learned inductively.

Second, why, given Universal Grammar, are all languages not the same? Here the standard answer is that universals are parametrized, thereby resulting in linguistic variation (Chomsky 1981). So, for instance, all languages have heads (e.g., the verb in the VP, adjective in the AdjP), but the Head Placement Parameter stipulates that there will be two possible orderings, cross-linguistically, of a head in relation to its complement: head-initial and head-final. The child learning the language is able to "set" the parameter correctly on the basis of exposure to a small amount of data. The general assumption is that both the parameters and the choice of settings for them are innate.

Third, how is exceptionality handled? That is, one would not want to say that a particular feature found in only one or two languages is handled by an innate parameter. In some cases, relatively superficial operations are said to mask the effects of parametrization. For instance, *wh*-movement generally moves the *wh*-phrase to the clause-initial position, and complementizers (closed-class words introducing subordinate clauses, such as *whether* in English) are also positioned to the left of the clause. In some languages, such as Vata, *wh*-movement is indeed to the clause-initial position, but a question particle fills the clause-final position, suggesting that Vata has final complementizers. One way to handle this exceptionality is an analysis in which the entire clause is moved leftwards, thus stranding the complementizer in the rightmost position (Kayne 1994). Similarly, while over 90% of languages are subject-initial, a few are object-initial, as noted above, such as the OVS language Hixkaryana spoken in Brazil. One analysis (Kayne 1994; Baker 2001) suggests that the language is underlyingly SOV, like Turkish or Japanese, but the [OV] constituent moves to the left of the subject, giving the superficial appearance of OVS order.

Handling the relative frequency of each variant within generative grammar is much more difficult than in the surface approach, though a few linguists have attempted to build variation directly into the theory of parameters (Baker's Parameter Hierarchy; Baker 2001). The essential problem is that it is very hard to predict what is probable and what is improbable in generative models, and why certain morphosyntactic features are more widespread than others. However, it may simply be the case that explaining variation is not part of what such models of grammar encompass.

One obvious question concerns whether the "surface" and "deep" approach to language universals can in any way be reconciled. A possible approach along these lines comes from the work of John A. Hawkins (1990, 1994, 2004), which proposes parsing-based functional explanations for the observed frequency of certain word orders. This predicts that in languages with consistent head placement, parsing will be easier because there is less distance between the heads of each phrase within a constituent. For instance, in a VP containing a head verb, a direct object and a PP, the heads will be "aligned" if the order is consistently head-initial, giving VO and preposition–NP order, or alternatively if the order is consistently head-final, giving OV and NP–postposition order. Such languages are assumed to represent steady states. On the other hand, languages with inconsistent head placement (e.g., VO but postpositional) have "nonaligned" heads, and may represent transitional states. Another possible approach is that taken by Cinque (1999) and Julien (2002), which is generative but takes seriously the results of the typological program.

For any proposed universal, then, many possible explanations exist at varying levels of generality, some syntactic and others not. Constraints on movement may be due to the nature of the short-term memory system. The general restriction of clauses to a maximum of four participants (most verbs take no more than three arguments) may in evolutionary terms be due to the primate system of visual representation, which cannot subitize more than four items (Hurford 2007). Much work remains in the untangling of the various factors underlying each universal.

Despite the fact that the universals proposed by generative grammarians have not, to date, been shown to be instantiated neurologically, many grammarians feel that abstract, high-level principles can guide empirical analysis in important ways. To illustrate, we first consider the hypothesis that tree structures are exclusively binary branching. This hypothesis has immediate analytical consequences which can be tested. Consider, for instance, VSO languages. If these appear to lack a surface VP constituent, one might assume that they should be analyzed as having a flat, ternary-branching structure (i.e., with the verb, the subject and the object as separate constituents). However, it turns out that VSO languages exhibit the same kind of structural asymmetries between subject and object that are observed in SVO and SOV languages, which have an obvious hierarchical structure with a subject–predicate bifurcation (Rizzi, this volume). It therefore seems reasonable to postulate an underlying binary-branching (SVO) order for verb-initial languages too, thus accounting for the subject–object asymmetries by proposing that such languages also have the subject–predicate division. The surface verb-initial order is then derived by head-movement (Rouveret 1994; Roberts 2005), with the verb head raising over the subject. Not all theoretical models would support such an analysis; see Borsley et al. 2007 for some discussion. Nonetheless, the facts remain to be explained, so alternative explanations must be sought by such models; for instance, the asymmetries observed in VSO languages might be due to discourse asymmetries, rather than to binary branching (cf. Van Hoek's 1997 analysis of English).

Let us also consider the syntactic "empty categories" which are postulated within various models of generative grammar. As Bickerton (this volume) notes, the referential properties of empty categories cannot be learned inductively, but depend on quite complex algorithmic processes obeying nonsuperficial, abstract principles, and therefore constitute strong evidence for at least some form of innateness. Intriguing evidence for the existence of the empty categories traditionally known as *pro* and *wh*-trace comes from a phenomenon in Welsh known as syntactic soft mutation. Soft mutation is a morphophonological process which changes the initial consonant of a word; most triggers for the process are lexical, and not of concern here, but one environment is purely syntactic. According to one analysis (Borsley and Tallerman 1996; Tallerman 2006), a phrasal category (XP) triggers the mutation on the initial segment of any following constituent which is a complement. Basic examples are shown in (16): the triggering phrasal category is bracketed, the element bearing the mutation is underlined, and its canonical form is given in parentheses:

(16) (a) Gwelodd [y ddynes] <u>gath</u>. (*cath*)
see.PAST.3S the woman cat
"The woman saw a cat."
(b) Mae [yn yr ardd] <u>gath</u> ddu. (*cath*)
be. PRES.3S in the garden cat black
"There is a black cat in the garden."

In (16a) for instance, the subject nominal phrase is the XP trigger, and the object bears the mutation (*cath* > *gath*); in (16b), the PP *yn yr ardd* ("in the garden") is the XP trigger, and the predicate nominal *cath ddu* ("a black cat") bears the mutation, which shows up as usual on the initial segment of the constituent (*cath* > *gath*). What is interesting is that both *pro* and *wh*-trace are also XP triggers for syntactic soft mutation: compare (16a) and (17):

(17) (a) Pwy welodd *wh-t* <u>gath</u>. (*cath*)
who see.PAST.3S cat
"Who saw a cat?"
(b) Gwelodd *pro* <u>gath</u>. (*cath*)
see.PAST.3S cat
"He/she saw a cat."

Although other explanations are of course possible, the XP trigger hypothesis neatly captures the facts and offers a strong generalization which holds across all available data. The data in (17a) also illustrates the fact, mentioned in the Introduction, that in a language with fronted *wh*-expressions, the *wh*-phrase appears to occur in two distinct places within the clause, namely its surface position and its canonical, "underlying" position.

Finally, the question arises as to the biological (im)plausibility of complex abstract structures. Is it reasonable to propose that the Coordinate Structure Constraint in (13), for example, is innate? What does it mean for movement constraints to be evolvable biologically?

One might suggest that we regard the principles which have been proposed within generative grammar as in some sense metaphorical, and thus are awaiting some more biologically plausible explanation. However, it must be stressed that throwing out abstract principles does not obviate the need for an explanation for the observed phenomena. We need a way to express, for instance, which long-distance dependencies are possible and which are not, and it is equally clear that there must in fact be some biological explanation for the phenomena observed. The research question that remains for linguists and biologists is to frame the observed properties of language in some way that makes them amenable to a biologically realistic explanation. Unifying the results found within the two disciplines (and other natural sciences) in this respect is the ultimate goal.

The Genesis of Syntax: What Can We Conclude about Early Human Syntax from Syntactic Phenomena Observable Today?

What Can Pidgins and Creoles Tell Us about Syntactic Phylogeny?

Pidgins arise whenever it is necessary for people to communicate without a common language. Pidgin is characterized by the following properties:

- absence of complex sentences,
- lack of consistent serial ordering,
- virtual absence of functional category words,
- frequent and random omission of both N and V with no systematic means of recovering the missing elements.

A pidgin may remain in this state indefinitely if it is used solely by adults. However, when children acquire a pidgin as their primary language, a creole arises. A single generation suffices for children to produce a full natural language, with all the usual features that this entails.

Typical features of creoles include (but are not limited to) serial verb constructions and a uniform TMA (tense, modal, aspect) system, where the ordering of elements (T > M > A) is invariant in all cases. Serial verb constructions share a single subject, but are characterized by having two or more verbs under the same intonation contour, with no subordinating or coordinating conjunctions. The object is also shared by both verbs in the serial construction and is never repeated. The following example is from Seselwa (spoken in the Seychelles Islands):

(18) il pran ti lisyen tuye
he take small dog kill
"He killed the small dog."

There are notable parallels with constructions involving empty categories in languages such as English, suggesting that there may be an innate algorithm for recovering the reference of empty categories in such constructions. Note the parallels between (19a) and (19b):

(19) (a) Mary wants someone [___ to talk to ___].
(b) il pran ti lisyen [___ tuye ___]

The parallels with other languages that we see in creoles cannot arise from anywhere except from Universal Grammar, since children forming creoles have no other language input to guide them.

Turning to their TMA system, we find that creoles typically have one marker for each of the three features: tense, aspect, and modality. These markers can occur either alone, or (modulo the fixed ordering noted above) combined with each other, in any of the eight logically possible ways, giving the following combinations of markers plus verb (including having no TMA marker at all): __ V, TV, MV, AV, TMV, TAV, MAV, TMAV.

It is notable that this same system occurs with only minor variations in two to three dozen unrelated creoles around the world. A reasonable conclusion is that this system is provided by Universal Grammar.

Finally, it must be emphasized that the similarities among creoles cannot possibly emerge from a common background, since the substrate languages spoken by pidgin speakers do not share these linguistic features.

What is the relevance of such languages to language evolution? While there are obvious differences between creole formation and the origins of human language, most notably in the rapidity of pidgin-to-creole development as compared with the likely scenario in phylogeny, some aspects of the pidgin–creole cycle may be analogous to the development of the human language faculty in evolution. In particular, an order of phylogenetic development is suggested with regard to the features which do, and which do not, emerge in creoles. For instance, creoles develop TMA markers but they do not develop case markers. This suggests that case marking may be a more recent development in language evolution. Moreover, creoles contain hierarchical structures, hypothetically formed by the Merge operation, but pidgins, which are much simpler, do not. In pidgins, and therefore most likely in the precursor to true human language, words occur singly as a linear string with no hierarchical structure.

What Can Diachronic Processes Tell Us about Syntactic Phylogeny?

Language has both an ontogeny and a phylogeny: it comprises both individual development in human infants and genetic development in the species. Both processes, of course, have clear precedents in biology, but a third is uniquely human: diachronic processes in individual languages (i.e., their changes in historical time). It is this biologically unprecedented developmental process which most directly shapes the morphosyntax of extant languages and also gives rise to typological diversity. Diachronic changes, unlike pure evolutionary developments, are mediated by the process of *cultural transmission*, as languages are passed from one generation to the next. The nature of this transmission also means that the processes of attrition (erosion, elimination, simplification, and loss) in language diachrony are starkly different from the corresponding process of simplification and restructuring in biological evolution.

In biology, due to genetic coding, evolutionary changes are virtually irreversible. Organs may be simplified, reduced, or altogether eliminated in extant adult structures. However, the process of both their innovation and elimination is still coded, in that order, in both the genome and, consequently, in ontogenesis (Gould 1977). Whales do not skip their terrestrial mammalian genes and embryology because they are now back in the water with fish. Both their genome and their embryology bear testimony to (a) their emergence from the water and (b) their subsequent return.

In contrast, in the absence of hard-wired genetic and ontogenetic coding, the attrition of linguistics structures may be absolute; once the morphology is eroded, syntactic constructions can "decay" to the point of utter functional inefficacy. This leads to an eventual renovation process, whereby new structures are recruited to pick up the slack; this is known as the diachronic cycle (Givón 1979). The diachronic cycle does not involve a reversal of directionality, merely the termination of one unidirectional process and the start of another cycle, in the same general direction, from scratch. As another cycle starts, a language may choose to pursue other structural options to perform the same communicative function(s). At such points a language, or some functional domains within a language, may change in terms of structural type.

Despite the differences between diachronic change and biological evolution, there do appear to be adaptively driven linguistic processes. For example, the motivation for historical change is typically local, but the results are often global. Micro-variation in idiolects and dialects leads to macro-variation in languages and language families.

In concrete terms, diachronic change creates grammatical structure in languages via a bundle of processes known as *grammaticalization*. One major trend is that functional elements are derived from lexical elements, with the lexical words coming to play a more "grammatical" role, and also becoming more abstract in meaning. Along with this, phonetic reduction typically occurs. Consider, for instance, the (putative) diachronic development of the future marker *be going to* in English.

(20) Stage 1: I am going to Durham. (Literal meaning involving movement from place to place)
Stage 2: I am going to sit in a chair all day. (Future marker, no movement necessary)
Stage 3: I am going to go to Durham next week. (Future marker plus verb of movement)

Note that at stages 2 and 3, phonetic reduction to *gonna* is commonplace, but this is not possible at the initial stage 1.

Some general examples of typical grammaticalization processes are shown below: the lexical elements are on the left-hand side in each case, with the corresponding derived functional elements shown on the right. The diachronic process seen in (20), for instance, illustrates the development of a lexical verb (*go*) into a TA marker in English.

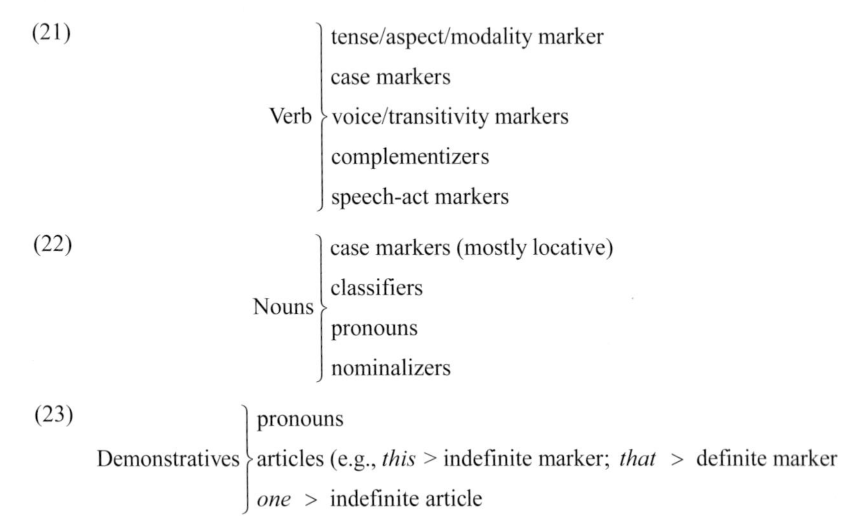

Another kind of grammaticalization process is *syntacticization*, meaning the creation of new syntactic structure. For instance, complex sentences (i.e., those in which one clause is embedded inside another clause) emerge in diachrony from conjoined clauses, so that two clauses initially occurring one after the other turn into a complex sentence. A good example of syntacticization may be argued to have occurred in the recent history of colloquial French, turning an emphatic "left dislocation" construction such as (24a) into a simple nonemphatic clause (24b):

(24) (a) Moi, je (ne) sais pas.
me I NEG know NEG
"As for me, I don't know."
(b) Moi je-sais pas.
me AGR.1s-know NEG
"I don't know."

In (24a), the optional, left-dislocated strong pronoun *moi* ("me") adds emphasis and is associated with a specific intonation, while the main clause has its own distinct subject, the weak pronoun *je* ("I"). In the later stage, (24b), the pronoun *moi* is now the obligatory subject pronoun and is amalgamated into the main clause in terms of intonation, while the erstwhile subject *je* is now even further reduced phonetically and is affixed to the verb itself as a first person singular agreement marker. This example also illustrates the diachronic cycle mentioned above. In French diachrony, the verb endings have undergone attrition, becoming phonetically opaque in the modern language: *(je) sais* ("I know") is pronounced the same as *(il) sait* ("he knows"). The affixation of the weak pronoun to the verb starts the cycle over again, producing an entirely new first person singular marker. Note also that this is a prefix, whereas the former agreement markers (the endings with a distinct pronunciation in Middle French) were suffixes: colloquial French could thus be considered to have undergone the kind of change in structural type noted above.

One school of thought (e.g., Heine and Kuteva 2002) considers that these same types of grammaticalization processes were also operative in the course of evolution, so that the earliest syntactic categories in protolanguage or in full language were solely lexical, with functional elements being created from these. It might be objected that the full cycle from lexical category to functional category to inflection can take place in 1000 years or less, whereas human language has probably been in existence for 100 times longer than this. However, given that inflectional markers are subject to quite rapid phonological decay, this cycle could have been repeated many times since the dawn of language. Moreover, no other plausible source for inflections and other grammatical markers has yet been proposed.

What Can Ontogenetic Processes Tell Us about Syntactic Phylogeny?

It is an interesting question whether child language development provides evidence about the origins and evolution of language. To review some basic facts, a pre-grammatical stage in infants lasts until roughly the end of the first year, followed by a one-word stage lasting until around 1;8 (i.e., one year 8 months); this stage is characterized by comprising around 95% nouns. The two-word stage (roughly from 1;6 to 2;4) consists of N–N and V–N combinations, with the latter slowly increasing in frequency. During this stage, constructions and morphology also begin to be acquired, gradually.

Verb complements are acquired quite early, beginning at about the age 1;8. Deontically oriented complements (e.g., those of "want" or "let") are acquired before those of epistemic verbs such as "know," "think," "say," and "see." Relative clauses are acquired much later (3;0–5;0), when the child begins to tackle communicative tasks of complex reference.

It is worth noting that there is almost certainly more grammatical structure in the child two-word stage than previously thought. For instance, in French the order of the negative marker with respect to the verb reflects exactly the same order as in adult speech, distinguishing between *pas manger* ("not eat") with a nonfinite verb, but *mange pas* ("eat not") with a finite verb. In other words, the correct adult order is produced even at this stage. Similarly, in German VO and OV order appears, at the same stage, in the correct places, the former occurring with finite verbs and the latter with nonfinite verbs. It may well be the case, then, that the superficially simple appearance of the infant two-word stage is misleading in terms of an underlying syntactic complexity.

What bearing might child language acquisition have on evolution? First, consider the domain of reference: children start with nondisplaced utterances and only later acquire displaced reference (i.e., the ability to talk about entities or events which are not observable in the here-and-now). We surmise that humans are the only species controlling displaced reference. Second, we know that animal communication is almost always manipulative and deontic, rather than declarative. The linguistic communication of human infants starts off the same way, with declarative sentences emerging later on.

It is possible, then, that linguistic features that emerge in ontogeny might be precursors of full language in the human species. In other words, like the development of pidgins to creoles, and the development of grammatical items in diachrony, the development of language in children ontogenetically might be a window on language evolution.

Such "recapitulationist" arguments are far from uncontroversial: see, for instance, the detailed critique of the "windows" methodology outlined by Botha (2006a, b, 2007). The problem is, however, that in the absence of other strong evidence for processes occurring in language evolution, recapitulationist arguments have become practically the default.

Conclusion

We have attempted to present an overview of the major syntactic phenomena requiring an explanation as part of the evolved biological capacities of our species. We outlined the main building blocks of syntax, concentrating specifically on lexical categories and functional categories, and examined hierarchical structure and recursion, particularly in light of proposals that this is the only uniquely human characteristic in the language faculty. We discussed the typology of the dependencies that exist between syntactic elements and considered the relationship between such dependencies and the needs of the human parser. Further, we considered syntactic universals of various kinds and their treatment within different grammatical traditions, and examined possible responses to exceptional constructions. Some evidence was presented in favor of abstract, high-level grammatical principles. Finally, we asked whether any evidence concerning the (unobservable) evolution of syntax itself can be gathered from an examination of different kinds of observable syntactic phenomena. In this regard, we considered the development of creoles from pidgins, the diachronic processes involved in language change, and the ontogenetic development of language in infants.

In terms of syntactic analyses, Occam's razor suggests that the most parsimonious solution consistent with the facts is always the best solution, an idea that has constrained linguistic theorizing (at least implicitly) for many decades. In many formal models of grammar, the evaluation metric is economy, simplicity, and elegance, thus leading to postulations that meet with Occam's razor. We have no evidence, however, that this same metric has any evolutionary validity. It may well be the case that less elegant theories are more in keeping with evolutionary precepts. Our present state of knowledge concerning the workings of the brain is not sophisticated enough to ensure that the solution considered to be most economical and elegant by syntacticians also reflects the workings of the human brain. Evidence from biology suggests that the most parsimonious explanation is not necessarily the correct solution, since systems display a great deal of redundancy. At the same time, there is little brain-based evidence to suggest that we should strive to build redundancy into our syntactic analyses.

Evolution

8

Possible Precursors of Syntactic Components in Other Species

Austin T. Hilliard and Stephanie A. White

Abstract

Language is a uniquely human phenotype. One might thus view language as a wholesale innovation and thereby limit inquiry into its evolutionary and neurobiological basis to noninvasive techniques. Alternatively, one may consider that other species possess subcomponents of language and that a comparative approach opens the door to controlled experimental investigations into the molecular, cellular and synaptic basis of vocal learning, a key subcomponent of language.

Introduction

Human linguistic syntax is the system of forming complex signals and the mapping of these signals onto conceptual/intentional representations. As such, syntax provides compositionality: it serves to combine a finite number of meaningful units to produce an infinite variety of sequences with larger meanings. This ability for limitless recombination appears uniquely human (Hauser et al. 2002). However, less sophisticated rule systems for recombining units and recognizing sequences of communication signals, or other stimuli, exist in animals and form the focus of this contribution.

Why Other Species?

How the human language singularity evolved remains a puzzle. We argue that a comparative approach examining animal communication and its underlying neural basis can inform studies of human language evolution. Biologists refer to the study of animals for the purpose of understanding the human condition as using an "animal model." While no animal model can fully capture any aspect of language, preadaptations for subcomponents of language, including syntax, most likely exist in nonhuman species. The alternative idea that syntax appeared *in toto*, via one fortuitous mutation, in a manner unique to humans, is not parsimonious. Rather, similar environmental and biological constraints on vocal communication have likely driven parallel solutions across multiple animal groups, examples of which we discuss below, with an emphasis on the meaningless syntax of songbirds. These preadaptations, coupled with the biological and cultural evolutionary interactions, discussed by Számadó et al. (this volume), may have combined in the hominid lineage to produce the unique language phenotype.

Any user of vocal communication signals must be able physically to transmit and receive these sounds. A co-requisite is the ability to convey (even passively, as in advertising the size of one's larynx simply by vocalizing) and decode behaviorally important information in vocal signals. Here, we focus on a narrower target, namely, vocal learners, or animals with the experience-dependent capacity to learn these production skills, and the neural basis that gives rise to this capacity. Vocal learning requires the ability to coordinate precisely and rapidly complex sequential movements of lingual, vocal, and respiratory musculature in order to mimic conspecifics or create new sounds. Although innate abilities may point to parallel mechanisms, from a biological perspective, neural circuitry that is capable of learning seems most likely to capture the unbounded features of language, a learned behavior. We sometimes broaden our perspective from the learning of vocal skills to the learning of any motor skill. The mechanisms which underlie vocal learning in more common neural circuits may form the evolutionary roots for the neural basis of speech.

Which Other Species?

Vocal learning depends upon hearing conspecific vocalizations as well as one's own. Comparison of these inputs determines whether neurobiological changes must occur for adaptive modifications of vocal output. Tests for the vocal learning capacity often rely on deprivation of these acoustic inputs and determination of whether the subsequent vocal output is abnormal. Deprivation can be drastic (such as deafening), dramatic (as in rearing in the absence of conspecific vocalizations), or refined (e.g., transient distortion of key auditory inputs). A noninvasive method examines whether changes in vocal output during normal development are more substantial than those expected due to physical maturation of the vocal apparatus (Fitch 1997) or are uncharacteristic of their species. Marine mammals, bats, and elephants are thereby considered vocal learners (Boughman 1998; Janik et al. 2006; Suzuki et al. 2006; Poole et al. 2005). By a majority of these tests, passerine birds of the oscine suborder, known as songbirds, are vocal learners as well.

Birdsong as a Unique Model System: Parallels between Song and Speech

About half (~4,500) of the extant avian species are songbirds; their phenotypic distinction is that they learn part of their vocal repertoire. This ability is shared with hummingbirds and parrots, which are in separate avian orders, raising the hypothesis that vocal learning emerged three times, independently, in the avian lineage. Among the identified vocal learners, songbirds are the most amenable to controlled experimentation. Certain species, such as white-crowned sparrows (*Zonotrichia leucophrys*), zebra finches (*Taeniopygia guttata*), and Bengalese finches (*Lonchura domestica*, also known as society finches), are small and breed in the laboratory. As a result, much about their song learning and its underlying neural bases is known. Songbird researchers divide song into songs, bouts, phrases, motifs, syllables, and notes (Konishi and Nottebohm 1969). Notes are the smallest unit, combining together to form syllables. Two or more syllables may group together to form a phrase. A motif is a sequence of notes and/or syllables that are repeated in a stereotyped order. One or more motifs or phrases followed by an interval of silence constitute a bout of song. (Brenowitz et al. 1997). Song phonation refers to the acoustic features such as amplitude, mean frequency, frequency modulation, amplitude modulation, and entropy. Song syntax refers to the temporal order of these features within a song (e.g., the order of syllables within a motif or phrases within a song).

Song and speech learning share key features, including dependence on hearing and on social interactions with conspecifics (Doupe and Kuhl 1999). Both occur during critical developmental phases beginning with an early perceptual phase where, in the case of songbirds, the song of an adult male is memorized. In humans, this corresponds to a time of universal speech perception when a baby listens to speech, but does not yet produce any learned vocal output. In a second phase, known as sensorimotor learning, both songbirds and humans practice and refine their own vocalizations in order to mimic adult sounds. In normal children, the onset of sensorimotor learning is marked by babbling at ~6 months and results in first words at ~one year, with two word combinations occurring at ~two years. Thereafter, word production and syntactical recombination takes off and can continue, to a degree, throughout life. In contrast, the degree of flexibility in mature songbirds depends upon the species.

By the time zebra finches reach sexual maturity, their previously variable songs have stabilized through a process termed *crystallization*. The unchanging song of adult zebra finches appears to contrast with the less limited capacity of human language. Yet other songbird species, such as mockingbirds, are capable of learning new songs throughout life. The maintenance of mature zebra finch song requires continuous auditory feedback, as it gradually deteriorates in deafened birds (Nordeen and Nordeen 1992; Brainard and Doupe 2000). Deafening-induced song deterioration is even faster in Bengalese finches (Woolley and Rubel 1997). Adult birdsong can be temporarily disrupted in intact birds exposed to abnormal auditory feedback (Cynx and Von Rad

2001). Similarly, adult speech depends on ongoing auditory feedback. Finally, as is evident to anyone trying to learn a new language after puberty, the human faculty for language peaks in youngsters. Thus, comparison of all of birdsong to human speech reveals both developmental constraints as well as relative openness to experiential input throughout life, and ongoing dependence on audition.

Beyond behavior, the neuroanatomical circuits underlying song and speech show additional parallels (Jarvis 2004). While the lateralized cortical regions classically known as Broca's and Wernicke's areas are arguably unique to humans (cf. Taglialatela et al. 2008 for chimpanzees), emerging evidence implicates subcortical brain structures, notably, the cerebellum and basal ganglia as critical for language and speech. The planning and execution of complex motor skills involves circuits that run through the cortex, basal ganglia, thalamus, and back to the cortex (Liebermann 2006). Relevant here, songbirds use similar loops during learning and production of song (Figure 8.1).

The collection of brain areas specialized for song is referred to as the song circuit and is well characterized, partly due to its sexual dimorphism in species such as the zebra finch (Nottebohm and Arnold 1976). Subregions within the cortical-like pallium, the basal ganglia, and thalamus are prominent and interconnected only in males. Neurons within these subregions are dedicated to song, lacking regular firing patterns during performance or perception of other behaviors or stimuli. The song circuit is comprised of two interconnected pathways that each stem from the cortical-like (pallial) HVC, analogous to association cortex in humans (Figure 8.1). In the first, known as the vocal motor pathway, auditory inputs enter the circuit at HVC. A subset of HVC neurons projects their axons to a region analogous to primary motor cortex. These neurons, in turn, make direct projections onto brainstem motor neurons,

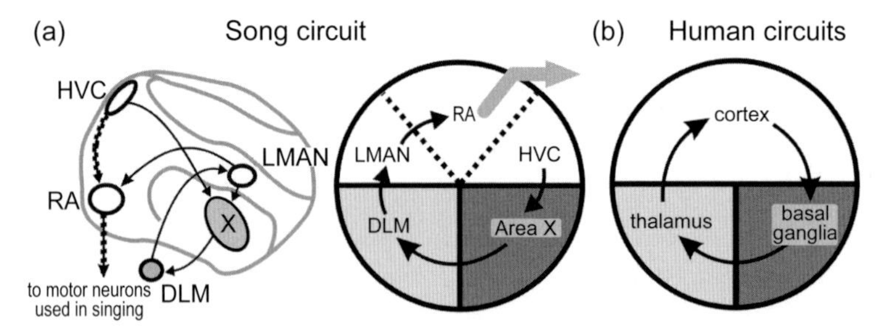

Figure 8.1 Schematic comparison of the avian song circuit and human cortico-basal ganglia-thalamo-cortical circuitry. The cortex is white, basal ganglia dark gray, and thalamus, light gray. (a) A composite sagittal view of songbird telencephalon is depicted on the left. Auditory input (not shown) enters the song circuit at HVC, the neurons of which contribute to two pathways: the vocal motor pathway (stippled arrows) and the anterior forebrain pathway (plain arrows; adapted from Teramitsu et al. 2004). (b) Human circuits. HVC: high vocal center; RA: robust nucleus of the archistriatum; DLM: dorsolateral thalamus; LMAN: lateral magnocellular nucleus of the anterior nidopallium.

which control the muscles used in singing (Nottebohm et al. 1982; Wild 1993). The direct connectivity between high-level cortical-like regions and the motor neurons that control the song organ is reminiscent of similar motor cortical projections in humans (and raccoons), which bypass brainstem way stations and directly contact finger motor neurons (Sereno 2005). This privileged access between high-level brain area and effector may underlie the enormous flexibility of motor output in each system. The second pathway indirectly links HVC to RA via forebrain structures and includes the striatal nucleus Area X, and the lateral portion of the magnocellular nucleus of the anterior nidopallium (LMAN). LMAN neurons rejoin the pathways by projecting to RA. This anterior forebrain pathway (AFP) is required for song modification, most prominent during sensorimotor learning (Bottjer et al. 1984; Scharff and Nottebohm 1991; Okuhata and Saito 1987; Sohrabji et al. 1990), but also during maintenance of adult song (Brainard and Doupe 2000; Williams and Mehta 1999). The essential point is that the pathway allowing for song modification and maintenance is similar to the cortical loops that underlie the planning and execution of complex sequential movements in humans.

The FOXP2 Puzzle Piece as a Genetic Example

A transcription factor known as FOXP2 was identified in 2001 as the monogenetic locus of a mutation causing an inherited speech and language disorder (Lai et al. 2001). Instantly, FOXP2[1] flashed as a key piece in the puzzle that, through interactions with other molecules, patterns the brain for language. Although the inherited disorder is not chiefly syntactical in nature, there is a morphosyntactic component (see below). Further, the FOXP2 story demonstrates how biologists, neurologists, and linguists can work collaboratively to uncover pieces of the neuromolecular puzzle underlying language.

The KE Family and Others

The FOXP2 discovery emerged from clinical characterization of the KE family. Of several expressive and receptive linguistic features, their most prominent deficits are in sequencing of orofacial movements especially those required for speech, referred to as verbal dyspraxia. Whether their cognitive deficits are in addition to, or a consequence of, their motor deficits is an area of ongoing investigation. Affected members have morphosyntactic difficulties as exemplified by their inconsistency in adding suffixes such as *s* for plurals or *-ed* for actions occurring in the past (Vargha-Khadem et al. 2005). Moreover, their gross

[1] By convention, human "FOXP2" is capitalized, mouse "Foxp2" is not, and "FoxP2" denotes the molecule in mixed groups of animals. Italics are used when referring to genetic material such as *FoxP2* mRNA (Carlsson and Mahlapuu 2002).

orofacial dyspraxia is relevant here, given our broad interest in the capacity to generate complex sequenced movements.

The identification of the genetic basis for the KE phenotype hinged on the discovery of an unrelated individual, CS, who exhibited similar deficits. Examination of CS's chromosomes revealed an easily detectable rearrangement, one end of which interrupted the gene encoding FOXP2, a transcription factor with unknown neural function. FOXP2 was known to be a transcriptional repressor in lung, binding to sequences in the noncoding region of its target genes, thereby decreasing their expression levels. Subsequent analysis of the KE family FOXP2 sequence revealed a single point mutation in the DNA binding domain of the molecule. Since 2001, additional cases of verbal dyspraxia linked to *FOXP2* have emerged, including a mutation that truncates the protein prior to the DNA binding domain (Macdermot 2005). These mutations impede FOXP2 from binding to its transcriptional targets (Vernes et al. 2006).

Imaging studies comparing affected to unaffected KE family members have revealed the neuroanatomical bases for their deficits. Affected individuals have bilateral abnormalities in the basal ganglia and cerebellum, in addition to cortical abnormalities including the Broca's area in the inferior frontal gyrus. Altered amounts of gray matter in these regions are accompanied by their underactivation during tasks of verbal fluency (Vargha et al. 2005; Belton et al. 2003; Liegeois et al. 2003). These findings suggest that a mutant copy of *FOXP2* during development results in the malformation of brain structures later used in the control of orofacial musculature important for speech.

Progress in Animal Models

Advances in understanding FoxP2 neural function have been made in songbirds and mice (White et al. 2006). As in humans, FoxP2 is expressed in the cortex/pallium, striatum, and thalamus of these animals during development, consistent with a role in forming these structures (Ferland et al. 2003; Lai et al. 2003; Takahashi et al. 2003; Teramitsu et al. 2004; Haesler et al. 2004). In songbirds, expression persists in adults when *FoxP2* mRNA and protein are actively regulated in the striatal song circuit region Area X when birds sing under certain social conditions (Teramitsu and White 2006; Miller et al. 2008). This implicates the molecule in the functional use of song circuitry. Beyond correlation, tests of molecular function rely on being able to manipulate molecular expression. The technology for manipulating genes in birds is not yet routine. Thus, Haesler et al. (2007) exploited viruses for their ability to enter cells and express foreign molecules. They used lentiviruses to express short hairpin sequences of RNA designed to knockdown FoxP2 expression. Virus was injected bilaterally into Area X of juvenile birds, which were then given the opportunity to learn their tutors' songs. Strikingly, in adulthood, their song copies lacked precision. The acoustic abnormalities were reported to resemble those exhibited by children with developmental verbal dyspraxia in that syllable structures and duration were abnormally variable. Regarding syntax, the jury is still out. No difference in the consistency with which knockdown birds ordered their syllables was detected, although the knockdown group had greater variability on this measure. Further, in the three sets of exemplar spectrograms, only the control pupils' syllables occur in the same order as those of their tutors. Perhaps more robust alterations in syntax would emerge with testing of more subjects, or with a consistently high percent of viral transduction (20 ± 10% of the Area X neurons were affected), or by testing other species with more sophisticated syntax (e.g., the Bengalese finch).

The technology for introducing inherited transgenes is well developed in mice, enabling functional tests of molecular manipulations. Thus far, four groups have used transgenics to alter Foxp2 in mice. In the first, Shu et al. (2005) produced a null mutation, creating intermediate Foxp2 levels in heterozygotes and undetectable levels in homozygotes compared to wild-type animals. As expected, the homozygous phenotype was the most dramatic with pups dying by postnatal day 21. Several behavioral tests were run, including examination of ultrasonic vocalizations. Although no spatial learning deficit was detected with the Morris water maze test, a major finding was that stranded heterozygous pups had reduced numbers of ultrasonic isolation calls, which typically cause the dam to retrieve the pup. The calls were reported to be acoustically normal, yet their decreased numbers in heterozygotes relative to wild-type pups was taken as evidence for a specific effect of Foxp2 on mouse vocalizations.

Subsequent studies did not aim to knock out Foxp2 but rather to mutate it. Groszer et al. (2008) generated lines of mutant mice by exposing founder males to N-ethyl-N-nitrosourea (ENU) which induces mutations randomly across the genome. The genomic DNA from >5,000 offspring was then screened. Incredibly, a mouse that carried the KE-like mutation in its Foxp2 gene was identified. While labor intensive, the ENU methodology does not introduce stretches of foreign DNA into the host, and backcrossing is used to remove nonrelevant mutations. In contrast to the heterozygous nulls described above, the heterozygote pups generated by Groszer et al. emitted similar numbers of normally structured isolation calls as their wild-type littermates. Analysis of different calls and call properties led the authors to conclude that severe Foxp2 mutations cause developmental delays. Rather than loss of a specific function of Foxp2, these generalized delays may underlie the lower number of calls in the heterozygous nulls. Assuming such calls are unlearned (which is likely given their function in newborn mice), the normal calls and call numbers of the KE-like heterozygotes are not surprising, as they reinforce the notion that FoxP2 is critical for learned skills such as speech in humans or other procedurally learned behaviors in nonvocal learners.

In line with this idea, the heterozygote KE-like mice showed deficits on two assays of motor skill learning: the accelerating rotorod and the tilted voluntary running wheel. Further, in these mice, neurons in the dorsal striatum, a region implicated in motor skill learning (Dang et al. 2006), lacked a form of synaptic plasticity known as long-term depression. Synapses in the cerebellar cortex also showed altered plasticity. Together, these findings depict a developmental role for Foxp2 in motor-skill learning, involving cortico-striatal and cortico-cerebellar circuitries, that extends to the learned speech of humans, but not to the unlearned calls of mice. Fujita et al. (2008), however, used targeted genetic recombination to generate KE-like heterozygote pups that had disrupted calls, raising a discrepancy between studies. This method necessarily leaves foreign genetic material in the animal, which may account for the different findings. Some resolution may emerge as other mice are generated with "conditional" mutations that can be switched on or off throughout life. Interestingly, while all three studies discussed thus far detected cerebellar abnormalities, only Groszer et al. (2008) found striatal ones.

Finally, rather than creating a loss of function phenotype, Enard and colleagues investigated whether the two amino acids from the FOXP2 sequence that, among primates, are unique to the human form (Asn at position 303 and serine at 325, Enard et al. 2002), would promote a new function when inserted into mice. At the time of this writing, the full results are not yet available, but a preliminary report suggests that the human-like sequence promotes neurite outgrowth and synaptic connectivity in the striatum (Enard et al. 2002). This observation, together with the disrupted striatal plasticity in mice with the KE-like mutation, the imprecise song copying of birds with lower striatal FoxP2 levels, and the altered striatal structure and activation in affected KE family members reinforce the importance of the striatum in skill learning. Perhaps FoxP2 function is critical for species-typical skills, be it mouse locomotor coordination, birdsong, or human speech.

FoxP2: Achieving Specificity

FoxP2 is expressed in most, if not all, body organs (Lu et al. 2002) and is present across the animal kingdom. The human mutant phenotype, however, appears restricted to neural function and, most generally, to learned motor control of the mouth and face, which does not extend to other body parts (Vargha-Khadem et al. 2005). How this specificity occurs requires consideration not just of the sequence that makes up FoxP2 itself, but also of the gene regulatory networks in which it participates.

FoxP2 Molecular Evolution

It is tempting to hypothesize that in certain phylogenetic lineages the molecular sequence of FoxP2 is critical for specialized acoustic motor function. However, only among primates is there an obvious correspondence between amino acid changes and vocal learning capacity. Accelerated evolution occurred in primates along the lineage between chimpanzees and humans (Enard et al. 2002), and in the human coding sequence, two amino acids are distinct (Zhang et al.

2002). Carnivores (who are not thought to learn vocalizations) share one of these two. Further confusing the picture, there is no correspondence between FoxP2 sequences and vocal learning abilities in birds (Webb and Zhang 2005) or acoustic motor abilities in bats (Li et al. 2007). Despite ignorance of the pressures driving accelerated evolution in primates, the finding of the modern human FOXP2 form from Neanderthal bones reopened the question of whether these archaic hominids possessed a protolanguage (Krause et al. 2007).

The difficulty in relating a single gene sequence to a complex behavioral phenotype highlights the fact that genes do not specify behaviors or cognitive processes (Fisher 2006). Instead, they make bits of biological machinery, things like signaling molecules, receptors, and regulatory factors such as FoxP2. These tiny machines interact in complex networks, and those networks are themselves cogs in larger hierarchical arrangements (Barabasi and Oltvai 2004). Thus, although *FOXP2* is implicated in speech and language impairment and neuroligins *NLG3* and *NLG4* in autism (Jamain et al. 2003), these genes cannot be considered "a gene for speech and language" or "genes for sociality." Coding variants of *FOXP2* have not been found in specific language impairment (SLI), and the relevance of *NLG3* and *NLG4* variants for common disorders remains open (Fisher 2006). The information carried in the genotype is simply the starting point for cascades of molecular, cellular, and systems-level interactions in the brain. As such, the discovery of the KE family and others with similar mutations is a fortuitous clue to the puzzle of the complex genetic basis of speech and language. The molecular targets of FoxP2's transcriptional regulation are the next set of interactors in the puzzle.

Gene Targets

Recently, the technique of chromatin immunoprecipitation coupled with gene promoter microarrays (ChIP-chip) was used by two groups to identify molecular targets of FOXP2 in human tissue (Spiteri et al. 2007; Vernes et al. 2007). This method uses an antibody to grab onto FOXP2 *in flagrante* (i.e., within a biological tissue and in the act of binding to the noncoding promoter regions of its target genes). After chemical dissociation, these DNAs are hybridized to microarray chips spotted with many samples of different known genes. A positive hybridization signal indicates that the spotted gene is likely a FOXP2 target. A variation of this method is referred to as ChIP-seq, in which, following dissociation, the target DNAs are directly sequenced to reveal their identities, an approach that we will return to at the end of this section.

Using ChIP-chip, Vernes et al. (2007) looked for targets in human SH-SY5Y cells, useful for studying neural processes while Spiteri et al. (2007) investigated FOXP2 targets in human fetal basal ganglia (BG) and inferior frontal cortex (IFC)—two main areas of dysfunction in people with FOXP2 mutations—and in human fetal lung. Spiteri et al. identified ~175 targets in each tissue, many of them overlapping. Most intriguing are the eight targets enriched in both brain areas but not in lung, three of which are involved in central nervous system development, including cortical patterning. Gene ontology and pathway analyses were used by both groups to investigate the strongest signals and revealed enrichment for molecules involved in ion transport, cell signaling, neurite outgrowth, and synaptic transmission. The Wnt/Notch signaling pathway was highlighted in the SH-SY5Y cells (Vernes et al. 2007). In mammals, *Wnt* genes are known to play a major role in forebrain patterning during development, consistent with the idea that FOXP2 targets in this pathway mediate structural changes in these regions. Additional targets have links to learning (Spiteri et al. 2007), which suggests a function for FOXP2 signaling cascades in activity-based sculpting of neural connections. This is intriguing in light of our own work, which shows down-regulation of FoxP2 in the striatal Area X of the song circuit when adult birds practice, but not when they perform their songs (Teramitsu and White 2006). As song practice is more acoustically variable, we posit a role for avian FoxP2 in adult neural and behavioral plasticity. Fourteen of the FOXP2 targets found by Spiteri et al. (2007) show accelerated evolution likely via positive selection, together comprising a genetic cohort potentially related to human cognitive specializations integrated by the BG and IFC, including speech and language. The significant overlap between the targets found in the BG and IFC and the targets identified in SH-SY5Y cells further validates both studies.

Significance of Gene Networks

Uncovering the genetic basis for a complex trait is a task rife with many challenges, yet just as many adaptive strategies are coming into use (Fisher et al. 2003; Felsenfeld 2002). One exciting strategy for investigating gene networks underlying complex traits was developed at our home institution, UCLA. Weighted gene co-expression network analysis (WGCNA) identifies groups of functionally related genes through statistical analysis of gene expression microarray data (Zhang and Horvath 2005). Until recently, the interpretation and analysis of microarray data has resulted in little more than lists of potentially interesting genes, left to be functionally validated via more traditional molecular techniques. That is because researchers have generally only considered genes in isolation, and whether they are expressed at different levels between control and experimental samples. A major drawback of this approach is the statistical difficulty inherent in comparing long lists of data points from insufficiently large sample sizes.

WGCNA takes a step back from the search for single differentially expressed genes, instead identifying groups of genes whose expression levels co-vary, called modules. The modules are defined via hierarchical clustering and are correlated to the trait of interest. In this way, multiple hypothesis problems are alleviated, and the investigator can be confident in the reproducibility of his/her results. By considering the functional activity of groups of genes,

WGCNA has thus far yielded robust and compelling findings for various traits, including identification of novel genes involved in brain cancer, gene pathways relevant to atherosclerosis, and genes that are key drivers of evolutionary change in humans (Horvath et al. 2006; Gargalovic et al. 2006; Oldham et al. 2006). Language may very well be the most complex of traits. Application of WGCNA or similar strategies to high throughput data on FOXP2 gene targets promises to help piece together the molecular puzzle underlying language.

The CNTNAP2 Connection

Discovery of gene networks using microarrays is technically limited by the number of genes that are printed on the arrays; often an incomplete set. An alternative approach is to identify target genes by direct, so-called "shotgun" sequencing. This unbiased approach was recently used by Vernes et al. (2008) to great effect. It identified the contactin-associated protein-like 2 gene (*CNTNAP2*) that encodes a neurexin protein, called CASPR2, as a direct target of FOXP2 transcriptional repression. In the brain, association of presynaptic neurexins with postsynaptic neuroligins is thought to be a key event in synaptogenesis (O'Connor et al. 1993; Dean et al. 2003). Accordingly, during human fetal brain development, *CNTNAP2* expression is enriched in areas of the cortex that give rise to language, while in developing rodents, its cortical expression is diffuse (Alarcón et al. 2008). The human *CNTNAP2* cortical pattern is opposite that of *FOXP2* (i.e., *FOXP2* levels are high where *CNTNAP2* levels are low), consistent with FOXP2 repression of this transcript. As mentioned above, FOXP2 variants have not been associated with SLI nor with common developmental disorders in which language is delayed or impaired, such as in autism. In sharp contrast, certain *CNTNAP2* variants in autistic children are associated with the age at first word (Alarcón et al. 2008). Further, Vernes et al. (2008) have found that genetic polymorphisms of *CNTNAP2* in children with SLI are correlated with their ability to do a nonword repetition task. The viewpoint that *FOXP2* is not a gene for language, but rather a key molecular piece that connects with many others in the language puzzle, is now directly supported by its transcriptional repression of *CNTNAP2*. This interaction connects FOXP2 with CNTNAP2-related language disorders in which FOXP2 was not otherwise implicated.

Syntax in Birdsong

Turning from the discussion on FoxP2, we now review what is known about "song syntax" in songbirds. In the following sections, we discuss birdsong syntax, its dependence on critical developmental phases for learning and auditory feedback, its role in perception, its social regulation, and its neural coding.

Phonetical Syntax

Birdsong syntax is defined simply as the temporal sequence in which discrete units of song (notes, syllables, phrases, motifs, bouts) are produced. These units do not carry independent meaning, and thus bird syntax has been called "phonological syntax" by analogy with the organization of independently meaningless phonemes into morphemes or words in language (Marler 1977). Since morphemes are symbolically meaningful, however, the analogy is not perfect. While birdsong can communicate species or individual identity and advertise mating or territorial ownership (Doupe and Kuhl 1999), it is not compositional; the position of a single note within a syllable, a single syllable within a motif, or a single motif within a bout does not by itself provide new semantic information. Thus, we adopt the terms "phonetical syntax" to refer purely to the sequence of sounds. Whatever vocabulary we use, in both birdsong and language, discrete acoustic units are produced and then combined according to specific sets of learned rules. In both, fine motor sequences are executed using the vocal motor apparatus, a task dependent on subcortical structures such as the basal ganglia and cerebellum.

Development

Similar to human speech and language learning, all aspects of birdsong development show critical constraints on acquisition as well as dependence on auditory feedback. In juvenile white-crowned sparrows syntactical and phonetical (i.e., the spectral features of song syllables) cues guide selective song learning during development (Soha and Marler 2001). Behavioral studies of zebra finch song development indicate that learning the spectral content of tutor notes and syllables occurs before learning how to order them in time (Tchernichovski et al. 2001). As development advances, zebra finches produce a greater variety of syllables delivered in sequence, reminiscent of the emergence in human infants of reduplicated babbling and the transition to variegated babbling. Multiple studies suggest that different neural pathways underlie the mimicry of individual syllables versus syllable order (Vu et al. 1994; Yu and Margoliash 1996; Hahnloser et al. 2002), and that aspects of song syntax can be acquired independently of spectral content (Helekar et al. 2003).

In 2003, Funabiki and Konishi used sustained white noise exposure to deprive zebra finches of auditory feedback during the sensorimotor phase of song development. Birds were reared with their biological parents until the onset of sensorimotor learning. Experimental birds were then moved into sound attenuation chambers with continuous loud noise playing in the chamber, depriving the birds of any auditory feedback. They were kept in noise for ~1–6 months before being released and allowed to sing under normal acoustic conditions. They learned phonetical features of the tutor song just as well as birds that did not endure noise exposure, with similar developmental trajectories.

Interestingly though, only those birds released from noise before 80 days of age were able to reproduce correctly the syntax of their tutor. All birds released after 80 days were unable to learn song syntax, perhaps indicating stronger developmental constraints on syntax than on phonation.

To examine the consequences of perturbing auditory feedback in real time, Sakata and Brainard (2006) interleaved trials of perturbed versus normal feedback to Bengalese finches during singing. They showed that, as for human speech, abnormal feedback disrupts both the sequencing and timing of song within tens of milliseconds. Specifically, altering auditory feedback reduced stereotypy in syllable transition probabilities and could elicit the production of completely novel sequences of syllables during what were otherwise completely stereotyped sequences.

Perception

Syntax is known to play a role in song perception, as evidenced by multiple studies using operant conditioning paradigms to train songbirds to discriminate between songs. Comparing results from different studies, Bengalese finches weigh syntax more heavily than zebra finches when doing discrimination tasks (Braaten et al. 2006), perhaps reflecting a functional importance of higher syntax variability in Bengalese finches. Gentner and colleagues used operant conditioning in European starlings (*Sturnus vulgaris*) to investigate the perceptual mechanisms underlying the vocal recognition of individual conspecifics (Gentner and Hulse 1998). After training, birds were able to discriminate individuals on the basis of novel song bouts, mediated by the memorization of specific motifs and the sequential ordering of motifs within bouts. A later expansion upon this study implicated HVC in the process of forming learned associations among conspecifics (Gentner et al. 2000) to show that HVC is important for the production of song syntax as well. The dual roles of HVC in song perception and production place it in an intriguing category of brain regions with "mirror like" properties, a topic that we will return to later.

Again using operant conditioning, starlings were trained to classify subsets of motif sequences in two different types of artificial starling "languages" (Gentner et al. 2006). Different types of typical starling motifs were used to generate a context-free grammar (CFG; A^nB^n) which entailed recursive center-embedding (i.e., a structure of one type, *AB*, is embedded in another instance of itself *AABB*) and a finite state grammar (FSG: $(AB)^n$). After thousands of training trials, nine of eleven birds were able to classify the CFG and FSG sequences accurately, as well as successfully classify novel sequences generated using the same rules. This study followed a similar experiment testing the syntactic capabilities of cotton-topped tamarin monkeys (*Saguinus oedipus*) wherein the authors concluded that the tamarins were able to master a FSG ($(AB)^n$), but not a CFG (A^nB^n) (Fitch and Hauser 2004). Humans trained on the same stimuli did master the CFG. This seems to substantiate an earlier claim (Hauser et al. 2002) that the capacity for recursion is unique to humans, but the questionable ecological relevance of the stimuli (recordings of human male and female voices) and small size of the training sets (only 2 samples per set) used in the tamarin study have prompted criticism. A major problem is the legitimacy, or lack thereof, of making conclusions based on generalizing from such small training sets to entire grammars.

The stimuli in the starling study were actual starling song motifs ("rattle" motifs were *A*, and "warble" motifs were *B*) and were varied such that multiple kinds of rattle and warble motifs were used, resulting in larger sets of ecologically valid stimuli. A series of probe tests provide compelling evidence that the successful starlings were not merely approximating the recursive structure of the CFG by learning an equivalent FSG, or using alternate discrimination strategies such as attending to only primary or terminal patterns or simply counting *A/B* transitions. This suggests that starlings can recognize strings formed using a recursive center-embedding rule and challenges the claim that only humans are capable of learning grammars with such a rule. If this is so, we now have at our disposal a well-established model system within which we can probe the neural systems encoding a learned CFG. However, it is important to keep in mind the immense amount of training required by the successful starlings (~9,400 to ~56,000 trials) and consider that even humans can fail to learn center-embedding grammar in artificial languages lacking semantic content, as shown by Perruchet and Rey (2005). Note, however, a general problem with these experiments is that animals in fact could solve the problem by simple counting (Corballis 2007b).

If even humans can fail to learn a CFG, where does this leave us? Note that the amount of center-embedding in the stimuli presented to Perruchet and Rey's subjects is no greater than that present in the simple sentence *Either they learned or they did not.* Hurford contends that these human subjects fared so poorly due to a lack of semantics in the artificial language (Hurford 2009). Perhaps they may have been successful if they had been sufficiently motivated to undergo exhaustive training like that endured by the starlings. Perhaps the difference between human and animal syntax can be accounted for by the lack of semantic interpretation in animal syntax coupled with stricter computational restraints in animal brains and is not so much a matter of FSGs and CFGs. It is possible that humans and animals all use some type of CFG, but humans simply exploit CFG to a greater depth, ultimately taking advantage of mechanisms such as recursion thanks to more formidable neural processing power combined with well-developed symbolic representational abilities. Clearly, birdsong lacks semantics, but if starlings or finches could map syllables or motifs of song onto conceptual or intentional representations of the world, perhaps a relatively simple (due to weaker neural processing power) compositional syntax would emerge in birds as well.

Social Context

Social regulation of syntax variability has been described in adult Bengalese finches (Sakata et al. 2008). Songbirds can sing in one of at least two social contexts: (a) directing song to a conspecific female, known as directed (FD) song, or (b) singing alone, known as undirected (UD) song. Bengalese finches were used to test specifically for social modulation of song syntax. Unlike zebra finch song, Bengalese song retains moment to moment variability in syllable sequencing (Okanoya 2004a). In zebra finches there is greater variability in syllable structure (mostly driven by variability in fundamental frequency) during UD song when compared to FD song (Kao et al. 2005). Analogous modulation was found in the syntax of Bengalese finch song, with greater variability in sequencing during UD over FD songs (Sakata et al. 2008).

Interestingly, social context regulation has been demonstrated at both the neural systems and molecular levels (Jarvis et al. 1998; Hessler and Doupe 1999; Hara et al. 2007). Acute down-regulation of *FoxP2* has been observed in Area X of zebra finches singing UD song when compared to FD song, indicating a possible role for FoxP2 in modulating behavioral variability (Teramitsu and White 2006).

Social context modulates dopamine (DA) levels in Area X, with levels being higher in FD song (Sasaki et al. 2006). The DA system is important for motivation and is thus a sensible candidate for driving social context modulation in the song circuit. In line with this notion, Sakata and Brainard (2006) propose a model of quick acting dopaminergic influence over the social modulation of song syntax variability found in Bengalese finches. The changes they observed occurred on a small enough time scale (1–3 minutes) which effectively rules out slower acting transcription factor-mediated processes as a biological mechanism. However, the medium spiny neurons expressing FoxP2 in Area X receive strong dopaminergic input. Thus the functional relationship between *FoxP2* and its gene targets, striatal microcircuits, DA, and social context regulation of song features like syntax variability requires further study.

Coding

The neural coding of syllable sequence production by the vocal motor pathway has been examined by making electrophysiological recordings from neurons in RA (analogous to primary motor cortex in mammals; projection neurons here directly innervate syringeal motor neurons) and HVC (analogous to premotor cortex in mammals, and presynaptic to RA). During singing, RA neurons generate a complex sequence of high-frequency spike bursts, reproduced precisely each time the bird sings a motif (Yu and Margoliash 1996). HVC neurons come in three basic types: interneurons (HVC_I), neurons projecting to Area X (HVC_X), and neurons projecting to RA (HVC_{RA}) (Hahnloser et al. 2002). By recording from HVC_{RA} neurons during singing, Hahnloser et al. showed that this specific subpopulation generates sparse patterns of bursts that are time-locked to the song, with each HVC_{RA} neuron emitting only one burst of spikes per song motif. The timing is extremely precise, with a jitter of less than 1 ms relative to the song. Recording simultaneously from HVC_{RA} and RA neurons in sleeping birds revealed that RA neurons generate complex burst sequences temporally locked to the sparse sequences in HVC as well as HVC_{RA} population activity correlated with bursting in RA. Thus, it is likely that sparsely firing HVC projection neurons are driving bursting activity in RA via direct feed-forward input (further discussed below in the section on Modeling). Population activity of HVC_{RA} neurons has since been highly correlated with syllable sequencing (Kozhevnikov and Fee 2007). Also, there is no correlation between the firing of HVC_{RA} neurons and spectral features of song syllables, or between the timescales of vocal dynamics and neural dynamics in HVC. This suggests that HVC codes for the temporal order of syllables in song and not for spectral structure, supporting the hypothesis of separate neural substrates underlying phonation and syntax in birds.

Quantification and Comparison

Birdsong syntax is typically analyzed in one of two subtly distinct ways: (a) using metrics of sequence stereotypy within the songs of a single bird or group to analyze sequence variability, or (b) directly comparing the song syntax of two birds or groups to obtain some measure of syntax similarity. At the core of both methods is the observation of syllable transitions and transition probabilities.

Metrics used in the first approach are sequence linearity and sequence consistency (Scharff and Nottebohm 1991). Sequence linearity quantifies the possible transitions that can be observed after each unique syllable of song and is calculated by dividing the number of unique syllables by the number of syllable transitions. In a completely linear song, each syllable has only one transition. Sequence consistency quantifies the frequency with which the dominant syntax occurs (i.e., how often the most common path through possible syllables is actually taken) and is calculated by summing the dominant transition probabilities for each syllable and dividing by the sum of all transition probabilities. Sequence stereotypy is defined as the average of the linearity and consistency scores. Newer versions of sequence stereotypy measure base scores on the entropy of the transition probability distribution for each syllable (Haesler et al. 2007; Sakata et al. 2008).

The direct comparison of the syntax of two birds/groups is a trickier proposition than comparisons of syntax variability. This approach must account for phonetic content of the song, assess sequence variability, and find a metric to combine these two measures. Funabiki and Konishi (2003) generated syntax similarity scores by first calculating the transition probability for every pair of syllables and then assessing the amount of overlap between the probability distributions for each syllable in the tutor and pupil song. These overlaps are summed and weighted by syllable frequency in the pupil's song. These scores are only valid if the experimenter is sure that syllable "X" in the tutor song is the same as syllable "X" in the pupil song. The authors addressed this by using a version of software called Sound Analysis Pro (SAP; Tchernichovski et al. 2000), developed for analysis of zebra finch songs. SAP calculates similarity scores based on spectral features of the syllables, including frequency modulation, pitch, pitch goodness, and entropy. Syllables in the pupil's song were scored for phonetic similarity against syllables in the tutor's song, and the above procedure performed on matched syllables. The latest version of SAP includes a function for measuring sequential similarity as well. We are implementing a variation of Funabiki and Konishi's procedure in our own studies. We start by generating transition probability distributions for each syllable, resulting in a matrix of probabilities for each bird/group. Each matrix row represents the distribution for a single syllable. The correlations between corresponding syllable distributions are calculated and weighted by the phonetical similarity score (from SAP) for the relevant syllable. These correlations are summed and divided by the number of syllables in the pupil's song to produce a similarity score. Syllables present in the pupil's song but not the tutor's, or vice versa, yield correlations of 0 and thus bring the score down. Again, all of these similarity measures depend on reliable phonetical scoring and may thus become problematic when studying phonetically disrupted song.

Distinct from quantification of variability or similarity, one group uses computational models to analyze the complex syntax of Bengalese finches (Kakishita et al. 2007). Multiple song units (motifs) are contained within a bout of Bengalese finch song. Each motif can be separated into small chunks of stereotyped sequences. Thus, Bengalese finch song can be said to possess a hierarchical structure. The authors use the term "double articulation" by analogy with language, but this term implies the organization of meaningless units (phonemes) into meaningful units (morphemes), which are then organized into meaningful strings (sentences). Based on the arguable assumption that Bengalese finch song is *k*-reversible (see Angluin 1982), the authors used an automata induction approach to model song syntax, with the final model represented as an N-gram of chunks. This seems an effective method for reducing complex syntax down to a relatively simple model, perhaps paving the way for novel analysis methods.

Modeling

A physiologically based associative learning model of the song circuit (Troyer and Doupe 2000) speaks to questions of network activity addressed above. Here we focus on findings relevant to syntax learning. Based on the model's behavior, song sequence generation results from a reciprocal sensorimotor interaction between the two populations of HVC projection neurons, with the motor component encoded in HVC_{RA} neurons, and the sensory component encoded in HVC_X neurons. Further, the AFP modulates activity in RA via a reinforcement signal, with RA activity biased by AFP input to more closely match syllable transitions in the tutor song. The AFP teaching signal is calculated based on information about the current motor output, received from HVC. The ability of the AFP to influence RA activity is predicted to be maximal during the peak period of sequence learning, with HVC_{RA} projections dominating RA input after learning has occurred.

Fiete and colleagues (2004) constructed a simple three layer feedforward model of HVC→RA→vocal output to investigate the role of HVC sparse coding on sequence learning. Essentially, the network produces some output that must be matched to some desired output through adjustments of the weights on connections between HVC and RA. The overall learning speed of the network decreased with increasing numbers of HVC bursts per motif under both gradient descent and reinforcement learning algorithms, due to increasing interference in the weight updates for different synapses. Thus, if HVC activity is sparse, synaptic interference is reduced. This interference is minimized if each HVC→RA synapse is used only once per motif, exactly what was observed biologically in awake behaving birds by Hahnloser et al. (2002).

A two-compartment physiological model of HVC_{RA} projection neurons has been used to investigate the role of intrinsic bursting in the generation of the sparse firing sequences observed in these cells (Jin et al. 2007). Simulations were first run without intrinsic bursting. While burst sequence generation occurred under these conditions, the network was highly sensitive to initial conditions of the simulation and could become unstable due to runaway excitation. After introducing intrinsic bursting properties into the HVC_{RA} neurons, runaway instability was abolished and the network became more robust. Intrinsic bursting in the model is driven by dendritic calcium spikes, a phenomenon yet to be observed in real HVC_{RA} neurons.

Other Species

This chapter has necessarily focused on songbirds for probing the neurobiological basis of vocal learning because so few other animal groups have the capacity for mimicry or the creation of new sounds via the vocal apparatus. Of those that do, fewer still are amenable to physiological experiments. In addition to songbirds, parrots and hummingbirds, certain species of bats, marine mammals such as whales, and elephants have been documented to possess this ability (Bougham 1998; Janik et al. 2006; Suzuki et al. 2006; Poole et al. 2005). Importantly, among primates, only humans are vocal learners. Changes in monkey or ape call structure are largely attributable to maturational factors, with a primary predictor of these changes being weight (Hammerschmidt and Fischer 2008). Some minor acoustic modifications not associated with weight have been documented. In the laboratory setting, cotton-topped tamarins exposed to loud white noise can make non-syntactical changes in the duration, timing, and amplitude of their calls to avoid the interference (Egnor et al. 2007). In natural settings, subtle modifications appear to enhance the match between vocalizations within given populations of Barbary macaques (*Macaca sylvanus*) (Fischer 2003). This latter phenomenon has been likened to speech accommodation in humans, where even subtle changes in syntax are observed as a person subconsciously adjusts their speech pattern to match that of his or her conversant (Giles 1984). How auditory feedback is used to accomplish vocal accommodation remains an open question. Again, these changes are on a much smaller scale than those observed in "bonafide" vocal learners.

Though limited in their ability to create new vocalizations, nonhuman primates can learn to associate certain calls with meanings, as can virtually all animals. The calls of vervet monkeys (*Chlorocebus aethiops*) appear innate: young vervets sound much like mature ones. However, a young vervet requires experience to use its vocalizations in the appropriate context and to interpret conspecific calls correctly (Seyfarth et al. 1980). A question of syntax occurs when combinations of calls convey new meaning. Putty-nosed monkeys (*Cercopithecus nictitans*) make two different alarm calls, termed "hacks" and "pyows," in response to distinct predators. A series of pyows appears to signify leopards, while hacks or hacks followed by pyows are a response to eagles. At other times, males produce 1–4 pyows followed by 1–4 hacks. These series reliably occur prior to movement of the group (Arnold and Zuberbühler 2008). The joining of two distinct alarm calls to convey a new meaning would appear to fit the bill for syntactic recombination. However, to be truly compositional, the meanings of the individual calls must be relevant to their combinatorial meaning (Hurford 2009), a finer but critical point that remains to be tested.

The possibility of compositionality arises from studies by Zuberbühler (2002) of Diana monkeys (*C. diana*) who listen in on sympatric Campbell's monkey calls (*C. campbelli*). Like vervet monkeys, Campbell's monkeys make alarm calls to nearby predators, and these are distinct for leopards versus eagles. When Diana monkeys hear the Campbell's calls, they make their own corresponding alarm calls. Campbell's males make a different sound when predators are at a distance, or when some other less critical disturbance occurs. In these situations, they emit a pair of low-pitched "boom" vocalizations. Strikingly, when these booms precede a Campbell's alarm call, Diana monkeys no longer make their own alarm calls. The addition of a boom appears to serve as a new, cross-species semantic signal, along the lines of "pay no attention to the man behind the curtain," or, as Zuberbhühler notes, something akin to human linguistic hedges such as "kind of" or "maybe." Though not proven, the latter interpretation raises the possibility of compositionality since the boom can be interpreted as adding a "not to worry" signal to the "there's a predator" one.

At the extreme of nonhuman primate call complexity is gibbon song. *Hylobates agilis* produce songs that are organized into complex sequences of several call phases. Females' calls differ acoustically between individuals, opening the potential for use in individual recognition (Oyakawa et al. 2007). At least in *H. lar* and *H. pileatus* however, such calls are inherited rather than learned. Hybrid offspring produce hybrid vocalizations in that the acoustic features are intermediate between those of each parent (Geissman 2000). Reminiscent of the putty-nosed monkey calls described above, certain components of gibbons' songs are differentially emphasized in different contexts. Specifically, white-handed gibbon songs were shorter, contained fewer sharp "wow" notes and shorter "hoo" notes when in a predatory context than when duetting. Differences in these song patterns appeared salient to the few members that temporarily left the group during the observation period. Upon returning, males responded with their own songs after hearing the groups' songs to predators, but not after hearing duets (Clarke et al. 2006).

As summarized by Hammerschmidt and Fischer (2008, p. 93–120), "While vocal production appears largely innate, learning does play a role in the usage and comprehension of calls, but this is not restricted to the primate order." Further, recombination of largely innate calls can be used to convey new meaning, but it remains unclear that the new "whole" is a manifestation of the sum of its parts, or indeed, has any shared meaning with them.

Returning to vocal learners, in a few cases, individual elephants (*Loxodonta africana*) have been observed to mimic nonspecies specific sounds (including human words!) (Poole et al. 2005). While the uniqueness of an elephant's trunk has been compared to singularity of human language (Pinker 1994), the discovery that elephants also possess the rare ability of vocal learning is quite recent. Investigation into the ontogeny of elephant's species-specific signals is even more nascent (Stoeger-Horwath et al. 2007). It will be interesting to learn whether elephants combine and respond to mixes of seismic and vocal communication signals in any sort of compositional manner (O'Connell-Rodwell 2007). Among marine mammals, humpback whales (*Megaptera novaeangliae*) produce some of the most elaborate songs in the animal kingdom. An individual whale sings its own distinctive song, with other whales in the same population singing similar ones. Songs recorded across 30 years reveal changes that are rapidly adopted within a given population. Changes occur mainly in the middle of the breeding season and are correlated with increased durations of the song sessions, suggesting that the song changes are part of a sexual display (Payne 2000).

Analysis of song structure in whales reveals a series of units referred to as a phrase and an unbroken sequence of similar phrases, called a theme (Payne and McVay 1971). Within a song there are variations on a phrase; slight changes occur across successive renditions. Themes are sung in an ordered sequence and compose a song which can last half an hour. Songs are strung together in sessions which, themselves, can last for hours. The use of information theory techniques (Suzuki et al. 2006) has confirmed the earlier proposal (Payne and McVay 1971) that the songs exhibit hierarchical structure. However, the claim that humpback whales can sing an infinite number of songs from a finite set of units, startlingly similar to human syntax, apparently overstates the case. As Hurford notes, Suzuki and colleagues combined the songs of several whales in their analyses. Simply put, a single humpback at any one time in its life only sings one song (Hurford 2009).

Finally, while marine mammals present obvious challenges for neurobiological study, neural activity in the brains of smaller mammalian vocal learners has been examined. Mustached bats (*Pteronotus parnellii*) can combine otherwise independently emitted syllables. Esser and colleagues (1997) used playback of these naturally occurring heterosyllabic composites as well as temporally destructured versions to determine the response properties of neurons. They uncovered regions in nonprimary auditory cortex that were sensitive specifically to the composite structure of communication calls. It has been proposed that the limited number of vocal gestures available to nonhuman primates has driven the ability of receivers to process signal combinations (see Számadó et al., this volume). Detection of combination-sensitive neurons in auditory association cortex in bats, also detected in songbird pallium (Margoliash and Fortune 1992), may provide the neural building blocks for more complex syntactical processing in humans.

Mirror Neurons

Any communication system requires individuals to attend to and understand the actions of others, be they gestural or vocal. A possible neural substrate for such has been well studied in primates and has recently been identified in songbirds. Mirror neurons were originally observed via neurophysiological recordings in monkey premotor cortex (Rizzolatti and Arbib 1998). These neurons fired not only when subjects grasped or manipulated objects, but also when the monkey observed the experimenter making a similar gesture. Some mirror neurons are highly specific, not coding only for an action, but also how that action is executed. For example, they fire during observation of grasping movements, but only when the object is grasped with the index finger and thumb. They can fire when the grasping action is performed with the mouth as well as with the hand (Gentilucci and Corballis 2006). In general, mirror neurons are proposed to code for representations of actions, which can then be used for imitating and understanding the actions of conspecifics. Evolutionarily speaking, they may have been instrumental in the transfer of the gestural communication system from the hand to the mouth (Gentilucci et al. 2001). Based on imaging studies of the KE family, it has also been speculated that *FOXP2* might play a role in the incorporation of vocal articulation into the mirror system (Gentilucci and Corballis 2006).

Along with substantial evidence for mirror neuron system (MNS) dysfunction in autism (Iacoboni and Dapretto 2006), the human MNS has also been hypothesized to play a role in apraxia, a cognitive motor disorder in which the patient loses the ability to perform learned, skilled actions accurately. The most common form, ideomotor apraxia, has been described as "an impairment in the timing, sequencing, and spatial organization of gestural movements"; patients with ideomotor apraxia cannot tell if someone else is performing an action correctly or not (McGeoch et al. 2007). Functional MRI has been used to investigate the role of the human MNS in representing hierarchical complexity during the observation of action sequences (Molnar-Szakacs et al. 2006). Observation of object manipulation sequences recruited classic MNS regions, and MNS activity appeared to be modulated by the perceived motor complexity of the action. These results support Arbib's theory of language evolution (Arbib 2005), wherein language is thought to have evolved out of the motor system for gestures, and provide a connection between developmental and neural evidence linking motor and language functions.

Neurons displaying precise auditory–vocal correspondence (i.e., mirror neurons) have recently been observed in the song system of swamp sparrows (*Melospiza georgiana*) (Prather et al. 2008), the first identification of auditory–vocal correspondence in single neurons. HVC_X neurons displayed highly selective auditory responses, typically activated by only one song type. These responses were sparse, occurring at a precise phase in a given syllable. The same neurons also fired selectively when the bird sang the preferred song type, also phase locked to a particular part of a given syllable. The singing-related activity of these neurons was motor related and not due to auditory feedback of the song. Interestingly, the auditory responses of HVC_X neurons extend to the songs of conspecifics with note sequences similar to that of the bird's own preferred song type.

Summary

For those interested in the biological origins and evolution of language, a daunting obstacle is the lack of neurobiological data on users of protolanguage and early language. Making matters more challenging, humans are the only current language users on the planet. Thus a comparative approach is necessary. While no complete animal model of human language exists, animal models can be used to investigate the neural systems underlying different aspects of language. This includes investigation of molecules identified in genetic studies of human language disorders. Animal models can also be useful in studying the biological basis of linguistic preadaptations, such as the motor and perceptual skills needed to learn sequential ordering of vocal utterances.

Songbirds provide an ideal system for such studies: they learn vocalizations under similar constraints to humans learning language, possess identified neural circuits underlying this ability that are similar to circuits in humans important for language, and share some genetic basis for vocal learning with humans (Jarvis 2004). From a practical standpoint, their behavior is easily quantifiable, and they are amenable to laboratory life. Admittedly, birdsong is far from being compositionally semantic. To stress the absence of meaning upon combination of its units, we have described the simple structure of their songs as exhibiting only phonetical syntax. However, the experience-dependent shaping of neural circuitry for song may point to building blocks for more complex micro- and macro-circuits that comprise human language centers. Perhaps just as similar selection pressures drove parallel evolution of the eye in dozens of distinct lineages (Land and Fernald 1992), a similar situation may likely hold for parallel biological solutions to the problem of learned vocal communication.

To date, FOXP2 is the only single molecule to be repeatedly linked to language (Marcus and Fisher 2003). Here, we argue that what we learn about FOXP2, from humans as well as animal models, can be leveraged as a molecular wedge into the networks underlying language. Studies in songbirds (Haesler et al. 2007) and the more genetically tractable mice (Enard 2002; Teramitsu and White 2008) suggest that FoxP2 is important for species-typical procedurally learned behaviors, such as locomotor skills in rodents and song in birds. These reports also highlight the role of the striatum in the experience-dependent neural changes underlying such skills. We predict that a comparative network analysis of the gene targets of FOXP2, identified via deep sequencing or microarray analyses, will reveal some shared connectivity related to basic skill learning, as well as connections unique to the species' skill sets including speech, and possibly language, in humans. In this light, it will be interesting to see whether CNTNAP2 is a FoxP2 target in species other than humans (Vernes et al. 2008).

Learned vocal motor control gets us only part of the way to human linguistic syntax. The additional capacity of conveying and processing semantic content is key to moving beyond the musical-like realm of birdsong to semantically compositional language. Discoveries on the biological basis and evolution of both vocal and nonvocal learners' capacity for symbolic representation and theory of mind must be joined with the findings reviewed here. In this vein, mirror neurons provide a hypothetical link between an action and its meaning, or intent, when committed by one's self or by others. Whether or not the mirror neuron system is part of this particular puzzle, somewhere along the hominid lineage, neural systems for complex meaning must have intersected with those for ordering of vocal output to lay the basis for human linguistic syntax.

Acknowledgment

The authors' research is supported by a grant from the National Institute of Health (NIH RO1 MH070712).

9

What Can Developmental Language Impairment Tell Us about the Genetic Bases of Syntax?

Dorothy V. M. Bishop

Abstract

Neuroconstructivist accounts of language acquisition have questioned whether we need to posit innate neural specialization for syntax, arguing that syntactic competence is an emergent property of the developing brain. According to this view, specific syntactic deficits in children are the downstream consequence of perceptual, memory, or motor impairments affecting systems that are implicated in nonlinguistic as well as linguistic processing. Genetic studies of developmental language disorders pose difficulties for this viewpoint; although syntactic deficits are highly heritable, they are not readily explicable in terms of lower-level perceptual or motor impairments, and are distinct from limitations of phonological short-term memory. Data do not support the notion of a single "grammar gene," but rather are compatible with an Adaptationist account, which postulates that humans evolved a number of neural specializations that are implicated in language processing. Cases of heritable language impairment may help us identify what these specializations are, provided we focus attention on those rare disorders that represent departures from normality, rather than the tail end of normal variation.

Introduction

Why do humans have language when other primates do not? One obvious answer is found in what may be termed the "Big Brain" theory: humans have bigger brains than other primates, and this gives them more computing power, hence making language possible. Such an explanation was, however, roundly rejected by Noam Chomsky, who regarded syntax as completely different from other cognitive abilities, requiring a qualitatively different neural substrate rather than just more computing power. He suggested that, in the course of evolution, a genetic mutation might have occurred that enabled a new kind of cognitive processing, making syntax possible; this is sometimes referred to as the Grammar Gene theory. In this chapter, I argue that, although the Big Brain and Grammar Gene accounts are often described as polarized positions in debates on evolution of language, in extreme form neither is tenable. Data from developmental language disorders join other sources of evidence in suggesting that language depends on multiple modifications to brain structure and function in humans.

The Grammar Gene Theory

Noam Chomsky put forward the proposal that human language is a "mental organ" that can be thought of as a species-specific characteristic. In his early writings, Chomsky showed little interest in speculating about how this human characteristic evolved, but appeared to suggest that it had arisen through a chance mutation. For instance, he wrote:

> …it seems rather pointless…to speculate about the evolution of human language from simpler systems….As far as we know, possession of human language is associated with a specific type of mental organization, not simply a higher degree of intelligence. There seems to be no substance to the view that human language is simply a more complex instance of something to be found elsewhere in the animal world (Chomsky 1968, p. 70).

Subsequently, he suggested:

> Perhaps at some time hundreds of thousands of years ago, some small change took place, some mutation took place in the cells of prehuman organisms. And for reasons of physics which are not yet understood, that led to the representation in the mind/brain of the mechanisms of discrete infinity, the basic concept of language and also of the number system (Chomsky 1988, p. 183).

A central plank in his argument for a "language organ" was the implausibility of a Big Brain account, together with evidence for the unlearnability of language, its universality in human populations, and absence in nonhuman populations. Although Chomsky derided the Big Brain account, he did not present any hard evidence. There would be considerable interest in studying the grammatical abilities of children with primary microcephaly, a genetic disorder in which brain size is dramatically reduced (Woods et al. 2005). If one could show syntactic competence in such cases, where brain size is comparable to that of chimpanzees, this would provide strong evidence against brain size as being a key determinant of syntactic skill. Unfortunately, as far as I am aware, there have been no systematic linguistic or neuropsychological investigations of such cases, though mental retardation and delayed speech have been noted in clinical accounts.

Although he was adamant that language had a biological basis, Chomsky did not discuss how linguistic functions were instantiated in neurobiology, but rather focused on analyzing the nature of syntactic knowledge. This is an obvious topic for a linguist to tackle, but his claims that such knowledge was innate had an unfortunate side effect of antagonizing large numbers of more biologically oriented scientists. In particular, most neurobiologists had difficulty in entertaining the idea that humans have an innate knowledge of syntax, especially when investigations of Universal Grammar led to formulations of that innate knowledge in terms of arcane concepts, such as Binding Principles, empty categories, move-α, and so on.

One apparent problem for a theory of innate grammar is that young children do not behave as if they have innate knowledge of grammar; they make numerous grammatical errors before they master the adult form. The solution to this dilemma was to conceptualize syntax acquisition as a process of specifying the settings of a small number of parameters (for an overview, see Bloom 1994). However, the idea that there is a specific set of parameters nestling in the brain waiting to be set by exposure to a target language was an anathema to most developmental psychologists, who could not see how this kind of representational knowledge could be prewired in the brain, and who found it incompatible with evidence that children's language acquisition is gradual rather than proceeding in quantum leaps (Bates and Carnevale 1993). It is unfortunate that a standoff resulted between Chomskyans and developmental psychologists, leading to a lack of debate of the central issues raised by Chomsky; namely, what it is that gives humans a unique capacity for language. Most developmental psychologists are happy to countenance the possibility that there might be innate constraints on how the human brain engages in face processing, social cognition (mind reading), or indeed the linguistic task of word learning (Markman 1992). However, the notion of innate brain specialization for syntactic processing is often derided or dismissed, because of its Chomskyan connotations of innate grammatical rules and parameter setting.

Another group that has had difficulty with Chomsky's position is the evolutionary biologists. How, they reason, could a function as complex as language arise in a single mutational step? Other complex characteristics may look qualitatively distinct from one species to the next, yet all the evidence points to the conclusion that complexity is achieved only by gradual adaptation through selectional pressures (Deacon 1997).

Alternatives to a Grammar Gene

Neuroconstructivism

Many linguists maintain that Chomsky effectively debunked the idea that syntax could be acquired by any kind of known learning mechanism (Hornstein and Lightfoot 1981). There are, however, several counterarguments to this position. First, our understanding of learning mechanisms has progressed enormously over the past few decades, with connectionist simulations throwing new light on processes of statistical pattern extraction that have considerable relevance for language acquisition. A simple associative net cannot learn the kind of long-distance dependencies that are seen in syntax, but it no longer seems reasonable to argue that learning by neural networks is a logical impossibility. The question, rather, is what constraints need to be incorporated in the learning mechanism to enable it to learn syntactic regularities from a noisy and underspecified input (see Briscoe, this volume). Chomsky's arguments, and mathematical evidence of the unlearnability of syntax, made fundamental assumptions about *what* is learned and these merit closer scrutiny. In particular, they assumed that syntax is independent from meaning, and that the task for the learner is to identify rules that generate legitimate strings of syntactic elements but do not generate illegitimate strings. Chomsky's arguments for the independence of syntax from meaning were based on armchair experiments concerned with adult syntactic competence—not from observations of children learning language. The fact that a competent adult can judge "colorless green ideas sleep furiously" as a legitimate sentence and "green furiously sleep colorless ideas" as nongrammatical was used to demonstrate that the grammatical rules of syntax are quite separate from meaning. However, it does not follow that there is independence of semantics and syntax in a child acquiring language. All the evidence we have from child language learners indicates that knowledge of syntactic categories is not present at the outset, when semantic and pragmatic considerations dominate children's utterances and comprehension. Children appear only to infer syntactic categories after learning chunks of language in a piecemeal fashion (e.g., Pine and Lieven 1997; Tomasello 2000). An alternative approach to language acquisition known as Neuroconstructivism (see Elman et al. 1996) uses findings from connectionist simulations and analyses of normal and abnormal development to argue against innate specialization for syntax. Instead, Neuroconstructivists conclude that functional localization of language in the brain of adults develops in the course of learning, and that abstract linguistic rules emerge as statistical regularities are extracted from specific learned instances of meaning–form relationships. The Neuroconstructivists are not opposed to there being some constraints on learning, but they are opposed to the idea of innate knowledge:

> ...knowledge is not innate, but the overall structure of the network (or subparts of that network) constrains or determines the kinds of information that can be received, and hence the kinds of problems that can be solved and the kinds of representations that can subsequently be stored (Elman et al. 1996, p. 30).

They question, however, whether any domain-specific language learning mechanisms need to be postulated.

The Adaptationist Account of Language Origins

Although sometimes depicted by psychologists as a Grammar Gene advocate, Pinker explicitly questioned the plausibility of such a view on evolutionary grounds (Pinker and Bloom 1990; Pinker 2003). According to his account, we should not attempt to reduce the human specialization for language to a single novel attribute, such as knowledge of grammar. Rather, we should conceive it as the culmination of gradual evolution of a whole set of traits, which, acting together, give us our language ability. According to this view, syntax does not take center stage, but is one feature of a complex new ability that emerged in humans. It is perhaps ironic that Pinker is regarded as an opponent by the Neuroconstructivists, as there are many points in common between their views and his approaches: both stress that language capability grows out of, and depends upon, a wide range of underlying sensorimotor and conceptual capabilities. The principal difference between the two viewpoints is that Pinker regards this growth as having occurred in the course of evolution to yield a human brain that is uniquely prewired to facilitate language acquisition, whereas the Neuroconstructivists focus on the process of language acquisition in the individual, noting how it depends on the integrity of nonlanguage faculties, rather than being modular from the outset.

In 2002, Chomsky coauthored a review in *Science* (Hauser et al. 2002), which stated a view of language evolution that bore a close relationship to the Adaptationist account. In stark contrast to Chomsky's earlier writings, this piece stressed the value of comparative studies on human and animal communication systems. Furthermore, nowhere did it make any mention of innate grammatical rules or parameter setting; instead, three hypotheses were put forward, varying in the extent to which they incorporated a specialized language system. In the third hypothesis, which came closest to the old Grammar Gene idea, the notion of a human specialization for syntax was discussed in terms of the evolution of brain systems that can do computations involving recursion. Once we couch the issue in terms of computational ability, rather than innate knowledge of principles that operate on syntactic elements, the notion of specialization for syntax becomes much more acceptable to neurobiologists (cf. Elman et al. 1996). This Recursive Brain account is amenable to experimental investigation and integrates readily with a more general research agenda that is concerned with discovering cognitive primitives.

Theoretical Positions: Similarities and Differences

From this review of contemporary viewpoints we can see that the polarization between the Grammar Gene and Big Brain account is rather a caricature of the current state of debate. There appears to be widespread agreement on several points:

1. Language is "a new machine built out of old parts," rather than a qualitatively distinct cognitive ability that arose "*de novo*" in humans.
2. Language depends on a range of cognitive processes, many of which are likely to have parallels with animal cognition.
3. Primates do not have anything that resembles syntax, and complex syntax cannot be learned by a simple associative net, or by an intelligent and rigorously trained chimpanzee.

A key question is what kind of underlying neurological structures are needed to create a brain capable of generating and understanding language, especially syntax. These structures must differ between humans and other primates, and it is an open question whether the differences will prove to be merely quantitative or qualitative. Also open is how many such differences will be found between humans and primates, and what their genetic bases will prove to be. A further question, implicit in the debate between Neuroconstructivists and Adaptationists, is how far the language-processing structures depend on language input to develop their specialized architecture.

My understanding of the theoretical positions outlined above is that the principal differences between them lie in their preferred answers to these three questions. Table 9.1 summarizes my interpretation of various theoretical positions in the field and emphasizes that there are plenty of possible positions other than the polarized Grammar Gene vs. Big Brain accounts with which Chomsky began. Included in this table are my interpretations of the three hypotheses formulated by Hauser et al. (2002): Hypothesis 1 regards human language as strictly homologous to animal communication, with only quantitative differences in the neurobiological basis. Hypothesis 2 seems equivalent to the Adaptationist account and likens the evolution of language to the development of the vertebrate eye, having arisen as a consequence of natural selection. As Table 9.1 points out, although couched in very different language, there are parallels between the original Grammar Gene account and the newer Recursive Brain hypothesis, in that both regard syntax as special and requiring a qualitatively different neural substrate from other cognitive functions.

Table 9.1 divides theories according to whether they postulate qualitative or quantitative differences between humans and other primates, but the difference between these is not always as clearcut as it may seem. For instance, Hauser et al.'s (2002) hypothesis 1 is similar in many ways to the Neuroconstructivist approach, except that the latter assumes that qualitatively new properties arise as neurobiological systems become more complex. A quantitative difference at one level (e.g., in number of brain cells) might make possible a qualitative change at another level (e.g., in ability to hold two elements in mind while operating on them). As noted by Gilbert et al. (2005, p. 584) "once brain size and structural complexity surpassed a certain threshold...cognitive abilities might increase disproportionately with physical improvements of the brain." Also, we might have a purely quantitative difference in brain size, but if this

Table 9.1 Number of language-related neurobiological differences between humans and other primates specified by different theoretical accounts.

Theory	Qualitative differences	Quantitative differences
Big Brain	None	One = brain size (though could be regionally specific)
Hypothesis 1 (Hauser et al. 2002)	None	Several, e.g., mirror neurons subserving imitation; systems for vocal learning; systems for categorical perception
Neuroconstructivist (Bates 2004)	None	Several (see Table 9.2), whose interaction could lead to qualitatively new behaviors
Grammar Gene	One: innate syntactic knowledge	(Not discussed; only syntax of interest)
Adaptationist (Pinker 2003) Hypothesis 2 (Hauser et al. 2002)	Several, including ability to use bidirectional symbols, ability to handle recursion, phonemic categorization; cortical control over articulation	Several, including spatial, causal and social reasoning; complex auditory analysis; vocal imitation
Recursive Brain Hypothesis 3 (Hauser et al. 2002)	One: ability to handle recursion	Several, including the nonsyntactic aspects of language regarded as qualitatively distinct by the Adaptationist account

is regionally specific, then this could lead to a different style of cognitive processing emerging (Passingham 2008). In a similar vein, Passingham (2008) noted that as brain size increases, there are associated changes in cerebral microstructure, with dramatic increases in branching complexity and dendritic spines. He makes the interesting point that as the neocortex increases in size, it becomes difficult to maintain the same proportion of neuronal interconnections, and this might lead to development of a larger number of smaller processing areas, each with strong local connectivity. The distinction between theoretical positions becomes decidedly blurred once we become aware of the possibility of nonlinear relationships between underlying structure and emergent function (Elman et al. 1996).

A related issue that is much debated by Neuroconstructivists is domain specificity of cognitive functions (see Elman et al. 1996). The notion of domain specificity has its origins in Chomsky's early writings, in which he emphasized the discontinuity between grammar and other kinds of knowledge. Fodor (1983) developed this idea with his writings on modularity, arguing that encapsulated brain systems evolved to perform a specific cognitive function and do not participate in other cognitive operations. To say that something is domain specific is thus closely allied to saying that it is qualitatively distinct from other functions. Domain specificity is a *bête noire* of the Neuroconstructivists, who have noted that just because a cognitive function is localized in a particular brain region and operates in a modular fashion this does not mean it is innately specified. For instance, reading or playing the piano would both meet this criterion, yet it would clearly be nonsensical to conclude that the human brain has evolved an innate reading or piano-playing organ. I suggest that domain specificity may be a red herring in debates about language evolution, because if a function *is* domain specific, this does not prove that it is innately specified, and if it is *not* domain specific, this is perfectly compatible with contemporary Adaptationist theories. In any case, logically, domain specificity has to be viewed as a continuum, ranging from cognitive operations that are implicated only in syntax and nothing else, to cognitive operations implicated in all types of mental processing. Most cognitive operations will fall somewhere between these extremes. It could be argued that, rather than having sterile debates about whether something is or is not domain specific, we should concentrate on specifying which linguistic and cognitive processes use overlapping neural circuitry

Evidence from Genetic Studies of Language

The condition of specific language impairment (SLI) is of particular interest when testing between theories, insofar as pathology may open a window on what happens when the normal genetic mechanisms go awry. SLI is diagnosed when a child's language acquisition is impaired in the context of otherwise normal development. As noted by Marcus and Rabagliati (2006, p. 1227), "developmental disorders are particularly well placed to yield insight into evolution of language by providing insight into both halves of the equation: that which is unique to language, and that which is not." The different theoretical accounts outlined above can to some extent be differentiated in their predictions about SLI.

Neuroconstructivists argue that any linguistic deficit in SLI (or in other developmental disorders) can be traced back to impairment of a system that is implicated in functions other than language. Bates (2004), for example, suggested that the capacity for language might be explicable in terms of the unique conjunction of sensorimotor, attentional, and computational skills that differentiates humans from other primates. Table 9.2 shows her proposals for necessary and sufficient prerequisites for language acquisition in humans. Two points are worth noting: First, all these abilities are present in nonhuman primates; humans differ only in having what Bates refers to as "exquisitely well-tuned" abilities in infancy. Second, the list does not include any special mechanism for acquisition of syntax.

The Adaptationist account would appear to make the specific prediction that we should be able to identify several genes that (a) differ between humans

Table 9.2 Functional infrastructure for language, and postulated impairments in developmental disorders, based on Bates (2004).[1]

	SLI	Autism	Williams syndrome	Down syndrome
Object orientation: peculiar fascination with small objects	intact	intact	intact	intact
Social orientation: tendency to orient towards faces and voices	intact	impaired	intact	intact
Cross-modal perception/sensorimotor precision/short-term memory	impaired	intact	intact	impaired
Computational power: rapid statistical induction	intact	variable	impaired	impaired

[1] Bates differentiates cross-modal perception and sensorimotor precision in her list of language prerequisites, but they are not clearly distinguished in her discussion of developmental disorders, so have been condensed here. Short-term memory is included here as it is mentioned as a key deficit in SLI and Down syndrome.

and primates, (b) have reached fixation in humans (i.e., do not vary in their form except in cases of pathology), and (c) can be associated with language impairment in cases of mutation. As Pinker pointed out, this does not mean that one should never find any other deficits associated with SLI; the genes associated with SLI might be expected to have pleiotropic effects on other systems. However, one should be able to identify allelic variations that have a distinct and disproportionate effect on language development.

The Recursive Brain hypothesis, like the Adaptationist account, would anticipate a range of genetic variants linked with language impairment, but would also predict the existence of some children who have a specific difficulty with recursion, and who will therefore have problems with complex syntactic operations. Let us now turn to examine the empirical evidence for these different positions.

The KE family and FOXP2

The KE family, whose pedigree is illustrated in Figure 9.1, has been described as presenting evidence for a Grammar Gene theory, but this interpretation has been hotly contested. The first published account of this family (Hurst et al. 1990) noted that affected members had a severe speech disorder, and commented on the striking pattern of inheritance that was consistent with a single defective autosomal dominant gene transmitted in Mendelian fashion. Subsequently, Gopnik (1990) drew attention to the grammatical deficits of affected members of the KE family, noting that these were not explicable in terms of low IQ or hearing difficulties, and concluding that these provided clear evidence for a genetic basis to syntactic knowledge. Researchers subsequently isolated a single

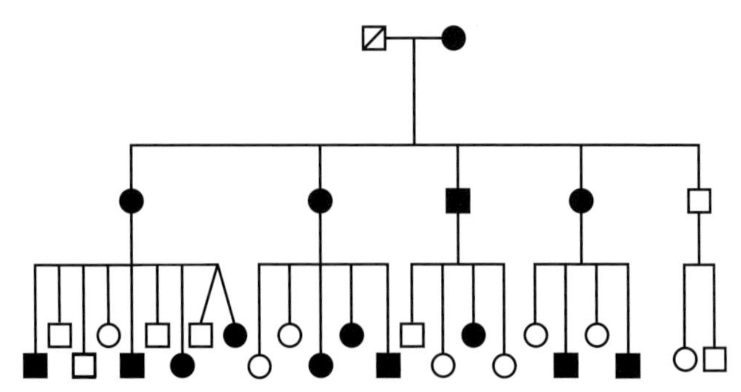

Figure 9.1 Pedigree of the KE family. Black symbols denote affected cases, and white symbols unaffected cases. Males are shown as squares, and females as circles. The first generation (an affected mother and unaffected father) are shown in the top row. The second row shows their five children, four of whom were affected. The third row shows the grandchildren, 10 out of 24 of whom were affected.

base mutation on chromosome 7 in a gene known as FOXP2. All affected individuals and no unaffected family member had this mutation; furthermore the mutation was not found in a large sample of normal adults who were screened, nor was any abnormality in FOXP2 detected in another sample of individuals with SLI who were unrelated to the KE family (for a review, see Fisher 2005; Marcus and Fisher 2003).

A comparative study of FOXP2 by Enard et al. (2002) identified only three differences between mouse and man in around 700 base pairs. Of particular excitement was the finding that two of these changes had occurred after the separation of lineages of chimpanzee and human some 5 to 6 million years ago. Studies of DNA diversity in regions of the genome adjacent to FOXP2 led to an estimate that the current form of FOXP2 became fixed in human populations around 200,000 years ago, consistent with estimates of when language first made its appearance.

So, does the KE family provide evidence for a genetic specialization for syntax? FOXP2 does seem to fit all three criteria for a gene specialized for language: (a) it differs between humans and primates; (b) except in cases of mutation, it takes a constant form in humans; and (c) language impairment occurs when there is mutation that affects function of the gene. Nevertheless, there have been arguments about the interpretation of this evidence, centering primarily on the conceptualization of the phenotype. Two related arguments have been put forward: the effects of the gene are not specific to syntax, and the language impairments in this family are explicable in terms of nonlinguistic factors.

The first point is clearly true. Watkins et al. (2002) noted that affected individuals did indeed have syntactic limitations, but they also had marked impairments of speech production with some associated orofacial dyspraxia (as evidenced by difficulties in imitating facial movements such as blowing out the cheeks and smacking the lips). Nonverbal IQ was lower in affected than unaffected individuals, though within normal limits in several of those who were affected. Furthermore, molecular studies indicate that FOXP2 is a regulatory gene that controls the expression of other genes which affect development of multiple systems, including heart, lungs and brain, and is highly conserved across species. I would argue that such evidence convincingly refutes the old Grammar Gene theory, but since nobody currently is advocating that theory, this is somewhat beside the point. The Adaptationist and Recursive Brain theories both explicitly argue that any gene involved in language will have evolved from a gene that is present in other animals and that such a gene is likely to serve many functions. The issue is not whether FOXP2 is involved in functions other than language, but whether it plays a key role in building a brain that can process language. Gene expression studies show that FOXP2 is involved in several brain regions, including some that are known to be important for language in humans (Fisher 2005).

Neuroconstructivists would argue that the KE family's deficits can be explained without postulating any innately specified language mechanism. The argument goes something like this: language is a highly complex function, requiring interaction between multiple systems; a complex system can go wrong for all sorts of reasons. An analogy can be drawn with the single-gene disorder, Duchenne muscular dystrophy. Affected children have great difficulty in walking, but we do not conclude that the mutated gene is a "gene for bipedal walking." Rather, we note that the mutation affects muscle function, with the large muscles in the legs being the first to show clear pathology, leading to weakness and functional incapacity. Similarly, a fairly low-level sensory or motor deficit could be responsible for the observed phenotypic profile in the KE family, without invoking any specialized language genes. However, if we are going to adopt this line of argument, then it behooves us to identify the underlying low-level nonlinguistic impairment that is responsible.

A key issue concerns the relationship between the linguistic difficulties and the problems with speech motor control. There are three ways of explaining this association. One possibility, compatible with Neuroconstructivism, is that speech motor problems make it hard for affected individuals to produce sentences and hence their syntactic skills are impaired (i.e., there is a peripheral explanation for the observed linguistic deficits). This is not a very satisfactory account, because we know that a peripheral difficulty with speaking does not by itself compromise syntactic development: a child who is anarthric (unable to speak because of damage to motor systems controlling the articulators) can do well on the kinds of test of syntactic comprehension that are failed by affected individuals of the KE family (Bishop 2002a). A second possibility is that there is a causal link between speech motor function and syntax, but this is fundamental rather than superficial. According to this view, the same brain systems that are important for motor sequencing are also implicated in syntax: thus, if they are disrupted, one will see both motor and syntactic deficits. This kind of argument fits with theories that see syntax as having evolved from motor specialization for speech and is entirely consistent with the Adaptationist view. It also meshes well with a theory proposed by Ullman and Pierpont (2005), who argue that humans have a neurobiological system responsible for procedural learning that is involved in automatization of motor skills and in mastering aspects of language form such as syntax and phonology. The main argument against this theory is that it would seem to predict a wider range of handicaps in the affected KE family members than is actually observed, given that skilled motor actions depend on the procedural learning system. For instance, Watkins et al. (2002) found no deficits on limb praxis in affected individuals. A final possibility is that the orofacial motor impairments seen in the KE family are associated deficits, not causally linked to the language deficits. Most genetic mutations are pleiotropic (i.e., they affect multiple systems) and thus it is not unreasonable to suppose that a defective gene will have a range of effects on development.

FOXP2 is not the only gene involved in language, and its influences extend well beyond the language regions of the brain. The role of this gene in speech production seems less controversial than its role in syntactic skill. Further studies using both linguistic and nonlinguistic tasks are needed to establish the nature of problems with recursion, syntax, and procedural learning in affected individuals, but so far I am not convinced by arguments that the syntactic deficits are simply secondary consequences of speech problems. Understanding how FOXP2 affects neurodevelopment will be an important step in throwing light on language-relevant differences between brains of humans and other primates.

Beyond the KE Family: Other Cases of SLI

Over the past twenty years, evidence has been mounting to suggest that SLI is a highly heritable disorder (Bishop 2002b). However, as noted above, genetic analysis of samples of people with SLI has failed to find any evidence of mutation in FOXP2. Furthermore, although SLI runs in families, one does not usually find the kind of clearcut Mendelian pedigree that characterizes the KE family. Rather, SLI is a complex disorder that aggregates but does not segregate in families.

There are several ways to approach the genetics of this kind of disorder. One possibility is to assume multifactorial causation, with many genes and environmental factors of small effect conspiring to depress language ability. This view of SLI was proposed by Plomin and Dale (2000), who suggested that rare mutations account for only a handful of cases of SLI, and that the disorder is best conceptualized as corresponding to the tail of a normal distribution of language ability, influenced by multiple genes of small effect (quantitative trait loci) rather than single mutated genes. Insofar as this is the case, the discovery of genes implicated in SLI is unlikely to throw light on the origins of human syntax. Consider an analogy with height: height in the human population is influenced by a host of small genetic and environmental influences; identifying allelic variants that lead to tall or short stature in individuals is unlikely to tell us anything about why, on average, humans are considerably taller than chimpanzees.

The kind of data reported by Plomin and colleagues do show that for many children, poor language development seems to differ from normal acquisition in quantitative rather than qualitative terms, and that whatever depresses language ability has a fairly global influence on many facets of both verbal and nonverbal development. To some extent the pattern of results may be seen as supportive of the Neuroconstructivist account in showing that there is nothing all that special about language. A supporter of an alternative account might argue, however, that such evidence in no way disproves the idea of genetic specialization for language, because the kinds of individual variation that are measured by psychometric tests of language ability may not be the crucial ones for understanding origins of language. If one accepts Chomsky's argument that (excepting cases of pathology) humans do not differ in their fundamental syntactic abilities, then such variation as can be seen in human children is bound to be relatively trivial, reflective of different rates of maturation or performance factors. The key question for those who postulate genes important for syntax is whether one can identify a subset of children who are qualitatively rather than just quantitatively different from their peers, in terms of lacking some aspect of syntactic competence.

A program of research headed by Mabel Rice and Kenneth Wexler is relevant to this question, as it focuses on one area of expressive syntax: morphological marking of finite verbs (for a review, see Rice 2000). A finite verb is one that is marked for tense and number with a grammatical morpheme such as past tense *-ed* or third person singular *-s*. In English, for example, the verb *go* is finite in a sentence such as *John went to school* or *he goes to school*, but is nonfinite in sentences such as *I want John to go to school*, or *I saw John go to school*. Before the age of about 4 years, typically developing English-speaking children do not reliably mark finite verbs and may say *John go to school* rather than *John goes to school*; after this age, verb morphology is usually mastered, with very few errors being made. In individuals affected by SLI, however, problems with this aspect of syntax are often seen well beyond 4 years of age. The problem is not just one of formulating utterances, because similar limitations are seen in children's grammaticality judgments (Rice et al. 1999). Rice and colleagues interpreted this pattern of results as evidence for a failure to set a specific parameter: one that specifies that finite verb marking is obligatory in English. Accordingly, mastery of finite verb marking is determined by biological maturation, with only a minimal amount of language exposure being needed to trigger the correct setting. However, in SLI, maturation of the relevant module is delayed, and children continue to treat the marking of tense and agreement on finite verbs as optional. Rice (2000) reviewed evidence for this Extended Optional Infinitive (EOI) theory, noting that it was supported by the distinctive pattern of grammatical errors seen in children with SLI, and by the lack of correlation between their use of verb morphology and vocabulary development. She noted that children who have problems with verb morphology often have other family members with language difficulties, and suggested impaired verb morphology might act as a phenotypic marker of heritable language disorder.

A rather different view of specific syntactic deficits has been proposed by van der Lely (2005), who has amassed evidence over several years for a specific subgroup of children who have, what she terms, grammatical SLI. These children are identified in terms of having grammatical deficits that are disproportionate not just to their nonverbal abilities but also to their other language abilities. Thus, while they do have deficits on tests of vocabulary or verbal memory, these are less striking than the impairments on tests of grammatical production and comprehension. The deficits described by Van der Lely include problems with marking verb morphology, but also encompass much broader difficulties with identification of thematic roles in complex sentences, and application of Binding Principles. Van der Lely (2005) interpreted these deficits in terms of a domain-specific deficit in computational syntax and noted that grammatical SLI runs in families in a way that is compatible with autosomal dominant inheritance.

These approaches to characterizing the phenotype of heritable language impairment have been highly controversial, with several researchers offering alternative interpretations of the data that challenge an account in terms of syntactic modularity (Freudenthal et al. 2007; McDonald 2008; Norbury et al. 2002; Tomblin and Pandich 1999). As argued above, however, debating whether deficits are domain specific or not may be of less importance than establishing whether there are genetic variants that are associated with relatively selective syntactic deficits. Thus, for instance, even if heritable problems with grammatical acquisition were due to processing limitations rather than domain-specific linguistic deficits, this would still be informative for our understanding of evolutionary origins of language. The evidence from family aggregation is not clearcut, because family members share environments as well as genes. To separate genetic and environmental contributions to individual variation, one needs a genetically informative design.

Twin Studies of SLI

Because monozygotic (MZ) twins share all their genetic material, whereas dizygotic (DZ) twins share, on average, 50% of segregating alleles, one can estimate the genetic contribution to a trait by seeing how far the correlation between twins and their co-twins differs in relation to zygosity. Twin studies have shown that there is higher concordance for SLI in MZ than DZ twins, but these studies have based the diagnosis of SLI on clinical tests that do not allow us to study separate components of language functioning (for a review, see Bishop 2002b). In my own program of research, I have moved away from the global clinical category of SLI to study the role of genetic influences on underlying cognitive processes implicated in this disorder. In this work, I have used a modification of the twin study method developed by DeFries and Fulker (1985). The starting point of the method is to take a quantitative measure of language impairment and identify as probands those children whose scores fall below a certain cutoff (e.g., the lowest 15% of the population). Suppose that performance on the language test were simply determined by random factors, so that two members of a twin pair bore no resemblance to one another. (In behavior genetics terminology, the trait is influenced solely by a non-shared environment). In that case, the mean score of the co-twins of probands would be at the population mean, and one could not predict a co-twin's score from that of the proband. Now consider an alternative scenario, whereby performance on the language test is solely determined by environmental factors shared by the two members of a twin pair (e.g., the amount of TV exposure at home or the quantity of maternal talk to the children in infancy). In this case, we would be able to predict a co-twin's score from that of the proband, regardless of zygosity. The average co-twin score would be impaired; indeed, in the theoretically implausible case where such environmental influence were the *only* factor affecting test scores, then means for probands and co-twins should be equivalent. The third possibility to consider is that ability to do a language task is solely determined by genetic makeup. In that case, probands and co-twins would be identical in the case of MZ twins, but only 50% similar for DZ twins. Using this logic, DeFries and Fulker (1985) showed that one could obtain estimates of the relative importance of genes, shared environment, and non-shared environment in determining whether a person had a deficit by doing a regression analysis in which one predicts the scores of co-twins from the scores of probands and a further term denoting the genetic relationship between the twins (1.0 for MZ and 0.5 for DZ). Figure 9.2 shows that, after transforming the data so that the mean proband score is 1.0 and the population mean is zero, one can get direct estimates of heritability of impairment by a comparison of mean scores for MZ and DZ co-twins. Heritability estimated in this way is known as group heritability (h^2_g), and is a measure of the extent to which differences between impaired and unimpaired children are caused by genetic variation.

Phonological short-term memory as a behavioral marker of heritable SLI. In my first studies using DeFries–Fulker analysis, I focused on two non-syntactic skills that had been postulated as underlying deficits in SLI: auditory temporal processing and phonological short-term memory. A low-level auditory perceptual deficit was first proposed by Paula Tallal and her colleagues as a key deficit in SLI (Tallal 2000). Elman et al. (1996) and Bates (2004) explicitly suggested that such perceptual impairments could cause downstream syntactic deficits. An alternative theory was advanced by Gathercole and Baddeley (1990), who maintained that the core deficit was not in auditory perception, but rather in a short-term memory (STM) system that was specialized for retaining

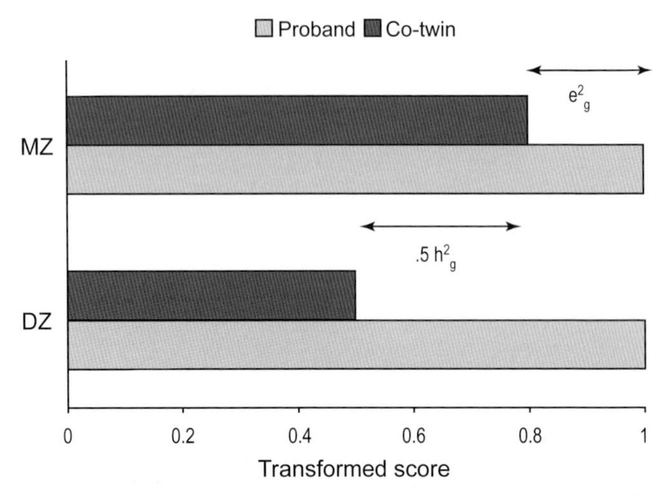

Figure 9.2 Illustration of DeFries–Fulker analysis. Data are transformed so that the population mean = 0 and the proband mean = 1. The effect of unique influences (non-shared environment) on impairment (e^2_g) is estimated as MZ proband mean – MZ co-twin mean. The effect of genes on impairment (h^2_g) is twice the difference between MZ and DZ co-twin means. The effect of shared environment (c^2_g) is $1 - h^2_g - e^2_g$.

phonological sequences for brief periods of time. A popular way to assess phonological STM is by asking the child to repeat nonsense words of increasing length (e.g., hampent, dopelate, perplisteronk). Gathercole and Baddeley (1990) showed that school-aged children with SLI could do this task provided the nonsense words were no more than 2 syllables in length; this indicated that they had the ability to perceive speech sounds and program the articulators to produce them in a novel sequence. However, as syllable length increased, the difference between SLI and typically developing groups was magnified, which suggests that the problem is in maintaining a sequence of novel sounds in memory. Baddeley et al. (1998) went so far as to suggest that phonological STM may have evolved to support language learning; note that their proposal is for a domain-specific system, but one that is very different from Chomksy's "language acquisition device," in that it has no syntactic content.

A twin study using measures relevant to both theoretical accounts gave sharply contrasting findings for the two measures (Bishop et al. 1999). Auditory temporal processing did distinguish children with SLI from unaffected cases, but there was no evidence of any genetic influence. Twins tended to resemble one another, regardless of zygosity, indicating that the similarity was due to shared environment, rather than to genetic similarity. Deficits in phonological STM, on the other hand, were highly heritable, and showed no significant influence of shared environment.

Marking of verb inflectional morphology. Bishop (2005) studied children's use of verb inflections for past tense and third person singular in a sample of 173 pairs of 6-year-old twins, selected to be over-representative of children

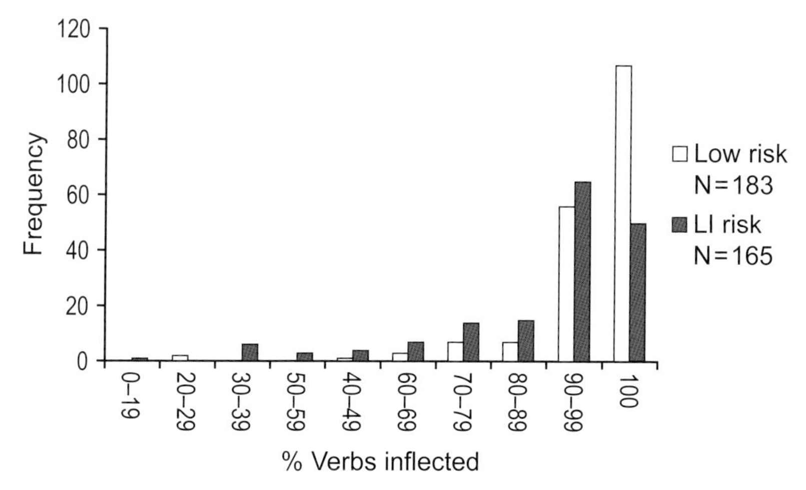

Figure 9.3 Percentage of inflected verbs produced by 6-year-old twins on a test designed to elicit past tense and third-person singular endings. Frequencies are shown separately for a subset of children who were identified at 4 years of age as being at risk of language impairment (LI risk) and from a comparison group of low risk twins. See Bishop (2005).

with language impairments. As predicted by Rice (2000), scores were strongly skewed, with most children near ceiling, but a minority omitting tense markings (see Figure 9.3). The DeFries–Fulker analysis gave high estimates of group heritability for impaired verb marking. The more extreme the cutoff used to select probands, the higher the heritability estimate. In a series of simulations, Bishop (2005) showed that the pattern of results was consistent with a model in which an allele of a single major gene of large effect depressed ability on this task. The prevalence of the deficient genotype in the population was estimated at between 2.5–5%.

Phonological short-term memory and syntax. The finding of high heritability for both nonword repetition (Bishop et al. 1999) and for verb morphology (Bishop 2005) raised the question of whether phonological STM is a key cognitive skill implicated in syntax acquisition, as suggested by Baddeley et al. (1998) and Bates (2004). It is noteworthy that poor nonword repetition is a correlate of syntactic impairment in both the KE family (Watkins et al. 2002) and the G-SLI cases described by Van der Lely (2005).

To test this possibility, Bishop et al. (2006) analyzed data from a nonword repetition task and a battery of other language tests, obtained with the sample of 6-year-old twins studied by Bishop (2005). Three points emerged from this study:

1. When working with children this young, it is important to be aware of two factors that contribute to nonword repetition performance. Some children did poorly on the test because they did not articulate the items

accurately, even when short nonwords were used. (This contrasted with data from older children, who tend to score close to ceiling on short nonwords.) Others could repeat two-syllable nonwords, but their performance deteriorated when longer sequences were used. If just the overall score on the test was used, then heritability of poor performance was unimpressive. However, if a measure which gave a purer index of memory was used, based on the mismatch between performance on short and long nonwords, then this was highly heritable ($h^2_g = .61$).

2. Apart from nonword repetition, only language measures with a syntactic component were significantly heritable. Group heritability was high not only for the test of verb morphology, based on Rice and Wexler's work ($h^2_g = .74$), but also for a measure of syntactic comprehension ($h^2_g = .82$). In contrast, vocabulary showed no genetic influence and appeared largely determined by environmental factors common to both twins. Poor performance on the syntax tasks was not due to low nonverbal IQ, and heritability estimates were unaffected when the analysis was re-run excluding children with significant articulation problems.
3. The most striking finding emerged when bivariate DeFries–Fulker analysis was conducted to see whether these different measures of language impairment had common genetic origins. There was no evidence of any genetic overlap between deficits in phonological STM and the two syntax measures; estimates of bivariate heritability were close to zero. However, there was evidence to suggest that the two syntax measures (receptive syntax and verb morphology) had genetic influences in common, indicating that any factor influencing syntactic functioning may exert an effect on a range of grammatical features, and not just verb morphology (cf. Van der Lely 2005).

Overall, the behavior genetic analysis indicated that there are two good linguistic markers of a heritable phenotype in SLI: measures of syntax (including marking of verb morphology) and phonological STM. Both do a good job of discriminating between cases of SLI and typical development, and both give high estimates of group heritability. They are not, however, different measures of the same underlying trait, but rather appear to be independent deficits.

Molecular Genetic Studies of SLI

We are still a long way from identifying genes that are associated with the heritable linguistic deficits identified by twin studies. Newbury and Monaco (2008) provide a useful review of what is known to date. These authors have worked with the SLI Consortium, which focuses on three main measures of the phenotype: scores from expressive and receptive composites of the Clinical Evaluation of Language Fundamentals, a widely used clinical assessment, and a test of nonword repetition, previously shown to show particularly high heritability in twin studies by Bishop et al. (1996, 2006), and regarded as a measure of phonological STM. Linkage was found between a region on the long arm of chromosome 16 and the nonword repetition phenotype, and between a region on the short arm of chromosome 19 and the expressive language score (SLI Consortium 2002). Both linkages have been replicated in additional samples, though the specific language traits linked to chromosome 19 are not consistent from study to study (SLI Consortium 2004). Interestingly, Falcaro et al. (2008) found that chromosome 19 had significant linkage to a measure of use of verb morphology similar to the one used by Bishop et al. (2006). Neither of the SLI linkage sites was, however, found to be significant by Bartlett et al. (2002), who instead reported significant linkage to chromosome 13. Lack of agreement in results of genome scans is not uncommon, and raises concerns about false positives, despite statistical attempts to control for these. However, other explanations are plausible: different methods of sampling, of phenotypic measurement, or statistical analyses. In general, these studies emphasize that most genetic influences are small in size and quite unlike the large effect seen with a mutation of FOXP2.

Implications of Genetic Data for the Evolution of Syntax

Chomsky emphasized that all normal humans had syntactic competence. In this chapter, the focus has been on those children who fail to develop normal syntactic competence, but it is worth stressing that even in the severely affected members of the KE family, we do not see people with *no* syntax, but rather people with *impaired* syntax. The abilities of young people and adults with specific grammatical deficits fall well below age level, but they are nevertheless typically well above that of a 2-year-old child, which is the level of comprehension achieved by Kanzi, the star of primate language learning. This in itself suggests that syntax is not a monolithic skill, but rather involves a complex set of interacting components, only some of which may be deficient in those with SLI.

How, then, should we proceed in our quest for the genetic origins of SLI? If our main goal is to study the etiology of SLI, then it makes sense to continue to work with traditional language measures that yield a normal distribution of scores, and to look for genetic correlates of scores in the tail of the distribution, because it seems that most cases will correspond to the tail of a normal distribution of language ability. If, however, our goal is to identify mutations that may have been important for the transition from a pidgin status to complex language, then we may make more progress if we focus on analyzing the SLI phenotype to identify those heritable deficits that are *not* part of normal variation, which may have their origins in rare allelic variants. To date we know of one genetic mutation, in FOXP2, that affects both speech and syntax. Furthermore, strong linkage to chromosome 16q for nonword repetition creates optimism that we may find another locus strongly associated with deficient phonological short-term memory (SLI Consortium 2004). The behavioral analyses suggest that yet another gene may be associated with pure syntactic deficits affecting verb morphology, and a preliminary linkage study indicates this has a different locus from the nonword repetition deficit (Falcaro et al. 2008). The evidence to date is very preliminary but it suggests that data from children with SLI will add to the evidence that humans have evolved not just one brain specialization for language, but many, which act together to give a robust communicative system.

A final intriguing question is whether a child who had SLI in English would have developed SLI if Chinese was the native language. Alas, this is likely to remain an armchair experiment! It does, however, get to the heart of the question of whether genetic influences are general for syntax (in which case the ambient language would be immaterial), affect acquisition of specific components of syntax (in which case a child with poor ability to master verb morphology would do fine in Chinese), or affect nonlinguistic perceptual or motor skills that impact on language (in which case the perceptual or motor demands of the language will affect whether SLI is manifest). If we do identify genes that are associated with risk for different SLI phenotypes in English, it will be of considerable interest to study their impact in individuals learning languages with very different grammatical and perceptual characteristics.

Acknowledgment

The author is supported by a Principal Research Fellowship from the Wellcome Trust.

Overleaf (left to right, top to bottom)
Szabolcs Számadó, Stephanie White, Kazuo Okanoya
Dorothy Bishop, Terry Deacon, Eva Hoogland
Group discussion, Eörs Szathmáry
Francesco d'Errico, Julia Fischer, James Hurford

10

What Are the Possible Biological and Genetic Foundations for Syntactic Phenomena?

Szabolcs Számadó, James R. Hurford, Dorothy V. M. Bishop, Terrence W. Deacon, Francesco d'Errico, Julia Fischer, Kazuo Okanoya, Eörs Szathmáry, and Stephanie A. White

Abstract

Syntax is a system for forming complex signals and mapping these signals onto conceptual/intentional representations. This system is unique to humans, and no animal species has a system of comparable complexity. The evolution of any complex organ can present a serious puzzle and the issue of the evolution of syntax is even more complex, given that one has to consider the interplay of three major systems: biological evolution (and thus inheritance), cultural transmission, and individual learning. To complicate things further, the comparative method—the basic tool of biologists—cannot be readily used given the uniqueness of human language. Yet, it still can be useful to start investigations from the bottom up. Accordingly, a survey of knowledge about the syntactic abilities of animals is made. Then a brief summary of knowledge on how and what genes have an influence on syntax or on language, in general, is provided. Potential evolutionary constraints are discussed as well as some potential mechanisms for the evolution of syntax, like the Baldwin effect, genetic assimilation, masking, and unmasking. Finally, a review is provided of the fossil and archaeological evidence that may be relevant to any detailed scenario considering the evolution of syntax and that might help to formulate testable predictions.

Introduction

Discussions in our group were based on the understanding of a number of shared starting viewpoints. First, our focus was on spoken language, the primary natural form of language. Written representations of language can be deceptive, especially in matters such as segmentation into words. Second, English should not be assumed to be a typical natural language. There is enormous variety and diversity in languages, and many language structures differ significantly from those exemplified by English. Third, the classic distinction between competence and performance is indispensable, but our stance on this differs somewhat from that of Chomsky, who first drew attention to this contrast. Competence is the tacit knowledge of an adult native speaker of his/her language. It determines the form of sentences and relations between form and meaning that a speaker can produce and understand. Performance is determined by two types of factors: accidental and constant, or permanent. Examples of accidental factors are distraction, drunkenness, and sporadic confusion. Constant performance factors include short-term memory and processing speed, which do not change significantly from utterance to utterance. In this sense, and as conceived by Chomsky, the causal relationship between competence and performance is one-directional: Performance factors affect the realization of well-formed sentences delivered by competence, but competence itself is not affected by accidental factors. We would argue that this viewpoint is unduly influenced by an idealized view of the adult language user. In development, it seems likely that the formation of adult competence is constrained by performance factors, some operating during ontogeny, others throughout a lifetime. We would therefore argue against a marginal role for performance factors, but instead suggest there can be a causal relationship from performance factors to competence that is not commonly recognized.

A problem for anyone attempting to investigate the biological bases of syntax is the plethora of different conceptualizations of syntax. There are several theoretical approaches to the problem of describing the complexity of syntax, and they range from complex context-free generative grammars to structurally simpler cognitive and discourse-based grammars. The dominant paradigm for over a generation has been the generative approach (e.g., Generalized Phrase Structure Grammars, GPSG). Recently, there has been a return to more Minimalist approaches (e.g., Culicover and Jackendoff 2005).

It might seem tempting to favor intuitively simple models of syntactic structures, especially in a biological context, because Minimalism offers attributes that are more easily assimilated into biological paradigms, as it makes less exotic claims about what must be contributed by unique human characteristics. However, the appeal of apparently very simple ways of putting things may be misleading. Certain types of complexity are indisputably inherent in the structure of natural languages, so that any model aiming at empirical adequacy must express them one way or another. Consider, for instance, the set of phenomena that linguists often refer to under the rubric of "movement." It is sometimes said that certain models are simpler because they avoid movement operations, but the movement/non-movement debate tends to hide the fundamental fact that any model trying to deal with natural languages must express the fact that there are long-distance dependencies in language. For instance, in interrogative constructions like *What book do you think I should read* ___ *?* one must express the fact that the phrase *what book* is understood as the thematic object of *read*, even if it is pronounced in a position which can be quite distant from *read*. Thus, any theory must possess a device to establish a relation between two positions, let us say an antecedent and a gap (but many other terminologies can be used), which can be indefinitely far away. Technical discussions on the exact formal nature of the device (e.g., classical generative grammar and government and binding approaches involve "displacement" from one position to another in the tree; GPSG and Minimalism do not) should not obscure the simple fact, which is generally accepted, that natural languages admit long-distance dependencies, and that any model must include a device to capture them.

Similarly, certain models look simpler in that they use very shallow, essentially flat tree structures (e.g., a ternary tree for clauses: [SVO]) rather than more articulated and "deeper" binary structures like [S[VO]]. Again, the "simpler" character of flat structures may be misleading if we look at the whole picture. If a model is intended to express properties of natural languages, it must state somewhere that subjects are systematically more prominent than objects, in that, for instance, a subject can bind a reflexive in object position but not vice versa; in fact, there are innumerable examples of such asymmetries between subjects and objects involving anaphoric and pronominal dependencies, the licensing of polarity items, agreement phenomena, etc. One approach is to express the asymmetry directly in the tree structure with the binary branching representation [S[VO]]. If a model uses flat structures, it must express the asymmetry in some other component, which shifts the question to what notation is the most perspicuous and useful, but the fact remains that the asymmetry must be expressed somewhere. Thus, to consider a flat model "simpler" can be misleading: the "simplification" in the tree structure forces the complication of some other component.

A note is in order about the role of Minimalism and evolution in general. Biological phenomena did not appear as the result of rational design, but rather due to tinkering. This means that components of the preexisting machinery, which may also serve other functions, have been recruited and modified to serve new functions, often in an idiosyncratic temporal order. No engineer would have designed crossing food and air channels between the upper vocal tract and the chest. This is a legacy of evolutionary tinkering. By the same token, genetic regulatory networks, so basic to complex traits, are by no means minimal. The moral is that formal Minimalist accounts of language may misguide us when we treat the historical complexification of language. However, unnecessary expressive power in theories is equivalent to excessive degrees of freedom, and weakens the theory, just as much as inadequately expressive power.

Another issue on which theories diverge is the mapping between syntax, on the one hand, and semantics and processing, on the other. It is important to realize that the emphasis in theoretical linguistics on syntactic theory does not imply contempt for, or ignorance of, the way such theories map onto semantics and processing. In fact all syntactic theories, from Chomsky (1965b) on, have been applied to sentence processing, and all of them come equipped with a compositional semantics.

The reason why syntacticians have not talked much about these interfaces is simply because they are painfully aware that the processors and semantics are the least psychologically plausible aspects of the theory. In particular, there is general agreement that psychologically real semantics is biologically embedded in ways that are mostly unknown. Researchers working in this field hope that neurocomputation and machine learning will deliver something better, so that they can cheerfully abandon Montague semantics. Meanwhile, syntax is presently seen as the best guide for the form that a biologically grounded theory of meaning will take, for the simple reason that evolution and child language acquisition could hardly allow syntax to be other than simply and compositionally related to semantics.

In general, certain differences in analytic details tend to be overemphasized by linguists compared to nonlinguists, giving the impression of a proliferation of syntactic theories and frameworks that are fundamentally different and inconsistent with each other. While there is genuine disagreement on how best to analyze certain phenomena, this should not obscure the fundamental agreement on a core of basic syntactic generalizations that has emerged over the last half century of syntactic research with precise formal tools. Identifying a consensus on the core syntactic structures would be particularly useful for promoting interdisciplinary collaborations to explore the cognitive and biological bases of language.

Syntax is a system for forming complex signals and mapping these signals onto conceptual/intentional representations. This system is unique to humans, and no animal species has a system of comparable complexity. Assuming that human language is like any complex organ evolved by natural selection (Pinker and Bloom 1990), one would expect language to have some sort of genetic profile. Yet we know very little about how and what genes have an influence on syntax or on language in general. (Present knowledge is reviewed in the section, Genetic Influences on Syntax). Thereafter, the biological grounding of syntactic theory and the potential constraints that such grounding can offer are discussed.

The evolution of any complex organ can present a serious puzzle, and we spent considerable time discussing the evolutionary processes that can render such evolution more probable (see section, Evolutionary Mechanisms for the Evolution of Syntax). Relatively simple processes, such as exaptation and self-organization, are highlighted first. The issue of the evolution of syntax is even more complex, however, given that one has to consider the interaction of three major systems: biological evolution (and thus inheritance), cultural transmission, and individual learning. The interplay of these systems is crucial to the evolution of syntax. There are at least two sources of regularities and generalizations in syntax: historico-social processes, including grammaticalization, and innate biases affecting language acquisition during ontogeny. Some intriguing interactions, like the Baldwin effect, genetic assimilation, masking, and unmasking are discussed. Finally, any serious theory of syntax evolution must be able to give testable predictions that can be linked to the paleontological record. While this may sound a daunting task, it cannot be avoided. We conclude with a review of the fossil and archaeological evidence that may be relevant to predictions about the evolution of syntax.

Syntax in Animals?

To shed some light on the evolutionary roots of syntactic abilities, researchers have examined the vocal utterances of several species. Candidate species include birds and some whale species, which are known to produce elaborate vocalizations and which can be described in terms of rules, as well as nonhuman primates because of the specific interest in shared features of nonhuman and human primate communication.

General Characteristics of Animal versus Human Communication

A striking feature of most, if not all, animal communication is the lack of a symbolic structure (Deacon 1997). Most of the complexity in animal communication can be explained by the fact that listeners are apt at extracting information from signals, while the sender does not always intend to provide that information (Seyfarth and Cheney 2003; Fischer 2008). Further analyses of the structure of animal communication need to take into account that both the acquisition and the performance of vocal behavior differs substantially between different taxa. In terrestrial mammals, the structure of the utterances is generally considered to be innate, while songbirds have to learn (based on innate biases) their species-specific songs. Some animals produce series of repetitions of the same sound (e.g., the croaking of a frog), whereas others utter strings of different notes, often composed into higher-order structures. The structure of both birdsong and humpback whale songs has been explored. One of the most elaborate singers among the songbirds, the nightingale, commands up to 200 song types (Hultsch and Todt 2004), with each song consisting of a succession of several elements or notes. Altogether, the song of a typical nightingale may have up to 1000 different elements. Thus, the number of combinatorial signals is effectively smaller than the elements which make up the signal (Hurford 2009). The same appears to be true for humpback whales, and is strikingly different from human languages, in which the number of words is orders of magnitude less than the number of possible sentences. The most elaborate bird and whale song exploits two main devices: repetition of syllables or phrases, and sequencing of up to about seven separate units (perhaps iterated) into a single phrase, itself perhaps iterated. Most significantly, bird and whale songs are *combinatorial but not semantically compositional in the sense that the elements that make up the utterance carry specific meaning.*

Perplexingly, the utterances of nonhuman primates are much less elaborate than that of songbirds or whales, with the notable exception of gibbon song, despite the fact that nonhuman primates do not simply utter single calls, but rather bouts of several calls. The question is (a) whether such sequences can be described in terms of syntactical rules, and (b) whether they allow listeners to attribute differential meaning based on the combination of different call units. The first point can be largely refuted as sequences do not follow fully predictable patterns (e.g., Crockford and Boesch 2003, 2005; Arnold and Zuberbühler 2006); instead, signal combinations can be described more appropriately in probabilistic terms. There is, however, good evidence for the second point (Arnold and Zuberbühler 2008; Arnold et al. 2008; Zuberbühler 2002). Since most monkey and ape species have relatively small repertoires, this constraint may have favored listeners' abilities to process signal combinations. On the production side, it remains unclear whether the processes that give rise to heterotypic call sequences (i.e., successions of different call types) are fundamentally different from those that lead to series of the same call (Hammerschmidt and Fischer 2008).

Phonological and Syntactical Complexity Need Not Equal Lexical Syntax

Consider an example based on a series of studies in Bengalese finches (Okanoya 2004b), which demonstrate that phonological and syntactical complexity need not result in lexical syntax. Bengalese finches are the domesticated strain of the wild white-rumped munias. In contrast to the wild white-rumped munias, Bengalese finches have been domesticated for over 250 years, during which time the courtship song became phonologically and syntactically complex. Acoustic and syntactical analyses of the two strains reveal marked differences between them. Bengalese finches use wide varieties of song note types, including the harmonic stack, frequency modulation, narrow-banded tone, wide-banded noise-like elements. White-rumped munias use primarily wide-banded noise-like elements. Bengalese finch song syntax is expressed by finite-state syntax with loops and returns, whereas white-rumped munia song syntax is simple and linear. Laboratory study of cross-fostering between the two strains revealed that white-rumped munias are more specialized in accurately learning their own-strain phonology, whereas Bengalese finches learned the phonology of both strains equally but less accurately, suggesting that Bengalese finches lost a species-specific bias to learn their own species' phonology accurately. However, using a nest-building assay, Okanoya (2004) found that females of both strains work more when stimulated with complex songs as opposed to simple songs, suggesting that bias to prefer the complex song exists in females even in the wild strain. Breeding experiments also suggest a preference for complex songs in Bengalese finches. Males reared in a nutritiously competitive environment tend not to develop longer song bouts. Since longer song bouts give more opportunity for demonstrating song complexity, males in less competitive environments have room to develop syntactically complex song. Furthermore, when two song tutors with different degree of song complexity were provided, male chicks were more likely to learn from complex singers.

Field observation in Taiwan suggests (Okanoya, pers. comm.) that syntactical complexity does not develop under the pressure for species recognition. Populations of white-rumped munia show a gradient of song syntactical complexity, but nevertheless songs are generally simpler in white-rumped munia than in Bengalese finches. In wild populations of white-rumped munia, when there are more sympatric species (in this case, the spotted munia), the population shows less syntactical complexity and vice versa. In addition, song phonological complexity, as examined by the degree of song sharing, suggests a similar tendency: the more sympatric species, the fewer song note variations were observed.

These behavioral/ecological studies are supported by neural and molecular studies (Okanoya, pers. comm.). When the sizes of song control nuclei are compared, Bengalese finches have a larger relative neural volume than white-rumped munias. Gene expression profiles of the two strains revealed higher levels of the neurotrophic factors in the auditory areas in Bengalese finches. In Bengalese finches, auditory selectivity of the neurons in the forebrain area for the bird's own song negatively correlated with song complexity of the subject bird. This suggests that neurons are less exactly tuned for the bird's own song in singers with complex songs.

For over 250 years, Bengalese finches have been artificially selected for their plumages and breeding capacity, not for their songs (Okanoya 2004b). Still, artificial selection may have influenced sexual selection of song complexity by female preference through indirect maternal control (Soma et al. 2009). Specifically, females may be changing their maternal investments to the siblings depending on the attractiveness of their mates: when breeding with a poor-quality singer, the female may reserve their investments for future opportunities. Results reviewed thus far suggest this possibility (Okanoya, pers. comm.).

Taken together, it appears likely that phonological and syntactical complexity in Bengalese finch songs evolved because (a) domestication freed them from pressure for species recognition based on song characteristics, and then (b) sexual selection advanced the syntactical and phonological complexity. This is consistent with the masking/unmasking idea which we discuss later. However, sequential variability only quantitatively stimulates females; different sequences of song phrases do not convey different meanings nor set different behavioral contexts (Okanoya, pers. comm.). Thus, song syntax in birds should be regarded as phonological syntax, but not lexical syntax.

Experimental Studies of Grammatical Capabilities of Nonhuman Species

Researchers have conducted experimental studies to explore animals' abilities to process different types of grammars. There is some evidence that tamarins and starlings can distinguish between sequences that were generated using different types of "grammars" (Fitch and Hauser 2004; Gentner et al. 2006). These studies examined whether the animals were able to distinguish between finite state grammar (FSG; exemplified by $(AB)^n$) and context-free grammar (CFG), exemplified by A^nB^n, with n = 2–4 (see Figure 10.1a and 1b, respectively). Syllables were drawn from a pool of spoken syllables or starling song types (warbles and rattles). Tamarins habituated with *abab* or *ababab* sequences were able to distinguish *aabb* and *aaabbb* sequences from the former, but not the other way around. Thus, Fitch and Hauser (2004) concluded that the animals were unable to process CFG grammars. In contrast, starlings (after exhaustive training) were able to discriminate between these two types of stimuli. Although one study maintained that starlings can master CFG grammar (Gentner et al. 2006), the evidence was considered inconclusive since they could resolve this task either by counting or a combination of attendance to the first stimuli in the sequence (primacy effect) plus counting (see Peruchet and Rey 2005; Corballis 2007b).

Experimental Studies on the Neurobiological Correlates of Syntax in Humans

Although few neurobiological studies have been performed on animals, Friederici et al. (2006a) have investigated humans using functional magnetic resonance imaging (fMRI) after they habituated to string sets of the same type as used in the Fitch and Hauser study. In the Fitch and Hauser study, elements (i.e., syllables) of the A category and the B category were marked by pitch (i.e., male vs. female voice); syllables in the Friederici et al. (2006a) study

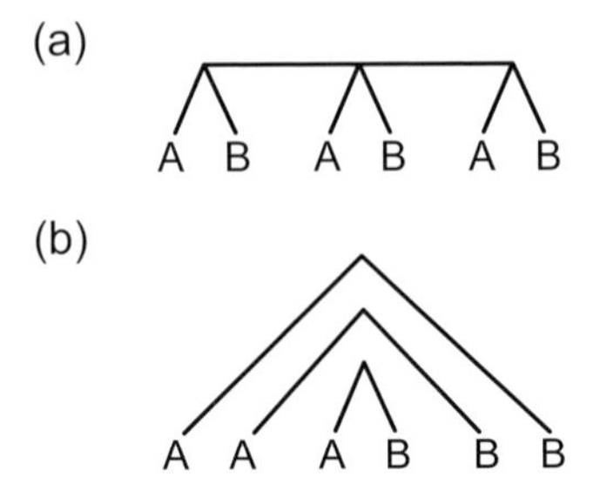

Figure 10.1 (a) FSG and (b) PSG (or CFG); n = 3.

were marked phonetically (i.e., high vs. low vowels). Brain imaging data revealed that the "FSG" stimuli, of the form *abab* and *ababab*, activated the left frontal operculum (fOP), in the inferior frontal gyrus, while "CFG" stimuli, of the form *aabb* and *aaabbb*, activated Broca's area (BA 44/45) in addition to the fOP.

The CFG used in all these studies does not necessarily require the construction of embedded hierarchical structure (see Figure 10.2) but can be processed based on a simple counting strategy. Therefore, a novel CFG was devised which required the construction of a structure, as in Figure 10.2 (Bahlmann et al. 2008). In this novel grammar, the relation between syllables in the sequence was marked by voiced–unvoiced consonants; that is, B-syllables began with consonants (p, t, k), which are the unvoiced counterparts of the onsets of the related A-syllables (b, d, g). For this CFG, again, Broca's area (BA 44/45) was found to be the crucial area recruited for the processing of CFG compared to FSG.

Using a natural language (German), Friederici's group demonstrated a similar finding. In this study, center-embedded sentences were used; the relation between A and B in a structure was marked by subject–verb agreement. Activation in the inferior portion of BA 44 of Broca's area increased as a function of syntactic hierarchy (number of embedded (nested) structures). In this experiment, the distance between dependent elements (subject, noun, and verb) was varied systematically. It was found that the factor of distance (i.e., number of words between subject-noun and verb) activated the left inferior frontal sulcus independent of the syntactic hierarchy. The interaction of the factors syntactic hierarchy and distance (i.e., general verbal working memory) was seen in the superior portion of BA 44. Functional and structural connectivity analyses revealed strong connections between the inferior BA 44 and the inferior frontal sulcus during the processing of multiple embedded sentences. These data indicate that syntactic hierarchy is represented separately from general verbal working memory in different brain areas, but that the two respective areas work together to process syntactically complex structures—successful processing that has only been demonstrated in humans.

The question then arises whether the ability to process hierarchically complex structures with nested dependencies is possible in humans due to specific neuroanatomical aspects of BA 44 either in the microstructure of BA 44 (e.g., receptoarchitectonics or microcircuitry), the relative size of BA 44 (which is

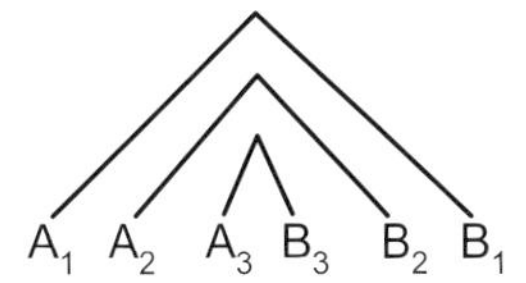

Figure 10.2 Novel PSG with a relation between syllables, where subscript denotes corresponding pairs.

defined cytoarchitectonically) in humans compared to nonhuman primates, or the macrostructure, be it the structural connectivity within the prefrontal gyrus or between the prefrontal and temporal gyrus. The available data on this question are sparse, but a recent structural connectivity study suggests different connectivity patterns in the human and nonhuman primate brain (Rilling et al. 2008). Moreover, the relative size of BA 44, when defined recepto-architectonically, appears to be somewhat different between the human and the macaque monkey (Petrides and Pandya 2002b). Additional studies comparing the neuroanatomy of the two species are necessary to evaluate whether these might explain the functional ability to process nested hierarchical dependencies that clearly characterize natural human language.

Finally, it is interesting to mention that Fedor, Ittzés, and Szathmáry (this volume) present a semi-realistic neural network that is able to parse input with center-embedding. It relies on the observation that CFG requires some implementation of a stack, with the necessary *pop* and *push* operations. Their proposed network is simpler than the previous solutions, since it rests on the assumption that gating of synaptic connections is critical for complex cognitive processes. There are four main components of the neural network: the input layer, the stack, the comparing layers, and an end-of-sentence neuron. A perceptron learns the dependencies between classes of words, and the neuronal stack takes care of the long-range dependencies that appear in a sentence. The capacity of parsing is limited only by the depth of the stack. This strongly suggests that to produce and parse CFG, the evolution of a neuronal implementation of "stack" is a necessary step.

Genetic Influences on Syntax

To discover genes that may be involved in syntax, a major approach is to identify humans with language impairments and then determine any underlying genetic anomalies. Specific language impairment (SLI) is diagnosed when a child has problems with acquisition of language production and/or understanding for no apparent reason. In the 1980s the results of twin studies were published showing that SLI is a highly heritable disorder (for a review, see Bishop 2002b). However, with one exception, which we discuss more fully below, single genes have not been found that have a strong causal association with SLI. It may seem odd that more genes have not been identified for such a highly heritable disorder, but this is actually to be expected, given that SLI is relatively common. In this regard SLI resembles other complex multifactorial disorders, such as heart disease and diabetes. If a single mutation were responsible, then we would expect it would have been selected against and gradually removed from the population. As noted by Keller and Miller (2006), where a common disorder is heritable, it is likely to be polygenic; it may either be influenced by the additive effects of many common allelic variants, each of which contributes only a small amount to the variance, or it may be heterogeneous, with many rare mutations, each of which explains only a small proportion of cases. In general, this kind of explanation seems increasingly likely for common developmental disorders.

An exception to this general rule was a three-generational British family in whom SLI appeared to be inherited in an autosomal dominant fashion. The first report by Hurst et al. (1990) stressed the speech difficulties in this family (the KE family), but later reports focused more on the associated language difficulties (see Bishop, this volume). Fisher and Marcus (2006) describe the history of discovery of the mutation in the KE family and its relevance for the evolution of language. Initially, linkage was found to a region of chromosome 7. Subsequently, a case of chromosome translocation was found with a similar phenotype, and this allowed geneticists to home in on a more precise location on the chromosome, eventually identifying a point mutation in the FOXP2 gene. This gene had not previously been thought to be implicated in brain function, but was subsequently shown to play a role in the development of regions of the frontal lobe, subcortical structures, and cerebellum. It is not, however, a language gene; rather it is a transcription factor that affects the function of many other genes, and it is involved in the development of the lung, heart, and other organs.

The mutation in the KE family is a clear case of speech and language disorder caused by a single gene mutation, but the mutation is very rare and only a handful of cases have been discovered, largely by searching for other cases with similar phenotype. The nature of the phenotype has been a matter of some debate. It is clear that affected individuals have problems with sequential speech production, which can lead to major problems with intelligibility. They also have language difficulties, evident in their written language and comprehension, which cannot simply be explained as secondary to the speech difficulties. It is unclear whether a higher-level problem with nonlinguistic cognition can be discovered to account for such problems.

One current hypothesis considered by Tomblin et al. (Christiansen, pers. comm.) is that sequential learning is a key underlying skill important for language learning, involving extraction and further processing of discrete elements that occur in complex temporal sequences. Rather than studying cases of rare mutation of FOXP2, Tomblin et al. asked whether common allelic variations in a FOXP2 promoter region might be associated with differences in sequential learning and language. A serial-response time task (SRT) was used as a measure of sequential learning in a sample of adolescents with and without language impairment. Associations were tested between SRT learning and variations in six single nucleotide polymorphisms (SNPs) selected from the major haplotype blocks within FOXP2. Two SNPs were associated with SRT learning. The association between genotypic status and language status was also found to be significant. These results suggest that FOXP2 influences systems that are important to the development of both sequential learning and language, supporting the hypothesis that language may be subserved by underlying mechanisms for sequential learning. Why should have common allelic variants with a detrimental effect on language persisted in the population? Is it possible that this aspect of the human genome is still undergoing evolutionary change?

Because FOXP2 shows such strong Mendelian effects—in the KE family a single mutant copy in humans reliably produces a severe speech and language phenotype—we should expect direct and specific control of a single gene over language. Yet, there is a problem: FOXP2, the gene itself, is far from specific. In addition to the brain, it is expressed in lung, heart, liver, gonads—most organs of the body. Further, it is expressed not only in humans but also in fruit flies and many other organisms which do not learn language. How then, can its mutation have such specific effects? For example, while affected KE family members exhibit language impairment, they also have difficulty performing manual sequential movements or in breathing.

Current techniques using microarray hybridization or deep sequencing to identify gene-level expression in different cells or tissues, followed by a statistical analyses of the covariance of these levels offer the next ideas for conferring specificity of FOXP2's actions. Since FOXP2 is a transcription factor, rather than its levels or mutation status being fully predictive of the language phenotype, the levels and state of key genes with which it interacts will be informative. These genes will differ across organs and animals, giving rise to specific gene networks. Indeed, an autism susceptibility gene known as contactin-associated protein-like 2 (*CNTNAP2*; Stephan 2008) has recently been discovered to be a transcriptional target of FOXP2 in humans (Vernes et al. 2008). Key pieces of evidence link *CNTNAP2* to language. These include the fact that certain *CNTNAP2* variants in autistic children are associated with the age at first word (Alarcon et al. 2008). Additionally, Vernes and colleagues (2008) found that genetic polymorphisms of *CNTNAP2* in children with SLI are correlated with their ability to do a nonword repetition task. As a member of the neurexin superfamily of transmembrane proteins, *CNTNAP2* is likely involved in cell–cell communication processes in the nervous system. Interestingly, its expression is enriched in areas of the human fetal cortex which give rise to language, while its cortical expression in rodents, who are not vocal learners, is diffuse (Alarcón et al. 2008).

Acceptance of FoxP2 as a "molecular entry point" to gene networks involved in language makes tests of its neurobiological function, and that of its key targets, of interest. As we cannot perform such tests in humans, what model organisms should we use? Mice are transgenically tractable, but they are not considered to be vocal learners. Nonetheless, mouse models of Foxp2 mutations (either with the normal human version of the gene, or the KE mutation; see Hilliard and White, this volume) have been generated and offer important insights. Although some of these data conflict as to whether there is an effect of specific mutations on vocal communication, in one report (Groszer et al. 2008) in which mice with the KE mutation lack vocal deficits, other learned motor skills are impaired. This tells us that FoxP2 in nonvocal learning animals may be more generally involved in other learned complex motor behaviors. This idea is compatible with the findings of Tomblin et al. (2007), mentioned above, and also suggests that detailed investigations of motor skill learning in humans with variants of the FOXP2 gene might be valuable.

In songbirds, many of which are vocal learners, experimentally induced diminishment of *FoxP2* levels in the specific subcortical neurons dedicated to song, results in imprecise song copying. Further, other studies show that when songbirds sing, *FoxP2* levels decrease naturally in these same neurons (White and Hilliard, this volume). Additionally, *CNTNAP2*, one of FoxP2's target genes mentioned above, is enriched in cortical-like pallial regions that give rise to song (S. A.White, pers. comm.), just as it is enriched in areas of the human fetal cortex that give rise to language. These exciting findings reinforce a specific role for FoxP2 in learned vocal communication signals in those species with this ability. Other comparative data on FoxP2 may have a bearing on the species-general contribution of this gene to vocal behavior. Each of the different clades of echolocating bats (Li et al. 2007) show extensive and diverse changes in the FoxP2 gene sequence. This stands out against a remarkably conserved profile of FoxP2 gene changes in other mammals. This unusual pattern of gene modification in a group of animals which rely on extremely precise vocalizations for predation suggests that FoxP2 may be playing a general role in highly demanding vocal behavior. In general, we are not in a position to be able to correlate specific sequence differences in this gene with specific vocalization consequences, either for animals or humans.

Together, the comparative data suggest that we can use nonhuman models to shed light on FoxP2 neural function. For example, work to detect FoxP2 gene targets and networks in songbirds can be compared to targets and networks in humans and thus highlight shared and unique subsets (see Hilliard and White, this volume). Shared gene networks are hypothesized to be involved in vocal mimicry and sequential learning in the vocal motor domain. Those unique to humans could highlight specializations for critical, semantic, and syntactic capacity. In tandem, the study of people with language disorders may have the possibility to throw light on the origins of language, but we do need to be careful in how we interpret the evidence. In many cases, language disorders are not going to tell us about specific genes involved in the evolution of language, but rather will result from combined effects of many genes of small effect. Even where large effect genes are found, such as FOXP2, they do not have syntax-specific effects. However, this does not mean they are uninteresting—they may indeed indicate just how genes involved in other functions developed new capacities.

How an Evolutionary Perspective Constrains Syntactic Theory

Those working on the evolution of language place new demands on any theory of syntax, namely the need for forms of syntax of intermediate complexity. In evolutionary theory the understanding of difficult transitions is much easier if the given transition can be broken down into a number of intermediate steps. For example, understanding of the origin of life is much easier under the assumption that genes came before protein synthesis and the genetic code. By the same token, the origin of the eukaryotic cell is more readily understandable if we accept that the origin of mitochondria and that of the cell nucleus did not have to happen simultaneously. This is why it is important to try to propose forms of syntax with intermediate complexity. Such intermediates will be far from perfect, but they still can be better than the previous stages. One example may be a first form of Merge that could operate only on words. Then in the next stage it could work on constructs from the previous stage. Such intermediate stages could be important for the modelers because they could then build specific models to try to simulate the evolutionary emergence of such stages. This could allow us to develop a feel for plausibility of alternative scenarios for syntactic evolution.

Another important contribution that the biological grounding of syntax can provide is a set of constraints that can narrow down the space of theoretical possibilities. Although it is often argued that there is an infinite space of possible forms of syntax, there are in fact many constraints on realizable forms that syntax can take. These arise from diverse sources, many of which are not explicitly linguistic in origin. Figure 10.3 (based on Deacon 2004) depicts how these many constraints collectively reduce the space of possibilities (see also Deacon 2003b). This will constrain the social evolution of language with respect to its learnability and transmissibility and likewise the biological evolution of adaptations supporting syntactic functions in language.

An additional class of constraints (not depicted) includes especially those associated with evolvability; the fact that evolution itself is a significantly constrained process in which certain restrictive conditions must be met for something to have the possibility of evolving. This includes having relevant evolutionary precursors that could be modified appropriately, the presence of appropriate selection pressures that are relatively stable over a very large number of generations, and some significant consequence for reproduction. Thus, for example, it is quite unlikely that primates would ever have evolved wings like angels. Consider words. The absence of any innate words from languages suggests that they are not evolvable. This begs a question about syntax. If innate words are not evolvable, why should we expect innate syntax to be evolvable? Are there some aspects of syntax that we might expect to be more evolvable than others? By focusing on the stability of the phenomenon across time and its relative invariant neural representation we gain some hints. As syntactic theories have matured during the past thirty years, there has been a

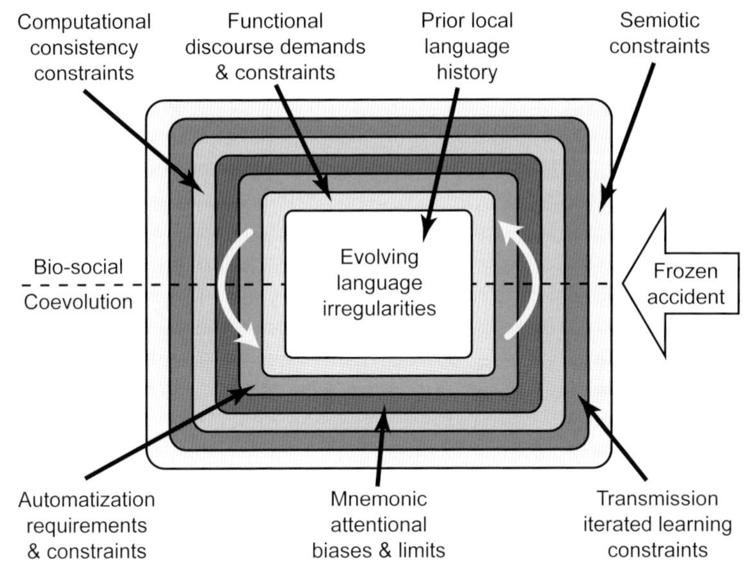

Figure 10.3 Nested hierarchy of constraints on possible syntactic forms. Higher level constraints are more general and set the boundary conditions for those lower down the list (indicated by the arrow). This diagram is meant to be exemplary of the logic of this constraint hierarchy, not a definitive list. The specific constraints are only intended as a suggestive list, and are not further explained. It shows how the space of "possible languages" (aka possible syntactic systems) may actually be quite reduced; both for evolution and acquisition. "Frozen accident" refers to the possibility that both social and genetic evolution can arbitrarily limit possible evolved forms simply by the fact that alternatives can disappear from the population by the chance effects of breeding and so simply reduce the space by accident. The inclusion of biosocial evolution refers to the fact that these constraints affect not only what can evolve biologically or in the transmission of language, but also how these processes interact over evolutionary time.

development from abstract algorithmic conceptions of syntax to more concrete and generalized conceptions. Whereas abstract and highly specific algorithmic capacities are unlikely to be evolvable for these reasons, more concrete and general processes (e.g., learning biases, working memory limitations, chunking strategies, hierarchical analysis, and automatization demands, to list few) that apply across syntactic operations are probably evolvable.

From an evolutionary perspective this introduces many additional constraints (not depicted in Figure 10.3) as well as prior biases, such as the visual specialization of primate ancestry, biases and constraints on learning and transmission, and the amount of evolutionary time and selection pressures supporting language. Generally, the briefer the time of language evolution, the less modified the brain should be for language processes and the more fragile this capacity should be to insult (e.g., by developmental damage), whereas the longer language has been around the more linguistically modified the brain should be and the more language-specific constraints should matter.

Evolutionary Mechanisms for the Evolution of Syntax

Exaptation of Nonlinguistic Capacities for Syntax

The concept of exaptation is crucial for understanding the origin of evolutionary novelties. Exaptation (previously called preadaptation) is the phenomenon where an adaptation that has gone to fixation in some selective environment turns out to be useful in a new one. The more complex the phenomenon to be explained, the greater the role exaptation is likely to have played in its appearance. It is the rule that after the initial stage of recruitment of the old adaptation into the new function, it gets refined through genetic evolution by natural selection. The bacterial flagellum is perhaps the most spectacular example of a series of exaptations and selective fine-tuning, producing a complex mobility apparatus based on a rotating electromechanical nano-machine (see Figure 10.4). This machine has several components, including (among others) the external filament, the motor and the stator, a secretory system and a chemotaxis apparatus. All of these components have proteins that are in part homologous to some other proteins in the cell. It is crucial that none these components was involved in mobility before. For example, the motor is homologous to a transport channel using a hydrogen ion gradient. The various components had been recruited (exapted) from previous functions, and they had undergone further evolution to varying degrees. It is very likely that the evolving structure

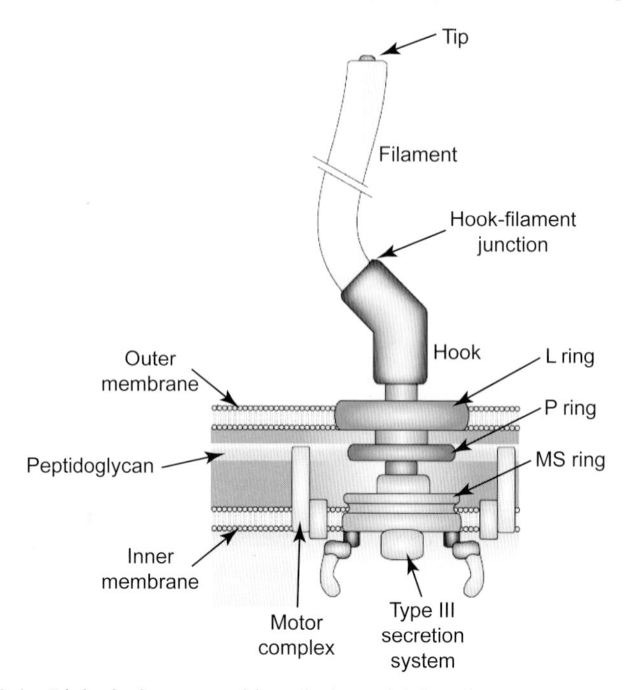

Figure 10.4 Biological nano-machine of a bacterial flagellum.

served some intermediate functions before motility, such as targeted secretion and adhesion.

We believe that recruitment of functionally different exapted modules played a crucial role in the evolutionary origins of language as well. For instance, it is possible that hierarchies got processed first in the domain of tool making and then was refined in the context of language; this refinement, in turn, fed back favorably on tool use and tool making. Humans have a suite of complex cognitive functions. It is likely that they are dependent on a set of intermediate phenotypes that act as processing modules in more than one complex adaptation. One important intermediate phenotype could be the ability to handle hierarchical representations in the brain. It is thus natural to expect that several mutations affecting human cognitive behavior will have pleiotropic effects. This shows that functional modularity does not imply the lack of pleiotropy. This may imply to some that evolution of natural selection of these traits could have been retarded. New studies in population genetics indicate, however, that pleiotropy can in fact speed up adaptive evolution, provided the population is far from its adaptive peak. This is likely to apply at the beginning of hominine evolution (for further discussion, see Szathmáry and Számadó 2008a).

Language functions are the result of emergent synergies that have subsequently come under selection for these higher-order interaction effects. A process that could lead to the emergence of such synergies could be important for the transition to language.

Coevolution of Genes and Social Transmission with Regard to Syntax

The notion that the evolution of syntax appeared by virtue of a very lucky accident (aka "hopeful monster" mutation) is considered by most biologists to be implausible. They would argue that the human facility of language syntax most probably arose from many changes in genetics and neurology, which collectively contributed appropriate processing biases. But how did this novel functional synergy arise in the first place?

The interaction of biological evolution and cultural transmission makes the evolution of human syntactic abilities more complex than other evolutionary processes (Boyd and Richerson 1985, 1996; Deacon 1997; Laland et al. 1999). The interplay between these processes has motivated considerable interest in evolutionary mechanisms that consider the influence of behavior on the direction of evolution. In recent years many researchers have suggested that a Darwinian variant of something like a Lamarckian process might have been involved; in other words, that learned syntax in our ancestry, before the hominid brain was in any way adapted for language, created conditions for this acquired behavior to become progressively controlled innately. Theoretical variants of evolutionary processes that take into account this multilayer interaction include the Baldwin effect, genetic assimilation, and niche construction.

Genetic assimilation has been demonstrated by experiments in various contexts (Pigliucci et al. 2006). It is important, however, to realize that in some cases, it leads to simplification while in others it can lead to more complexity. In the classic Waddington-type experiment, it led to simplification. The crucial concept here is the norm of reaction of a genotype: how is the phenotype affected by environmental change (see Figure 10.5). Due to selection in the new environment there is no selection against the loss of plasticity.

It is important to look at a mechanism where genetic assimilation can create adaptive complexity. We begin with a general model of this process, as developed in quantitative genetics (see Figure 10.6). First, phenotypic plasticity can be essential to survival in a changed environment (Figure 10.6a). Note that without a plastic response the population will go extinct. Second, peak shifts on an adaptive landscape can be produced by plasticity on an unchanging adaptive surface (Figure 10.6b). The bold line shows the adaptive surface. Third, the combination with genetic evolution is as follows. If plasticity is too low, peak shifts can be impossible. If it is very broad, adaptation is possible without genetic change. An intermediate degree of plasticity is the most interesting case, because the population can start shifting due to plasticity followed by genetic change through natural selection in the same direction. Complexity could be gradually built up due to a series of such plastic and genetic peak shifts in succession. The process just described is sometimes called the Baldwin expediting effect.

There are many attractive features of this view which might help explain the origins of syntax; however, the process requires that learned syntactic behavior was already available in some form prior to being assimilated to genetic control. Another problem is that there is often confusion between the Baldwin effect, genetic assimilation, and niche construction (and sometimes a tendency for people to treat these as Lamarckian processes, which they are not). Thus, we must first carefully distinguish between them. The Baldwin effect and genetic assimilation processes are actually quite different processes—in some ways inverses (Deacon 2003a), while niche construction typically will involve both in complex interactions.

Baldwinian Process

Baldwin (1896) originally called this hypothetical mechanism "organic selection" because it was ultimately due to factors intrinsic to the organism rather than extrinsic, as in Darwinian natural selection. The same basic idea was simultaneously presented by James Mark Baldwin (1896), Conwy Lloyd Morgan (1896), and Henry Osborne (1896), though their proposals differed in details. In fact, Osborne argued that the mechanism showed that evolution could proceed by this mechanism in the absence of natural selection (he was an anti-Darwinian).

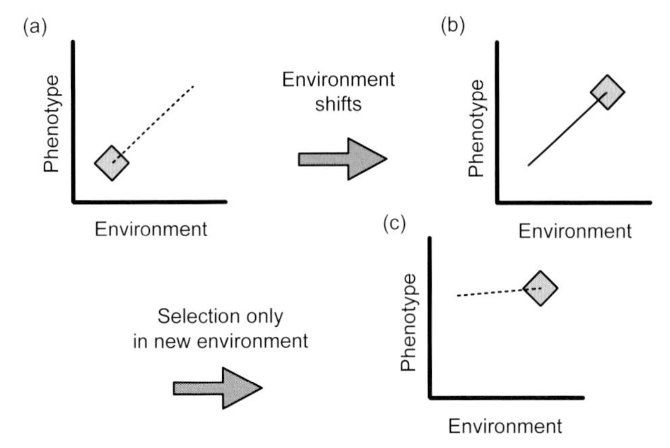

Figure 10.5 Genetic assimilation. Norm of reaction can drift in the new environment resulting in loss of the phenotype typical for the old environment (after Pigliucci et al. 2006)

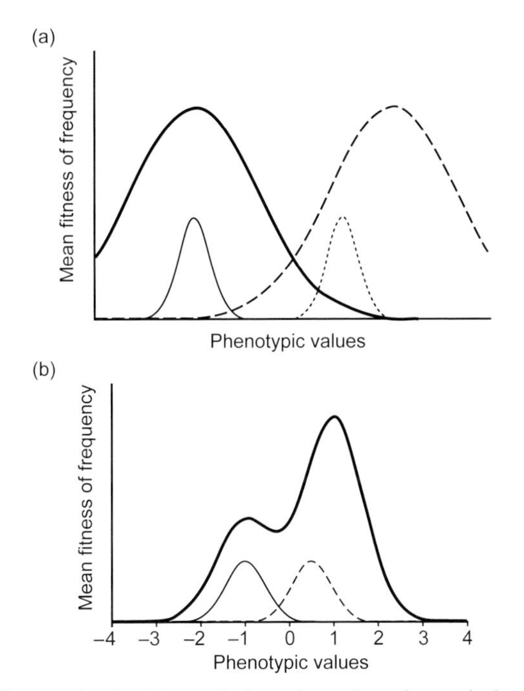

Figure 10.6 Phenotypic plasticity and adaptation, where the x axis denotes phenotypic values and the y axis denotes both mean fitness and the frequency of a given phenotype in the population. Bold lines are mean fitness; the dashed line represents mean fitness in the new environment. The thin solid line represents the population distribution in the old environment and the thin dotted line represents the population distribution after a plastic response to the new environment. (a) Without a plastic response the population will go extinct. (b) Peak shifts on an adaptive landscape can be produced by plasticity on an unchanging adaptive surface.

In this hypothetical process, plasticity and/or learning (i.e., epigenetic effects) shield a lineage from elimination by natural selection, and thereby allow more variants to emerge and be retained in the population. This increases the chance that at some point in that lineage a more appropriate congenital (not requiring plasticity) variant can emerge and come to replace the plastic/conditional phenotype. Individuals who could adapt without suffering the costs associated with plasticity or learning would presumably out-compete the rest and in this way enable this variant to become dominant in the population.

This mechanism has been difficult to demonstrate empirically, and to many it appears only to be a claim that behavioral plasticity allows a species to shift between niches and distinct adaptive optima. A more serious limitation is that being able to adapt to conditions due to flexibility reduces the selective value of any alternative less plastic phenotypes that might arise, and so would resist being so replaced (though this does not exclude selection favoring adaptations that add support to this plasticity). Cost/benefit factors are critical determinants of the process, with the high cost of plasticity/learning contributing to the relative selective advantage of an innate variant. Of course, being too inflexible for the context is also a problem. There will likely be a cost of inflexibility relative to the changeability of the environment to consider as well.

Deacon (2003a) describes the process presumed to initiate this change as the "masking" of selection on specific traits, due to relaxation of the adaptive demands. For example, redundant extra-genomic support for a phenotypic function (such as the need to produce vitamin C endogenously, see below) relaxes selection on the maintenance of genetic information to produce this effect and thus allows mutations to accumulate and degrade the genetic information. Degradation at the genomic level comes easy, due to random spontaneous point mutations, and is hard to reverse because this requires precisely inverse mutations.

Waddingtonian Process

In the 1940s and 1950s, Conrad Waddington ran breeding experiments on fruit flies to determine if acquired traits could be bred to become ineluctable. He showed that breeding for traits that were expressed only if specific environmental conditions held could lead to their expression irrespective of environment. He called this process "genetic assimilation" (see Figures 10.5).

This mechanism differs from Baldwin's in two ways. First it is a consequence of the unmasking of traits that are below the threshold for selection and thus are otherwise neutrally varying; second it was an empirically demonstrated effect. The mechanism remained only theoretical until quite recently (Rutherford and Lindquist 1998). The mechanism involves many genes with weak effect (thus requiring environmental support) being progressively grouped together because selective mating will be much more probable between carriers due to their reproductive advantages. After many generations of breeding between carriers, Waddington showed that the initially conditional trait tended to be produced irrespective of environmental influence. Waddington described this increase in the probability of epigenetic expression as an increased "canalization" of the process.

Niche Construction

Recently, a more complex variant of this has been invoked implicitly and explicitly by a number of people (Deacon 1997; Laland et al. 1999; Odling-Smee et al. 2003). It can be exemplified by the beaver dam effect, which also shows its link to the "extended phenotype" of Dawkins (1982). Beavers actively modify their niche, such that over evolutionary time beavers have become increasingly adapted to this artificial aquatic niche. It is important to note that niche construction improves the fitness of the organisms. Many effects that populations of organisms have on their environment are negative; this can be referred to as niche destruction. Niche construction involves a combination of Baldwinian and Waddingtonian processes. It is often a response that functions to reduce exposure to the negative influences of natural selection, as in beaver dams (in this way it is Baldwinian); however, by creating new conditions, previously neutral phenotypic variants that are relevant to taking advantage of or fitting into this niche become exposed to selection (unmasked). In this way, if niche construction produces a relaxation of selection, it can mask selection on some traits and, at the same time, create conditions for unmasking others, related in complementary ways to the masking process. More importantly, the unmasked traits will involve a wide variety of previously unrelated genetic loci and epigenetic processes that just happen to contribute. So even if the Baldwinian mechanism is merely degenerate, it may set up conditions for indirectly correlated functional mechanisms to be collectively modified in linked ways with respect to the masked traits.

This is important to the evolution of the syntax problem because it suggests an evolutionary mechanism whereby previously unlinked neural mechanisms may be drawn together in such a way that they become collectively selected with respect to some synergistic consequence. Whatever evolutionary modifications to brain function made syntactic operations more easily acquired and effortlessly employed in our communication, they were inevitably a highly diverse set of neural structures and mechanisms that were not originally evolved for language. Explaining how they came to cooperate synergistically and become progressively modified, with respect to each other to more efficiently serve this novel collective function, has long been one of the more insuperable problems for biologists attempting to explain the evolution of our extensive language adaptations.

Deacon has provided a number of simpler analogical cases of this process to help exemplify how it works. Some particularly clear and analogous cases involve the degradation of endogenous ascorbic acid biosynthesis in primates, the evolution of gene duplication in the evolution of hemoglobins and in homeotic genes, and, by reference, to work by Okanoya on song control in a domesticated finch (Deacon 2003a, b, 2004).

For example, the loss of endogenous ascorbic acid synthesis and the redistribution of selection to distributed genetic loci and diverse phenotypes involve the following steps:

1. Fruit eating, due to facultative (flexible) adaptation (i.e., arboreal feeding), supplements vitamin C (ascorbic acid) synthesis, and masks selection for the benefits of endogenous production.
2. Masking allows degradation of endogenous vitamin C synthesis in primate lineage and eventually complete degradation of the gene (GULO) responsible for its production.
3. This results in something analogous to addiction to environmentally supplied dietary vitamin C.
4. Since this is an extrinsic source that will be competed for, the addiction produces selection for means to better guarantee this extrinsic contribution.
5. This will redistribute selection to other genetic loci and phenotypic features that support access to vitamin C, including, for example, 3-color vision for ripeness identification, changes in teeth and digestive tract, changes in taste preferences, changes in metabolic handling of other substances (e.g., alcohols) potentially present in fruits, etc.
6. These independently unmasked features will collectively and synergistically contribute toward this common end (e.g., via a Waddingtonian process) and thus make the adaptation more stable by widely distributing the selection load.

In parallel ways, relaxation of selection, due to what might be described as linguistic niche construction, likely contributed to the evolution of the complex synergistic brain functions associated with language syntax. The role it may have played in the emergence of these novel interdependencies among cortical areas is to increase the ease of recruitment of diverse brain structures to aid language. For this to have led to efficient utilization of this higher-order integration of cortical functions, however, subsequent selection had to reinforce this synergy and refine the match between these unmasked capacities and the unique functional demands imposed by language processing. In other words, analogous to the case of recruited adaptations for obtaining vitamin C, unmasking of these many independent neurobiological contributions to syntactic processing would have initiated Waddingtonian selection to stabilize this synergy and further hone the fit, respectively. The extent to which the sort of interactions resulting from masking effects are predisposed by prior functional relationships and epigenetic fine-tuning to bias the direction of evolutionary change is subject to debate and, more importantly, open for future detailed investigation of the relevant parameters. But the capacity of this mechanism to expose unprecedented interrelated clusters of traits to the effects of natural selection on their synergistic contributions to a complex trait makes it an attractive mechanism for explaining many aspects of syntactic evolution. In sum, classical natural selection, genetic assimilation and relaxation from selective constraints are all likely to have contributed to the evolution of the complex of adaptive changes supporting syntax, but their relative contributions are unknown and need to be explored.

The very fact that language is culturally transmitted could lead to masking of at least some of the genes that influence the acquisition and facility with syntax, ultimately giving rise to a weakening of innate constraints. Smith and Kirby (2008) present a model of the coevolution of learning biases and languages transmitted by learners possessing those biases innately. Using a Bayesian model of learning, they show that although the nature of the bias provided by genes is crucially important, evolution will favor learners in which the *strength* of the bias has no effect on the outcome of cultural evolution of language. This is a case of masking, because natural selection cannot "see" the strength of the innate bias. They hypothesize, in line with Deacon's argument, that strong biases will therefore erode. Ultimately, this model leads us to expect that once a language can be culturally transmitted with some reliability, then evolution will lead to the precise set of circumstances where these cultural processes are increasingly important and where selection will favor agents with traits that help maintain this process.

It may be worth considering how these issues relate to Lachlan and Slater's (1999) "cultural trap hypothesis" for the evolution of socially acquired birdsong. In their model, learned birdsong may be maintained by selection even when the average fitness of a population of learners may be lower than a population with an innate song. This is due to an evolutionary trap where, once there are individuals in a population able to produce variable behavior, more constrained behavior is selected against it if it reduces the effectiveness of social transmission. We can think of this as a kind of *addiction* to culture. Once we are reliant on cultural transmission of behavior then it is hard to shake the habit.

The interaction of genetic evolution and cultural transmission can have a profound effect on the nature of the genetic contribution to the acquisition and neural processing of syntax. According to a prominent view, genetic constraints specify a "universal grammar" (UG): a system of arbitrary linguistic principles that explain the speed of language acquisition, language universals, and why language is uniquely human. But could such genetic constraints co-evolve with language? Deacon (1997, 2003) reviewed a number of requirements for the evolvability of language adaptations, arguing that in principle a biologically evolved UG is unlikely, as is the evolution of innate words. Christiansen et al. (2006) investigated this hypothesis with a theoretical model, implemented in computer simulations, analyzing when genes encoding UG could coevolve with language. Their results suggest that genes for UG could only coevolve with stable aspects of the linguistic environment. Yet, prior to the existence of putative language genes, the only linguistic constraints are cultural conventions, which typically change much faster than genes. A fast-changing linguistic environment does not provide a sufficiently stable target for biological adaptation: so an innate UG could not have evolved by genetic assimilation from a culturally transmitted syntax (even if we allow for niche construction). This is in disagreement with Baldwinian scenarios described, for example, by Pinker (1996) and Jackendoff (2004) and strongly suggests that UG is an evolved cultural product dependent on preexisting cognitive mechanisms that have been modified to complement, but not replace, constraints arising from learning, communication, and social transmission. However, this does not mean that there could not be genetic assimilation of a wide variety of functional supports for social acquisition and ease of implementing syntactic algorithms (such as a more efficient working memory) as Christiansen et al. (2006) demonstrated in another set of simulations. Most importantly, there are procedural neurobiological capacities, such as operation on hierarchical structures (among many others), that would be selectively favored in a communication scenario, and thus collectively constituting language-specific innate adaptations even though they do not determine anything that looks like UG.

All these mechanisms assume some version of gene–culture coevolution, such that genes and culture relate to each other a little like a symbiosis between two species (e.g., Deacon 1997). Genes evolve by mechanisms that are reasonably well understood, but cultural evolutionary processes have only recently been systematically explored. Beginning in the 1970s evolutionary biologists began to apply Darwinian principles to cultural evolution (Cavalli-Sforza and Feldman 1981; Boyd and Richerson 1985; Henrich and McElreath 2003). Early work was mostly based on theoretical models but more recently the number of experimental and field studies has begun to increase (Mesoudi 2007; Henrich and Henrich 2007; see also Jaeger et al., this volume). Cultural variation is subject to some of the same processes that operate on genes. An analogous process to natural selection can operate on cultural variation. For example, if an individual somehow culturally acquires an aberrant ideolect, their ability to communicate will be handicapped and selection will act against the propagation of the ideolect. Cultural evolution also includes a number of processes with little parallel in the genetic system. Labov (2001) reviews the longish list of processes that are involved in dialect change. For example, locally prestigious women often inspire others to imitate the dialect innovations they model. The picture that emerges is that culture evolves under the influence of a complex concatenation of forces which shape it over time.

This overview of evolutionary processes suggests that coevolutionary approaches may hold the key to answering some of the apparently paradoxical problems this unprecedented language function poses. Specifically, it overcomes the difficulty of explaining the evolution of the complex interdependence of brain functions recruited for this capacity, as well as explains the powerful role played by cultural transmission in structuring syntactic conventions in ways that are learnable and well-suited to functional brain architecture. Because of the many levels of interactions involved, these processes are remarkably complex and convoluted, and are not easily understood without the aid of simulation research to test the contributions of many relevant parameters and boundary conditions. Although these theoretical innovations are still new and exploratory, they offer methodological approaches that can make the insights of biological evolution more compatible with the insights of syntactic analysis.

Correlates of Syntax in the Fossil Record

One of the paradoxes of the origins of syntax lies in the fact that while archaeologists and paleoanthropologists are probably not the best scientific community to ask what syntactical language is and when it arose, they probably retain in their hands the best, not to say the only, information to answer the latter question. Language does not fossilize. Complex technologies, regional trends in the style and decoration of tools, systematic use of pigments, abstract and representational depictions on a variety of media, burials, grave goods, musical instruments, and personal ornaments are among the more common long-lasting human creations that may be considered, at one degree or another, as nonlinguistic phenotypes associated with the emergence of language (Barham 2002; Klein 2000; d'Errico et al. 2003; Henshilwood and Marean 2003; Coolidge and Wynn 2004; Henshilwood et al. 2004; Conard 2003; Vanhaeren and d'Errico 2006; d'Errico and Vanhaeren 2009; Henshilwood and Dubreuil 2009). They could, if their significance in this respect was more precisely evaluated, provide valuable information on the origin and, if any, on the evolutionary steps that have led to syntactical language. Recent discoveries have dramatically changed our knowledge on the chronology of the emergence of these traits and the fossil human populations to which they were associated. We know now that composite tools using elaborated hafting techniques were used since at least 200 ky and are found among both Anatomically Modern Humans (AMH) in Africa and Neanderthals in Europe and the Near East (Mazza et al. 2006). A systematic use of pigments, probably used for body decoration, is attested in Africa at archaeological sites dated to 160 ky (Marean et al. 2007) and possibly at sites dated to 280 ky (McBrearty and Brooks 2000). In the Near East the oldest evidence for a systematic use of pigments dates back to ca. 100 ky (Hovers et al. 2003). Pigments are sporadically used by Neanderthals in Europe since 300 ky (Marshack 1981) but their use becomes systematic only after 60 ky (d'Errico 2003; Soressi and d'Errico 2007; see Figure 10.7). Burial practices go back to 120 ky among both AMH and Neanderthals (Pettitt 2002; d'Errico et al. 2003). Fully shaped bone tools (spear points, awls, spatulas, harpoons) are found in Africa since at least 75 ky, possibly 90 ky (d'Errico

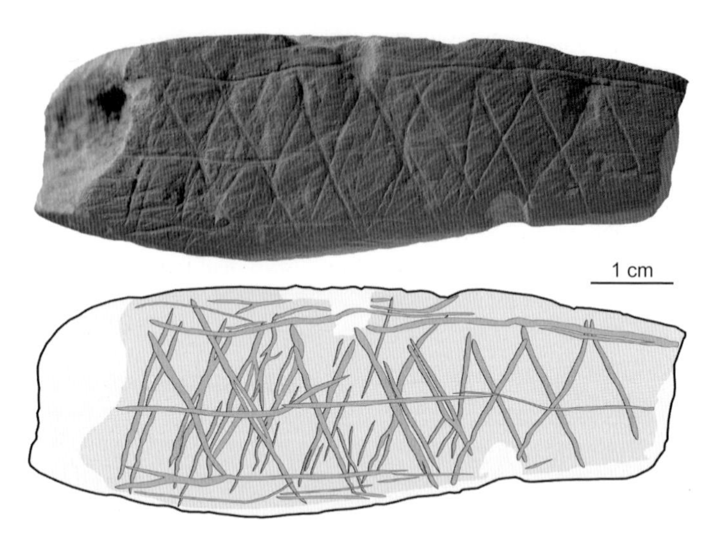

Figure 10.7 Ochre slab engraved with an abstract pattern found in Middle Stone Age layers of Blombos cave, Cape Province, South Africa, dated to ca. 75,000 years (photo F. d'Errico).

and Henshilwood 2007; Yellen et al. 1995). They disappear between 70 ky and 50 ky. Neanderthals, in contrast, produced complex bone tools only after 40 ky, just before their extinction (Villa and d'Errico 2001; d'Errico et al. 2003). Convincing evidence for the use of personal ornaments, consisting of perforated marine shell belonging to a single species, is found at sites from South Africa (d'Errico et al. 2005), North Africa (Bouzouggar et al. 2007), and the Near East (Vanhaeren et al. 2006) dated to between 100 and 70 ky. Beads disappear in these regions between 70 ky and 40 ky (d'Errico and Vanhaeren 2009) and reappear almost everywhere in Africa and Eurasia after this time span. 40 ky beads from Europe are associated to both Neanderthals and AMH (d'Errico 2003). They differ from their 100–70 ky antecedents in that they take the form of hundreds of discrete types identifying regional patterns (Vanhaeren and d'Errico 2006). The earliest abstract designs, engraved on bone and ochre, are found in South Africa and are dated to ca 75 ky (Henshilwood et al. 2002). Figurative representations, consisting of painted, engraved and carved animals, appear much later, at ca 28 ky in Africa and 32 ky in Europe, Asia, and the Near East (McBrearty and Brooks 2000; Conard 2003). The oldest carved musical instruments, consisting of flutes made of bird bone and mammoth ivory, are found in Europe and date back to 32 ky ago (d'Errico et al. 2003; Conard et al. 2004). No convincing musical instruments are associated to Neanderthals (d'Errico and Lawson 2006).

The arguments put forward to link these innovations to syntactical language are of various natures. The uniformitarianist approach to this issue postulates that societies of the past, which show in their social and cultural systems the same degree of complexity recorded in historically known societies speaking syntactical languages, must have had means of communications of comparable complexity (Binford 1983; Renfrew 1996; Tschauner 1996; d'Errico and Vanhaeren 2009; Henshilwood and Dubreuil 2009; but see Botha 2008). Deacon recalls that knapping techniques requiring a complex sequence of actions to produce fully shaped symmetrical tools have been considered viable correlates for the presence of hierarchical structures in contemporary communication systems on the grounds that those tools required a structured "syntax" of actions to be produced (Dibble 1989; Deacon 1997; Stout et al. 2000, but see Stout and Chaminade 2007). For example, early Oldowan stone tools, which date to 2.4 million years ago, required only a few simple strikes of stone to produce their crude cutting edge (clearly nonhierarchical). Acheulean hand axes associated with *Homo erectus* required multi-step preparation and a shift in methodology to prepare the stone core and refine the cutting edge (demonstrating the dependence of nested production techniques). Some Mousterian and Upper Palaeolithic tool technologies (associated with Neanderthals and modern humans) often required careful preparation of a core so that precise blades could be chipped off, as well as hafting of points to wood. Both of these procedures show evidence of multi-level nesting of significantly different techniques suggesting some facility with hierarchic thinking. Again, whether this correlates with syntactic hierarchical ability is unclear, but it is important to note that the manufacturing today of the more complex hand axes by archaeologists activates their right-hemisphere analogue of the Broca area, whereas no such brain activity is elicited by manufacturing the simpler tools (Stout et al. 2008).

Symbolic manifestations have been repeatedly mentioned as the best evidence for the emergence of language on the grounds that only a communication system with complex symbolic functions can create symbolic codes, embody them in material culture, and unambiguously transmit their meaning from one generation to another (Donald 1991; Deacon 1997; d'Errico et al. 2003, Henshilwood and Marean 2003).

Complex techniques and symbolic artefacts, however, as observed by Bickerton, are not in themselves conclusive with respect to syntax: a structureless protolanguage may create and maintain complex technological traditions and symbolic codes without the need of syntax. According to Bickerton, a way for dating the emergence of syntax would be that of using convergent lines of evidence. Since evolutionary novelties are more likely to appear during a speciation event than in the midlife of a species, the emergence of modern humans in Africa could set the commencement date for syntax at ~ 160 ky BP. Since the laying down of universal grammar most probably preceded the *H. sapiens* Diaspora from Africa, dated to ~ 60 ky, and most of the behavioral novelties appear now to date between 160 ky BP and 110 ky BP, these three lines of arguments converge on, and are fully consistent with, the emergence of syntax in this 100 ky period. D'Errico suggests that in order to test this scenario we should gain better insight into the pertinence of each category of symbolic artefacts as correlates of language, in general, and syntax, in particular. Correlates of functions constitutive of syntax, such as hierarchical organization, Merge, recursion, and links between distant elements (see Rizzi, this volume) are found in complex symbolic nonlinguistic codes. The discovery in the archaeological record of symbolic artefacts displaying codes of such a complexity may identify past populations with cognitive abilities compatible with the development of syntactical language. This hypothesis testing approach should not be restricted to the African AMH considering that Neanderthals developed a number of behavioral innovations (e.g., burials, pigment use, ornaments, composite tools) suggesting their capacity of creating symbolic codes (d'Errico et al. 2003; Zilhão 2006).

In conclusion, languages with some form of syntax probably emerged before the appearance of complex symbolic artefacts in the archaeological record. This may have occurred in Africa, in conjunction with the origin of our species in that continent, or independently at different times and places among different fossil populations. The latter scenario is consistent with disputed (Coop et al. 2008) recent genetic evidence (Krause et al. 2007), indicating that the FOXP2 variant of modern humans was already present in the Neanderthal genome and that its appearance predates the common ancestor (dated to around 300–400 ky) of modern humans and the Neanderthals. This is because the complete replacement of the ancestral form with the modern form implies the presence of a selective pressure for whatever these mutations have contributed (i.e., possibly the increased fluency of syntactic speech). So the presence of FOXP2 in Neanderthals would suggest that this syntactic speech demand was present at least at the point that our two lineages shared a common ancestor.

Finally, diverse fossil evidence concerning such features as brain and vocal tract evolution may also aid in determining the time course of the evolution of syntactic speech. For example evidence for changes in brain size and morphology—beginning roughly 2 million years ago and reaching modern proportions as early as 400,000 years ago—may be relevant (Bruner et al. 2003). In addition, evidence for the time course of vocal evolution has been supplied by discoveries of modern appearing hyoid bones that predate AMH (Martínez et al. 2008) and changes in the cranial and spinal openings associated with neural innervations of tongue and diaphragm in hominids, which may correlate with the timing changes in voluntary control of these muscles associated with vocalization.

Conclusions

We are very far away from being able to answer all of the relevant questions regarding the biological grounding and evolution of syntax. However, progress can be observed on almost all grounds and some conclusions can be made.

First, one can safely say that no known animal communication system has the complexity of human language (syntax). Combinatoriality, which can be observed in animal communication, does not equal compositionality, which as far as we know has not been observed. Neuronal implementation of stacks for language processing (i.e., the ability to move beyond level 1 recursion or to process CFG) might be one of the key features lacking in animals. It is important for research to continue exploring animal's abilities to process and produce different types of grammars. Neurobiological studies on animal cognition are currently severely lacking in this regard, and thus this represents a very promising and interesting field for future studies.

As far as the genetic background of human language is concerned, we have come a long way from the simplistic view of FOXP2 as a gene for grammar. To determine how it truly acts with regard to language requires a careful characterization of the phenotypes associated with its variation in humans and in animal models. In addition to further probing its linguistic effects, nonlinguistic phenotypes associated with the KE mutation should be evaluated. If we could unveil a subtle motor deficit in the general domain of sequential learning, this would fit with the altered motor learning exhibited by mutant mice bearing the KE-like Foxp2. Perhaps a nonlinguistic phenotype is shared across mammals.

With regard to vocal communication, careful characterization of the impact of diverse Foxp2 mutations on mouse ultrasonic vocalizations is warranted to rule out deficits due to general developmental delays. Similar care in the construction of mutant animals can avoid unintended effects of the sheer mechanics of gene manipulation. Conditional mutants in which the expression of the altered gene can be activated at different stages of maturity will help identify which aspects of the full phenotype are the result of altered brain structure and which are due to altered real-time function.

All of the above issues apply to the study of FoxP2 in songbirds, the most experimentally tractable animal model to share the vocal learning capacity with humans. Following alteration of FoxP2 levels in the brains of songbirds, we should carefully pursue (a) any syntactic phenotype, (b) any nonvocal motor deficits such as serial response time, and (c) the relative effects of the mutation on brain formation versus ongoing function during singing. Finally, comparison of the gene networks that FoxP2 participates in across several animal groups and humans holds promise for providing valuable information about what is shared and what is unique. This will likely point us to other genes that contribute to acoustic communication and vocal learning, including more common developmental disorders of language.

The exploration of evolutionary mechanisms that are capable of producing such a complex set of adaptations like human language should also continue. Functional recruitment of old adaptations to new functions (exaptation followed by fine-tuning) could have been a major factor in language evolution. Relaxation of selective constraints is an important source of variation. It may have allowed the creation of synergistic circuits of previously decoupled or loosely coupled brain regions which then could generate more complex behavior as a raw material of further adaptive evolution.

The coevolution of genes and social transmission probably played a crucial role in the evolution of syntax. Breaking down this process into subprocesses (i.e., Baldwin effect, genetic assimilation, the evolution of learning rules and biases) and specifying the role of these subprocesses could be a huge step forward. Of course this assumes a better understanding of gene–culture coevolution which underlines the importance of the modeling approach (see Jaeger et al., in this volume).

Finally, to test our theories of syntax it is crucial to identify the cultural and the biological correlates that we may find in the fossil record. There are several promising ways to carry out such research, from identifying the symbolic, syntactic, and cognitive requirements of the complex traditions that left their traces in the fossil record, to identifying the genetic, neuronal, and morphological correlates of modern human language.

Here we tried to show the usefulness of biological grounding of potential theories of syntax. While many questions remain, we are encouraged by the apparent progress. The flourishing cooperation between scientists from very different backgrounds offers hope that knowledge from various fields will be integrated, eventually enabling us to resolve the "hardest problem of science": the origin of human language and that of syntax.

Acknowledgments

We would like to thank the insightful and inspiring contribution of the following people: Derek Bickerton, Denis Bouchard, Ted Briscoe, Morten Christiansen, Angela Friederici, Tom Givón, Simon Kirby, Pete Richerson, Luigi Rizzi and Mark Steedman.

Brain

11

Brain Circuits of Syntax

Angela D. Friederici

Abstract

Against the background of possible neuroanatomical differences of the human and nonhuman primate brain, brain circuits for the processing of syntax are considered. Evidence from event-related potential studies as well as from functional brain imaging studies is presented indicating that local phrase-structure processing and the processing of hierarchical dependencies can be mapped onto two separable neural networks: the former involves the frontal operculum and the anterior superior temporal gyrus; the latter involves Broca's area and the posterior superior temporal gyrus. Neuroanatomical data indicate that the respective brain areas of each of these networks are connected by different fiber tracts. These findings suggest that the two syntactic brain circuits are separable, functionally as well as structurally.

Introduction

The evolution of language has been discussed for decades. The discussion has ranged from a view that specialized brain mechanisms specific to language acquisition have evolved over long periods of natural selection (Pinker and Bloom 1990) to a view that language has adapted to the nonlinguistic constraints deriving from language learners and users, i.e., their brains (Christiansen and Chater 2008). Both views admit the relevance of the brain to the evolution of language, either with respect to domain-specialized brain systems or to its domain-general processing constraints. It may therefore be worthwhile to consider the human brain in more detail.

In this chapter, I will describe the brain systems underlying syntax in the contemporary human and discuss their possible phylogenesis.

Most views on the evolution of language agree with the notion that syntax is the crucial aspect that differentiates human natural languages from other communication systems. Chimpanzees, for example, are able to learn words (i.e. the arbitrary relation between a symbol and an object or an action), but they are unable to learn syntactic structures beyond the level of a Subject–Verb or Verb–Object relation (Terrace et al. 1979; Ristau and Robbins 1982).

Hauser, Chomsky, and Fitch (2002) claim that recursion is the core of human syntax and the ability to process recursive structures is a major aspect differentiating human from nonhuman primates. This claim has triggered a number of empirical studies in humans and nonhuman animals, and has launched an intensive discussion which is still ongoing.

Behavioral Studies in Humans and Nonhuman Animals

While investigating cotton top tamarins (*Saguinus oedipus*), Fitch and Hauser (2004) found that these animals were able to learn a finite state grammar (FSG) of the AB^n type, but not a simple phrase structure grammar (PSG) of the A^nB^n type. Humans, in contrast, were able to learn both equally well. It was concluded that humans, but not nonhuman primates, can process hierarchical structures. This finding seemed to provide preliminary support for the claim put forth by Hauser et al. (2002) although, clearly, the PSG used in the empirical study did not include recursive structures, but only hierarchical structures.

This finding and the conclusions have been challenged both on theoretical and empirical grounds. It was argued that the PSG used in this study did not necessarily require the processing of hierarchical structures, but could be accommodated by a simple counting strategy (Perruchet and Rey 2005; de Vries et al. 2008). Thus it was not too surprising when it was demonstrated that this PSG could be learned and processed by starlings (Gentner et al. 2006), animals that are able to count.

Neural Structure in the Human and Nonhuman Primate

Apart from triggering comparative behavioral studies, the argument put forward by Hauser et al. (2002) also led to the consideration that if there was a principled difference between human and nonhuman primates, this might be evidenced in a specialized brain system fully developed in the human brain, but not in the nonhuman primate (Friederici 2004b). A review of the available literature (Friederici 2004b) revealed that complex syntax processing in humans was subserved by Broca's area, BA 44/45, whereas local syntactic dependencies (similar to an AB^n relation) involved the ventral premotor cortex (vPMC) and the frontal operculum (fOP).

This seemed interesting insofar as the vPMC/fOP is cytoarchitectonically different from Broca's area (Amunts et al. 1999) and in some views considered to be phylogenetically older than the Broca's area (Sanides 1962). The human vPMC is cytoarchitectonically characterized as agranular cortex, whereas BA 44 is dysgranular and BA 45 granular cortex. This characterization is based on the observation that the six-layered cortex differs with respect to the presence and amount of granular cells in layer IV. Brodmann (1909) used this characterization to define different cortical areas, the so-called Brodmann areas (BA). According to Brodmann's analysis, the vPMC (BA 6) is cytoarchitectonically clearly differentiated from BA 44 and BA 45, respectively. Using modern methods of cytoarchitectonic analysis, this differentiation between BA 6, BA 44, and BA 45 has found support (Amunts et al. 1999). A final analysis of the cytoarchitectonic structure of the fOP, however, is still missing, but underway (Amunts et al., in preparation).

A direct comparison between humans and our evolutionary next relatives, the chimpanzee, with respect to cytoarchitectonic structure of the cortex, is not possible as the respective cytoarchitectonic data from the latter species are not available. Cytoarchitectonic data are, however, available on the macaque brain (Petrides and Pandya 1994, 1999). A visual comparison of the relative extension of BA 6, BA 44, and BA 45 in the human and the macaque PFC, as displayed in Petrides and Pandya (1994), does suggest differences. It appears that the relative size of BA 44 (in relation of BA 6 and BA 45) in the macaque is smaller compared to the relative size of BA 44 in the human. However, the search of homologies and nonhomologies between the human and the monkey brain is difficult (Arbib and Bota 2003). There seems to be agreement that human BA 44 resembles area 44 in the macaque, as described by Petrides and Pandya (1994, 2002b), and F5, as defined by Matelli et al. (1985), although human BA 44 and macaque area 44 differ cytoarchitectonically. In contrast to human BA 44, which is dysgranular cortex, area 44 in the macaque is agranular and thus cytoarchitectonically comparable to human agranular BA 6. Functionally, the macaque agranular area 44 is involved with orofacial musculature (Petrides et al. 2005), whereas human dysgranular BA 44 subserves the processing of grammatical sequences (for a review, see Friederici 2004a). Orofacial movements in the human brain are controlled by the agranular BA 6. Thus the PFC in the human and the macaque brain differ in its macro- as well as its microstructure. Whether this observation holds once a direct and systematic comparison between the human and the macaque cortex is available, must await further studies.

Another possible difference between the human and macaque brain may lie in the structural connectivity between brain areas known to be relevant for syntactic processing. In humans, it has been reported in a number of different studies that BA 44/45 together with the posterior portion of the superior temporal gyrus (STG) support the processing of syntactically complex sentences (for a review, see Grodzinsky and Friederici 2006). Structural connectivity analysis in the human suggests that these two brain areas are connected by the superior longitudinal fasciculus (SLF) and the articulate fasciculus (AF) as the connecting fiber tracts between the prefrontal and the posterior temporal regions (Catani et al. 2003; Friederici et al. 2006a).

Recent analyses of the macaque brain indicate that in these animals the prefrontal cortex is not primarily connected to temporal regions, but to parietal regions (Schmahmann et al. 2007). On the basis of these data, Schmahmann et al. called into question the existence of a prefrontal-to-temporal connection as reported for the human brain. This conclusion is based on the preassumption of a structural identity of the human and the nonhuman primate brain. This preassumption, however, has been called into question by a recent analysis of structural connectivities comparing the human, the chimpanzee, and the macaque brain (Rilling et al. 2008). This study indicates phylogenetic differences in the structural connectivity pattern in the human and the macaque brain. Whereas in humans, there are strong connections from BA 44, BA 45, and BA 47 to the STG and the MTG, connections from the frontal lobe in macaques terminated in parietal and most posterior temporal regions.

Ontogenetically, it appears that the SLF and AF are among the last fiber tracts that myelinize (Pujol et al. 2006), suggesting their special role in brain maturation. The SLF is not fully myelinized even at the age of 3½ years (Pujol et al. 2006), an age at which children still demonstrate problems with the processing of complex syntactic structures, such as passive sentences (Fox and Grodzinsky 1998) or case-marked noncanonical object-first sentences (Dittmar et al. 2008; MacWhinney et al. 1985; Slobin and Bever 1982). Thus it is tempting to speculate that the processing of complex syntactic structures is related to the maturational stage of the SLF and AF.

Against this admittingly speculative neuroanatomical background, let us turn to an evaluation of the function of the different brain regions in the PFC and the STG for syntactic processing, and the structural connectivity between these.

Syntactic Circuits in the Human Brain

Recently, we proposed two different brain circuits supporting syntactic processing (Friederici et al. 2006a; Grodzinsky and Friederici 2006). One network consists of the fOP and the anterior STG involved in the buildup of local phrase structures (local syntax); the other network consists of Broca's area (BA 44/45) and the posterior STG that comes into play additionally when dependency relations between constituents of a sentence (complex syntax) are to be computed. The empirical evidence that led to this proposal is multifold. Here, I will mainly present evidence from my laboratory.

Early vs. Late Syntactic Processes

Starting from the theoretical consideration that parsing involves two processing stages—a first processing stage during which local phrase structures are built up on the basis of word category information, and a second stage during which grammatical relations between constituents are assigned (Frazier 1987b)—we conducted a number of experiments to evaluate these different processing stages (for a review, see Friederici 1995, 2002). In our first experiment, we focused on local phrase-structure building (the first processing stage). We used a violation paradigm in which the processing of correct sentences is compared to the processing of sentences containing a phrase-structure violation (i.e., a violation of the obligatorily required word category). Measuring event-related brain potential (ERP), we identified two ERP components in response to a phrase-structure violation: an early left anterior negativity (ELAN) and a late centro-parietal positivity around 600 ms (P600; Friederici et al. 1993, 1996; Hahne and Friederici 1999).

The P600 component had been observed in response to a number of different syntactic anomalies, including syntactic violations and syntactic ambiguities (Hagoort et al. 1993; Osterhout and Holcomb 1992; Osterhout et al. 1994) as well as syntactically complex sentences (Kaan et al. 2000), and is widely taken to reflect syntactic processing. We interpreted the P600 to reflect a late processing phase during which the ultimate syntactic relations are assigned and during which, if necessary, syntactic revision and repair takes place.

The ELAN which occurs with outright syntactic phrase-structure violations was taken to reflect an initial processing phase of local phrase-structure buildup; that is, the inability to build a local phrase structure due to an element with the incorrect word category (e.g., verb instead of noun; correct: *The pizza was in the fridge*; incorrect: *The pizza was in the eaten*). The ELAN effect was reported in a number of studies in different languages (Friederici et al. 1993; Neville et al. 1991; Ye et al. 2006). Dipole localization analysis of this early syntactic effect in magnetoencephalography (MEG) data revealed two dipoles in each hemisphere with maxima in the left hemisphere: one in the anterior STG and one in the inferior frontal region (Friederici et al. 2000).

The involvement of the anterior temporal region and the inferior frontal region during the process reflected in the ELAN was confirmed by lesion data. In patients with lesions in the anterior temporal lobe, the ELAN component was absent; the same was true for patients with lesions in the inferior frontal cortex (Friederici et al. 1998, 1999; Friederici and Kotz 2003).[1] Patient data additionally revealed that only left inferior frontal cortical lesions, but not left subcortical lesions in the basal ganglia, lead to an absence of the ELAN, indicating that the left anterior cortical structures are relevant for the processes reflected by the ELAN (Friederici et al. 1999).

An fMRI study using the same sentence material was conducted to specify the brain regions involved in more detail. The fMRI data showed increased activation in the temporal cortex when comparing syntactically and semantically

[1] The specific functions of the temporal and the inferior frontal region in the process reflected by the ELAN still need to be defined. The process clearly should involve several subprocesses, such as the prediction of the upcoming word category (based on the prior syntactic information), the recognition of the upcoming word category, and the check for a match of the predicted and perceived word category. The prediction is likely to involve the IFG as the PFC is known to predict upcoming elements in structured sequences (Schubotz and von Cramon 2002). The recognition is likely to involve the temporal region as the temporal cortex is taken to house the lexicon (e.g., Mummery et al. 1996; Binder et al. 1997). It is debatable, however, whether the checking process necessarily involves the IFG or whether this checking process can be thought of as a template-matching process which could be based in the anterior STG assumed to be involved in combinational processes (Hickok and Poeppel 2007).

incorrect sentences to their correct counterparts. While the middle portion of the left STG showed increased activation for both violation types, the anterior STG and the fOP displayed more activation for sentences with syntactic phrase-structure violations compared to correct sentences and to sentences with semantic violations. The posterior STG was seen active in both violation conditions, suggesting its involvement in syntactic and semantic sentential processes (Friederici et al. 2003). Thus it appears that the fOP, anterior STG, and posterior STG are supporting syntactic processes. Based on the MEG localization data for the ELAN effect, which showed an involvement of the anterior STG and the frontal brain region and by logical exclusion, one might hypothesize that the posterior STG supports processes reflected by the P600.

Local Phrase Structure vs. Complex Syntactic Structure

It is interesting to note that syntactic phrase-structure violations do not seem to activate Broca's area, but only the fOP. Broca's area comes into play when syntactically complex sentences (e.g., object-first sentences) are processed. This was evidenced by the present data as well as by other fMRI studies on syntactic processes in different languages, such as German, English, and Hebrew (for a review, see Friederici 2004a).

Using German as a testing ground, we demonstrated that the inferior portion of the left BA 44 parametrically increased its activation with increasing syntactic complexity operationalized as the number of objects moved in front of the subject (Friederici et al. 2006b). In earlier studies with nonparametric designs, the left IFG had been reported to show increased activation for syntactically more complex compared to less complex sentences in German (Roeder et al. 2002), in English (Stromswold et al. 1996), and in Hebrew (Ben-Shachar et al. 2003).[2]

In these studies, however, the factor syntactic complexity was confounded with the factor working memory, as the syntactically more complex sentences also increased working memory demands—at least under the common assumption that object-first structures are noncanonical (derived from canonical subject-first structures) whose parsing requires the reconstruction of the underlying canonical form. Studies that varied syntactic complexity and working memory as independent factors only recently managed to separate the neural basis of these. Earlier fMRI studies reported a recruitment of the left inferior frontal cortex for both object-relative clauses and long-distance syntactic dependencies (Cooke et al. 2001) or for object-first *wh*-questions and the distance of the syntactic dependency (short vs. long) (Fiebach et al. 2005). The latter study reports a post hoc cluster analysis of the IFG activation, which suggests

[2] Note that these studies also report additional brain regions for the critical comparisons between more complex and less complex structures. For critical reviews of these studies, see Caplan (2006a) and Caplan et al. (2007b).

two functionally distinct areas within BA 44: the superior portion correlated with the factor distance and the inferior portion correlated with the factor syntactic complexity.[3]

In a more recent fMRI study, we were able to segregate syntactic complexity from verbal working memory in a design which crossed the two factors systematically (Makuuchi et al. 2009). Syntactic complexity was operationalized as the number of center-embedded sentences while working memory was operationalized as the number of words between syntactically dependent elements in the sentence. Different subregions in the left IFG were shown to be recruited as a function of these two factors. Syntactic complexity affected the inferior portion of BA 44, whereas working memory demands affected the superior frontal sulcus dorsally located to BA 44.

From these fMRI studies, we may conclude that Broca's area is recruited when complex syntactic structures are processed, whereas local phrase-structure building does not necessarily recruit this area, but rather a phylogenetically older brain region: the fOP.

Functional Segregation of Broca's Area and fOP

A functional specification of those brain areas supporting local syntactic processes and complex syntactic processes was achieved by a direct comparison of brain activation elicited by local phrase-structure processing and the processing of more complex, hierarchical structures in fMRI studies using an artificial grammar paradigm (Friederici et al. 2006a; Bahlmann et al. 2008). In a first fMRI experiment (Friederici et al. 2006a), we investigated the processing of an FSG of the AB^n type and of a PSG of the A^nB^n type similar to the grammars used by Fitch and Hauser (2004). In the Fitch and Hauser study, category membership was marked by pitch (i.e., male vs. female voice). In our study, category membership was marked by phonology of the consonant-vowel syllables (Figure 11.1).

Participants in this study were divided into two groups: one group learned the FSG, the other learned the PSG. During the test phase in the MR scanner, each group was presented with correct and incorrect sequences of the respective grammar which they had previously learned. A comparison of brain activity between incorrect and correct sequences revealed an increase in activation in the fOP for the incorrect FSG sequences. For the processing of the PSG, the same comparison showed activation not only in the fOP, but, moreover, in Broca's area; that is, BA 44 extending to the posterior portion of BA 45 (Figure 11.2). From this finding, we concluded that Broca's area only comes into play when complex, hierarchical structures are processed.

[3] Using English as a testing ground, Santi and Grodzinsky (2007) also varied syntactic complexity (number of movements) and distance between dependent elements and found an interaction of the two factors in Broca's area, here BA 45.

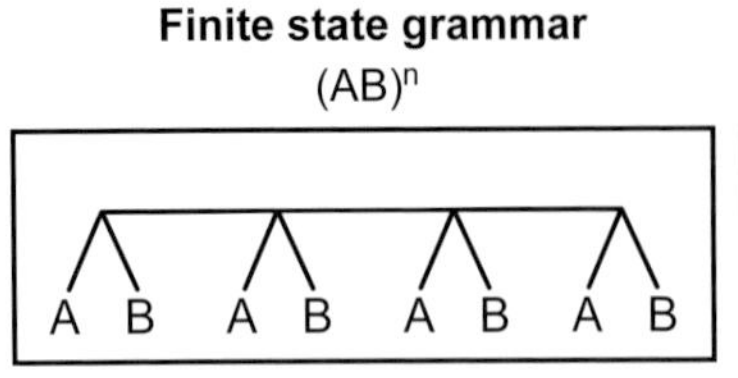

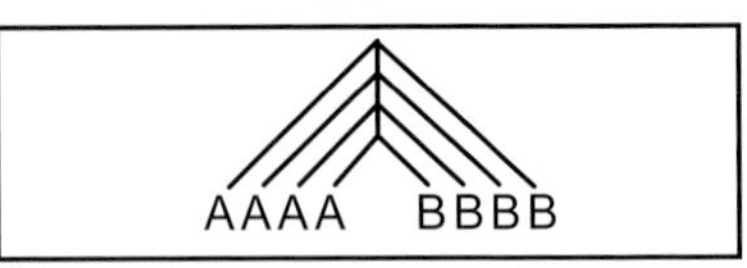

cor/short: A B A B de bo gi fo
viol/short: A B A **A** de bo gi **le**
cor/long: A B A B A B A B le ku ri tu ne wo ti mo
viol/long: A B A B A B A **A** le ku ri tu ne wo ti **se**

cor/short: A A B B ti le mo gu
viol/short: A A B **A** ti le mo **de**
cor/long: A A A A B B B B le ri se de ku bo fo tu
viol/long: A A A A B B B **A** le ri se de ku bo fo **gi**

syllables of category A: de, gi, le, mi, ne, ri, se, ti
syllables of category B: bo, fo, gu, ku, mo, pu, to, wu

Figure 11.1 General structure and examples of stimuli in the FSG and PSG. Members of the two categories (A and B) were coded phonologically with category "A" syllables containing the vowels "i" or "e" (de, gi, le, ri, se, ne, ti, and mi) and with category "B" syllables containing the vowels "o" or "u" (bo, fo, ku, mo, pu, wo, tu, and gu). The same syllables were used for both types of grammar. The positions of the violations in the sequences were systematically changed. Examples of correct (corr) and violation (viol, in bold) sequences are shown for the short and the long condition in each grammar. After Friederici et al. (2006b).

However, the criticism that was raised with respect to the Fitch and Hauser (2004) study—namely that the A^nB^n grammar used does not necessarily require hierarchical processing, but may be processed by a simple counting strategy—also applied to our study. Indeed, this criticism was raised quite specifically, challenging our conclusions (Perruchet and Rey 2005; de Vries et al. 2008).

In a second fMRI study (Bahlmann et al. 2008) we designed a new PSG that forced necessarily hierarchical processing. In this PSG, dependency of the related elements was marked by a voiced–unvoiced consonant relation. This led to subcategories within the category of A elements (i.e., A_1, A_2, A_3, etc.) and B elements (i.e., B_1, B_2, B_3, etc.) (Figure 11.3). To avoid item learning, each of these subcategories had more than one member. In order to process a sequence $A_3 A_2 A_1 B_1 B_2 B_3$, the parser had to build up a hierarchical relation between the respective embedded elements (i.e., $A_1 B_1$ embedded within $A_2 B_2$, etc.).

In this study, one single group of participants learned both the FSG (AB^n) and the PSG (as defined above) in order to allow for direct comparisons of FSG and PSG and of correct and incorrect sequences separately. In the test phase, participants were confronted with incorrect and correct sequences and were required to judge these for grammaticality with respect to the implicitly learned rules. The comparison between grammar types revealed a selective increase of activation in BA 44/45, both when comparing all sequences (collapsed over correct and incorrect items) (see Figure 11.4), but also, importantly, when only comparing correct sequences of PSG and FSG (see Table 2 in Bahlmann et al. 2008). Thus this second study confirmed our conclusion from the first artificial grammar study (Friederici et al. 2006a) by demonstrating that the BA

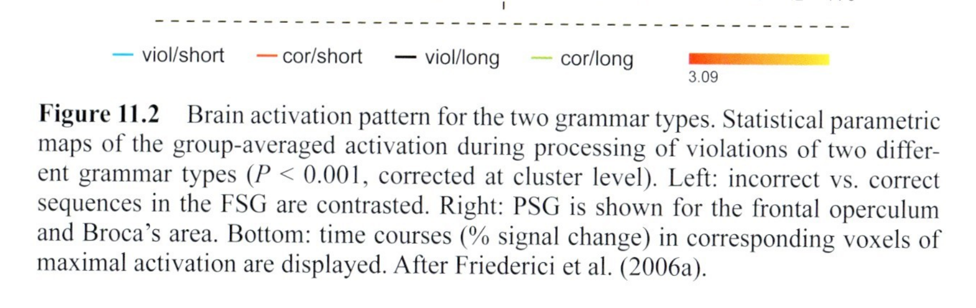

Figure 11.2 Brain activation pattern for the two grammar types. Statistical parametric maps of the group-averaged activation during processing of violations of two different grammar types ($P < 0.001$, corrected at cluster level). Left: incorrect vs. correct sequences in the FSG are contrasted. Right: PSG is shown for the frontal operculum and Broca's area. Bottom: time courses (% signal change) in corresponding voxels of maximal activation are displayed. After Friederici et al. (2006a).

44/45 is recruited for the processing of hierarchical structures compared to the processing of local phrase structures.

Thus, there appears to be some evidence for a functional segregation of the Broca's area from the fOP. To provide additional support for a possible segregation of the fOP and Broca's area, we sought structural differences between

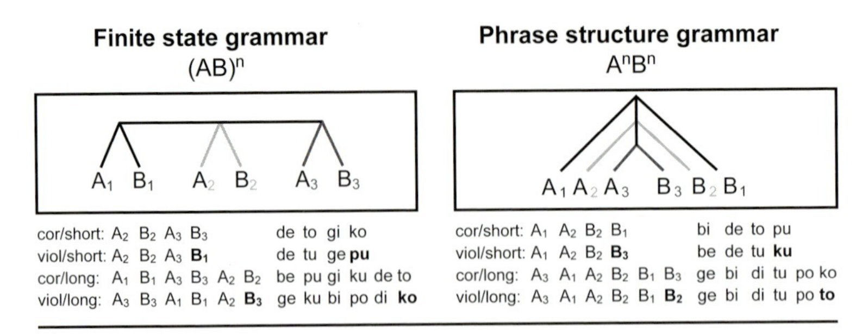

syllables of category A: be, bi, de, di, ge, gi
syllables of category B: po, pu, to, tu, ko, ku
Relation between A_n - B_n: voiced - unvoiced consonant

Figure 11.3 General structure and examples of the two rule types. The local dependency rule was generated by simple transitions between categories of consonant-vowel syllables. The hierarchical rule was produced by embeddings between the two syllable categories. Short and long sequences were applied. Violations of the structure were situated at the last 3 or 4 positions (short sequences) and at the last 4, 5, or 6 positions (long sequences). In the given example, the violations are placed at the fourth position for short sequences and at the sixth position for long sequences (bold letters). After Bahlmann et al. (2008).

the two brain areas. We reasoned that if the two brain areas were functionally different, they should each be part of different structural networks.

Our hypothesis was that given the MEG data (Friederici et al. 2000) and the fMRI data (Friederici et al. 2003) on local phrase structure violation, one network might involve the fOP and the anterior STG, while another network

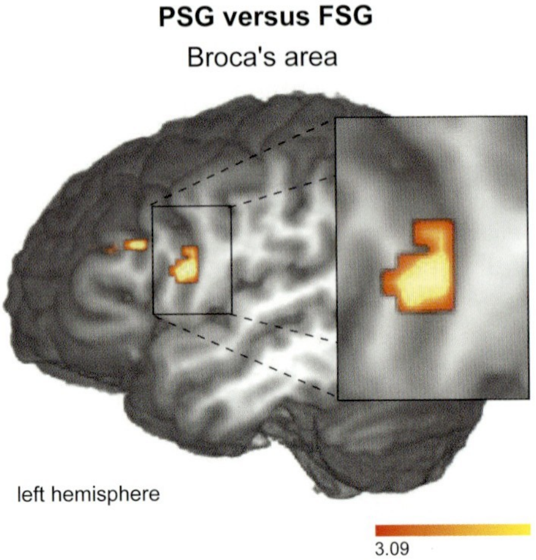

PSG versus FSG
Broca's area
left hemisphere
3.09

Figure 11.4 Brain activation pattern for PSG minus FSG rule. Statistical parametric maps are shown of the group-average activation ($P < 0.001$, corrected at duster level) for Broca's area. After Bahlmann et al. (2008).

supporting the interpretation of complex sentences might include Broca's area (Stromswold et al. 1996; Caplan et al. 2000b; Roeder et al. 2002; Ben-Shachar et al. 2003) and possibly the posterior STG, as this region was found to activate more for noncanonical object-first sentences (Cooke et al. 2002; Constable et al. 2004; Bornkessel et al. 2005).

Structural Connectivity Pattern for Broca's Area and fOP

In search of information about the structural underpinning of two separable networks, we used diffusion tensor imaging (DTI) techniques which allows the fiber tracts connecting different brain areas to be analyzed. We set the respective starting points (seeds) of the analysis in those brain regions that were identified as showing the maximal functional activation in our first artificial grammar fMRI study (i.e., the fOP and BA 44/45; Friederici et al. 2006a). The analysis suggested two separate networks: one connecting the fOP via the fasciculus uncinatus to the anterior STG and further into the temporal cortex (local syntax); the other connecting Broca's area via the SLF and AF to the posterior STG expanding further into the temporal cortex (complex syntax) (see Figure 11.5).

Functionally, the local syntax network appears to support local phrase-structure building (cf. MEG study, Friederici et al. 2000, and fMRI study, Friederici et al. 2003). The complex syntax network, however, seems to subserve the processing of syntactically complex sentences. While the former network with its frontal and temporal part is well evidenced, the same does not hold for the latter network. With respect to Broca's area and its involvement in the processing of complex sentences, the fMRI studies discussed above provide good evidence (Stromswold et al. 1996; Caplan et al. 2000b; Roeder et al. 2002; Ben-Shachar et al. 2003; Friederici et al. 2006b; Santi and Grodzinsky 2007b). With respect to the posterior STG and its involvement in the processing of syntactically complex sentences, the supportive database is sparse (cf. Cooke et al. 2002; Constable et al. 2004; Bornkessel et al. 2005).

Although the left posterior STG has been shown to be involved in sentence processing, its specific role still needs to be defined. This area is activated when the parser encounters ungrammatical strings of natural language (Schlesewsky and Bornkessel 2004), when processing syntactically complex object-first compared to subject-first sentences (Cooke et al. 2002; Constable et al. 2004; Bornkessel et al. 2005). This area, however, has also been reported to activate when processing selectional restrictions of verbs (Friederici et al. 2003), as a function of verb complexity (Ben-Shachar et al. 2003), and of verb-based argument hierarchies (Bornkessel et al. 2005). Thus it appears that the function of the posterior STG is to promote the integration of syntactic and verb-based syntax-relevant information during sentence comprehension.

Tentative support for this view comes from ERP work. Late centro-parietal distributed positivity (P600) has been correlated with processes of syntactic

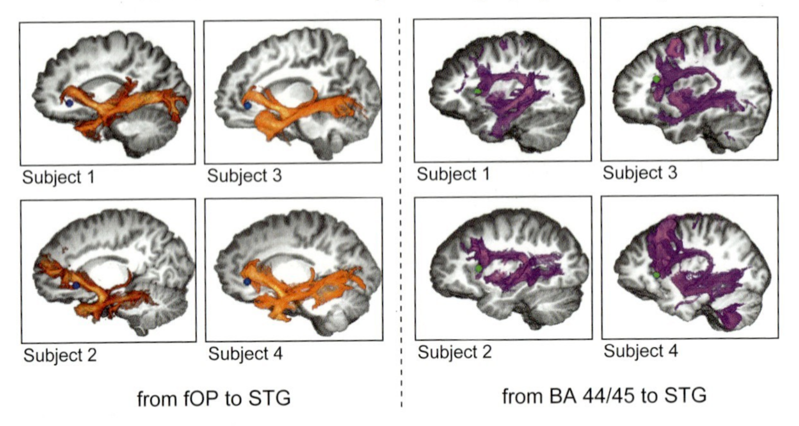

Figure 11.5 Tractograms for two brain regions: fOP and Broca's area. Three-dimensional rendering of the distribution of the connectivity values of two start regions with all voxels in the brain volume (orange, tractograms from fOP; purple, tractograms from Broca's area). Four representative subjects of the FSG group with their individual activation maxima in the fOP (blue) in the critical contrast incorrect vs. correct sequences ($P < 0.005$). The individual peaks of the functional activation were taken as starting points for the tractography. Four representative subjects of the PSG group with their individual activation maxima in Broca's area (green) in the critical contrast incorrect vs. correct sequences ($P < 0.005$).

integration (Kaan et al. 2000), and has been observed during the processing of garden-path sentences (Osterhout and Holcomb 1993) and syntactically complex sentences (Kaan et al. 2000). It has also been observed with violations of the verb-argument structure (Friederici and Frisch 2000; Frisch et al. 2004), with combined syntactic and lexical-semantic violations (Gunter et al. 2000), and with semantically based violations of verb-argument relations (Kuperberg 2007). Unfortunately, however, no data concerning the localization of the P600 are available. Further empirical support for the view that the posterior STG is involved in processes of syntactic integration is needed before any firm conclusions can be drawn.

The functional relevance of the connection of Broca's area and the posterior STG via the SLF and AF certainly has yet to be determined in more detail. Based on the available ontogenetic and phylogenetic data, we may hypothesize that the SLF and AF play a crucial role in the neural network supporting the processing of complex syntax.

Ontogenetically, the SLF and AF appear to be among the latest fasciculi to myelinize fully (Pujol et al. 2006). Even by the age of 6 years, children still show significant differences to the myelinization status compared to adults (Brauer et al., in preparation). Behavioral studies in children show that they only begin to process complex sentences (e.g., passive sentences; Fox and Grodzinsky 1998) between 3.5 and 4 years of age and demonstrate problems with the processing of object-first case-marked sentences beyond the age of 5 years (Dittmar et al. 2008). These observations suggest that there might be a relation between functional development and structural maturation. However, this has not yet been demonstrated and requires further correlational studies.

Phylogenetically, it has been shown that the SLF in the macaque brain does not project to the STG, but to the parietal cortex (Schmahmann et al. 2007). A recent DTI study comparing the fiber connections between the frontal and the temporal cortex in the human and macaque brain reported substantial differences between the two species, in particular with respect to their strength and outbranching in the temporal cortex (Rilling et al. 2008).

Conclusion

The evidence discussed here allows us to separate two different functional brain circuits: one responsible for local phrase-structure building involving the fOP and the anterior STG; another subserving the processing of hierarchically structured complex sentences. A crucial part of this latter circuit is clearly Broca's area (BA 44) and most likely the posterior STG. The available data, moreover, allow us to define two separable pathways connecting the respective brain regions in each of these circuits. It is hypothesized that the connection between Broca's area and the posterior STG relevant for complex syntactic processes may develop late phylogenetically as well as ontogenetically. This hypothesis, however, needs future empirical evaluation.

12

Neural Organization for Syntactic Processing as Determined by Effects of Lesions

Logic, Data, and Difficult Questions

David Caplan

Abstract

The two sources of information about the way the brain is organized to support syntactic processing are (a) analysis of effects of lesions on syntactic processing (the lesion approach) and (b) analysis of changes in neural activity that are associated with syntactic functioning (the activation approach). This chapter reviews results using the first of these approaches. It will become clear from the discussion that I believe that the effort to understand aphasic performances is very demanding; in my view, it is only recently that some of the challenges in understanding these performances have begun to become at all clear, and these insights have only occurred in the study of comprehension, not production. For this reason, I focus solely on comprehension.

Terms

In this chapter, the terms *function*, *operation*, *process*, and similar terms refer both to (a) basic, elementary, "atomic," psychological processes that cannot be further divided into smaller functions (e.g., inserting a lexical item into a phrase marker, or moving a node to another location in a phrase marker) as well as to (b) natural (i.e., scientifically legitimate) groupings of these elementary processes (e.g., all syntactic movement, or all syntactic movement of a particular type, could be referred to as a function, operation, process, or by a similar term). A *deficit* is a disturbance in a function, operation, process, etc. "Deficit" can refer to any abnormality of a function, including the absence of a representation, the inability to apply an operation, or inefficient application of an operation. Terms such as *neurological feature*, *neurological entity*, and *area* refer to natural (i.e., scientifically legitimate) divisions of the brain. These terms will be used to refer to subcellular elements (e.g., receptor profiles), microscopically identified features of areas of the brain (e.g., cytoarchitectonic areas), macroscopically identified features of areas of the brain (e.g., gyri), physiological processes (e.g., neuronal spikes), and others. In some places (e.g., when developing the basis for drawing inferences from data to models of brain organization), the broadest possible range of referents of these terms is appropriate; in other places (e.g., when discussing particular models), specific senses are intended.

Logical Issues in Identifying Neural Features that Are Necessary and Sufficient for a Function

The two approaches to studying the functional organization of the brain involve different logical relations between data and models.

Beginning with lesion data, two conclusions can be drawn from the effects of a lesion. If it can be established that a function cannot be performed normally after a lesion (i.e., that there is a deficit in a function), it follows that the area of the brain that is lesioned is necessary for the normal exercise of that function. This logical inference applies to lesions and deficits at a particular point in time. It is possible that, over time, other brain areas become capable of performing a cognitive function that was deficient at the time a patient was first tested. Available knowledge points to recovery of normal functions after lesions, especially early in life. In such cases, the areas that were necessary for the function at the time they were lesioned were not irreplaceable; other areas have a latent capacity to perform the function, which emerges at some point after a lesion to the primary areas in which the function is accomplished. If it is reasonable to assume that a deficit is permanent, it is valid to conclude that the features of the brain that are necessary for the function that is deficient cannot be replaced; no other brain areas have the latent capacity to perform the function.

Conversely, if a function is spared by a lesion, the features of the brain that are affected by the lesion are not necessary for the normal performance of that function, at the time of the lesion. If a second lesion affects the function if and only if the first lesion is present, the implication is that combination of the features of the brain affected by the two lesions is necessary for the function.

In contrast, the logic underlying activation approaches is that if a feature of the normal brain is affected by the performance of a function, that feature of the brain is part of the neural system sufficient for the performance of that function. The term *affected* should not be interpreted as meaning "increases its activity." Any neural change associated with the exercise of a function is a candidate for a feature of the brain that is part of the neural system sufficient for that function; the exact role of a neural change must be examined by systematic exploration.

Activation and lesion approaches are logically related. Phrased in terms of areas of the brain, in any one individual, the totality of all the areas of the brain that are individually necessary for a function must be the same as the set of areas that are jointly sufficient for the performance of that function. There can be discrepancies, however, between the results of lesion and activation studies for many reasons.

Neurologically, in the best of circumstances, we can only assume that the activation approach reveals at most a subset of the brain areas sufficient for the exercise of a function (e.g., better technology may lead to more activation); areas that are missed may be ones that are necessary for the function. Therefore, it is possible for a brain area to not be activated by a function and for a lesion in that area to lead to a deficit in the function. Conversely, not all lesions in a brain area affect the neural elements necessary for a function. It is commonplace that slowly growing infiltrative lesions, such as brain tumors, can leave functions (apparently) undisturbed even though acutely developing necrotic lesions, such as strokes in the same areas, lead to deficits. Also, any given psychological measure may not be sensitive to the effects of a lesion on a function. For both of these reasons, an area may be activated by a function, but the function may not be affected by certain lesions in that area, as measured by specific tests of the function. Psychologically, experiments virtually never activate only the functions under study; some activation usually reflects other cognitive operations that accompany the operation of the function under study. Thus a lesion in an area which is activated in a study designed to activate a function may have no effect on the function because the lesioned area supports these other concurrently active functions.

However, if a study can be designed so that any areas activated are due in part to the function under study, then those areas are a subset of the neural system sufficient for the performance of the function. Therefore, it is possible to check on the validity of inferences about functional neural organization by comparing activation and lesion studies. Converging results—lesions in activated areas leading to the expected deficits—point to the areas that are individually necessary and jointly partially sufficient for the exercise of a function. Divergent results require that one or another assumption be abandoned. If an area of the brain is activated in a study and a lesion in that area does not lead to a deficit in the function which the activation study was designed to activate, then either the activation study was imperfect and cognitive operations other than the function under study were activated, or the measure of the deficit is inadequate and further studies will reveal a deficit in the function in question, or the features of the brain that are lesioned are not ones that are necessary for the function, or the model of cognition is incorrect and the postulated function does not exist, or the function is duplicated in several areas of the brain or its neural basis varies across the population. Given this broad range of possibilities, it is clear that, unless results are convergent, the implications of the combination of lesion and activation studies for functional neuroanatomy are not clear.

Activation studies performed after a lesion yield information about plasticity; the areas that are sufficient for a function after those that were necessary for it at one point in time have been lesioned. These areas may be ones that were activated before the lesion, or ones that are entirely newly recruited to perform the function. Lesions to such areas should result in the reemergence of the deficit if those areas were in fact activated by the function, and not by concurrent psychological operations.

The basic logic of lesion and activation studies holds if functions are duplicated or vary in their neural basis, but the inferences are much more complex. For instance, if functions are duplicated in different areas, the set of areas that are individually necessary for a function is disjunctive, and the effect of a lesion in an area activated in an ideal experiment must be restated to say that all the areas activated by a function must be lesioned to produce a deficit. If the function varies in its location (and thereby leads to multiple areas of activation in a group of participants in an experiment), the effect of lesions in areas activated by a function must be considered on a case-by-case basis (obviously impossible with naturally occurring lesions, but possibly possible with transcranial magnetic stimulation) or becomes probabilistic in a group.

Deficit Lesion Correlation Studies

As noted above, if a function cannot be performed normally after a lesion, the feature of the brain that is lesioned is necessary for the normal exercise of that function. To apply this approach, we need to characterize deficits, lesions, and the relations between them.

Characterization of Deficits in Syntactically Based Comprehension

Since the seminal paper of Caramazza and Zurif (1976), it has been appreciated that evidence for a deficit affecting syntactic operations in a patient consists of the combination of abnormally low performance in understanding sentences that require a syntactic analysis to be understood and the retained ability to understand sentences with the same syntactic structures in which meaning can be inferred from the meanings of the words in a sentence and knowledge about likely relations between them. Sentences that require a syntactic analysis to be understood are called *semantically reversible sentences* because the nouns can be reversed around the verbs and the sentences remain plausible (e.g., *The boy who the girl pushed is tall*). Sentences in which meaning can be inferred from the meanings of the words in a sentence and knowledge about likely relations between them are known as *semantically irreversible sentences* (e.g., *The book that the girl read is long*). Caramazza and Zurif (1976) were the first researchers to document this pattern of performance, which they found in four Broca's aphasics and four "mixed anterior" aphasics. Models of deficits use the pattern of reversible sentences that can and cannot be understood to characterize deficits. For instance, if a patient cannot understand sentences such as *The boy who the girl pushed is tall* normally, but shows normal comprehension of sentences such as *The girl who pushed the boy is tall*, the range of operations that are deficient is restricted to some degree.

Two basic views have been articulated regarding the nature of deficits in syntactically based comprehension. The first is that individual parsing or interpretive operations are selectively affected by brain damage. The second is that patients lose the ability to apply what have been called "resources" to the task of assigning and interpreting syntactic structure. The first of these deficits may be likened to a student not being able to calculate π to 8 decimal places in his/her head because s/he does not know the formula for calculating π. The second may be likened to a student knowing the formula but not being able to hold the intermediate products of computation in mind.

Researchers who advocate "specific deficit" accounts of aphasic disturbances have claimed that certain patients' patterns of performance indicate that specific representations and processes specified in linguistics and in psycholinguistic models of parsing and sentence interpretation are deficient. For instance, the *trace deletion hypothesis* (Grodzinsky 2000) maintains that individual patients cannot process sentences which Chomsky's theory maintains contain a certain type of moved items (trace deletion hypothesis refers to an earlier version of Chomsky's theory in which these items were moved and left a "trace"). The claim is that some patients have lost the ability to connect certain types of moved (in modern version of syntactic theory, copied) items to their points of origin.

Evidence for such deficits would come from the finding that a patient had an impairment restricted to processing the sentences that required that structure or operation to be understood. Proponents of these models have argued that there are data from aphasia of this sort. For instance, Grodzinsky (2000) reviews studies in several languages which he claims show that Broca's aphasics have abnormally low performances on sentences in which these movement/copy operations have taken place, and where simple heuristics cannot lead to a normal interpretation. Such patients have chance performance on sentences such as in (1), but above chance performance on the sentences in (2).

(1) (a) The boy$_i$ was pushed t_i by the girl.
(b) It was the boy$_i$ who$_i$ the girl pushed t_i.
(c) The boy$_i$ who$_i$ the girl pushed t_i was tall.

(2) (a) The boy pushed the girl.
(b) It was the boy_i who_i t_i pushed the girl.
(c) The boy_i who_i t_i pushed the girl was tall.

According to the trace deletion hypothesis, in both (1) and (2), the trace (indicated by *t*) is absent. Therefore, the noun to which it is normally related (shown by the subscript) is not connected to it in the usual fashion. In these circumstances, Grodzinsky argues, a heuristic assigns the thematic role of Agent to this noun. Therefore, in (1), the aphasic constructs a representation of the sentence in which there are two agents, and chooses one or the other at random when asked to match these sentences to pictures. In (2), only one agent is assigned and the patients assign the normal meaning of the sentence. The correct interpretation is achieved by the wrong means: the heuristic, not the usual parsing process.

Although a considerable body of data is consistent with the trace deletion hypothesis, there are serious problems in this analysis, as well as in all other proposals that aphasic deficits affect specific parsing and interpretive operations.

First, adequate linguistic controls have not been run which would show that the deficit is restricted to the structures claimed. For instance, the trace deletion hypothesis maintains that patients with the deficit in question are able to co-index items other than traces, such as pronouns and reflexives. Thus, a patient who has this deficit would not be able to connect *the boy* to the trace in *The boy who the man pushed t bumped him*, but should be able to connect *the man* to *him* in this same sentence. Accordingly, the patient should not know who pushed whom, but should know that the boy bumped the man. None of the papers in the literature that have been taken as supporting the trace deletion hypothesis have reported patients' performance on both sentences with traces and sentences with reflexives (or other referentially dependent items, such as pronouns; see Caplan 1995; Caplan et al. 2007b).

This brings up another issue. Much of the evidence for specific syntactic deficits is not based upon the performance of *individual* patients but rather on the performance of *small groups* of patients with certain diagnoses drawn from the traditional clinical literature on aphasia, such as Broca's aphasia or agrammatic aphasia. It has been claimed that the objection raised above is answered by these group data. For instance, Grodzinsky (2000) has argued that some agrammatic patients have shown integrity of processing pronouns, answering questions about the adequacy of linguistic controls. In my view, these studies do not address the issues raised above. For example, although some agrammatic patients have shown normal performances on sentences with pronouns (Grodzinsky et al. 1993), these patients have not also been tested on sentences with traces; thus, we do not know if they show the deficit specified by the trace deletion hypothesis. Empirical data show that not all patients with a clinical diagnosis of Broca's aphasia or agrammatism have problems with sentences containing traces (Swinney and Zurif 1995; Zurif et al. 1993; Blumstein et al. 1998; see Berndt et al. 1996; Drai and Grodzinsky 1999, Caramazza et al. 2001; Caplan 2001a, b) so there is a need to verify the presence of this problem on a patient-by-patient basis.

A second problem with specific deficit analyses is that the data in support of the analysis overwhelmingly consist of measures of accuracy in one task, usually sentence–picture matching. However, it has been well documented that performance may dissociate over tasks (Cupples and Inglis 1993; Caplan et al. 1997). An inability to perform accurately on a set of sentences in one comprehension task cannot be taken as a reflection of an impairment of a syntactic operation if the patient can perform accurately on those sentences in another comprehension task.

Analysis of the largest series of single cases currently available (Caplan et al. 2006, 2007b) provides data regarding the frequency of occurrence of specific deficits in syntactically based comprehension in aphasia. Forty-two right-handed native English-speaking aphasic patients with single left hemisphere strokes and 25 age- and education-matched controls were tested in object manipulation, sentence–picture matching, and grammaticality judgment tasks; the latter two with whole sentence auditory and word-by-word self-paced auditory presentation. In each task, three syntactic operations were tested: passivization, object relativization, and co-indexation of a reflexive. Two constructions instantiating each structure were presented in each task. Deficits were identified as either below normal or chance performances on sentences containing specific structures or requiring specific operations, and normal or above chance performance on baseline sentences that did not contain those structures or require those operations. Adjustments were made for speed-accuracy trade-offs, and deficits were determined by response times as well as accuracy criteria. Considering only the two comprehension tasks (sentence–picture matching and enactment), four types of deficits could be observed: task-independent, structure-specific deficits (i.e., those affecting the experimental and not the baseline sentences in both constructions in both tasks); task-independent, construction-specific deficits (i.e., those affecting the experimental and not the baseline sentences in one construction in both tasks); task-dependent, structure-specific deficits (i.e., those affecting the experimental and not the baseline sentences in both constructions in one task); task-dependent, construction-specific deficits (i.e., those affecting the experimental and not the baseline sentences in one construction in one task). Two patients had task-independent, structure-specific deficits, and two others had task-independent, construction-specific deficits; all had other deficits. No patient had task-dependent, structure-specific deficits. Thirty patients had task-dependent, construction-specific deficits. Task-dependent, construction-specific deficits were found in both sentence–picture matching and enactment. These results strongly point to most deficits affecting the ability to assign and interpret syntactic structure in one construction in one task. Specific deficits affecting a syntactic representation or a parsing/

interpretive process postulated by theories such as the trace deletion hypothesis are very rare, if they exist at all.

As noted above, the alternative view is that deficits of aphasic syntactic comprehension consist of reductions of processing capacity. Four arguments have been made in support of this suggestion:

1. Some patients can understand sentences that contain certain structures or operations in isolation but not sentences that contain combinations of those structures and operations (Caplan and Hildebrandt 1988; Hildebrandt et al. 1987).
2. In large groups of patients, as patients' performances deteriorate, more complex sentence types are affected more than less complex ones (Caplan et al. 1985, 2007b).
3. In factor analyses of performance of such patient groups in syntactic comprehension tasks, first factors on which all sentence types load account for the majority of the variance (Caplan et al. 1985, 1996, 2007b).
4. Simulations of the effect of such reductions on syntactic comprehension in normal subjects through the use of speeded presentation (Miyake et al. 1994), concurrent tasks (King and Just 1991), and other methods mimic aphasic performance.

These arguments are not iron-clad. The argument that some patients can understand sentences which contain certain structures or operations in isolation but not sentences containing combinations of those structures and operations suffers from the same limitations of the database discussed above: it is based on a single performance measure (accuracy) in a single task (enactment). Testing the second result (i.e., as patients' performances deteriorate, more complex sentence types are affected to a greater degree than less complex ones) risks circularity unless the effects of resource reduction are modeled and measured, and how to do this is a point of contention. Two studies have addressed this issue (Caplan et al. 1985, Caplan et al. 2007b) and found this pattern; a third, smaller, study did not (Dick et al. 2001). The data regarding interference effects in normal subjects are complex. Interactions of load and syntactic complexity, and of these factors with subject groups that differ in processing resource capacity, are critical pieces of evidence that would support this model. However, these interactions occur only under special circumstances (Caplan and Waters 1999, Caplan et al. 2007a), making this argument suggestive at best. The finding that first factors on which all sentence types load account for the majority of the variance is an extremely robust finding, regardless of the task over which factors are extracted or whether they are extracted over several tasks (Caplan et al. 2007b).

Additional evidence regarding the nature of aphasic deficits in syntactic comprehension comes from Rasch models of the performance of the 42 patients reported in Caplan et al. (2007b) on sentence–picture matching and enactment tasks. In a Rasch model, the probability of a correct response to a sentence is modeled as a function of the difficulty of that sentence and the ability of a patient. We explored models in which (a) sentences were grouped by task (i.e., the difficulty of sentences was modeled separately for sentence–picture matching and enactment), (b) sentences were grouped by structural type (i.e., the difficulty of sentences was modeled separately for sentences of different sorts), and (c) patients were clustered on the basis of different clustering algorithms. Seen in terms used above, adding a factor is the equivalent of postulating that difficulty or capacity differs for the sentence types or patients who differ along the lines established by that factor. We found that the introduction of a task factor and a patient grouping factor significantly improved the goodness of fit of the model, but that all sentence groupings either had no effect or made the goodness of fit worse. These results indicate that sentences exert different levels of demand as a function of task but not as a function of specific syntactic structure. This argues against the importance of structure-specific deficits as a major determinant of aphasic performance in syntactic comprehension tasks.

If deficits in aphasic syntactic comprehension consist of a reduction in the processing resources used to support the construction of syntactic representations in particular tasks, a critical question is: What, exactly, are resources? Seen in the most general terms possible, resources are features of a model of a cognitive system that allow certain operations to occur and set limits on their occurrence, but do not themselves enter into computations and are not representations. There are a variety of ways of conceptualizing resources, which are tied to models of the operations that underlie syntactic comprehension itself.

One way to conceive of resources is to see them as intrinsic aspects of the parsing/interpretive process. There are currently two different types of models of this type. In the first type, the operations that underlie syntactic comprehension are algorithms (procedures) which apply to symbolic representations, such as production rules. At least one model that views parsing this way models resources as a cost of each computation and each item stored in a memory system, and sets a limit on the total costs that can be incurred at any point in a computation (Just and Carpenter 1992). In the second type, the operations that underlie syntactic comprehension are adjustments to weights in connectionist models, and resources are modeled as the number of units in critical positions of the model or as aspects of the weighting process. An example is the presence and number of hidden units in a Boltzmann machine, whose existence extends the computational power of one-level perceptrons and whose number affects the types of generalizations that the system achieves. Though these types of models differ in fundamental ways, from the point of view of modeling resources, they share the feature that resources, and their limitations, are specified within the models. This is more obviously the case for connectionist models, where the computational units themselves determine the capacity of the system, than in procedural models, in which the limits on computational capacity are separate from the operations themselves, but in both types, resources are not set by other systems and the features of the model that determine capacity are imposed arbitrarily.

A second approach to resources is to see them as the effect of cognitive capacities other than the intrinsic operation of the parser/interpreter. Cognitive functions that have been suggested as resources that support parsing and interpretation include various type of short-term memory (phonological short-term memory: Baddeley 1986, Caramazza et al. 1980; working memory: Miyake et al. 1994) and speed of processing capacity (rates of activation and decay: Haarmann and Kolk 1991, 1994; Haarmann et al. 1997). None of these capacities have been definitely shown to play this role (for discussion, see Caplan and Waters 1990, 1999).

Finally, it is possible that deficits affecting operations that provide input into parsing and interpretation appear as resource reductions in parsing and interpretation. Lowered efficiency of lexical processing can affect syntactic processing by producing delays between the construction of partial parses and interpretations and the presentation of the input to the processes that create these representations, especially in the auditory modality. Again, this possibility has not been adequately explored to know to what extent it accounts for aphasic performance.

All things considered, in my opinion it is reasonable at present to conceive of resources very abstractly.

Further insight into the nature of deficits comes from online measures of syntactic processing. These studies have shown that behavioral measures differ for sentences that are correctly and incorrectly interpreted. For instance, Caplan et al. (2007b) found that the difference in self-paced listening times for syntactically demanding segments in more versus less complex sentences were normal when patients responded correctly to the sentences and deviated from normal (either faster or slower) when those patients responded incorrectly to the sentences. This suggests that the sentence comprehension process provides correct representations on some trials and incorrect representations on others. The deficit—whether in resource availability or in the ability to apply a specific operation—occurs intermittently, not in a fixed fashion. Additional evidence for a random factor in generating responses is the finding that, in 7% of comparisons of experimental and baseline sentences, performance on "experimental" sentences was significantly better than performance on baseline sentences (Caplan et al. 2006, 2007b). Assuming that the experimental sentence in the pair requires all the parsing and interpretive operations required in the baseline sentence, plus one or more additional operations, this reversed complexity effect can only be due to random factors affecting a patient's ability to comprehend a sentence (and to demonstrate that comprehension on a given task).

In summary, the view that deficits in aphasic syntactic comprehension consist of a reduction in the processing resources used to support the construction of syntactic representations in particular tasks seems to me to be the best supported analysis of aphasic deficits in this cognitive domain. From the perception of psycholinguistic models, this is a possible deficit: a growing body of research that finds that the mapping of comprehension onto perceptual experience interacts incrementally with the mapping of syntactic structure onto meaning (Tanenhaus et al. 1995). Different levels of resource availability, upon which noise is surperimposed, lead to differences in how often sentences with particular features are affected in a given task. Resource availability varies on a very short timescale. The physiological basis for a deficit of this sort remains to be understood.

Characterization of Lesions

There are two issues that have to be faced when characterizing lesions. The first is to specify the units of the brain that a lesion affects. The second is to specify the aspects of the lesion that affect those units.

Neural tissue can be characterized at many levels, from subcellular neural elements and processes to large brain regions. At present, researchers can only see macroscopic features of the brain (e.g., gyri, sulci, fissures, subcortical gray matter nuclei) via *in vivo* structural neuroimaging and physiological activity related to large aggregates of neurons in physiological recordings (a few studies of recordings from selective neurons have been done intra-operatively and DSI offers new data). As far as I know, since Caramazza and Zurif (1976), no studies have examined brains of patients with syntactic comprehension disorder at postmortem, where cellular features of an area could be directly examined.

At the level of the individual patient, there are a number of difficulties in identifying many of the gross neuranatomical structures that researchers relate to syntactic processing. Many gyri are not completely defined by sulci, but require researcher-based completion of the course of a sulcus to its intersection with another, and/or researcher-drawn "limiting planes," to be delineated. The boundaries of a lesioned brain area adjacent to cerebrospinal fluid are often interderminate and must be estimated. Both these factors make the identification of the gross neuroanatomical regions affected by a lesion difficult.

Though macroscopically defined areas of the brain are what can be observed *in vivo*, this level of characterization of the brain is inadequate in basic ways. Macroscopically defined areas of the brain have their functional capacities by virtue of their cellular and subcellular features, including the type and distribution of neurons in each area, the receptors and neurotransmitters of these neurons, the secondary messenger systems that are active in them, etc., and the connectivity of the neurons in the area. Macroscopically defined areas that play roles in cognition thus contain cellular and subcellular features that determine their ability to encode, transform, store, and retrieve psychologically pertinent information.

For the most part, researchers in the area of syntactic processing have focused on cytoarchitectonic analysis as specifying the cellular features that are relevant to functional specialization. The most widely used division of the brain into regions that are considered the loci of syntactic operations is the Brodmann cytoarchitectonic map (Brodmann 1909). Most researchers consider that Brodmann areas (BAs) in the perisylvian association cortex are critical components of the neural system that supports parsing and interpretation. The most conservative view is that this area includes BA 45, 44, 22, 39, and 40; more expansive conceptions include all or parts of BA 46, 47, 21, 37, and even 10 and 11, although these last two areas, especially, are usually considered too far forward to be part of this area. Many researchers collapse some of these areas into larger entities: Broca's area, consisting of BA 44 and 45, sometimes BA 46 and even less frequently BA 47; and Wernicke's area, consisting of BA 22 and possibly part of BA 21; the inferior parietal lobe, consisting of BA 39 and 40.

There are, however, many difficulties in identifying the Brodmann's areas that are affected by a lesion. Brodmann's areas are not isomorphic with macroscopically determined regions, such as gyri, sulci, or lobes (Roland et al. 1997). Cytoarchitectonic areas show individual variability in their mapping onto gyri and sulci (Amunts et al. 1999). Probabilistic maps of cytoarchitecture-gyral mappings, such as the MNI atlas, have not been applied to the analysis of lesions in studies of syntactic processing. It is also important to appreciate that Brodmann's map is only one of many cytoarchitectonic maps. This map may not always divide the cortex into functionally distinct regions or be the only cytoarchitectonic map that does so. Finally, ways to divide neural tissue in terms of features that are potentially relevant to psychological operations, such as neurotransmitter systems, cross-cut cytoarchitectonic maps (Mazziotta et al. 2001). It is not clear at this point whether these ways of characterizing neural tissue are more appropriate for purposes of localization of syntactic comprehension (or many other cognitive functions). Thus, the use of cytoarchitectonic areas is both too weak and too strong for purposes of empirical neuropsychological study. Their use is too weak because these regions are not unique among ways to divide neural tissue into areas that are potentially relevant to information-processing; their use thus does not capture other major physiological divisions of the neuraxis, such as into areas that share a neurotransmitter profile. On the other hand, their use is too strong because these regions are not reliably discernable at present; that is, they do not directly and invariantly align with macroscopic landmarks that are currently visible in *in vivo* images.

In my view, these problems preclude the use of lesion data as the basis for claims about the effects of lesions in very small areas of the brain. For instance, Friederici (this volume) presents a model of the neural basis for syntactic comprehension that attributes specific types of syntactic operations to parts of Broca's area and the superior temporal gyrus. In my view, lesions cannot now be reliably analyzed at the level of anatomical specificity that is needed for data from patients to confirm or contradict hypotheses phrased in terms of such small brain areas. On the other hand, lesion data are more than adequate, at least in principle, to confirm or contradict hypotheses phrased in terms of larger brain areas. For instance, they are adequately reliable to provide data relevant to the question of whether the correct model of neural organization for syntactic processing is invariant location, distribution, variable localization, etc. Finding, for example, that lesions confined to (roughly) STG affect particular aspects of comprehension in some patients and lesions confined to (roughly) IFG do so in others is inconsistent with invariant location of the operations which are deficient in those patients in only one of these areas.

In practice, the problems raised above regarding the analysis of brain regions affected by lesions do not arise in the vast majority of studies of syntactic processing, because these studies do not examine lesions quantitatively. In fact, many studies that make claims about the neural basis of parsing on the basis of patient data do not examine lesions at all, but simply classify patients as Broca's or nonfluent patients and Wernicke's, fluent, conduction and anomic patients. The assumption made in these studies is that patients in the first group have anterior lesions and those in the second have posterior lesions, whereas the reality is far more complicated (Mohr et al. 1978; Vanier and Caplan 1989). A large number of studies summarize radiological reports and/or display lesions, usually on a single transverse section of the brain imaged with CT or MR. The extent of lesions in these cases is unknown. A small number of studies have reported qualitative measures of lesions seen on CT and MR scans. A number of these images have been analyzed by purely subjective techniques, based on templates applied by eye to CT or MR cuts (Naeser and Hayward 1978; Naeser et al. 1979; Dronkers et al. 2004). These approaches typically yield inter-observer reliability correlations of about 80% for analyses that assign the quartile involvement of large regions of interest by a lesion (Naeser and Hayward 1978). One may legitimately assume that inter-observer correlations are lower for smaller regions of interest (ROIs) and for smaller percentages of ROIs. These data are at best rough estimates of lesion location and of the extent of a lesion in selected ROIs. In what follows, I discuss the few studies in which lesions were analyzed by quantitative and semi-quantitative methods. Even here, the delineation of gray-white matter and lesion-CSF boundaries, tracing sulci, estimation of the continuation of sulci, and other factors related to image analysis introduces uncertainties into the characterization of the grossly defined areas of the brain affected by lesions. No study has reported lesions in white matter tracts or subcortical nuclei, which are very likely critical in determining deficits.

The second issue that needs to be addressed when considering the use of lesions is what aspects of a lesion are measured. MR and CT imaging provides information about certain aspects of a lesion, such as areas that have undergone necrosis in stroke. However, lesions have widespread effects that are not measured by one imaging modality. For instance, strokes are associated with metabolic and perfusion effects outside areas of necrosis; these have rarely been investigated for their contribution to syntactic comprehension deficits.

In summary, lesions can be described *in vivo* in terms of macroscopically visible brain features, visualized radiologically. With some imprecision, it is possible to identify ROIs corresponding to cortical gyri and to determine the extent of a lesion in these ROIs in a single patient. Inferring the cytoarchitectonic regions that are affected by a lesion introduces considerably more uncertainty into the measure of a lesion. In most studies, only necrotic areas are measured, not areas of hypometabolism, hypoperfusion, or abnormal oxygen utilization. The resulting available database allows for tests of gross localization of function, but is unable to provide data relevant to hypotheses about the role of small areas of the brain, and risks missing the effect on syntactic processing of many aspects of lesions.

Characterization of the Relation between Lesions and Deficits in Syntactically Based Comprehension

Terms

Localization and distribution are relative terms. Almost all models assume that cognitive functions are supported by populations of neurons, not single cells; in this sense, all functions are distributed over some part of the brain. Whether a function is said to be localized or distributed depends upon how big the area that supports it is, and different researchers may use different terms for the same relation. (It is just a matter of terminology whether one says a function that is supported by pars opercularis, pars triangularis, and pars orbitalis is distributed over those three areas or localized in Broca's area.) I shall refer to *areas* as parts of the brain within which a function is *localized* and *regions* as parts of the brain within which a function is *distributed.* The real issue, as far as I can see, is the extent to which functions are supported by cell assemblies that have particular properties.

The physiological basis of a function is dynamic on a very short temporal scale. For instance, the somatotopic map in area 3b of monkeys can be changed for short periods of time as a result of repeated intracortical microstimulation of a given cortical site: neurons close to the stimulated site (within ~ 500 μm) come to have receptive fields closely matching those of the stimulated neurons (Recanzone and Merzenich 1988). Thus localization refers either to the area in which stimulation affects the function (or in which the function leads to physiological responses) in the absence of particular temporospatial contextual factors or to the entirety of the area in which this occurs under all circumstances. The same is true of distribution.

Models of Regional Functional Neuroanatomy

There are three basic models of the relation of functions to neural areas and regions: (a) localization, (b) distribution, and (c) duplication (degeneracy). Figure 12.1 presents these models in a visual form.

Localization hypothesizes that a small area of the brain supports a function. Localization can be invariant (Figure 12.1a: a function can be localized in the same location in all individuals) or variable (Figure 12.1c: a function can be localized, but in different locations in different individuals). If localization is variable, it can vary in an unconstrained fashion (i.e., there is no area that is more likely to be the locus of the function than another) or the variation can be constrained (i.e., a function is localized more often in one area).

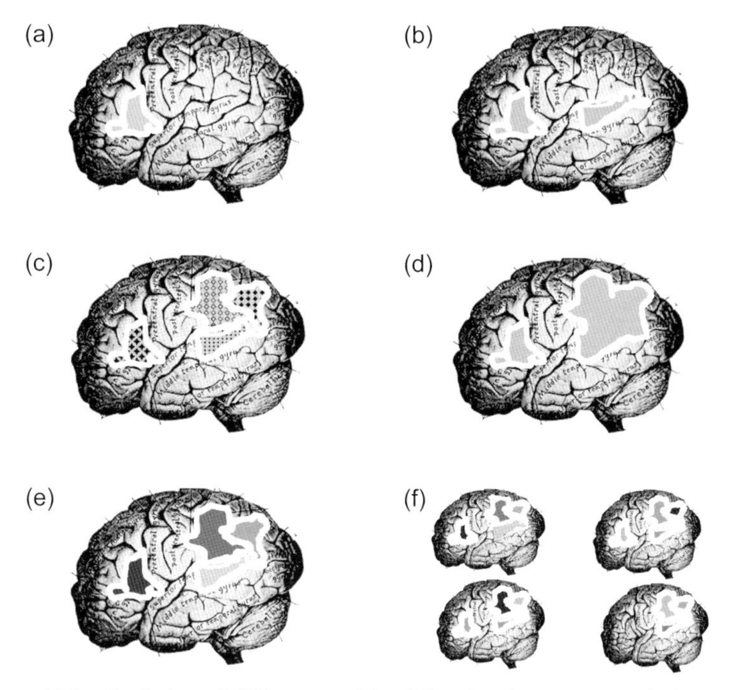

Figure 12.1 Depiction of different models of functional neuroanatomical organization. (a) Invariant localization: only one area performs an operation. (b) Degeneracy: several areas can perform the operation on their own. (c) Variable location: the same operation is performed in different areas in different individuals (indicated by variation in texture). (d) Even distribution: several areas contribute equally to performing the operation (no area can perform the operations on its own). (e) Constrained uneven distribution: several areas contribute unequally to performing the operation (indicated by variation in color saturation). The contribution of each area is the same in all individuals. (f) Unconstrained uneven distribution: Several areas contribute unequally to performing the operation. The contribution of each area differs in different individuals (indicated by different patterns of variation in color saturation).

Distribution hypothesizes that a region of the brain supports a function. If a function is evenly distributed throughout a region (Figure 12.1d), there can be no individual variability in its neural basis. If it is unevenly distributed throughout a region, particular parts of the region are more important to the function than others. In that case, the unevenness can be constrained (Figure 12.1e: the areas within the region that are more important to the function are the same in all individuals) or un-constrained (Figure 12.1f: there are no constraints on which areas within the region are more important in any given individual).

Degeneracy hypothesizes that more than one area independently supports a function. Invariant degeneracy (Figure 12.1b) hypothesizes that the areas which support the function are the same in all individuals. Unconstrained variable degeneracy hypothesizes that the multiple small areas of the brain which support a function differ in different individuals. Constrained variable degeneracy hypothesizes that certain areas are more often the sites of the function than others. (The last two models are not depicted.)

Of these theories of gross functional neuroanatomy, two major types of models—localization and distribution—have clearly been proposed regarding syntactic processing. Localizationist models are represented by Grodzinsky (2000), who claims that Chomskian traces are co-indexed in Broca's area; distributed models are exemplified by Dick et al. (2001) and Damasio and Damasio (1992). Within these models, a constrained variable localization model has been proposed (Caplan 1994) as well as an invariant unevenly distributed model (Mesulam 1990). It is not clear to me whether there is an advocate of a duplication model of syntactic operations. The model proposed by Friederici (this volume) might be such a model. Friederici argues that certain syntactic operations are supported by the parts of left IFG and left STG. If her view is that the integrity of *both* these areas is necessary for the functions, she is advocating a complex form of invariant localization or even distribution; if it is that the integrity of *either* of these areas is necessary for the functions, she is advocating a duplication (degeneracy) model.

These models make specific predictions regarding the effects of lesions on functions. One set of predictions relates lesion size in areas and regions of the brain to the magnitude of a deficit. Assuming that there is a monotonic relation between lesion size in an area or region that supports a function and the magnitude of a deficit of that function (an assumption made by all models), and that a sufficiently large patient sample is tested, these predictions include the following: Invariant localization predicts that larger lesion size in only one area will lead to an increase in the magnitude of a deficit. Unconstrained variable localization predicts that the relation of lesion size and deficit will be the same for all areas within the region in which a function is variably localized. Constrained variable localization predicts that the (positive) relation of lesion size and deficit will differ for the areas within the region in which a function is variably localized. Even distribution predicts that the relation of lesion size and deficit will be the same for all areas within a region in which a function is distributed. Unconstrained uneven distribution predicts that the relation of lesion size and deficit will be the same for all areas within a region in which a function is distributed. Constrained uneven distribution predicts that the relation of lesion size and deficit will differ for areas within a region in which a function is distributed.

As can be seen, several models generate identical predictions; however, they do so for different reasons. The prediction that the relation of lesion size and deficit will be the same for all areas within the region in which a function is (a) evenly distributed, (b) variably unevenly distributed in an unconstrained fashion, or (c) variably localized follows from even distribution because *every individual* has the same, even distribution of function over the region as well as from variable uneven, unconstrained distribution and variable localization because, *over the entire population*, functions are equally distributed over the region. Therefore, these models generate different predictions regarding the effects of lesions in single cases. In individual cases, even distribution predicts that lesions of the same size in different areas within the region in which a function is distributed will result in equivalent deficits; variable uneven distribution and unconstrained variable localization predict that lesions of the same size in different areas within the region in which a function is distributed will result in different magnitudes of deficits. Variable uneven, unconstrained distribution predicts that lesions anywhere in the region will produce *some* deficit in any patient, however small (because the function is distributed throughout the region); unconstrained variable localization predicts that some lesions will not affect performance at all in some patients. This difference in predictions is also true of variable constrained uneven distribution and constrained variable localization models.

The predictions made by degeneracy are the same as those made by localization, except that they apply to more than one area.

Relation of Performance on Syntactic Comprehension Tasks to Lesions

To my knowledge, there are five studies in the literature in which radiological images have been analyzed and related to sentence comprehension in aphasics.

Kempler et al. (1991) reported CT and PET data from a group of 43 aphasic patients. Comprehension performance was correlated with lesion size in Broca's and Wernicke's areas as measured by CT and with hypometabolism throughout the temporal, parietal, and occipital lobes, as measured by FDG-PET. A problem with this study is that Kempler et al. used the Token Test, which seriously confounds syntactic processing with short-term memory requirements. Thus performance may reflect short-term memory capacity rather than parsing and interpretive abilities.

Tramo et al. (1988) reported three cases where comprehension of reversible active and passive sentences was studied in a sentence–picture matching task.

Comprehension of reversible passives improved to above chance levels in two patients with anterior lesions, but not in the patient with a posterior lesion. This speaks to plasticity and suggests that the posterior perisylvian region has a latent ability to perform the syntactic functions of the anterior region.

Caplan et al. (1996) obtained CT scans in 18 patients with left hemisphere strokes. Scans were normalized to the Talairach and Tournoux atlas, and five perisylvian regions of interest defined following the Rademacher et al. (1992) criteria: the pars triangularis and the pars opercularis of the third frontal convolution, the supramarginal gyrus, the angular gyrus, and the first temporal gyrus excluding the temporal tip and Heschl's gyrus. Lesion volume was calculated within each ROI. Syntactic comprehension was assessed using an object manipulation task which presented 12 examples of each of 25 sentence types, selected to assess the ability to understand sentences containing only fully referential noun phrases, sentences containing overt referentially dependent NPs (pronouns or reflexives), and sentences containing phonologically empty NPs (PRO, NP-trace and *wh*-trace; Chomsky 1986, 1995).

Neither overall accuracy on the 25 sentence types, nor a syntactic complexity score, nor 19 separate measures corresponding to particular syntactic operations differed in groups defined by lesion location. None of the 168 correlations between overall accuracy on the entire set of 25 sentence types, overall syntactic complexity score, or the 19 separate measures of particular syntactic operations with normalized lesion volume in the language zone, normalized lesion volume in each of the five ROIs, or normalized lesion volume in the anterior and posterior ROIs were significant. These correlations remained insignificant when the effect of overall lesion size was partialled out by using the residuals of regression analyses in which the normalized lesion volume in the language zone was regressed against the overall accuracy scores and the syntactic complexity scores. The results all remained unchanged in ten patients who were studied and scanned at about the same time relative to their lesions. Detailed analysis of single cases with small lesions of roughly comparable size, who were tested at about the same time after their strokes, indicated that the degree of variability found in quantitative and qualitative aspects of patients' performances was not related to lesion location or the size of lesions in the anterior or posterior portion of the perisylvian association cortex. Caplan et al. concluded that the syntactic operations examined in this study were not invariantly localized in one small area, but were either distributed or showed variability in localization.

This study has many limitations. Only 18 subjects were scanned, making the sample potentially unrepresentative and limiting the use of regression analyses. The study was based on CT scans and did not measure hypoperfusion or hypometabolism in cerebral tissue. CT scans were normalized along a single linear dimension, introducing distortions in volumes of ROIs. ROIs and lesions were identified subjectively. The boundaries of lesions and CSF spaces were identified subjectively. Though many syntactic structures were examined, syntactic comprehension was only tested on a single off-line task.

Dronkers et al. (2004) studied 64 patients with left hemisphere strokes as well as 8 right hemisphere stroke cases and 15 controls. Patients were scanned using CT or MR. The patients were tested on the Western Aphasia Battery (WAB) and the Curtiss-Yamada Comprehensive Language Evaluation (CYCLE) for sentence comprehension. A voxel-based lesion-symptom mapping (VLSM) approach to analysis of the data was reported. A radiologist identified areas on a template of the brain (consisting of 11 transverse slices) which corresponded to lesions in each patient's brain. Patients were divided into groups with and without lesions at each voxel. For every voxel for which there were at least 8 patients with lesions and 8 without lesions, the groups were examined for differences in their performance on the CYCLE and on each of its subtests, using Bonferroni-corrected *t*-tests.

The authors created regions of interest based upon the presence of significant *t* values. The performance of patients with and without lesions in five areas of the left hemisphere (MTG, the anterior STS, the STS and the angular gyrus, midfrontal cortex (said to be in BA 46), and what was said to be BA 47) differed on the total CYCLE score. Dronker et al. also divided patients into groups based upon the presence of lesions in "the bulk" of an ROI, and compared the scores of the resulting groups on the CYCLE and its subtests using uncorrected *t*-tests. The results were largely similar. The process was repeated for each subtest of the CYCLE. The authors interpreted the pattern of subtests that were different in each of the regions in terms of psycholinguistic processes that were affected in patients with and without lesions in these areas. They suggested that the MTG was involved with lexical processing, the anterior STS with comprehension of simple sentences, the STS and the angular gyrus with short-term memory, and the left frontal cortex with working memory required for complex sentences. Dronker et al. concluded that neither Broca's area nor Wernicke's area contributes to sentence comprehension and that the apparent involvement of these regions in previous studies is "epiphenomenal," due to the role that adjacent cortex plays in these processes.

This study suffers from many limitations. Beginning with the treatment of neurological data, the identification of areas and lesions—including lesion-CSF boundaries—was entirely subjective, and only structural measures of lesions were considered. Unlike the Caplan et al. (1996) study, where ROIs were defined according to the Rademacher system, in the Dronkers' study, no indication is given of how ROIs were defined. It is not clear what "the bulk" of an (undefined) ROI consisted of, or whether this measurement differed for different ROIs. The decision to eliminate voxels in which 7 or fewer patients did or did not have lesions may have eliminated both Broca's area and Wernicke's area (as well as the insula) from consideration, since these regions may have been spared in fewer than 8 patients. The approach of evaluating the effects of lesions at each voxel or at each ROI independently makes no allowance for interactions of lesions in different locations; in particular, effects of total lesion volume were not considered.

With respect to the psycholinguistic analyses, the differences in performance that formed the basis for deficit analyses were those between groups of patients, not differences between patients with lesions in an area and normal subjects. Normals performed at ceiling on all subtests of the CYCLE; accordingly, all the lesions were associated with abnormal performance on all tests. The deficits that underlie these abnormal performances are unclear. CYCLE is a sentence–picture matching test with both syntactic and lexical foils for many items. Errors in which lexical foils are selected provide evidence for lexical deficits and errors in which syntactic foils are selected provide evidence for sentence-level deficits; however, Dronker et al. did not separate different error types in their analyses. Therefore, the psychological data are inadequate to support any of the conclusions made, beyond the claim that they had abnormal language comprehension.

The most noticeable result of this study is the fact that lesioned voxels in Broca's and Wernicke's areas were not associated with poor performance compared to patients without lesions in these voxels. There are several possible accounts for this finding. One, mentioned above, is that voxels in Broca's and Wernicke's areas may have been eliminated because they were spared in too few subjects. Another possible explanation of the failure to find effects of lesions in voxels in Broca's and Wernicke's areas and effects of lesions in adjacent voxels is the uncertainty associated with the normalization process. Even if these artifacts are not the source of the finding, the finding only shows that patients with lesions in these areas are not more affected on language comprehension than patients whose lesions spare these areas, not that patients with lesions in these areas are not impaired relative to normals (with whom they were not compared, as noted above).

Caplan et al. (2007c) imaged 32 of the patients and 13 controls reported in Caplan et al. (2007b), using MR and FDG-PET scanning. First factor scores of principle components analyses of performance on each task were taken as reflections of resource availability of each patient in each task, and differences in accuracy, reaction time, and listening times for words in critical positions (corrected for word length and frequency) in experimental and baseline sentences served as measures of syntactic processing ability for particular structures in particular tasks.

The relation between lesion size and deficits was investigated by regressing MR and PET measures of the extent of the lesion in each of seven ROIs against the measures of syntactic processing mentioned above. Because of the small number of cases, regressions were done in series, using general factors and lesion size in larger regions as independent measures in the first analyses and progressing to the effects of lesions in smaller areas. Time since lesion and age did not affect performance. Percent lesion volume on MR and mean PET counts/voxel in a variety of small regions accounted for a significant amount of variance in performance measures, after total lesion size was entered into stepwise regressions. For instance, percent MR lesion in the inferior parietal lobe, the anterior inferior temporal lobe and superior parietal lobe, and PET counts/voxel in Broca's area accounted for a significant amount of variance in first factor scores for all tasks combined and for object manipulation. Similar patterns (i.e., significant effects of lesion size in several areas on performance measures) were found for other dependent variables. Some areas, such as the anterior inferior temporal region, were significant in several tasks, implying that each of these cortical areas is necessary for multiple functions. Finally, some dependent variables were predicted by lesions in multiple areas, implying that multiple, unrelated cortical areas are necessary for some functions.

The data were also examined on a case-by-case basis to look for evidence of distribution of the functions assessed. The range of performance on five measures (first factor scores for sentence–picture matching, enactment and grammaticality judgment; syntactic complexity scores for accuracy in sentence–picture matching and enactment) was examined in patients with lesions within .25 SD of the mean lesion size in four regions (the entire left hemisphere, the left hemisphere cortex, the perisylvian association cortex, and the combination of the perisylvian association cortex, the inferior anterior temporal lobe and the superior parietal lobe). In each case, performances of patients with lesions restricted to this small proportion of the total range of lesion sizes covered a wide range of total performance; in some cases almost the entire range, and in one case the entire range, of performance was found. The converse was also considered. The range of lesion sizes in these four regions was examined in patients whose performance fell within .25 SD of the mean performance on these measures. In each case, percent of region lesioned in the selected cases covered a wide portion of the range of percent of region lesioned in all cases. These results argue against models that maintain that the functions which these performance measures assess are evenly distributed across large contiguous brain areas, such as the left hemisphere cortex or the perisylvian association cortex. If this were the case, lesions of equal size in these regions should have led to similar magnitudes of a deficit. The data are consistent with the view that localization of these functions varies across individuals. They are also consistent with the idea that the functions being measured are unevenly distributed throughout large areas, either with the same pattern of unevenness in all individuals (invariant uneven distribution) or with different patterns of uneven distribution in different individuals (variable uneven distribution).

A previously unreported analysis pertains to deficits and spared functions in patients with lesions that primarily occupy small areas of the brain. Table 12.1 shows these results for patients with lesions that primarily occupy the posterior or anterior portion of the perisylvian association cortex. The pattern of spared and affected sentence types differs considerably for patients with similar lesions. In the posterior cases, patient 50017 shows a lesion occupying 50% of the posterior perisylvian area and almost no abnormal performances

Table 12.1 Performance of patients with lesions primarily located in anterior and posterior perisylvian areas on sentence–picture matching (SPM) and object manipulation (enactment: OM) tasks. Broca's area: Rademacher parcellation units F3o, F3t, and FO; posterior perisylvian area: Rademacher parcellation units T1p, AG, SGa and SGp. Shaded cells indicate performance 1.79 SDs below the normal mean (the cut-off for abnormal performance with $n = 25$). Low scores and large SDs make this cut-off low in some sentence types (e.g., SO). Similar results are found when performances at or below chance, and above chance, are considered.

Patients with primarily posterior lesions:

Patient	% Broca's area lesioned	% Posterior perisylvian area lesioned	SPM task: % correct by sentence type										
			A	PF	PT	RG	RGB	RP	RPB	CO	CS	SO	SS
50013	0.0	36.5	78	89	78	80	100	90	90	100	100	50	100
50017	16.4	50.7	78	100	89	100	60	100	90	100	70	80	80
50027	5.5	52.2		67	22	70	70	90	50	50	90	60	70
			OM task: % correct by sentence type										
			A	PF	PT	RG	RGB	RP	RPB	CO	CS	SO	SS
50013	0.0	36.5	100	100	100	100	100	90	100	100	100	70	90
50017	16.4	50.7	100	100	100	100	100	100	90	100	100	50	60
50027	5.5	52.2	80	80	90	50	20	70	30	60	70	20	0

Sentence types: A: active; PF: full passive (with *by* phrase); PT: truncated passive (without *by* phrase); RG: reflexive with genetive subject NP; RGB: control for reflexive with genetive subject NP (third NP in place of reflexive); RP: reflexive with possessive subject NP; RPB: control for reflexive with possessive subject NP (third NP in place of reflexive); CO: cleft object; CS: cleft subject; SO: subject object relative; SS: subject subject relative

Table 12.1 *continued*

Patients with primarily anterior lesions:

Patient	% Broca's area lesioned	% Posterior perisylvian area lesioned	SPM task: % correct by sentence type										
			A	PF	PT	RG	RGB	RP	RPB	CO	CS	SO	SS
50005	100.0	10.1	100	100	100	100	100	100	100	100	100	100	100
50009	19.5	0.0	100	100	100	100	100	100	90	100	100	80	100
50018	39.4	4.7	56	44	78	60	50	60	70	20	60	50	70
50022	41.3	0.0	100	100	100	100	90	80	100	90	100	90	90
50025	55.8	11.7	89	89	100	70	60	50	30	60	80	70	70
50026	78.3	1.1	89	100	100	50	60	90	100	70	90	70	90
50043	33.2	0.0	100	100	100	70	70	90	100	90	100	50	100
			OM task: % correct by sentence type										
			A	PF	PT	RG	RGB	RP	RPB	CO	CS	SO	SS
50005	100.0	10.1	100	100	100	100	100	100	100	100	100	100	100
50009	19.5	0.0	100	90	100	100	100	100	100	100	100	70	70
50018	39.4	4.7	90	100	0	50	50	80	70	80	70	0	22
50022	41.3	0.0	100	100	90	100	100	90	100	100	100	60	90
50025	55.8	11.7	100	100	90	70	70	80	80	20	100	0	10
50026	78.3	1.1	90	80	70	30	30	60	40	90	100	20	10
50043	33.2	0.0	100	100	100	100	90	90	100	90	100	20	100

in enactment; patient 50027, with a lesion whose perisylvian extent is very similar to that seen in patient 50017, shows abnormal performance on almost all sentence types in enactment. In the anterior cases, patient 50005 shows a complete lesion of the anterior perisylvian area and no abnormal performances; patient 50018, with a lesion that occupies 39% of the anterior perisylvian area, has virtually no normal performances. Since poor performance could be due to any aspect of the lesion in a patient, the most interpretable results are those in which a lesion is associated with normal performance, which imply that the aspects of the brain that are lesioned are not necessary for the functions that are spared. These data indicate that the total integrity of particular areas of the brain is not necessary for particular functions in some patients, though it appears to be necessary for those functions in others (though deficits could be due to lesions in other areas). A larger series of cases is needed to determine the extent to which lesions in these and other areas occur without disrupting particular functions.

The implications of these results for the neural basis of syntactic processing depends upon the nature of the functions that the behavioral/performance measures measure. As noted above, deficits affecting syntactic comprehension appear to reduce the resource system that is used to support parsing and interpretive operations in a task, not the operations or resource systems that are used to assign and interpret syntactic structures in an amodal, abstract fashion. The operations in question do not simply map propositional representations onto perceptual and motor functions; deficits that resulted from deficits in such operations would affect all propositions with similar semantic values (e.g., with the same number and type of thematic roles) in a task, whereas the deficits seen in aphasia affect sentences differently as function of their syntactic form. Rather, the resources that appear to be reduced in aphasia support mapping propositional content onto the products of the comprehension process at a point in processing at which the syntactic structure of a sentence influences the mapping (i.e., during the online computation of the structure and meaning of a sentence). This has significant consequences for the interpretation of the effects of lesions.

If this deficit analysis is correct, the neural basis for domain-independent (amodal) parsing and interpretive operations is not addressed by the studies reviewed here (the same is true of all activation studies of which I am aware). The aphasia data raise the question of whether the brain is organized to segregate parsing and interpretive operations from perceptual, motor, and mnemonic functions. As noted above, there is considerable behavioral evidence that parsing and interpretation is mapped onto task performance incrementally, and the possibility exists that the brain is thoroughly interactive—that all areas of the brain that support parsing and interpretation are also involved in these incremental interactions. Regardless of whether this is correct or not, to date, I would argue that the data from aphasia provide information about areas that are involved in these interactions, not any that may support amodal parsing and interpretation.

Summary and Conclusions

The study of aphasic disturbances of syntactic comprehension has been taken as providing a window into the organization of the brain for parsing and sentence interpretation. Recent studies provide strong reasons to believe that the view through this window is blurred: it is only visible through the screen of task performance. Deficits affecting syntactic comprehension affect the interfaces between the comprehension process and task performance. The little data available suggest that lesions in different sites affect different comprehension-task interfaces, and that some areas support several such interfaces. This may be because some of these interfaces are variably localized or variably unevenly distributed, but the results are also consistent with invariant localization. If there are many aspects to these interfaces, each of which is invariantly localized, and lesions in each aspect of an interface affect performance, lesions in different areas will be related to performance on a given task. If some of these interfaces are the same across comprehension-task pairings, some areas will be involved in several mappings. Practically nothing is known about the interface of task demands and comprehension and how these integrated functions break down in aphasia. A great deal of research into these questions—both in terms of model development and in terms of empirical investigation—is needed if we are to use the performances of patients with cerebral lesions as a basis for information about the neural basis for syntactic processing.

13

Reflections on the Neurobiology of Syntax

Peter Hagoort

Abstract

This contribution focuses on the neural infrastructure for parsing and syntactic encoding. From an anatomical point of view, it is argued that Broca's area is an ill-conceived notion. Functionally, Broca's area and adjacent cortex (together *Broca's complex*) are relevant for language, but not exclusively for this domain of cognition. Its role can be characterized as providing the necessary infrastructure for unification (syntactic and semantic). A general proposal, but with the required level of computational detail, is discussed to account for the distribution of labor between different components of the language network in the brain. Arguments are provided for the *immediacy principle*, which denies a privileged status for syntax in sentence processing. The temporal profile of event-related brain potential (ERP) is suggested to require predictive processing. Finally, since, next to speed, diversity is a hallmark of human languages, the language readiness of the brain might not depend on a universal, dedicated neural machinery for syntax, but rather on a shaping of the neural infrastructure of more general cognitive systems (e.g., memory, unification) in a direction that made it optimally suited for the purpose of communication through language.

Introduction

Recent years have seen a growing number of studies on the neural architecture of language, using both electromagnetic methods (EEG, MEG) and hemodynamic methods (PET, fMRI). These studies have added to, but also changed, previous views on the brain's infrastructure for language, which were based primarily on patient studies. Before discussing the relevant issues in detail, here is what I believe to be the major conclusions that can be drawn from the overall body of literature on the neurobiology of language:

1. The language network in the brain is more extensive than the classical language areas (Broca's area, Wernicke's area). It includes, next to

Broca's area, adjacent areas in the left inferior frontal cortex (LIFC), as well as substantial parts of superior and middle temporal cortex, inferior parietal cortex, and parts of the basal ganglia. In addition, homologue areas in the right hemisphere are often found to be activated to a lesser extent.

2. In contrast to classical textbook wisdom, the division of labor between Broca's area (frontal cortex) and Wernicke's area (temporal cortex) is not language production vs. language comprehension. LIFC is strongly involved in syntactic and semantic unification operations during comprehension. Wernicke's area is involved in language production, at least at the level of word form encoding (Indefrey and Levelt 2004).

3. None of the language-relevant areas and none of the language-relevant neurophysiological effects are language specific. All language-relevant ERP effects (e.g., N400, P600, (E)LAN) seem to be triggered by other than language input as well (e.g., music, pictures, gestures; see Kaan, this volume).

4. For language, as for most other cognitive functions, the notion of function-to-structure mapping as being one-area-one-function is almost certainly incorrect. More likely, any cortical area is a node that participates in the function of more than one network. Conceivably, top-down connections from supramodal areas could differentially recruit such a cortical node into the service of one network or another (Mesulam 1990, 1998).

5. Despite the syntacto-centrist perspective in most of modern linguistics, in terms of processing there is no evidence for a privileged status of syntactic information. Language comprehension, beyond the single word level, happens in accordance with the immediacy principle, which states that all relevant information types (e.g., syntactic, semantic, extra-linguistic information) are brought to bear on language interpretation as soon as they become available, without giving priority, on principled grounds, to the syntax-constrained combination of lexical semantic information. The immediacy principle does not apply to language production, which requires a conceptual specification that precedes syntactic and phonological encoding, at least to some extent.

In this chapter, I provide further background and additional reflections on the biological foundations of syntax. I begin with a closer look at Broca's area, from a neuroanatomical perspective. The focus on Broca's area will enable some general conclusions about the relevant features of language-relevant brain areas.

Deconstructing Broca's Area

Classically, Broca's area has been considered to be a key site for syntax. Despite some disagreement in the literature (Uylings et al. 1999), most authors agree that Broca's area comprises Brodmann areas 44 and 45 of the left hemisphere. In classical textbooks, these areas coincide at the macroscopic level with the pars opercularis (BA 44) and the pars triangularis (BA 45) of the third frontal convolution. However, given anatomical variability, in many brains these two parts are not easy to identify (Uylings et al. 1999), and clear micro-anatomical differences (see Amunts and Zilles 2006) have been missed when macro-anatomical landmarks are used (Tomaiuolo et al. 1999). Furthermore, cytoarchitectonic analysis (Amunts et al. 2003) shows that BA 44 and BA 45 do not neatly coincide with the sulci that form their boundaries in macro-anatomical terms. More fundamentally, one has to question the justification for subsuming these two cytoarchitectonic areas under the overarching heading of Broca, rather than, say, BA 45 and BA 47. Areas 44 and 45 show a number of clear cytoarchitectonic differences, one of which is that BA 45 has a granular layer IV, whereas BA 44 is dysgranular. In contrast, like BA 45, BA 47 is part of the heteromodal component of the frontal lobe, known as the granular cortex (see Figure 13.1; Mesulam 2002). In addition, BA 44 and BA 45 have clearly distinct postnatal developmental trajectories and show a difference in their patterns of lateral asymmetry (Uylings et al. 1999). Using an observer-independent method for delineating cortical areas, Amunts et al. (1999) analyzed histological sections of ten human brains. They found a significant left-over-right asymmetry in cell density for BA 44, whereas no significant left–right differences were observed for BA 45. However, BA 44 and BA 45

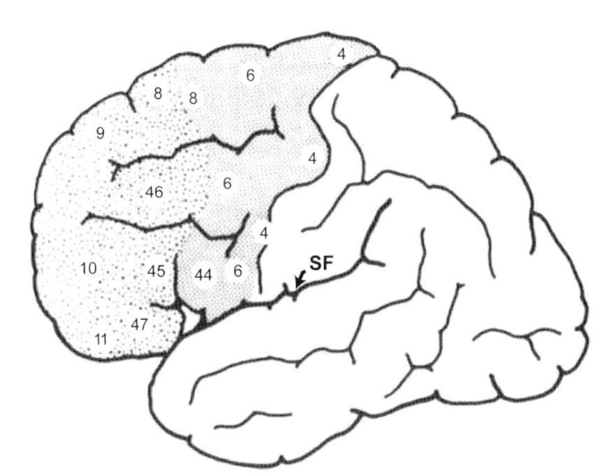

Figure 13.1 Lateral view of the frontal lobes. The numbers refer to Brodmann areas. Hashed markings: motor-premotor cortex; dotted markings: heteromodal association cortex. SF: Sylvian fissure; CS: central sulcus. After Mesulam (2002).

are cytoarchitectonically more similar to each other than BA 44 and BA 6, or BA 45 and BA 6 (Amunts and Zilles 2001).

Studies on corresponding regions in the macaque brain (Petrides and Pandya 2002a) have shown that BA 44 receives projections primarily from somatosensory and motor-related regions, the rostral inferior parietal lobule, and supplementary and cingulate motor areas. There is input from portions of the ventral prefrontal cortex, but only sparse projections from inferior temporal cortex (Pandya et al. 1996). Conversely, BA 45 receives massive projections from most parts of prefrontal cortex, from auditory areas of the STG, and visually related areas in the posterior STS. In other words, the connectivity patterns of macaque BA 44 and 45 suggest clear functional differences between these areas. Differences in connectivity have also been found in human studies, using a technique called *diffusion tensor imaging* (Glasser and Rilling 2008). Friederici et al. (2006a) report syntax relevant differences in connectivity from Broca's area and from the frontal operculum to different parts of the temporal lobe.

Xiang, Norris, and Hagoort (submitted) performed a resting state functional connectivity study to investigate directly the functional correlations within the perisylvian language networks by seeding from three subregions of Broca's complex (pars opercularis, pars triangularis, and pars orbitalis) and their right hemisphere homologues. A clear topographical connectivity pattern in the left middle frontal, parietal, and temporal areas was revealed for the three left seeds in Broca's complex. These results demonstrate that a functional connectivity topology can be observed in the perisylvian language areas in the left hemisphere, in which different parts of Broca's area and adjacent cortex show a differential pattern of connectivity. This pattern is only seen in the left hemisphere and seems to be organized according to information type: semantic, syntactic, phonological (Figure 13.2).

Finally, studies on the receptor architecture of left inferior frontal areas indicate that functionally relevant subdivisions within BA 44 and BA 45 might be necessary (for more details, see Amunts and Zilles 2006). For instance, there is a difference within BA 44 of the receptor densities, for example of the $5HT_2$ receptor for serotonin, with relatively low density in dorsal BA 44 and relatively high density in ventral BA 44.

In short, from a cytoarchitectonic and receptor architectonic point of view, Broca's area, comprising BA 44 and BA 45, is a heterogeneous patch of cortex; it is not a uniform cortical entity. The functional consequences of this heterogeneity are unclear, since the degree of anatomical uniformity required for an inference of functional unity is unknown. Here, two different views exist about the functional relevance of architectural differences in brain structure, which can be made clear in connection to the Brodmann map.

A prime example of the contribution of neuroanatomy is the famous map by Korbinian Brodmann (1869–1918). This map consists of 47 different areas, usually referred to by expressions such as BA 44 for Brodmann BA 44.

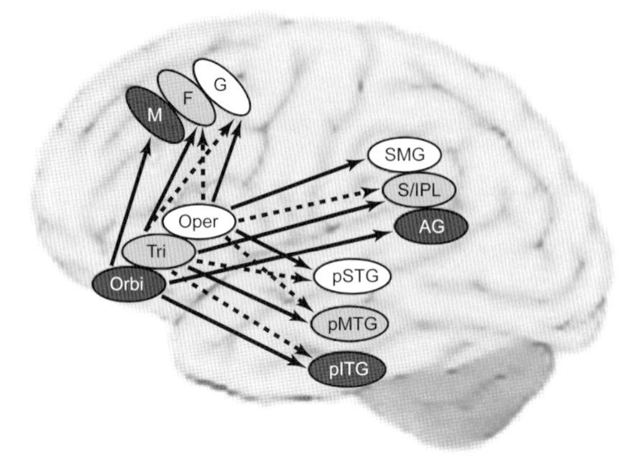

Figure 13.2 The topographical connectivity pattern in the perisylvian language networks. Connections to the left pars opercularis (oper), pars triangularis (tri) and pars orbitalis (orbi) are shown in white, gray, and black arrows respectively. The solid arrows represent the main (most significant) correlations and the dashed arrows represent the extending (overlapping) connections. Brain areas assumed to be mainly involved in phonological, syntactic and semantic processing are shown in white, gray, and black circles, respectively. MFG (middle frontal gyrus) is separated into three parts. SMG: Supramarginal gyrus; AG: Angular gyrus; S/IPL: the area between SMG and AG in the superior/inferior parietal lobule; pSTG: posterior superior temporal gyrus; pMTG: posterior middle temporal gyrus; pITG: posterior inferior temporal gyrus.

The numbers of the Brodmann areas were determined by the order in which Brodmann went through the brain, as he analyzed one area after the other. Brodmann's classification is based on the cytoarchitectonics of the brain, which refers to the structure, form, and position of the cells in the six layers of the cortex. Quantification was done by Brodmann on postmortem brains, which were sectioned into slices of 5–10 micron thickness that underwent Nissl-staining and were then inspected under the microscope. In this way, the distribution of different cell types across cortical layers and brain areas could be determined. Even today Brodmann's map (Brodmann 1909) is recognized as a hallmark in the history of neuroscience. Brodmann's work reveals that the composition of the six cortical layers, in terms of cell types, varies across the brain. Cell numbers can vary as well. The primary visual cortex, for instance, has about twice as many neurons per cortical column as other brain areas (Amaral 2000).

The classical view among neuroanatomists is that these architectural differences in brain structure are indicative of functional differences, and, conversely, that functional differences demand differences in architecture (Bartels and Zeki 2005; Brodmann 1905; Vogt and Vogt 1919; Von Economo and Koskinas 1925). Following the classical view, through different ways of characterizing brain structure (i.e., cyto- myelo- and receptor architectonics; Zilles and Palomero-Gallagher 2001), brain areas can be identified, for which differences in structural characteristics imply functional differences. Accordingly, it follows that one should look for the structural features that determine why a particular brain area can support, for instance, the processing of a first or second language.

In contrast to the classical view in neuroanatomy, more recent accounts have argued that from a computational perspective, different brain areas are very similar. For instance, Douglas and Martin argue that:

> The same basic laminar and tangential organization of the excitatory neurons of the neocortex, the spiny neurons, is evident wherever it has been sought. The inhibitory neurons similarly show a characteristic morphology and patterns of connections throughout the cortex….all things considered, many crucial aspects of morphology, laminar distribution, and synaptic targets are very well conserved between areas and between species (Douglas and Martin 2004, p. 439).

Functional differences between brain areas are, according to this perspective, due mainly to variability of the input signals in forming functional specializations. Domain specificity of a particular piece of cortex might thus not be determined so much by the heterogeneity of brain tissue, but rather by the way in which its functional characteristics are shaped through input. Moreover, findings of neuronal plasticity (e.g., the involvement of visual cortex in verbal memory in the congenitally blind; Amedi et al. 2003), suggest substantial plasticity in structure-to-function relations.

The above considerations result in a view of Broca's area that is different from the classical perspective. With respect to language areas in frontal cortex, it has become clear that, in addition to BA 44 and 45, at least BA 47 and the ventral part of BA 6 should be included in the left frontal language network. Recent neuroimaging studies indicate that the pars orbitalis of the third frontal convolution (roughly corresponding to BA 47) is involved in language processing (e.g., Devlin et al. 2003; Hagoort et al. 2004). From a functional anatomical perspective, it thus makes sense to use the term *Broca's complex* for this set of areas. Broca's complex is used here to distinguish it from Broca's area as classically defined, which is both too broad (since it comprises anatomically and functionally distinct areas, with differences in their connectivity patterns) and too narrow (since it leaves out adjacent areas that are shown to be crucial for language processing). Broca's complex, as defined here, is the set of anatomical areas in left inferior frontal cortex that are known to play a crucial, but by no means exclusive, role in language processing.

The Role of Broca's Complex

A particular cognitive function is most likely served by a distributed network of areas, rather than by one local area alone. In addition, a local area participates in more than one function. A one-to-one mapping between Broca's area and a specific functional component of the language system would thus be a highly unlikely outcome. Even for the visual system, it is claimed that the representations of, for example, objects and faces in ventral temporal cortex are distributed and overlapping (Haxby et al. 2001). Moreover, Broca's area has been found activated in imaging studies on nonlanguage functions, such as action recognition (Decety et al. 1997; Hamzei et al. 2003) and movement preparation (Thoenissen et al. 2002). Of course, all this does not mean that cognitive functions are not localized and that the brain shows equipotentiality. It only means that the one-area-one-function principle is in many cases not an adequate account of how cognitive functions are neuronally instantiated.

If Broca's complex is not a domain-specific area, what properties does it have that makes it suitable for recruitment for unification operations in the language domain?

The answer that I propose is based on (a) an embedding of this complex in the overall functional architecture of prefrontal cortex, and (b) a general distinction between memory retrieval of linguistic information and combinatorial operations on information retrieved from the mental lexicon. These operations are referred to as *unification* or *binding*. The notion of binding is inspired by the visual neurosciences, where one of the fundamental questions concerns how we get from the processing of different visual features (color, form, motion) by neurons that are far apart in brain space to a unified visual percept. This is known as the *binding problem*. In the context of the language system, the binding problem refers to an analogous situation, but is now transferred to the time domain: How is information that is incrementally retrieved from the mental lexicon unified into a coherent overall interpretation of a multi-word utterance? Most likely, unification needs to take place at the conceptual, syntactic, and phonological level, as well as between these levels (see Figure 13.3; Jackendoff 2002). In this context, binding refers to a problem that the brain has to solve, not to a concept from a particular linguistic theory.

Integration is an important part of the function of the prefrontal cortex. This holds especially for integration of information in the time domain (Fuster 1995). To fulfill this role, prefrontal cortex needs to be able to hold information online (Mesulam 2002) and to select among competing alternatives (Thompson-Schill et al. 1999). Electrophysiological recordings in the macaque monkey have shown that this area is important for sustaining information triggered by a transient event for many seconds (Miller 2000). This allows prefrontal cortex to establish unifications between pieces of information that are perceived or retrieved from memory at different moments in time (Fuster 1995).

Recent neuroimaging studies indicate that Broca's complex contributes to the unification operations required for binding single word information into larger structures. In psycholinguistics, integration and unification refer to what is usually called post-lexical processing. These are the operations on information that is retrieved from the mental lexicon. It seems that prefrontal cortex is especially well suited to contribute to post-lexical processing, since this includes selection among competing unification possibilities, so that one unified representation spanning the whole utterance remains.

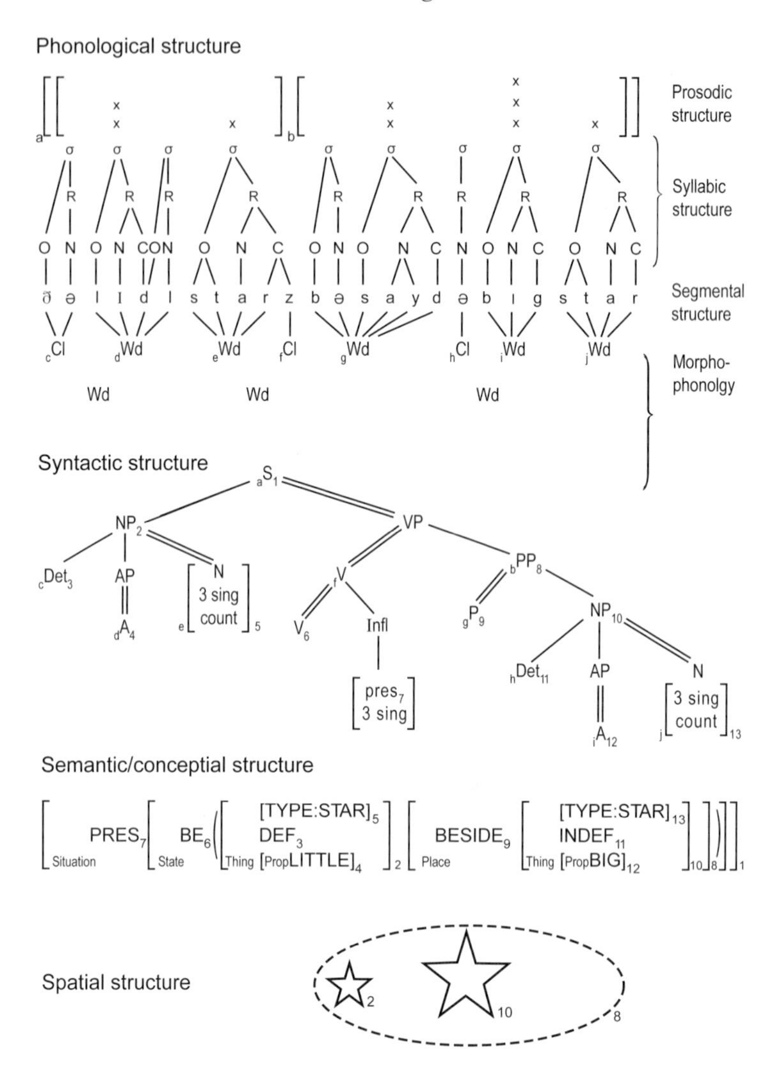

Figure 13.3 The phonological, syntactic, and semantic/conceptual structures for the sentence *The little star's beside the big star* (Jackendoff 2002). The unification operations involved are suggested to require the contribution of Broca's complex.

In short, the properties of neurons in the prefrontal cortex of macaques suggest that this part of the brain is suitable for integrating pieces of information that are made available sequentially, spread out over time, irrespective of the nature of the material to be handled (Owen et al. 1998). Clearly, there are interspecies differences in terms of the complexity of the information binding operations (Fitch and Hauser 2004), possibly supported by a corresponding increase in the amount of frontal neural tissue from monkey to man (Passingham 2002). With respect to language processing in humans, different complex binding operations take place. Subregions in Broca's complex might contribute to different unification operations required for binding single word information into larger structures.

A Unification Model of Parsing

Progress in the neurobiology of language suffers from the lack of (or the unawareness about the availability of) detailed, preferably computationally explicit models of language processing. In many ways, this situation holds for most domains of cognitive neuroscience research. If it comes to the neurobiology of syntax, the specification of a grammar does not suffice. If anything, all there is to be found in the brain is a system capable of parsing and syntactic encoding. Therefore, we need explicit models of parsing (and syntactic encoding) to guide our search and interpretation of results in ERP, MEG, or fMRI experiments. Here I offer a proposal for an explicit model of parsing. An interesting aspect of the model is that it also accounts for syntactic encoding, with the input–output relations reversed (from concept to phonology). This model is not to be taken as a final theoretical commitment. Instead, it is an illustration of the explicitness that is necessary to make progress in this domain of research. I offer an account based on the Unification Model for parsing (Vosse and Kempen 2000).

According to this model, each word form in the mental lexicon is associated with a structural frame. This structural frame consists of a three-tiered unordered tree, specifying the possible structural environment of the particular lexical item:

(1)

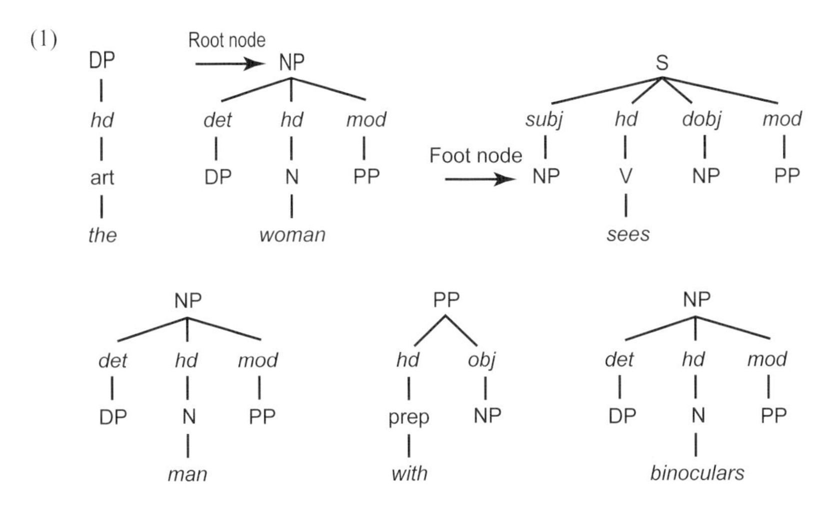

where DP: determiner phrase; NP: noun phrase; S: sentence; PP: prepositional phrase; art: article; hd: head; det: determiner; mod: modifier; subj: subject;

dobj: direct object. The top layer of the frame consists of a single phrasal node (e.g., NP). This so-called root node is connected to one or more functional nodes (e.g., subject, head, direct object) in the second layer of the frame. The third layer contains again phrasal nodes to which lexical items or other frames can be attached.

This parsing account is "lexicalist" in the sense that all syntactic nodes (e.g., S, NP, VP, N, V) are retrieved from the mental lexicon. In other words, chunks of syntactic structure are stored in memory. There are no syntactic rules that introduce additional nodes. In the online comprehension process, structural frames associated with the individual word forms incrementally enter the unification workspace. In this workspace constituent structures spanning the whole utterance are formed by a unification operation of two lexically specified syntactic frames:

(2)

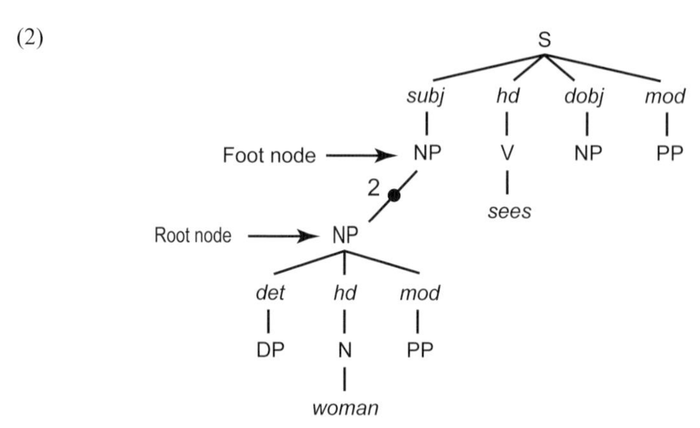

Unification takes place by linking the root node NP to an available foot node of the same category. The number 2 indicates that this is the second link that is formed during online processing of the sentence *The woman sees the man with the binoculars*. This operation consists of linking up lexical frames with identical root and foot nodes, and checking agreement features (e.g., number, gender, person). It specifies what Jackendoff (2002) refers to as the only remaining "grammatical rule": UNIFY PIECES.

The resulting unification links between lexical frames are formed dynamically, which implies that the strength of the unification links varies over time until a state of equilibrium is reached. Due to the inherent ambiguity in natural language, alternative binding candidates will usually be available at any point in the parsing process. That is, a particular root node (e.g., PP) often finds more than one matching foot node (i.e., PP) with which it can form a unification link (for examples, see Hagoort 2003).

Ultimately, this results in one phrasal configuration. This requires that among the alternative binding candidates, only one remains active. The required state of equilibrium is reached through a process of lateral inhibition between two or more alternative unification links. In general, due to gradual decay of activation, more recent foot nodes will have a higher level of activation than ones which entered the unification space earlier. In addition, strength levels of the unification links can vary as a function of plausibility (semantic) effects. For instance, if instrumental modifiers under S-nodes have a slightly higher default activation than instrumental modifiers under an NP-node, lateral inhibition can result in overriding a recency effect.

The Unification Model accounts for sentence complexity effects known from behavioral measures, such as reading times. In general, sentences are harder to analyze syntactically when more potential unification links of similar strength enter into competition with each other. Sentences are easy when the number of U-links is small and of unequal strength. In addition, the model accounts for a number of other experimental findings in psycholinguistic research on sentence processing, including syntactic ambiguity (attachment preferences; frequency differences between attachment alternatives) and lexical ambiguity effects. Moreover, it accounts for breakdown patterns in agrammatic sentence analysis (for details, see Vosse and Kempen 2000).

The advantage of the Unification Model is that (a) it is computationally explicit, (b) it accounts for a large series of empirical findings in the parsing literature and in the neuropsychological literature on aphasia, and (c) it belongs to the class of lexicalist parsing models that have found increasing support in recent years (Bresnan 2001; Jackendoff 2002; Joshi and Schabes 1997; MacDonald et al. 1994).

Further support for a distinction between a memory component (i.e., the mental lexicon) and a unification component in syntactic processing comes from neuroimaging studies on syntactic processing. In a meta-analysis of 28 neuroimaging studies, Indefrey (2004) found two areas that were critical for syntactic processing, independent of the input modality (visual in reading, auditory in speech). These two supramodal areas for syntactic processing were the left posterior STG and the left prefrontal cortex. The left posterior temporal cortex is known to be involved in lexical processing (Indefrey and Cutler 2004). In connection with the Unification Model, this part of the brain might be important for the retrieval of the syntactic frames that are stored in the lexicon. The unification space, where individual frames are connected into a phrasal configuration for the whole utterance, might recruit the contribution of Broca's complex (LIFC). Empirical support for this distribution of labor between LIFC and temporal cortex was recently found in a study by Snijders et al. (2008). They did an fMRI study in which participants read sentences and word sequences containing word category (noun–verb) ambiguous words at critical positions. Regions contributing to the syntactic unification process should show enhanced activation for sentences compared with words, and only within sentences display a larger signal for ambiguous than unambiguous conditions. The posterior LIFG showed exactly this predicted pattern, confirming the hypothesis that LIFG contributes to syntactic unification. The left posterior middle temporal gyrus was activated more for ambiguous than unambiguous conditions, as predicted for regions subserving the retrieval of lexical-syntactic information from memory.

A Unification Model Account of Syntactic ERP Effects

Since the discovery of the N400 effect in the beginning of the 1980s (Kutas and Hillyard 1980), a whole series of language-relevant ERPs have been observed. ERPs reflect the sum of simultaneous postsynaptic activity of a large population of mostly pyramidal neurons recorded at the scalp as small voltage fluctuations in the EEG, time locked to sensory, motor or cognitive processes. In a particular patch of cortex, excitatory input to the apical dendrites of pyramidal neurons will result in a net negativity in the region of the apical dendrite and a positivity in the area of the cell body. This creates a tiny dipole for each pyramidal neuron, which will summate with other dipoles provided that there is simultaneous input to the apical dendrites of many neurons, and a similar orientation of these cells. The cortical pyramidal neurons are all aligned perpendicular to the surface of the cortex, and thus share their orientation. The summation of the many individual dipoles in a patch of cortex is equivalent to a single dipole calculated by averaging the orientations of the individual dipoles (Luck 2005). This equivalent current dipole is the neuronal generator (or source) of the ERP recorded at the scalp. In many cases, a particular ERP component has more than one generator and contains the contribution of multiple sources. Mainly due to the high resistance of the skull, ERPs tend to spread, blurring the voltage distribution at the scalp. An ERP generated locally in one part of the brain will therefore not only be recorded at a nearby part, but also at quite distant parts of the scalp.

In connection to syntactic processing, two classes of syntax-related ERP effects have been consistently reported for over a period of more than ten years. One type of ERP effect related to syntactic processing is the P600 (Hagoort et al. 1993; Osterhout and Holcomb 1992). The P600 is reported in relation to syntactic violations, syntactic ambiguities, and syntactic complexity. This effect occurs in a latency range between roughly 500–800 ms following a lexical item that embodies a violation or a difference in complexity. However, the latency can vary, and earlier P600 effects have also been observed (Hagoort 2003; Mecklinger et al. 1995). Another syntax-related ERP is a left anterior negativity (referred to as LAN or, if earlier in latency than 300–500 ms, as ELAN; Friederici et al. 1996). In contrast to the P600, the (E)LAN has thus far (almost) exclusively been observed to syntactic violations. LAN is usually observed within a latency range of 300–500 ms. ELAN is earlier and has an onset between 100 and 150 ms. The topographic distribution of ELAN and LAN is very similar. The most parsimonious explanation is, therefore, that the same neuronal generators are responsible for LAN and ELAN, but the temporal profile of their recruitment varies.

How does the Unification Model account for these effects? In the Unification Model, binding (unification) is prevented in two cases: (a) when the root node of a syntactic building block (e.g., NP) does not find another syntactic building block with an identical foot node (i.e. NP) to bind to; (b) when the agreement check finds a serious mismatch in the grammatical feature specifications of the root and foot nodes. The claim is that (E)LAN results from a failure to bind, as a consequence of a negative outcome of the agreement check or a failure to find a matching category node. For instance, the sentence *The woman sees the man because with the binoculars* does not result in a completed parse, since the syntactic frame associated with *because* does not find unoccupied (embedded) S-root nodes with which it can bind. As a result, unification fails. However, this does not necessarily mean that no interpretation of the grammatically ill-formed input will result. There is good evidence that semantic unification and syntactic unification both occur in parallel and, to some degree, independently. Moreover, ERP recordings in aphasic patients have shown that agrammatic aphasics can reduce the consequences of their syntactic deficit by exploiting a semantic route in online utterance interpretation (Hagoort et al. 2003). They thus provide evidence for the compensation of a syntactic deficit by a stronger reliance on another route in mapping sound onto meaning (multiple-route plasticity).

In the context of the Unification Model, the P600 is related to the time it takes to establish unification links of sufficient strength. The time it takes to build up the unification links until the required strength is reached is affected by ongoing competition between alternative unification options (syntactic ambiguity), by syntactic complexity, and by semantic influences. The amplitude of the P600 is modulated by the amount of competition. Competition is reduced when the number of alternative binding options is smaller, or when lexical, semantic, or discourse context biases the strengths of the unification links in a particular direction, thereby shortening the duration of the competition. Violations result in a P600 as long as unification attempts are made. For instance, a mismatch in gender or agreement features might still result in weaker binding in the absence of alternative options. However, in such cases the strength and buildup of U-links will be affected by the partial mismatch in syntactic feature specification. Compared to less complex or syntactically unambiguous sentences, it takes longer in more complex and syntactically ambiguous sentences to build up U-links of sufficient strength. Thus, complex, syntactically ambiguous sentences result in a P600, as compared to less complex, syntactically unambiguous sentences.

This account is not without problems, in terms of the available data. For instance, a word category violation often triggers both (E)LAN and P600. The current version of the model does not specify a mechanistic account of simultaneous unification failure and ongoing U-link attempts. Thus, empirical data provide input for adaptations and improved versions of the model. However, this does not invalidate the conclusion that what we need is models beyond verbal description, with a sufficient level of computational explicitness to be able to characterize the immense body of empirical data on the electrophysiology of language.

The Immediacy Principle

The immediacy principle states that language comprehension beyond the single word level happens incrementally, in close temporal contiguity with information provided by the input signal. There is no principled priority for certain information types. This is not to deny that as a default there is a certain order in the flow of information. For instance, in hearing a word, its phonological information will be retrieved before its semantics; in speaking, the reverse relation holds. However, in the process of composing an interpretation from the lexical building blocks that make up a multi-word utterance, syntactic, semantic, and extra-linguistic information conspire and constrain the interpretation space in parallel. This view stands in contrast to a class of processing models that claim a priority for syntactic information (Frazier 1987a). The strong version of these serial syntax-first models of sentence processing assumes that the computation of an initial syntactic structure precedes semantic unification operations, because structural information is necessary as input for thematic role assignment. In other words, if no syntactic structure can be built up, semantic unification will not be possible. Recent electrophysiological evidence has been taken as evidence for this syntax-first principle (Friederici 2002). Alternative models (Marslen-Wilson and Tyler 1980; MacDonald et al. 1994) claim that semantic and syntactic information are processed in parallel and immediately used once becoming available.

Relevant empirical evidence for a syntactic head start in sentence comprehension is provided in a series of studies in which Friederici and colleagues found an ELAN to auditorily presented words, whose prefix is indicative of a word category violation. For instance, Hahne and Jescheniak (2001) and Friederici et al. (1993) had their subjects listen to such sentences as:

(3) (a) Die Birne wurde im *gepflückt.*
"The pear was being in-the *plucked.*"
(b) Der Freund wurde im *besucht.*
"The friend was being in-the *visited.*"

where the prefixes *ge-* and *be-* in combination with the preceding auxiliary *wurde* indicate a past participle and where the preposition *im* requires a noun. In this case a very early (between 100 and 300 ms) LAN is observed that precedes the N400 effect.

Although this evidence is compatible with a syntax-first model, it is not necessarily incompatible with a parallel, interactive model of sentence processing. As long as word category information can be derived earlier from the acoustic input than semantic information, as was the case in the above-mentioned studies, the immediacy principle predicts that it will be used as it comes in. The syntax-first model, however, predicts that even in cases where word category information comes in later than semantic information, this syntactic information will nevertheless be used earlier than semantic information in sentence processing. Van den Brink and Hagoort (2004) designed a strong test of the syntax-first model, in which semantic information precedes word category information. In many languages, information about the word category is often encapsulated in the suffix rather than the prefix of a word. In contrast to parallel models, a syntax-first model would, in such a case, predict that semantic processing (more in particular, semantic binding) is postponed until after the information about the word category has become available.

Van den Brink and Hagoort (2004) compared correct Dutch sentences (4a) with their anomalous counterparts (4b) in which the critical word (italicized in 4a/b) was both a semantic violation in the context and had the incorrect word category. However, in this case word category information was encoded in the suffix "-de."

(4) (a) Het vrouwtje veegde de vloer met een oude bezem gemaakt van twijgen.
"The woman wiped the floor with an old *broom* made of twigs."
(b) *Het vrouwtje veegde de vloer met een oude *kliederde* gemaakt van twijgen.
"The woman wiped the floor with an old *messed* made of twigs."

Figure 13.4 shows the waveform of the spoken verb form *kliederde* (messed). This verb form has a duration of approximately 450 ms. The stem already contains part of the semantic information. However, the onset of the suffix *-de* is at about 300 ms into the word. Only at this point will it be clear that the word category is a verb and not a noun as required by the context. We define this moment of deviation from the correct word category as the *category violation point* (CVP), because only at this time is information provided such that it can be recognized as a verb, which is the incorrect word category in the syntactic

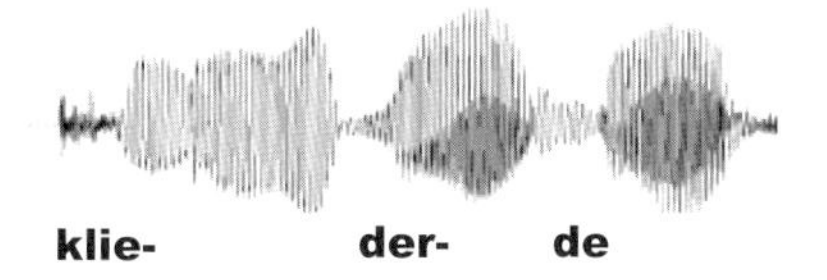

Figure 13.4 A waveform of an acoustic token of the Dutch verb form *kliederde* (messed). The suffix *-de* indicates past tense. The total duration of the acoustic token is approximately 450 ms. The onset of the suffix *-de* is at approximately 300 ms. Only after 300 ms of signal, can the acoustic token be classified as a verb. Thus, for a context that does not allow a verb in that position, the category violation point is at 300 ms into the verb (see text).

context. Although in this case semantic information can be extracted from the spoken signal before word category information, the syntax-first model predicts that this semantic information cannot be used for semantic unification until after the assignment of word category.

Figure 13.5 shows the averaged waveforms that are time-locked to the CVP for two frontal sites where usually the ELAN is observed, and two posterior sites that are representative of N400 effects. As can be seen, the N400 effect clearly precedes the ELAN in time. Whereas the ELAN started at approximately 100 ms after the CVP, the N400 effect was already significant before the CVP. This finding provides clear evidence that semantic binding/unification (as reflected in the N400) can start before word category information is provided. This is strong support for the immediacy principle: information available in the signal is immediately used for further processing. In contrast to what a strong version of the syntax-first model predicts, semantic binding/unification does not need to wait until an initial structure is built on the basis of word category information.

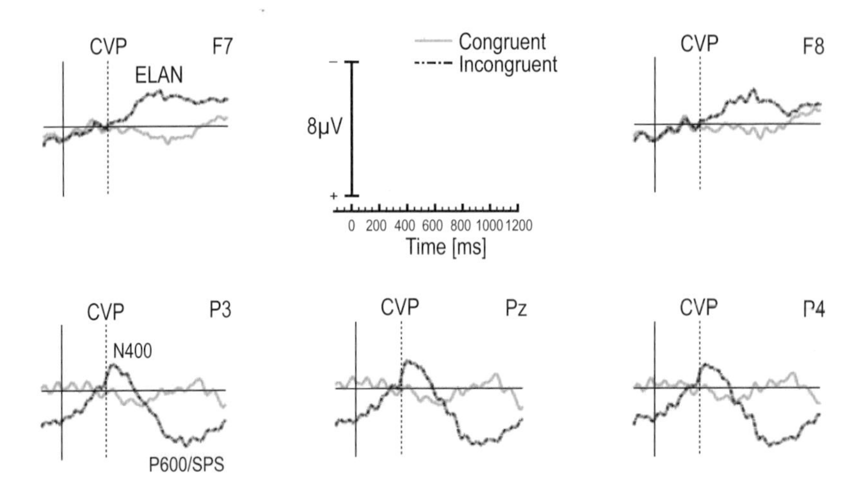

Figure 13.5 Connected speech. Grand average ERPs from two frontal electrode sites (F7, F8) and three posterior electrode sites (Pz, P3, P4) to critical words that were semantically and syntactically congruent with the sentence context (congruent: solid line), or semantically and syntactically incongruent (incongruent: alternating dashed/dotted line). Grand average waveforms were computed after time locking on a trial-by-trial basis to the moment of word category violation (CVP: category violation point). The baseline was determined by averaging in the 180–330 ms interval, corresponding to a 150 ms interval preceding the CVP in the incongruent condition. The time axis is in milliseconds, negativity is up. The ELAN is visible over the two frontal sites: the N400 and the P600 over the three posterior sites. The onset of the ELAN is at 100 ms after the CVP; the onset of the N400 effect precedes the CVP by approximately 10 ms (Van den Brink and Hagoort 2004).

Predictive Processing

One of the most remarkable characteristics of speaking and listening is the speed at which it occurs. Speakers easily produce 3–4 words per second; information that has to be decoded by the listener within roughly the same time frame. Considering that the acoustic duration of many words is on the order of a few hundred milliseconds, the immediacy of the ERP effects (discussed above) is remarkable. The ELAN has an onset on the order 100–150 ms, the onset of the N400 and the LAN is approximately at 250 ms, and the P600 usually starts at about 500 ms. Thus, the majority of these effects happen well before the end of a spoken word. Classifying visual input (e.g., a picture) as coming from an animate or inanimate entity takes the brain approximately 150 ms (Thorpe et al. 1996). Roughly the same amount of time is needed to classify orthographic input as a letter (Grainger et al. 2008). If we take this as our reference time, the earliness of an ELAN to a spoken word is remarkable, to say the least. In physiological terms, it might be just too fast for long-range recurrent feedback to have its effect on parts of primary and secondary auditory cortex involved in first-pass acoustic and phonological analysis. Recent modeling work on the *mismatch negativity* suggests that early ERP effects are best explained by a model with forward connections only. Backward connections become essential only after 220 ms (Garrido et al. 2007). The effects of backward connections are, therefore, not manifest in the latency range of at least the ELAN, since not enough time has passed for return activity from higher levels. In addition, LAN and N400 are following the word recognition points closely in time in the case of speech. This suggests that what transpires in online language comprehension is presumably based, to a substantial degree, on predictive processing. Under most circumstances, there is just not enough time for top-down feedback to exert control over a preceding bottom-up analysis. Very likely, lexical, semantic, and syntactic cues conspire to predict characteristics of the next upcoming word, including its syntactic and semantic makeup. A mismatch between contextual prediction and the output of bottom-up analysis results in an immediate brain response recruiting additional processing resources for the sake of salvaging the online interpretation process. As presented above, the Unification Model for parsing has prediction built into its architecture. Syntactic frames activated on the basis of an activated word form specify the local syntactic environment options and carry as such a prediction about the syntactic status of the next upcoming word; see Example (1). Recent ERP studies have provided evidence that context can indeed result in predictions about a next word's syntactic features (e.g., gender; Van Berkum et al. 2005) and word form (DeLong et al. 2005). Lau et al. (2006) have shown that the (E)LAN elicited by a word category violation was modulated by the strength of the expectation for a particular word category in the relevant syntactic slot. The role of structural predictions has, thus, found support on the basis of recent empirical findings. One possible source for these predictions is the language production system.

It might well be that the interconnectedness of the cognitive and neural architectures for language comprehension and production (Hagoort et al. 1999b) enables the production system to generate internal predictions while in the business of comprehending linguistic input. This "prediction-is-production" account has, however, not yet been tested empirically.

Afterthoughts on the Neurobiology of Syntax

Next to speed of language processing, another decisive characteristic of human language is its diversity. Decades of cross-linguistic work by typologists and descriptive linguists have shown "just how few and unprofound the universal characteristics of language are, once we honestly confront the diversity offered to us by the world's 6–8000 languages" (Evans and Levinson 2009). Languages vary substantially in sound, lexical meaning, and syntactic organization. Nevertheless, children acquire language faster, almost universally, and seemingly more automatically than, for instance, musical skills. This indicates that evolution has provided humans with a brain that is characterized by a certain language-readiness. This might have happened by optimizing the neural infrastructure for the ensemble of cognitive systems (e.g., systems for memory, unification, executive control) that collectively provided language-readiness to the brain, instead of forming a universal neural machinery dedicated to syntax. According to the former view, there might also be overlap and common recruitment of certain brain regions in the service of different cognitive functions with similar requirements. For instance, domain-specific memories for syntax, music, and action schemata might all recruit the domain-general unification capacities of Broca's complex to produce or decode the combinatorial aspects of language, music, and action (Patel 2003). In support of this account, a recent study with Broca's aphasics characterized by a syntactic deficit showed that these same patients also were impaired in processing musical syntactic (harmonic) relations in chord sequences (Patel et al. 2008). How we continue to interrogate the brain will, to a large extent, depend on the perspective taken on this issue.

Acknowledgment

I am grateful to Karl Magnus Petersson for discussions about and comments on a previous version of this contribution.

Overleaf (left to right, top to bottom):
Tatjana Nazir, group discussion, Balázs Gulyás
Csaba Pléh, Anna Fedor, Peter Hagoort
Angela Friederici, Group discussion
David Caplan, Jens Brauer, Wolf Singer

14

What Are the Brain Mechanisms Underlying Syntactic Operations?

Anna Fedor, Csaba Pléh, Jens Brauer, David Caplan,
Angela D. Friederici, Balázs Gulyás, Peter Hagoort,
Tatjana Nazir, and Wolf Singer

Abstract

This chapter summarizes the extensive discussions that took place during the Forum as well as in the subsequent months thereafter. It assesses current understanding of the neuronal mechanisms that underlie syntactic structure and processing.... It is posited that to understand the neurobiology of syntax, it might be worthwhile to shift the balance from comprehension to syntactic encoding in language production

The Main Questions

The focal question posed to our group can be answered on many descriptive levels. At the finest-grained level are processes that store and activate syntactic representations in neuronal networks. Here, relevant variables are the frequency of spikes, the temporal structure of spike sequences, changes in sensitivity and number of synapses, and modifications of excitability. To date, however, little is known about the physiological basis for syntactic processing at this cellular level. In our discussions we therefore concentrated on relating syntactic representations and processes to larger-scale neural features, such as event-related potentials (ERPs) and areas of the brain determined anatomically or functionally.

It is important to realize why issues related to the "localization"of cognitive skills in the brain, including the processing of language, are hotly debated. These issues have a long history in neuroscience, and the answers given were always dependent on the original assumptions, the methodology, and methods of the investigations. Nowadays it cannot be denied that the textbook notion of "Broca's area for syntax and Wernicke's area for semantics" is outdated. Localization of components of language, including syntax, shows variation both across and within individuals; the latter indicates that the problem has a strong ontogenetic component as well. The enormous plasticity of the developing human brain (as shown by recovered language after lesions early in life) demonstrates that the crucial involvement of Broca's area in syntactical processing in most people cannot be a genetically hardwired, rigid condition. It seems more correct to say that some areas of the normally developing human brain are more prone (in a quantitative sense) to host and process different components of language than others. This "hospitality" may be due to the particular features of the local microanatomy/physiology or connectivity to other areas, or both. Data support both alternatives. Ultimately, what we need is a detailed neuronal model of how syntactic operations can be implemented as well as how such neuronal structures could have been rigged by genetic influence. Apes do not talk and acquire syntax; by comparison, humans with even low IQs are linguistic geniuses. We need to gain an understanding of the neurobiological and correlated genetic changes in evolution; we are slowly but surely approaching this target. This chapter reports on the current level of understanding.

Syntactic Structure and Processing

In its essence, syntax is a means by which elements are arbitrarily combined; in the case of language, the elements are semantically interpreted symbols. The syntax of natural languages has a number of properties that have guided the study of its evolution and neurological basis. Several pertain to the nature of the meanings that it encodes; others to the forms of syntactic representations.

Three important features of semantic meanings encoded by syntactic relations are:

1. Meaningful symbols that are combined syntactically are both referential (reflected in, e.g., content words) and logical/formal (reflected in, e.g., function words and affixes).
2. Meanings conveyed by combinations of these symbolic elements, collectively referred to as propositional and discourse-level meanings, constitute a restricted set of semantic values and include the relation between items and their properties (modification and quantification), the participant structure of events (thematic roles), temporal and aspectual features of events, illocutionary force, discourse prominence, as well as others.
3. Propositional meanings can be combined with the referential meanings of individual items, as in the modification of referential elements by

propositions using relative clauses (e.g., *The boy who is wearing the blue shirt is my brother*). Propositions can also be combined with other propositions, as in complementation (e.g., *John believes that carrots are a good source of vitamins*). Meanings are therefore compositional, productive, and recursive.

The syntactic relations between constituents that determine aspects of propositional and discourse meaning have been the subject matter of intense inquiry on the part of linguists since the seminal work of Noam Chomsky in the 1950s. Models of these relations differ widely. The latest version of Chomsky's theory, for instance, maintains that binary branching nodes create hierarchical structures in which nodes "merge" "externally" to create local phrase markers that assign certain semantic values. This aspect of the syntax is the basis for the compositional, productive, and recursive features mentioned above. Second, in this model, constituents "merge" "internally" to create complex phrase markers, subject to language-specific and language-universal constraints (Chomsky 1995, 2004, 2006). Semantic relations between constituents are determined and constrained by particular asymmetrical dominance relations between nodes in these phrase markers ("c-command"). Other linguists view syntax differently. Goldberg (1995, 2006) and Croft (2001), for example, argue that the syntactic structure of a sentence is a direct reflection of its surface form, and that propositional meanings are determined by the "constructions" of the syntax (see Tallerman et al., this volume).

Regardless of how the syntactic structures that determine propositional and discourse meanings are represented, they must capture the fact that form-meaning relations are complex. Take, for example, two of the central features of syntax: *hierarchical structures* and *distant dependencies*. These features can be distinguished from each other. One can easily compose sentences with distant dependencies, where the distance between the connected words is made wider by inserting adverbs. Thus, in empirical studies, it is very important to ensure that it is indeed complexity per se (hierarchical structure including distant dependencies vs. linear structure including local dependencies only) and not merely greater distance between constituents that makes the understanding of one sentence more difficult than another. If the distance between constituents is too large (e.g., in the case of a very long sentence), it may be difficult to judge whether the sentence is grammatically correct, even if the structure is simple. It is therefore crucial to disentangle empirically the factors of syntactic complexity and working memory through a design that varies these two factors independently whenever possible (Makuuchi et al. 2009); see Figure 14.1 for examples.

For processing long-distance dependencies, a system is needed that carries information over time. In the human brain, the prefrontal cortex may be suitable to this task because of its involvement in working memory. Its ability to support short-term functions is known from neuropsychological studies in

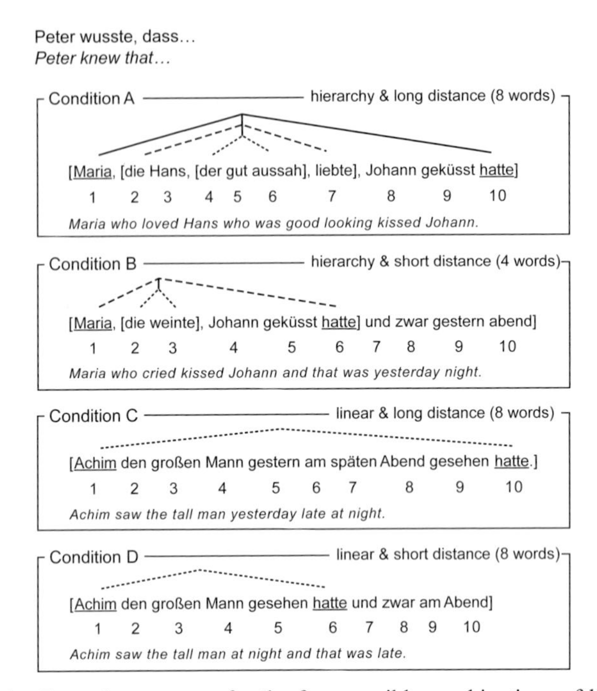

Figure 14.1 Example sentences for the four possible combinations of linear/hierarchical structure and short-/long-distance dependencies.

human subjects and animal experiments. Understanding the neural mechanism for processing hierarchical structures in sentences is, however, a much harder problem and, recent pioneering studies in this field notwithstanding, there are only speculative ideas about its implementation in the brain. A step toward understanding this issue are findings from brain imaging studies, which suggest that the effects of long-distance dependencies and those of hierarchical structures are processed in separable, but closely interacting areas in the inferior frontal gyrus (IFG; see Figure 14.2) (Makuuchi et al. 2009; Santi and Grodzinsky 2007b).

The example above demonstrates potential interactions between linguists, psychologists, and cognitive neuroscientists. In general, linguists and psychologists can provide theories that constrain the hypothesis space regarding syntactic structures and their processing, presenting testable predictions; neuroscientists can provide empirical evidence that constrains the theory space. Experimental evidence could help linguists decide between competing models of structure and processing. Neuroscientists, in turn, need from linguistics a certain level of generalization across the different versions of linguistic theoretical models. This would help define aspects of complexity and structure for translation into experimental manipulations aimed at identifying biological consequences.

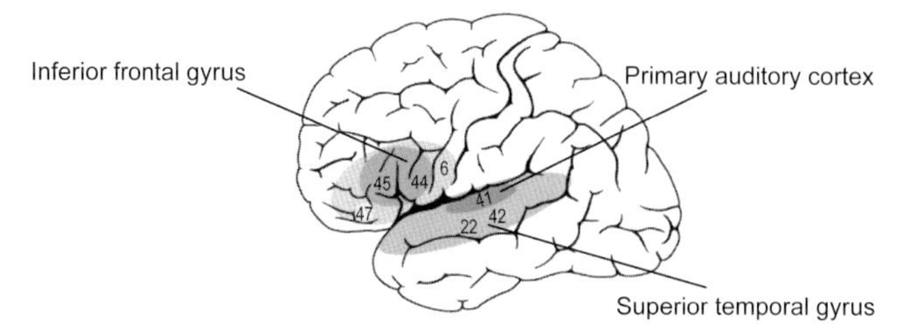

Figure 14.2 Left sagittal view of the human brain. The shaded areas indicate the main language-relevant perisylvian regions in the left hemisphere (LH). Broca's area proper (Broca 1863) comprises Brodmann area (BA) 44 and BA 45 of the IFG (Amunts et al. 1999). Further frontal areas (light gray), also involved in some language processes, are BA 47 of the IFG and the precentral gyrus (BA 6), and it has been suggested to label all four areas together as Broca's area extended (Hagoort 2005). The shaded region around 22 and 42 depicts the superior temporal cortex with the primary auditory cortex (dark gray) in its mid-portion.

To illustrate such an interdisciplinary interaction, consider the syntactic phenomenon that many linguistic theories specify as the *movement* or *displacement* property in many natural languages; that is, elements are often pronounced in positions distinct from those in which they are interpreted. For example, in a question like

(1) Which book do you think I should read?

the phrase *which book* must be interpreted as a thematic argument of the verb *read*. It, however, is not in the canonical object position adjacent to the verb as in *I should read this book*. Instead, it has been displaced to the front of the sentence and, in fact, can be indefinitely far away from its thematic position. Thus, the grammar and the parser must be able to relate two positions which can be quite distant in the tree structure. One may think that this operation has a certain computational cost which brain imaging studies may be able to highlight, in terms of accrued activation of the neural structures which are involved. As movement can be seen as a sub-case of Merge, the fundamental recursive structure-building mechanism, it can be applied several times in a structure. One might entertain the working hypothesis that the number of applications of movement provides a rough measure of complexity (Jakubowicz, unpublished). This would be difficult, however, to substantiate, because it is likely that different types of movement have different computational costs; for instance, the computational cost of some very local movements (e.g., head movement of the verb to the inflectional system or to the complementizer in verb second language) may be negligible if compared to the cost of long-distance phrasal movements. We can avoid these complications, if we restrict our attention to phrasal movements.

An important property of long-distance movement is that, in theory, it typically takes place through a number of successive local steps in compliance with the requirements of locality principles. There is strong and varied linguistic evidence suggesting that the movement of our sample sentence in (1) takes place in at least two steps: (a) from the thematic argument position to the embedded complementizer system, and (b) from the embedded to the main complementizer system. In (2), the blanks indicate the two "traces" of movement:

(2) Which book do you think [___ that [I should read ___]]

To test this hypothesis empirically, it would be interesting to compare this movement in terms of computational cost with the case in which two distinct constituents have moved once. A possible approximation of this in English may be a simple clause in which one verbal complement is passivized and the other is moved:

(3) To whom was the book given ___ ___ by John?

Another possibility might be to investigate this phenomenon in languages with scrambling (Germanic) or cliticization (Romance) structures in which two verbal complements are allowed to take free-word order to a certain degree, or where pronouns are moved to a higher position in the tree structure of the sentence. Testing scrambling in German subjects, Friederici et al. (2006a) found Broca's area to increase its activation parametrically as a function of object–noun phrases moved (scrambled) in front of the subject–noun phrase.

Movement is only a one-level complexity problem. If the sheer number of movements provides a rough measure of complexity, one would expect, if the hypothesis is correct, that two successive movements of the same element, as in (2), would roughly correspond to two local movements of two distinct constituents that have each moved once, as in (3). Finding similar neural effects of (2) and (3) relative to an appropriate control would support the view that movement occurs through a series of local steps, and demonstrates an important case where linguists and neuroscientists could benefit from each others' point of view.

Neural Organization for Processing Syntax in the First Spoken Languages of Adults

Neural Codes for Syntax

How are syntactic categories and their relationships coded in terms of physiological properties of neural tissue? A leading hypothesis is that they are related to the frequency of neuronal firing patterns and their correlations across space and time. Some data, collected during neurosurgical procedures, are available on extracellular recordings of single neurons in response to language stimuli; more data exists on such responses in subdural electrode grids used in preparation for resections (Quian Quiroga et al. 2008; Mormann et al. 2008). These studies raise the possibility of a neural code for syntax, but further research is needed to determine if there is any specificity for syntactic processes on this elementary level.

Lateralization

Since the early studies by Paul Broca (1863), language has served as an example of functional cortical lateralization in the human brain. The common notion of language lateralization to the LH does not, however, imply that the right hemisphere (RH) is not involved in language processing. Instead, it refers generally to a stronger activation in LH than in RH and to differences in the functions that the two hemispheres perform at each level of language processing. This has been studied fairly extensively with respect to auditory and phonological processing. One model of functional asymmetry in the auditory cortex differentiates the processing on a basic feature level of auditory information to the two hemispheres. In this model, temporal information is processed primarily in the LH and spectral information primarily in the RH (Zatorre and Belin 2001). The role of the two hemispheres in syntactic processing remains uncertain. Some studies have found bilateral activation associated with syntactic contrasts (e.g., Just et al. 1996; Cooke et al. 2002). Other models of language processing suggest a specialization of the perisylvian cortex of the LH for processing semantic and syntactic information (Friederici 2002), while the perisylvian cortex of the RH, in particular the posterior superior temporal gyrus (STG) and the frontal operculum, is regarded as responsible for the processing of prosodic information (Meyer et al. 2002; Meyer et al. 2004; Zatorre et al. 2002; Friederici and Alter 2004).

Consonant with functional hemispheric specialization, structural hemispheric asymmetries were found in language-related areas of, for example, the planum temporale in the perisylvian cortex (Geschwind and Levitsky 1968) and in the organization of intrinsic connectivity (Galuske et al. 2000). Amunts et al. (1999) reported a left-larger-right asymmetry for BA 44. The relation of these specific features of brain structure to aspects of syntactic processing remains unknown.

Hemispheric specialization is not unique to language. Besides the speech and language domain, cortical laterality has been reported for a variety of sensorimotor functions such as handedness or visual processing and other cognitive functions such as memory (Gainotti 2007; Habib et al. 2003). Morphological asymmetries have also been observed for cortical areas not related to language processing, e.g., the primary motor cortex (Amunts et al. 2000). From a comparative perspective, the morphological asymmetry of the planum temporale does not seem to be exclusively human; it has been found in nonhuman primates (Gannon et al. 1998). Functional hemispheric specialization is also not uniquely human; it has been observed in the auditory cortex of songbirds (George et al. 2004). Despite the fact that we know very little about the cortical organization of the brain of some other species, hemispheric organization can be inferred from lateral bias. Lateral bias (as opposed to cortical lateralization), referring to afferent or efferent bias (e.g., hand or eye preference), can be assessed behaviorally, and has been discovered for a variety of nonhuman species, including chicks, rats, and primates (Denenberg 2005; Hopkins and Cantalupo 2008; Rogers 2008).

Given its ubiquity among species and its very early presence in development, laterality appears to be a phylogenetically old phenomenon. It may underlie homologous phenomena across species or constitute a convergent characteristic that has been repeatedly favored in natural selection. However, the evolutionary advantage conferred by hemispheric specialization remains unknown, and the extent to which functional and/or structural hemispheric specialization might have promoted the evolution of language and syntax remains an open issue.

Regional Specialization

There is a general consensus that cognitive operations engage cortical macronetworks consisting of activated neuronal populations in various brain regions. To perform a certain cognitive task successfully, the involvement of a number of activated neuronal assemblies is required. A widespread view holds that a core network performs a set of operations that occur under all conditions of performance, and that this core network is complemented by other activated neuronal populations that map its products onto input and output systems, memory, decision making processes, etc. Different nonlinguistic systems are engaged as a function of varying modalities or stimulus and task conditions, and their involvement affects efficiency of task performance.

One goal of current research is to identify the areas of the brain responsible for parsing and interpretation. The areas and physiological events *necessary* for a function are revealed through the effects of lesions that occur either naturally or experimentally, as in transcranial magnetic stimulation: deficits that follow lesions imply that the lesioned aspect of the brain is necessary for the deficient function, while spared functions that follow lesions imply that the lesioned aspect of the brain is unnecessary for the function. The areas and physiological events that are *sufficient* for a function are revealed through the study of neural responses to linguistic tasks: areas or physiological events affected by psychological operations are a subset of those sufficient for the function that is required for the performance of those psychological operations. The implementation of these research approaches involves many decisions about methods, analyses, and interpretation of results that are subject to discussion.

The most widely accepted model of neural organization for cognitive processing, including syntactic processing, is *localization* (Friederici 2002; Grodzinsky 2000): the view that particular syntactic computations are processed in circumscribed areas of the perisylvian association cortex, which are defined by cytoarchitectonic or, more recently, receptotectonic criteria. These models do not deny macroscopic variation across individuals with respect to the grossly defined areas of the brain that support a language operation, but rather attribute this variability to variability in the mapping of cytoarchitectonic areas onto grossly defined brain areas, such as gyri (Amunts et al. 1999), and to individual differences in connectivity between areas (Anwander et al. 2007). These models postulate localization of syntactic functions in circumscribed brain regions, when brain regions are correctly described at the cytoarchitectonic, receptotonic, or other cellular or subcellular level. Models of this sort carry the implication that the computational capacities of a brain area are determined by its *unique* informationally relevant features.

The alternative is that computational capacities are determined by the neurological features that are *common* to a broader range of brain areas within a specified, but more extended brain region. If that is the case, a functional capacity such as language comprehension or, more narrowly, specific syntactic processes, could be distributed over different areas within the perisylvian cortex more or less evenly or unevenly (e.g., Mesulam 1998), localized in a number of areas that have the critical cellular features but which are otherwise unrelated (so-called "degeneracy"), or variably localized in different individuals (Caplan et al. 2007b). Borrowing a term from Lashley (1950), Caplan (1994; Caplan et al. 2007b) used the term "equipotential" to refer to an initial neural state in which many areas, or a large area, are capable of supporting a syntactic operation and the determination of which part of the potentially responsible neural system ultimately supports any given operation is determined by a variety of factors.

These possibilities have implications for mechanisms that underlie the evolution of the neural substrate for language and syntax. Since, as far as is currently known, all genes that are expressed in language-related cortex are expressed in more than one cytoarchitectonically defined area, invariant localization requires a confluence of genetic effects. On the other hand, initial equipotentiality and ultimate variable localization or distribution of a syntactic operation over a wide area of the brain is compatible with multiple patterns of gene expression resulting in cortex that could support a syntactic operation, coupled with a variety of effects of epigenetic factors, the developmental history of an individual, and other influences (including the presence of disease) .

Though they cannot both be correct for a single parsing or interpretive operation, the localized and the distributed/variable models each have their supporters. Support for a localized organization of the brain comes from data that report certain syntactic phenomena to be observed consistently in a specific brain region, such as scrambling in BA 44 or movement in BA 45 (Ben-Shachar et al. 2003; Santi and Grodzinsky 2007b; Friederici et al. 2006a). A particular model of localization can be found in Friederici's position paper (this volume) which claims that two pairs of cytoarchitectonically defined areas of cortex, which are functionally and structurally interconnected via different white matter fiber tracts, support specific functions in syntactic computations. According to this hypothesis, one network consists of the deep frontal operculum in the medial IFG and the anterior STG and is responsible for processing local phrase structures. The processing of complex sentences with a hierarchical structure is supported by a second network which consists of Broca's area (in particular BA 44) and the posterior STG. In Friederici's view, activation in these brain areas is not specific to language processing; they are also involved in other cognitive processes. The specific function of these areas for language processing emerges from their contributions to a broader network of perisylvian areas and develops during normal ontogenesis. In the case of a major disturbance, such as lesions of these areas early in development, plasticity of the maturating brain allows other parts of the cortex, especially homologous areas of the contralateral hemisphere, to take over these functions .

Evidence for other models comes from the study of aphasia as well as other neuroimaging results. Caplan (this volume) reviews evidence that strokes in either the posterior or anterior parts of the perisylvian cortex are associated with normal performances on tasks that require syntactic comprehension, indicating that different parts of this area are not necessary for syntactic operations in individual cases. He presents data that the severity of deficits in various aspects of syntactic processing varies greatly in patients whose lesions occupy the same proportion of the perisylvian cortex and that the proportion of the perisylvian cortex that is lesioned in patients with similar degrees of impairment of various aspects of syntactic processing also varies greatly. These results lead to the view that there are individual differences in the areas that are necessary for particular syntactic operations. Caplan et al. (2008) argue that BOLD signal correlates of the same or very similar syntactic contrasts often show multiple areas of activation within a single study and considerable variation across studies.

Hagoort (this volume) argues that a particular cognitive function, such as syntax, is most likely subserved by a distributed network of areas, rather than by a local area alone. A one-area-one-function principle is in many cases not an adequate account of how cognitive functions are neuronally instantiated. Connectivity is the major force in shaping the functional contributions of a particular piece of cortex.

These major disagreements result in part from differences in how the research has been conducted (i.e., choice of methods) and in how the results have been interpreted. Many questions arise about experimental design. The activation of brain areas during certain tasks is inferred on the basis of statisti cal models that usually employ a *subtraction* between experimental conditions to isolate a specific process under investigation. The resulting activation pattern refers to only those activations that are stronger in one condition than in the other. Using this approach, activation common to both conditions is invisible, which is intended according to the specific process under investigation. However, the choice of *contrasts* is critical in both aphasiology and functional neuroimaging. Baseline sentences need to test all relevant operations that are antecedent to the parsing and interpretive operations under study. For instance, testing for a deficit in the ability to relate a noun phrase to a particular position in a sentence (i.e., to a trace) requires showing that structures with that feature cannot be comprehended normally, that the patient can understand thematic roles in sentences that do not require the operation (e.g., in active sentences), and that a patient can understand sentences that require relating noun phrases to other referentially dependent items (e.g., reflexives).

Another reason for disagreement is that most neuroimaging studies are done on comprehension (parsing), whereas many patient studies have focused on language production (syntactic encoding). The few imaging studies that have been done on syntactic encoding and on comparing encoding and parsing suggest that in comprehension one might be able to bypass syntax, but not so in production. If there have been evolutionary selection pressures and precursors for language, they are very likely related to wiring up the brain for language production and the role of syntax as an intermediary between a nonlinearized conceptual representation and a linear string of speech sounds. To understand the neurobiology of syntax, it might be worthwhile to shift the balance from comprehension to syntactic encoding in language production.

With respect to what has yet to be done, many basic studies are needed to shed light on the neural organization for syntactic processing. On the psychological side, it is becoming apparent that, in both aphasiology and functional neuroimaging, parametric variation of a factor that affects an operation or repetition suppression paradigms are designs that have much promise, but they are only beginning to be used. Qualitative contrasts, which are more commonly used, are hard to interpret. In both aphasiology and functional neuroimaging, the study of the effects of a syntactic contrast in more than one task is critical, but very few studies examine task effects on syntactic determinants of aphasic performance or neural activity in normal subjects.

In addition to these issues, there are several sources of individual variability in functional neuroimaging studies other than distribution, duplication, or individual differences of the areas that support syntactic processing. These sources of differences across studies include variance in the statistical methods used, normalization algorithms, the effects of the tasks used upon syntactic processing, differences in task performance when participants process different sentence types, differential use of ancillary (sometimes strategic) cognitive mechanisms such as subvocal rehearsal in different sentence types, and others. There are differences in parsing and interpretive processes as a function of language, which raises the possibility that different patterns seen in different languages might reflect invariant localization of different syntactic and/or parsing operations. Thus, understanding the reasons for variable results in different studies requires detailed examination of methods, and progress can be expected as this examination reveals limitations of existing methods, leading to better and more detailed experiments. Other analyses of existing data may also be useful, such as conjunction analyses of the effects of particular contrasts over tasks, languages and participant groups, meta-analysis of existing studies.

The extent of individual variability in the neural organization for syntactic processing remains an open, basic question. It is worth noting that variability is very common at every level of biology. Phenotypic variation results from genetic, environmental, and developmental variation. Thus, even in genetically identical individuals who grow up in the same environment, some phenotypic variation can be found due to subtle differences in the factors that affect the individuals during development. Thus, the reasons for the various results found in different studies constitute an important area for future study.

Differences between Languages

Friederici proposed that, in general, the same brain areas may be recruited for different tasks depending on the functional significance of different linguistic markers. One crucial example is thematic role assignment, which seems to be done by the LH, but the cues used for this computation might differ greatly across languages. Investigating different cues across languages, Bornkessel and Schlesewsky (2006) argued that Broca's area supports thematic role assignment independent of the particular cue (word order, case marking, or animacy). They suggest that this region is responsible for mapping the linear sequence of cues regarding the position of a noun on a thematic hierarchy onto the thematic role assigned to each noun.

The idea that areas of the LH become responsible for the efficient use of whatever cue is most reliable or available in a language is supported by developmental studies. Studies by Slobin and Bever (1982), MacWhinney et al. (1985), and Pléh (1989, 1990) reveal interesting and intricate differences between languages in the use of different structural cues in agent assignment, such as case marking or word order. Interesting temporal developmental differences follow: In Turkish, due to its clear case marking (the accusative is coded by vowels), children become proficient in agent assignment ("who-did-what-to-whom") by 2 years of age. Hungarian children also use case marking, but it is less transparent in this language; thus, children are tuned to the use of exclusive case marking around 3½ years of age, whereas English-speaking children stabilize the order-based strategy around the age of 5. In Serbo-Croatian, where there is an interaction between animacy and case marking, children use a combined word-order and case-marking strategy to assign thematic roles.

In English, Bever (1971) showed that the increase of effective LH dominance in language processes correlates with increasing use of the "first noun is the agent" strategy in sentence interpretation. In Hungarian, Pléh (1982) showed that children of the same age (4–6 years) displayed the opposite tendency; namely that with the increase of LH lateralization, a decrease was observed in the use of order-based strategies. It seems to be that the LH is tuned to the most valid cues of the given language. Therefore, the same brain areas recruited to support order-based strategies in English support morphology-based strategies in Hungarian.

Another much studied example is the domain of prosody. Suprasegmental intonational patterns are mainly represented and computed by the RH (Meyer et al. 2002) whereas in tonal languages, where (Gandour et al. 2002; Gandour and Dardarananda 1983) prosodic cues (i.e., tones) have lexical consequences, these are computed and represented by the LH.

Clearly, there is a need to do systematic cross-linguistic comparisons using identical methods, both behaviorally and in regard to neuroscience data gathering, to reveal the exact nature of neural network activation in typologically different languages.

Other Brain Features: Receptor Structure

Theories of regional specialization raise many issues that must be considered from both a psychological and neural perspective. The most fundamental is: What is the proper decomposition of function and structure? We have briefly discussed what we consider to be important aspects of syntactic representations (for further discussion, see Tallerman et al., this volume). From a neural perspective, we need to go beyond both macroscopic landmarks and cytoarchitectonic classifications, such as that of Brodmann (Figure 14.3), to identify brain areas that could be related to syntax.

Receptor structure is a good candidate for future research. In addition to the most widely distributed major neuroreceptors and neurotransmitters (glutamate, GABA-benzodiazepine, monoamines, acetylcholine, endocannabinoid, opioid), there are several hundred known neuroreceptors as well as neurotransmitter systems in the primate brain. The fine balance between these systems (i.e., the brain's receptor fingerprint) is determined by genetic and epigenetic factors, as well as by ongoing brain activities and the brain's interactions with its social or physical environment. Individual cortical areas have typical receptor structures which are usually characteristic for each cytoarchitectonically defined area (Brodmann area; see Figure 14.4) (Zilles et al. 1991, 2002, 2004; Amunts et al. 1999; Amunts and Zilles 2001; Eickhoff et al. 2007a, b).

The human brain's receptor fingerprint is changing continuously, in response to continuous challenges, including normal or physiological challenges (maturation, aging, habituation, sensory processes, social interactions, diurnal rhythms, etc.), pathological challenges (schizophrenia, depression, epilepsy, etc.), internal (endogenous neurotransmitter release due to, e.g., thinking or meditation) or external pharmacological challenges (medicines, drug abuse). The human brain's receptor fingerprint has enormous individual variations and thus could underlie variability in regional specialization. At the same time, the human brain's receptor fingerprint shows a close correlation with personality types, behavioral traits, and temperament. For instance, the level of dopamine shows a close correlation with extroverted or introverted behavior or

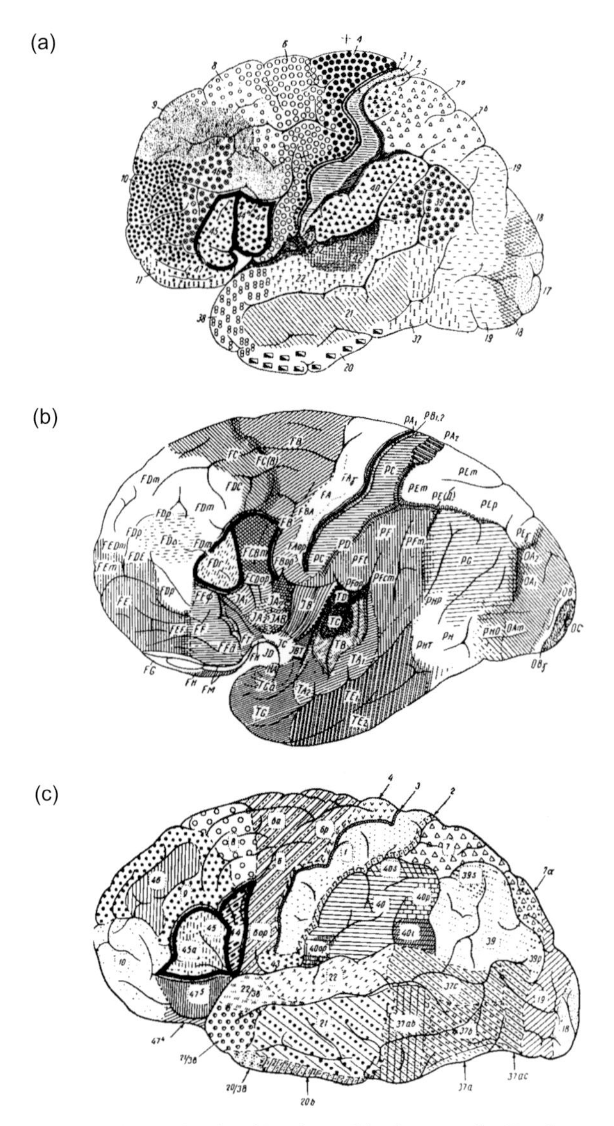

Figure 14.3 Cytoarchitectonic classification of brain areas by Brodmann (reprinted with permission from Amunts et al. 1999).

with novelty seeking, whereas serotonin correlates with harmony or challenge seeking, with proneness to transcendental experience, or depression (Gulyás 2007; Farde et al. 1997; Zald et al. 2008).

Thus receptor profiles could contribute to stable neural systems that underlie syntactic processing. An open issue is whether the involvement of the perisylvian cortex in syntactic processing might be based on certain properties in its receptor architecture. To date, it is unclear whether receptor

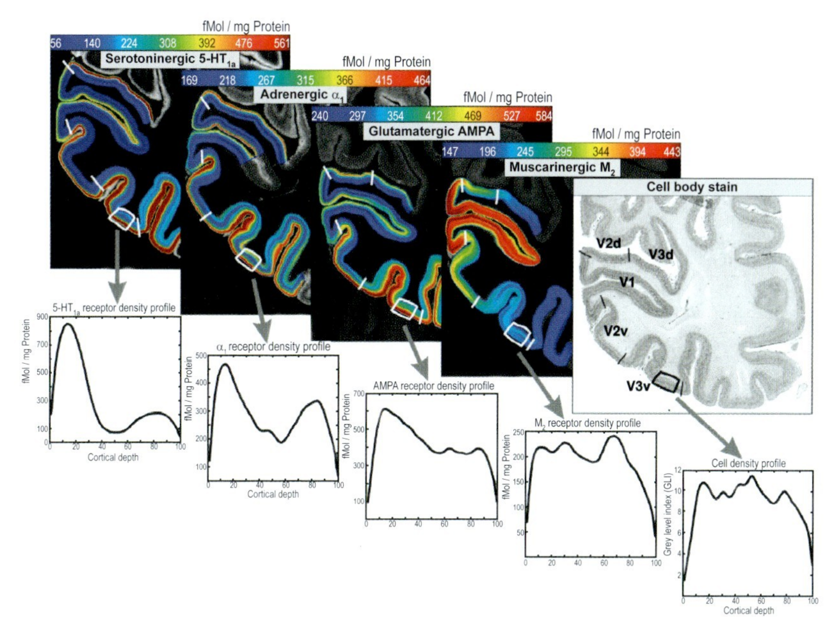

Figure 14.4 Receptor structure of brain areas (after Eickhoff et al. 2007b).

architecture is a precondition or a consequence of shared functionality between two brain areas.

At the same time, many crucial aspects of morphology, laminar distribution, and synaptic targets are very well conserved between areas and between species (Douglas and Martin 2004). Functional differences between brain areas might, therefore, be mainly due to variability of the input signals in forming functional specializations. Domain specificity of a particular piece of cortex might thus not so much be determined by heterogeneity of brain tissue, but rather by the way in which its functional characteristics are shaped by the input. Moreover, findings of neuronal plasticity (e.g., the involvement of visual cortex in verbal memory in the congenitally blind; Amedi et al. 2003), suggest substantial plasticity in structure-to-function relations (see Hagoort, this volume).

The Temporal Dimension of Syntactic Processing

Electrophysiological measures provide data of the temporal aspects of brain mechanisms underlying syntactic processing. The most intensely studied ERP components, often studied using violation paradigms and by presenting sentences with increased processing difficulty, are briefly described here.

Based on the time course of language-related ERP effects, one can say that the major syntactic and semantic processing events happen between 150 and 800 milliseconds. In connection with syntactic processes, two classes of syntax-related ERP effects have been consistently reported over a period of more than ten years. Examples of sentence material that would elicit the four ERP components are:

(4) (a) *The boy of eats ice cream ELAN, P600
(b) *The boys eats ice cream LAN, P600
(c) The boy eats socks N400

One type of ERP effect related to syntactic processing is the P600 (Hagoort et al. 1993; Osterhout and Holcomb 1992). P600 is reported in relation to syntactic violations, syntactic ambiguities, and syntactic complexity. This effect occurs in a latency range between roughly 500–800 ms following a lexical item that embodies a violation or a difference in complexity. However, the latency can vary, and earlier P600 effects have also been observed (Hagoort 2003; Mecklinger et al. 1995). Another syntax-related ERP is a left anterior negativity, referred to as LAN or, if earlier in latency than 300–500 ms, as ELAN (Friederici et al. 1993). In contrast to the P600, (E)LAN has thus far been (almost) exclusively observed in syntactic violations. LAN is usually observed within a latency range of 300–500 ms; ELAN is earlier, with an onset between 100–150 ms. The topographic distribution of ELAN and LAN is very similar. The most parsimonious explanation is, therefore, that the same neuronal generators are responsible for LAN and ELAN, but the temporal profile of their activation varies (for an alternative view, see Friederici, this volume). A negativity around 400 ms (N400) with a central topography is related to semantic processes, for example, at semantic violations and corresponding lexical search processes.

ERP data provide an example of the feedback from neurological studies to models of language processing. One of the most remarkable characteristics of speaking and listening is the speed at which it occurs. Speakers easily produce 3–4 words per second; information that has to be decoded by the listener within roughly the same time frame. Considering that the acoustic duration of most words is in the order of a few hundred milliseconds, the immediacy of the ERP effects is a highly salient feature. The ELAN has an onset of 100–150 ms, the onset of the N400 and the LAN is approximately at 250 ms, and the P600 usually starts at about 500 ms. The majority of these effects happen well before the end of a spoken word. Classifying visual input (e.g., a picture) as coming from an animate or inanimate entity takes the brain approximately 150 ms. If we use this as our reference time, the early brain response reflected in the ELAN to a spoken word is remarkable, to say the least. In physiological terms, it might be just too fast for feedback to have an effect on parts of primary and secondary auditory cortex involved in first-pass acoustic and phonological analysis. This suggests that what happens in online language comprehension is substantially based on predictive processing. Under most circumstances, there is just not enough time for top-down feedback to exert control over a preceding bottom-up analysis. Very likely, lexical, semantic, and syntactic cues are used to predict characteristics of the next upcoming word, including its syntactic and semantic makeup. A mismatch between contextual prediction and the output of bottom-up analysis results in an immediate brain response activating additional processing resources used in revising the products of the initial online interpretation process (see Hagoort, this volume).

Neural Organization for Processing Syntax in Children and Individuals with Impaired Development

Apart from the individual variability in performance and, possibly, the neural substrate for syntactic processing that can be found in healthy, monolingual adults with a normal ontogenetic background, it is important to investigate how great the divergence can be in individuals whose language systems differ from that of adult monolinguals, such as children, bilinguals, and individuals who have experienced brain damage early in their development. Studies of these groups suggest that different brain areas can be recruited during ontogeny to support language, with similar efficiency to that seen in adult monolinguals.

Children

Early investigations of language acquisition in children by functional brain imaging focused mainly on language lateralization during rather broad task requirements, such as word generation or text comprehension (Gaillard et al. 2001; Lee et al. 1999). Results of these early experiments showed that children make use of the same network of perisylvian cortical areas in the inferior frontal and superior temporal cortices during language processing as adults. However, in some studies, the organization of language in the developing brain seemed to be less left-lateralized than in adults (Brauer and Friederici 2007, Gaillard et al. 2000; Holland et al. 2001; Ulualp et al. 1998); other studies did not find any differences in language laterality between adults and children (Balsamo et al. 2006; Lohmann et al. 2005). The degree of lateralization appears to depend on age, task, and cortical area (Holland et al. 2007). Nevertheless, the developing language comprehension system shows some hemispheric specialization for certain aspects of language: e.g., the specialization of the RH for prosodic information (Meyer et al. 2003) was observed in young children (Wartenburger et al. 2007) and infants (Homae et al. 2006).

With respect to the intra-hemispheric organization for language, children show less neural specification than adults (Brauer and Friederici 2007). Using a violation detection task with its attendant limitations, Brauer and Friederici (2007) found that children used the entire perisylvian network during sentence comprehension, as opposed to the regional specialization for syntactic and semantic processes seen in adults. A specification for syntactic processes was observed only in Broca's area in the IFG, where there was more activity associated with syntactic processing than is seen in adults. Interestingly, this is the same area that is also more strongly engaged in adult second language learners and in adult native speakers when processing more complex sentence structures. The inference is that this pronounced involvement of the IFG most probably reflects higher processing demands (Bornkessel et al. 2005; Rüschemeyer et al. 2005).

With respect to the timing of syntactic and semantic processes during sentence comprehension, electroencephalic (EEG) investigations in very young children have shown ERPs that resemble those seen in adults during syntactic processing (ELAN, P600) and during semantic processing (N400). These ERPs components are, however, slightly later and more sustained (Friedrich and Friederici 2004, 2005; Oberecker and Friederici 2006; Oberecker et al. 2005).

The anatomical bases for differences in intra-hemispheric organization for syntactic processing in children and adults have begun to be explored. Connectivity through the subcortical white matter connections revealed by diffusion tensor imaging shows differences between adults and children in fiber tract integrity in exactly those perisylvian regions where the functional differences between them are observed (i.e., IFG and STG) (Brauer et al. 2008). Presumably, this is based on lower myelination of these fiber pathways in the immature brain (Paus et al. 1999). The main fiber tract connecting IFG and STG shows continuous maturation during development until adolescence (Giorgio et al. 2008). Other white matter connections, such as those underlying the sensory and motor cortices, mature much earlier than those in temporal language areas, arguing for a particularly slow maturation of language-related pathways (Pujol et al. 2006). During maturation, increasing integrity and myelination of white matter fiber tracts permits faster and more accurate information transmission between the cortical structures involved in the network of language comprehension. Simultaneously with changes in white matter, gray matter maturation progresses during childhood and adolescence.

Thus, the development of the human brain is accompanied by a general pattern of progressive and regressive adjustments, including cortical and subcortical brain structures (Toga et al. 2006). Brain maturation is characterized by changes in gray matter with a reduction of cortical thickness and density (Giedd et al. 1999; Sowell et al. 2003). Pruning out of synapses and reduction of neuropil might be responsible for this (Staudt et al. 2000). Simultanuously, white matter gain occurs due to ongoing myelination of the fiber pathways, a process which lasts until young adulthood (Barnea-Goraly et al. 2005; Lebel et al. 2008). These maturational processes are likely to be related to, if not underlie, functional development which, in turn, might affect structural maturation.

The study of the neural basis for syntactic processing in children is linked to the question of the neural basis of the capacity to learn a syntactic system that has features found in human languages. Clearly, only learning systems with particular properties, likely coupled to cognitive systems with particular properties, are able to acquire natural language syntax, and these learning and related cognitive systems must have evolved in humans. Behavioral studies have made claims about these learning and cognitive systems on the basis of the study of syntax acquisition in children. Findings such as those presented above are relevant to the ontogenetic neural changes that lead to changes in the nature and efficiency of syntactic processing, but only provide very general information about the features of neural systems that have evolved in such a way so as to allow syntax to be acquired. A number of studies of adults learning syntactic systems have, however, been interpreted as providing more information on this topic. Opitz and Friederici (2004), for example, demonstrated that learning a grammar is initially supported mainly by the hippocampus and that, as learning proceeds, the activation in the hippocampus decreases while activation in Broca's area increases. Studies by Tettamanti et al. (2002) have been interpreted as showing that Broca's area is responsible for the ability to apply principles of Universal Grammar to learning artificial languages, but this interpretation has been questioned (Caplan 2007b).

We do not know the minimum amount of input that is required for a child to be able to learn a language without being negatively affected. However, it seems that it is much less than is usually expected. Surprising examples include hearing children with deaf parents for whom acoustic input is confined to the media and occasional encounters with relatives and yet still develop good language skills. Conversely, deaf children who communicate with their parents using a home-style signing language can readily learn real, grammatical sign language when going to school; even without proficient teachers, they develop grammatical signing themselves. Similarly, in groups of children who hear only pidgin at home, Creole languages emerge.

A well-established aspect of language development is that it is usually less efficient after a certain age—the "critical period." The critical period for language acquisition seems to differ for different components of language (Hakuta et al. 2003). The reasons why language cannot be perfectly acquired after this critical period are not well understood and may be due to several mechanisms. It has been shown that long-distance growth of axons is not possible after a certain age. Most pruning also takes place at an early age; thereafter plasticity may even decrease (Huttenlocher 2002). Myelination may also play a role, as discussed earlier.

Individuals with Impaired Development

The effects of lesions on the areas that support syntactic processing reveal additional aspects of neural organization for these processes. Before a certain age, focal damage to LH language regions does not necessarily lead to impaired linguistic capacities in the adult state (Bishop 1988; Bates 1999). In epileptic children, where Broca's area has been damaged, the center of language production "moves" to the RH homologue of that area, and these children exhibit good linguistic abilities, although language development is slow (Vargha-Khadem et al. 1997). These new areas in the RH are also often activated in unaffected individuals during language tasks, although their particular functions are not yet known. The importance of RH has also been emphasized by Bates et al. (1997), who examined children that suffered brain injury to either the LH or RH before 6 months of age. Results from 10–17 months suggest that children with RH injuries are at greater risk for delays in word comprehension and in the gestures that normally precede and accompany language onset. The relevance of the RH during early childhood might reflect the fact that early language development is based to a large extent on prosodic processes represented in the RH. Likewise, following early surgical interventions in the LH, RH activity can be observed in linguistic tasks but not entirely over homologue areas. Using fMRI, Liégeois et al. (2004) showed that the age of lesion effects the neural basis for language: following early LH damage, the RH takes over linguistic functions, but if the damage occurred after the age of 5, language remains based in the LH.

Potential Relations between Nonlinguistic Functions and Syntax

One hypothesis of how syntactic capacities evolved is that they are due to exaptations: brain areas that evolved under selection pressures for other functions became capable of supporting syntactic processing. Several hypotheses of this sort have been suggested.

By modernizing the classical gestural theory of the origin of language, Corballis emphasizes that modern spoken language was modeled on simple elementary gesturing (Corballis 2002). He proposes that autonomous speech may have arisen only as recently as 50,000 years ago, shortly before the cultural explosion. Spoken language allowed humans to use their hands for technology, which led to the development of more sophisticated cultural products.

The discovery of mirror neurons (neurons that fire when someone performs a specific action as well as when they observe someone else performing the same action) in the ventral premotor cortex has led to theories about potential links between motor gestures and language. Mirror neurons provide a direct link between sender and receiver by which parity (what counts for the sender of a message also counts for the receiver) and direct comprehension becomes possible (Rizzolatti and Arbib 1998). Parity is seen as a key to the initiation of language evolution.

Other capacities have also been suggested as evolutionary precursors of syntax. One is a nonlinguistic basis for representing events. When speakers of separate languages that differ in their predominant word order are asked to describe an event by using gestures without speech, they uniformly use a fixed actor–patient–act order (e.g., girl–box–cover; Goldin-Meadow et al. 2008). This order, which is analogous to the subject–verb–object pattern found in many languages of the world, is also the order found in the earliest stages of newly evolving (gestural) languages. This observation suggests that when humans initially created language, they may have exploited a natural disposition for presenting events nonverbally.

Social and internal cognitive capacities may also be a precursor of syntactic abilities. They are believed to have provided adaptive pressures for further evolution of capacities such as anticipation and working memory because language enriches the ability to plan for the future (Arbib 2008). The ability to "plan ahead" is seen in many nonlinguistic functions, such as motor planning (e.g., as in animal foraging). An action like pushing a door, for example, is a function of situations where the door is closed (or open) to situations where it is open (or closed). A plan to push a (closed) door, and then go through it, is the composition of two such functions:

(5) "Push, then go through."

Such a plan is already (weakly) hierarchical and the definition of planning itself is recursive, thus a plan is an elementary action, or a plan is the composition of two plans. "Plans" of nested relations might therefore have preceded and triggered the evolution of language.

Hierarchical computations involved in motor planning and language do not have to be supported by the same brain regions because the aptitude for such computations may not be restricted to one specific region in the brain. Hence, while Broca's area (BA 44/45) is involved in hierarchically organized sentence processing or in the processing of artificial grammar (Friederici et al. 2006b; Bahlmann et al. 2008), processing hierarchical structures of nonlinguistic nonsense shapes, seemed to involve the pre-supplementary motor area in addition to Broca (Bahlmann et al., submitted). It is worth noting that the majority of studies that investigated manipulation of *sequential* information documented the involvement of Broca's area independently of the nature of information that was manipulated. Gelfand and Bookheimer (2003) showed that the posterior portion of Broca's area responded to sequence manipulation tasks, independent of whether the stimuli were composed of phonemes or hummed notes. Similarly, the invention of novel motor sequences in musical improvisation recruits a network of brain regions that includes Broca's area (Berkowitz and Ansari 2008). Ullman (2004), therefore, suggested that the rule-governed combination of lexical items into complex representations evolved from a network of frontal, basal ganglia, parietal, and cerebellar structures, which supports the learning and execution of any skills that involve sequences.

Though planning abilities may be one evolutionary precursor for language, some aspects of language, including features of syntax, appear to have a more complex structure than planning in animals. An example is long-distance dependency of the kind exhibited by the relative clause construction. A person who has the goal state of having a banana, and constructs a plan:

(6) GRASP(GO_THROUGH(PUSH))
"Push, then go through, then grasp."

which is applied to a banana on the other side of a door, has in some sense established a long-distance (three-step) dependency between the present state and the banana.There is some evidence, however, that animal planning does not have this character. Animal planning is more like reactive search through reachable situations (Koehler 1925), and in that sense is probably expressible by finite-state machines. Most likely, the intrinsically recursive character of language has other origins. One that has been suggested is the additional distinctively human involvement of propositional attitude verbs, and the nature of the associated specifically recursive concepts of other minds (Tomasello 1999).

Phylogenetic Development of the Human Brain and the Evolutionary Neural Basis for Syntax

Language is a recent achievement, but not as recent as other cognitive capacities of the brain, such as arithmetic, reading and writing, for which there has been no time for selection to drive genetic evolution. In contrast to these late cultural innovations, language has been around much longer, making it likely that some genetic evolution has accompanied language evolution. If so, it is legitimate to ask how genes can influence our language faculty by affecting the brain. What is the novelty in the brain compared to that of apes that allows us to handle language?

Increased brain size and specialized brain areas are two commonly cited features of the brain that may be the basis for this capacity. However, there are problems with both explanations. Microcephalics suffer from several cognitive deficits, but they are usually still much better at language than nonhuman primates (Woods 2005), arguing that increased brain size cannot be the (only) evolutionary change in the brain that led to language. We should also bear in mind that there is great variation in brain size within the general population. By the same token, early lesions of the LH can be compatible with the development of good language, suggesting that the emergence of specific neural areas is also not the sole basis for language evolution. Taken together, these facts suggest that the human brain is able to represent and process language, albeit with less efficiency, even when it falls outside of the normal range of brain size and has encountered major changes in the areas normally involved in language functioning. We do not know what the computational correlates of these differences are, but we can pose the following inferences: Synaptic dynamics, functional connectivity between cells and between cortical regions based on fiber tract connections, propagation of spike packets can, in principle, all be crucially different. In particular, the immature human brain could have a widespread latent capacity to handle operations on hierarchical structures, which is an empirically testable question. This is compatible with the statement that in normal brains, Broca's area is in fact specialized for some syntactical operations.

One possibility is that the emergence of language and syntax depended upon the evolution of a greater capacity for neural specialization during ontogeny than was found in nonlinguistic ancestors (see Szamado et al., this volume). Accordingly, an "equipotential" initial state (in the sense of the word used above) would be transformed into one of a number of adult specializations, such as a specialization of Broca's area for some syntactical operations. In the young human brain, this specialization could be more quantitative than qualitative. In contrast, in the mature brain, the difference becomes qualitative due to ontogenetic progressive modularization. Changes in the extent to which this ontologenetic specialization can occur may be one step in the evolution of a brain that supports language and syntax. We note that this suggestion places less importance on phylogenetic analogies to language development during ontogeny than the view that the evolution of language, or syntax, requires highly specific, human cortical microcircuitry. The idea of ontogenetic maturation repeating phylogeny is hard to evaluate, since many changes in brains across species (e.g., brain size, the relative amount of white matter, functional and anatomical connectivity) are correlated. Qualitative differences could be an emergent property of quantitative changes (as in phase-transitions in physics).

Acknowledgments

Our discussion benefitted from the participation of many others at this Forum. The contributions of Derek Bickerton, Dorothy Bishop, Terrence Deacon, Tom Givón, James Hurford, Frederick Newmeyer, Luigi Rizzi, Mark Steedman, Luc Steels, Eörs Szathmáry and Stephanie White are specifically acknowledged here.

Modeling

15

Syntax as an Adaptation to the Learner

Simon Kirby, Morten H. Christiansen, and Nick Chater

Abstract

This chapter considers the implications of an evolutionary approach for the idea that human language syntax can be explained by an appeal to strongly constraining domain-specific linguistic nativism. Three sources of evidence that appear to support this particularly strong nativist position are examined: universals, the appearance of design, and the poverty of the stimulus. By taking seriously the fact that the cultural transmission of language has its own adaptive dynamics, it is shown that each of these three motivations is undermined, drawing on evidence from mathematical, computational, and experimental studies. It is suggested that a truly explanatory account of the origins of syntactic structure needs to tackle the interactions between culture, biology, and individual learning—interactions that are perhaps uniquely complex in the case of human language.

Introduction

One of the most obviously striking features of human language, especially in comparison with all other communication systems in nature, is syntax. More precisely, language is unique in providing an open-ended system for relating signals and meanings, one which has its own internal structure. The particular structure of the mapping between meanings and signals varies from language to language, and for many researchers, the central challenge for linguistic theory is an explanation of the constraints on this variation. In other words, linguistics seeks an explanatory account of the universals of syntactic structure.

A hugely influential approach to this explanatory challenge has involved a direct appeal to biology. In this view, syntax arises from our species-specific biological endowment which is specific to language. We have the languages we do because an innately given "language faculty" has a particular structure that constrains the possible types of language (e.g., Hoekstra and Kooij 1988).

In particular, Chomsky (1975) suggests that it is a set of innate constraints on language acquisition that determines the nature of syntactic universals.

This view directly relates universal properties of syntax on the one hand, with a universally shared biological trait on the other.[1] One issue with this attempt at explanation (for a discussion, see, e.g., Hurford 1990) is that it simply replaces one explanatory challenge with another. Although it appears to answer the question of why we have the particular language universals we do, it immediately poses another: Why is our language faculty constrained in the way it is? In a landmark paper, Pinker and Bloom (1990) directly address this question in an attempt to support the nativist approach to explanation. They set out what might be called the orthodox evolutionary approach to language (see Figure 15.1). In this approach, our innate language faculty shapes the structure of language and is in turn shaped by biological evolution driven by natural selection for communicative function. This is motivated by the observation of the apparent adaptive nature of syntactic structure:

> Grammar is a complex mechanism tailored to the transmission of propositional structures through a serial interface....Evolutionary theory offers clear criteria for when a trait should be attributed to natural selection: complex design for some function, and the absence of alternative processes capable of explaining such complexity. Human language meets this criterion (Pinker and Bloom 1990, p. 707).

Pinker and Bloom (1990) provide an influential recasting of Chomskyan nativism in evolutionary terms, one that takes us from observed universals of syntactic structure, through an inferred innate Universal Grammar (UG), grounded firmly in standard mechanisms of evolutionary biology. To critically assess the foundations of this view, it is worth unpacking some of the motivations for assuming this kind of evolutionary nativism. In this paper we will consider three in the light of recent research on the adaptive mechanisms underlying human language:

1. *Universals*. Languages vary, but that variation is constrained. The nativist approach provides a simple and compelling account of this: the constraints on cross-linguistic variation *directly* reflect the languages we can acquire.
2. *The appearance of design*. This is the point made in the quotation from Pinker and Bloom (1990) above. Language structure is adaptive—natural selection of innate constraints appears to be the only available explanation.

---

[1] Note that there is a presumption here that the language faculty is uniform across members of our species, or at least it is uniform with respect to the constraints on cross-linguistic variation. This is a reasonable assumption to make in that there is no obvious evidence that some individuals find particular types of language harder to acquire than other individuals. However, it has recently been challenged as a result of large-scale statistical analysis of genetic and linguistic variation (Dediu and Ladd 2007).

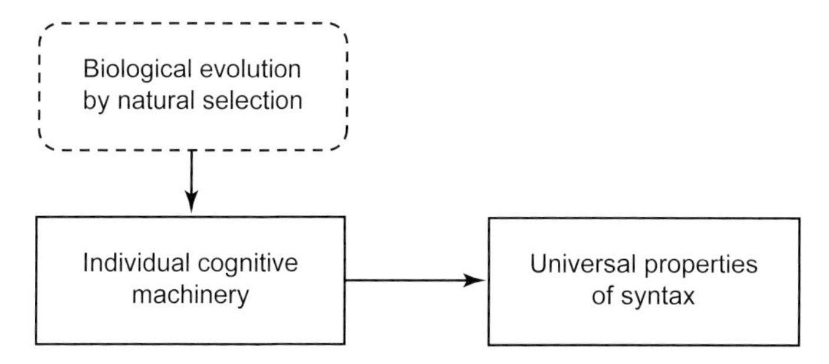

Figure 15.1 The orthodox evolutionary approach to explaining syntactic structure. Biological evolution by natural selection shapes our innate cognitive mechanisms for acquiring language which directly determine the universal properties of syntax.

3. *Poverty of the stimulus.* For many linguists, this is the most familiar reason for assuming innate constraints. Children have access to only limited and degraded evidence that underdetermines the language they are attempting to acquire. Nevertheless children robustly converge on the correct language. Language acquisition therefore appears to be impossible without significant innate knowledge about the languages children may face.

These motivations seem well-founded and reasonable, and appear to provide solid ground on which to build an evolutionary account of the origins of syntactic structure. However, in this paper we wish to argue that there is something missing from the orthodox evolutionary approach sketched in Figure 15.1 that undermines each of these motivations. A key unstated assumption underlying Pinker and Bloom's framework (and indeed the standard nativist position more generally) is that there is a straightforward link between our innate language faculty and universal properties of language structure. This assumption seems reasonable on the face of it, but on closer inspection it is problematic. After all, these are two very different kinds of entities: a genetically determined universally shared part of our cognitive machinery; and constraints on the variation of internalized patterns of linguistic behavior shared within speech communities.

What is the mechanism that bridges the gap between an individual-level phenomenon (the structure of a language-learner's cognitive machinery) and a population-level phenomenon (the distribution of possible languages)? As Kirby et al. (2004) argue, the solution to this problem is to explicitly model the way in which individual behavior leads to population effects over time. Language emerges out of a repeated cycle of language learning and language use, and it is by studying this socio-cultural process directly that we will see how properties of the individual leave their mark on the universal structure of language (Figure 15.2).

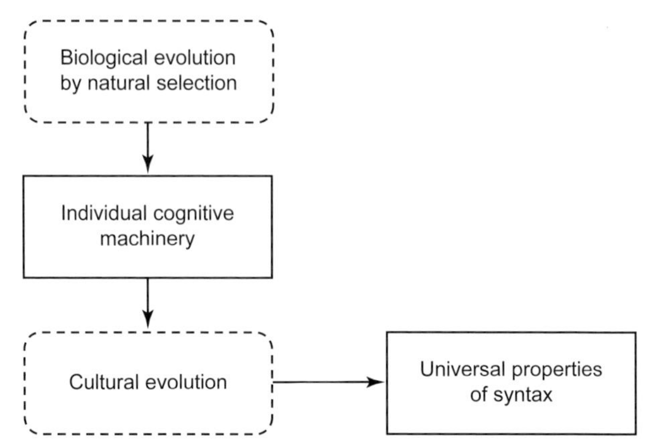

Figure 15.2 The place of cultural evolution in determining the universal properties of syntax. Biological evolution shapes our individual cognitive machinery, but this is only indirectly connected to the object of explanation. Individuals influence a process of social transmission and cultural evolution that eventually leads to emergent universals.

Of course, it is not a priori obvious that the extra box in Figure 15.2 will add anything substantial to the picture—that considering the role of *cultural* as well as biological evolution will change anything. The goal of this chapter is to argue the contrary. By ignoring or downplaying the importance of cultural evolution, evolutionary linguistics risks coming to the wrong conclusions. It is crucial that researchers interested in language evolution do not make the mistake of assuming that evolution is a purely biological process that can be studied in isolation from the dynamics operating at shorter timescales. Although it would be convenient if we could say that the study of social transmission and cultural evolution is purely the realm of historical linguistics and therefore evolutionary linguistics can essentially ignore these mechanisms to focus purely on natural selection, this is not how evolutionary systems work.

By taking the role of cultural evolution seriously, we will show that the motivations for a *strongly-constraining domain-specific linguistic nativism*[2] are undermined. Throughout we will stress the importance of taking an empirical approach to language evolution. Three complex systems are involved in the emergence of syntax: individual learning, cultural transmission, and biological evolution. We cannot reasonably expect our intuitions about the interactions of these to be sound. One response is to build models, both in the computer and in the laboratory, which allow us to explore in miniature how the processes underlying language evolution work, and then apply what we learn from the models to better understand the real object of enquiry: human language.

---

2 It is important to stress that our arguments in this chapter apply to a particular nativist stance: one which infers innate constraints on language that are both specific to language and map directly onto language universals. It is a common misunderstanding that this position is synonymous with generative approaches to language, but this is not necessarily the case (for extensive discussion, see Kirby 1999).

The Logical Problem of Language Evolution

Before turning to models of cultural evolution, we first wish to explore some of the issues underlying the second motivation listed above—that the existence of language structure implies an explanation in terms of natural selection of innate constraints. We argue that advocates of a richly structured, domain-specific, innate UG confront a "logical problem of language evolution" (Christiansen and Chater 2008). To see this, we begin by noting that, as for any other putative biological structure, an evolutionary story for UG can take one of two routes. One route is to assume that brain mechanisms specific to language acquisition have evolved over long periods of natural selection (e.g., Pinker and Bloom 1990). The other rejects the idea that UG has arisen through adaptation and proposes that UG has emerged by nonadaptationist means, (e.g., Lightfoot 2000).

The nonadaptationist account can rapidly be put aside as an explanation for a domain-specific, richly structured UG. The nonadaptationist account boils down to the idea that some process of *chance variation* leads to the creation of UG. Yet the probability of randomly building a fully functioning, and complete novel, biological system by chance is infinitesimally small (Christiansen and Chater 2008). To be sure, so-called "evo-devo" research in biology has shown how a single mutation can lead, via a cascade of genetic ramifications, to dramatic phylogenetic consequences (e.g., additional pairs of legs instead of antennae; Carroll 2001). But such mechanisms cannot explain how an intricate and functional system can arise, de novo.

What of the adaptationist account? UG is intended to characterize a set of universal grammatical principles that holds across all languages; it is a central assumption that these principles are arbitrary. This implies that many combinations of arbitrary principles will be equally adaptive, as long as speakers adopt the *same* arbitrary principles. Pinker and Bloom (1990) draw an analogy with protocols for communication between computers: it does not matter what specific settings are adopted, as long as everyone adopts the same settings. Yet the claim that a particular "protocol" can become genetically embedded through adaptation faces three fundamental difficulties (Christiansen and Chater 2008).

The first problem stems from the spatial dispersion of humans, which occurred within Africa, and ultimately beyond Africa, before and during the period (100–200 ky) within which most scholars assume language emerged. Each sub-population would be expected to create highly divergent linguistic systems. But, if so, each population will develop a UG as an adaptation to a *different* linguistic environment—hence, UGs should, like other adaptations, diverge to fit their local environment. Yet modern human populations do not seem to be selectively adapted to learn languages from their own language groups. Instead, every human appears, to a first approximation, equally ready to learn any of the world's languages.

The second problem is that natural selection produces adaptations designed to fit the *specific* environment in which selection occurs, i.e., a language with a specific syntax and phonology. It is thus puzzling that an adaptation for UG would have resulted in the genetic encoding of highly abstract grammatical properties, rather than fixing the superficial properties of one specific language.

The third, and perhaps most fundamental, problem is that linguistic conventions change much more rapidly than genes do, thus creating a "moving target" for natural selection. Computational simulations have shown that even under conditions of relatively slow linguistic change, arbitrary principles do not become genetically fixed. Chater et al. (2009) illustrate this problem in a series of computer simulations. They model the specific evolutionary mechanism to which Pinker and Bloom appeal to explain the evolution of innate knowledge of language. This mechanism is the *Baldwin effect*: information which is initially acquired during development can become gradually encoded in the genome (for review and discussion, see Briscoe 2003, this volume, and Deacon 1997). Chater et al. (2009) assume the simplest possible set-up: that (binary) linguistic principles and language "genes" stand in one-to-one correspondence. Each gene has three alleles—a neutral allele, and two alleles, each encoding a bias for a version of the linguistic principle. Agents learn the language by trial-and-error, where their guesses are biased according to which alleles they have.[3] The fittest agents (i.e., the fastest learners) are allowed to reproduce, and a new generation of agents is produced by sexual recombination and mutation. When the language is fixed, there is a selection pressure in favor of the "correctly" biased genes, and these rapidly dominate the population (Figure 15.3).

However, when language is allowed to change gradually (e.g., due to grammaticalization-like processes or exogenous forces such as language contact), the effect reverses—biased genes are severely selected against when they are inconsistent with the linguistic environment, and neutral genes come to dominate the population. The selection in favor of neutral genes occurs even for low levels of language change (i.e., the effect occurs, to some degree, even if language change equals the rate of genetic mutation). Of course, linguistic change (prior to any genetic encoding) is likely to have been much faster than genetic change.[4]

It remains possible, though, that the origin of language did have a substantial impact on human genetic evolution. The above arguments only preclude

[3] Because learning biases are probabilistic, learners are always able to learn the language eventually, even if their genetic biases are in the wrong direction (which will make them slow learners).

[4] It may be tempting to object that UG principles do not change, and hence provide a stable environment over which adaptation can operate. However, such an objection would be circular because it presupposes what an evolutionary theory of UG is meant to explain. That is, an innate UG is supposed to explain language universals and thus it cannot be assumed that language universals predate the emergence of UG.

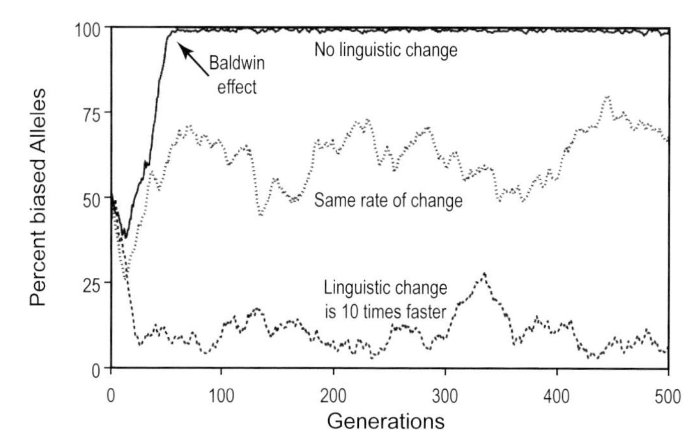

Figure 15.3 The effect of linguistic change on the genetic encoding of arbitrary linguistic principles. The simulation has a population of 100 agents, a genome size of 20, survival of the top 50% of the population, and starts with 50% neutral alleles. With no linguistic change, a Baldwin effect occurs (i.e., alleles encoding specific aspects of language emerge rapidly). But when the language changes, biased alleles are no longer advantageous and are selected against. The results are typical of those obtained using a wide range of parameters (Chater et al. 2009).

biological adaptations for *arbitrary* features of language. There might have been features that are universally stable across linguistic environments that might lead to biological adaptation (such as the means of producing speech, the need for enhanced memory capacity, or complex pragmatic inferences; see Kirby and Hurford 1997, and Christiansen et al. 2006, for computational models that look at nonarbitrary adaptation). In addition, the situation becomes more complex when we look in more detail at interactions between cultural evolution and biological evolution of *weak* constraints. We will return to this issue later.

Language and Cultural Evolution

The problem with the straightforward application of the arguments from biological adaptation to theories of UG lie principally in our poor understanding of exactly how the process of cultural evolution works for language. Specifically, we need to move towards a general theory of how particular kinds of UG constraints or biases lead to language structure when mediated by a process of cultural transmission (Figure 15.2). Only once we have this can we hope to disentangle the precise roles of the different adaptive processes involved.

The *iterated learning model* (e.g., Kirby et al. 2004; Brighton et al. 2005) aims to provide a general solution to this problem. The idea is simple: to build idealized models of the process of cultural transmission that show how global effects emerge from the repeated process of individuals learning and producing linguistic behavior. The simplest iterated learning models consist of a chain of *agents* (individuals modeled in simulation, or in an experimental setting) each of which observes the linguistic behavior of the previous agent in the chain, attempts to learn the underlying linguistic system, and then goes on to produce observable behavior for the next agent down the chain. Like the parlor game, *Telephone*, this produces a cultural dynamic whereby the behavior produced by agents may change over time purely by virtue of being passed-on by an iterated cycle of learning and production. In general, we define iterated learning to be a cultural process whereby an individual learns a behavior by observing another individual's behavior, *who acquired it in the same way*.[5]

This general approach—modeling the way in which linguistic behavior is repeatedly transmitted between individuals—has been studied extensively in the literature, using everything from dramatically idealized simple models (e.g., Kirby et al. 2007) to extremely sophisticated models involving realistic populations of agents interacting socially and grounded in a real environment (e.g., Steels 2003). A thorough review of the result of this modeling work is well beyond the scope of this article, but one of the recurrent observations relates to the importance of what have been called *transmission bottlenecks*. Specifically, if a learner is given imperfect information about the language they are trying to acquire (i.e., where there is some kind of bottleneck on the transmission of language from one individual to another, be it in terms of noise, processing constraints, or simply not hearing all the relevant data) then cultural transmission becomes an *adaptive system*. What this means is that language will adapt so that it appears to be designed to fit through whatever bottleneck the experimenter imposes.

A classic example of this kind of result is provided by several studies into the emergence of compositional syntax (for a review see, Brighton et al. 2005). The existence of compositional structure in the mapping between meanings and strings is an apparently adaptive feature of human language syntax—it is a crucial part of what enables us to have open-ended expressivity, an assuredly adaptive trait.[6] However, computational models of iterated learning which start

[5] Note that this does not limit iterated learning to purely vertical transmission. Indeed, one of the earliest models of this process (Batali 1998) employed purely horizontal transmission (i.e. with individuals learning, producing and then learning again in a completely mutually interacting population). Batali's results bear striking similarities to the quite different models with only vertical transmission. It is the similarity of results across a range of models that has led researchers to attempt to understand the dynamics of iterated learning in as general terms as possible (e.g., Griffiths and Kalish 2007).

[6] Of course, compositionality is only one aspect of the uniquely human structure of syntax (and a very basic one at that). A language with just compositionality and none of the other features of human syntax might arguably be described as "protolanguage," so more work is needed to see if similar processes as those described here can take us further. Nevertheless, it is important to understand that these results do have significant implications as they stand for the particular kind of nativism we are discussing. Furthermore, the Bayesian model we turn to later is completely general and not reliant on a particular view of what constitutes syntactic structure.

from random noncompositional initial languages, or with no language at all, show that this property emerges from the repeated cycle of production and learning without any biological evolution of the agents. The reason is straightforward: compositional structure improves the stability of languages transmitted through a bottleneck. To put it another way, compositionality is an adaptation by language to improve its own survival. There is nothing mysterious or teleological about this. Rather, it is the inevitable consequence of the process of cultural transmission. As Hurford (2000) puts it succinctly, "social transmission favors linguistic generalization."

To check the generality of the conclusions from the computational models, we developed an experimental framework for iterated learning (Kirby et al. 2008). In our experiments, human participants were faced with an artificial language learning task in which they were required to learn to associate strings of written syllables with pictures of colored moving shapes. Each picture was either a square, triangle or circle, was colored either red, blue or black, and was depicted as bouncing, spiraling, or moving horizontally. Although, in the testing phase, participants were asked to produce strings for all 27 different possible pictures, they were only actually trained on a random subset of 14 of these.

The crucial aspect of these experiments that makes them relevant here is that the language a participant is trained on is actually a random sample of the output of the previous participant in the experiment at test, with the very first participant being trained on a randomly constructed language (i.e., one which exhibits no compositional structure). With this experimental set up we are able to observe in the laboratory exactly how a simple language like this one evolves culturally. Two questions present themselves: Will languages adapt to be increasingly learnable? Will structure emerge?

The answer turns out to be "yes" to both questions, but the exact kind of structure that emerges depends in an interesting way on the nature of the bottleneck. Figure 15.4 shows quantitative results for the experiment outlined above (with the lines marked "unfiltered"). Clearly, the languages become more learnable and more structured over time, purely as a result of being transmitted repeatedly from individual to individual. We start with a language that is impossible to learn in the sense that there is no way of accurately guessing what an unseen meaning might be called, and end with a language where participants do extremely well in generalizing accurately to unseen examples.

What does the emergent structure that makes this possible look like? It turns out that, in this version of the experiment, what emerges is a kind of structured lexical underspecification. The number of distinct strings in the language plummets from 27 at the start to a handful after 10 "generations" (the exact number varies from replication to replication). These remaining strings thus refer to a set of meanings, rather than a single one. What is fascinating is that these sets show distinctive structure (picked up by the quantitative measure in Figure 15.4). For example, in one run of the experiment, a single word emerged to

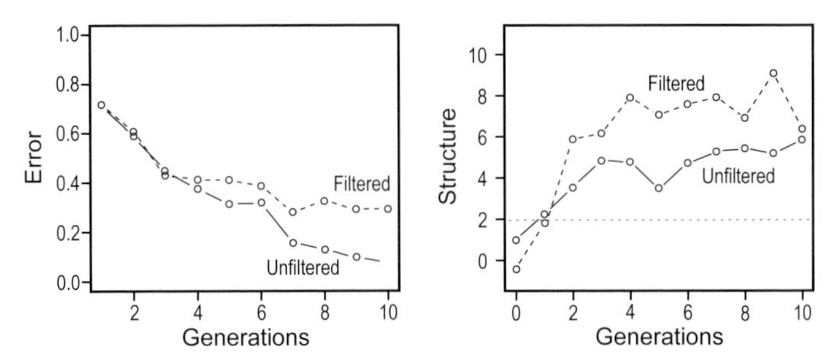

Figure 15.4 The average of four cultural transmission chains in each of two experimental conditions showing languages evolve to be more learnable and more structured over time. Each generation here is an individual experimental participant who learns an artificial language produced by the previous participant in the experiment. The graph on the left shows the average error in learning the language (a score of 1 means that the strings produced were completely dissimilar to the target, a score of 0 means they were exactly correct). The graph on the right shows a measure of structure in the mapping between strings and meanings (for more information, see Kirby et al. 2008). Wherever this structure measure is above the dotted line, the language is nonrandom at the 95% confidence interval. The "unfiltered" condition indicates that the training data is passed directly to the next participant, whereas in the "filtered" condition ambiguous strings are removed (see text).

refer to all the horizontally moving objects. This kind of nonrandom underspecification of meanings in the language allows learners to generalize accurately to unseen meanings.

Why weren't we seeing the kind of compositional structure that was apparent in the computational models? One difference is that the simulations typically built-in some motivation for the agents to maintain expressivity and avoid collapsing all meaning distinctions to a single string, for example. We wanted to show that a similar result could be achieved in the human experiment just by a minimal change to the transmission bottleneck. Accordingly, we reran the experiments with a single alteration: before giving the training data to a participant, we scanned it for any underspecification. If the same string was used for more than one meaning, we simply filtered all but one of those instances out of the training data. This filtering step corresponds to the pressure in real language use to maintain expressivity, such that distinct meanings tend to be assigned distinct signals.[7] Note that participants were not aware we were doing

[7] Filtering underspecification from input is not necessarily a particularly realistic way of achieving this, although it is likely that something like filtering based on communicative utility is a real mechanism in language transmission. In the experimental model it should be seen as a stand-in for a more complex suite of communicatively motivated pressures. An alternative might have been to set up the experiment within the frame of an overtly communicative task. However, it was crucial for our purposes to demonstrate that participants were not intentionally and intelligently *designing* a communication system (for example on analogy with their own language), as they may well have done if this became the overt goal of the experimental setup. Rather, we wanted to show that the cultural transmission process alone is all that is required to generate adaptive structure.

this (indeed, in neither version of the experiment did any participant ever guess that the experiment involved cultural transmission in any case). However, the difference in outcome was dramatic. Figure 15.4 shows that the quantitative results showed the same trend, albeit revealing this was a more difficult task. The big difference was in the particular structure of the language.

With filtering in place, the kind of structured underspecification we saw previously no longer could take hold. Nevertheless, a different adaptation emerged which led to both an increase in learnability *and* the maintenance of expressivity. This adaptation was exactly the same as the one that emerged in the computational models: compositionality. The examples below (from data presented in Kirby et al. 2008) show how three "morphemes" emerged in one chain encoding color shape and motion respectively (note the hyphens are included for clarity only, they are not present in the participants' output or input):

(1)	(a)	n-eke-ki "black-triangle-horizontal"	(b)	n-eki-pilu "black-triangle-spiral"
	(c)	l-aho-ki "blue-circle-horizontal"	(d)	l-aho-plo "blue-circle-bounce"
	(e)	r-e-plo "red-square-bounce"	(f)	r-e-pilu "red-square-spiral"

This result was not invented by one particularly smart individual in the experiment, but rather appeared cumulatively and without deliberate design on behalf of the participants. The participants were not trying to construct a perfect language to fit through the bottleneck (which would have been impossible given that they could not know the constraints we were placing on the bottleneck). They were simply trying their best to give us back what we gave them. Many participants did not even realize that we were asking them to generalize to unseen meanings. Nevertheless the language underwent cumulative cultural adaptation, just as predicted by the computational models. Clearly, these are adult participants that already have a native language and as such we need to be aware that the biases they bring to bear on the learning task are a combination of biologically basic ones and those that arise from their existing specific cultural inheritance (i.e., their native language). Of course, the close fit of the experimental results with those predicted by the simulation models speaks against the idea that acquired biases are the primary driver. More importantly, however, the primary purpose of these models is not as a discovery procedure for our biological biases but rather as a way of determining how a culturally transmitted language responds to whatever biases and transmission pressures are placed upon it.

This result, and others like it, cast doubt on all the motivations for strongly constraining domain-specific innateness listed in the introduction. Firstly, and most importantly, it demonstrates that there is more than one mechanism capable of delivering the appearance of design. Natural selection (in the biological sense) is no longer the only possible explanation for adaptive structure in language—the mere fact that language is transmitted culturally induces adaptation by language itself. Secondly, it recasts the so-called "poverty of the stimulus" problem in a new light. As others have argued (e.g., Zuidema 2003) these results show language structure does not exist in spite of the impoverished stimulus available to the child, but rather precisely because of the stimulus poverty. When more, and better, data is provided to learners in the models, the languages that emerge exhibit less structure.[8]

What of the remaining motivation: universals? In some sense, the experiments with compositionality already address this issue. If we were to look only at the end result of the simulations, compositional language, we might be led to the wrong conclusion that the learners were equipped with a mechanism that constrained them only to learn compositional languages. However, this would be a mistake. In these models, even when learners do not reliably acquire compositional structure an exceptionless universal outcome can still be expected (Hurford 2000).

To understand the relationship between universals and UG better, we implemented a mathematical model of iterated learning using Bayesian agents (Kirby et al. 2007). This allows us to control very precisely the contribution of innateness and see what language universals emerge for a given transmission bottleneck. The innate contribution is represented in the model in terms of a prior bias over possible languages. That is, we are able to provide a probability distribution over languages that reflects the innate preference for one language over another. These innate biases can therefore be arbitrarily strong or weak, covering the spectrum of possibilities from hard constraints to slight tendencies.

By treating iterated learning as a Markov process in which the transition between languages is determined by the Bayesian model of learning, we are able to predict exactly what universals should emerge for a given model of innateness. The most striking result from this work is that, given reasonable assumptions about how learners select hypotheses, innate biases are *not* reflected directly in language universals. Specifically, the strength of the language universals that emerge is independent of the strength of the innate bias and is instead determined by the nature of the transmission bottleneck. Simplifying somewhat, in conditions of data poverty, arbitrarily weak innate predispositions are amplified by cultural transmission. Indeed, the strength of innate bias makes absolutely no difference to the final distribution of language types in the model.

What this means is that we cannot infer strongly constraining innateness simply by looking at language universals. This observation actually has some

[8] Of course, this in itself does not provide a solution to learnability arguments, but note that iterated learning ensures that the training data provided for a learner will be the best possible data for the particular learning problem learners typically face (because cultural evolution will tend to maximize the learnability of the language).

empirical support. For example, there are cases where culturally transmitted birdsong for a particular species has an exceptionless universal, but where a bird of that species can nevertheless acquire an atypical song (e.g., Hultsch 1991). Similarly, Dediu and Ladd (2007) present evidence that there is genetic variation in prior disposition to acquire tone languages which result in a clear and strong skewing of language types in different populations. Nevertheless, it is clear that any normal individual can acquire any existing language whatever their genetic makeup. Both these cases suggest that whatever biases lead to the population-level effects, their effect at an individual level can be tiny.

Finally, this result has implications for the biological evolution of innateness discussed in the previous section. Smith and Kirby (2008) look at the co-evolution of Bayesian learners and the languages they transmit culturally. They argue that cultural transmission shields bias strength from the view of natural selection (see also Deacon 2003a), leading to the possibility that strong biases may be impossible to maintain against mutation pressure.

So, where does this leave the biological evolution of innate bias? The results of the models discussed in the previous two sections do not rule out the evolution of innate bias, but they narrow down the possible ways this evolution could take place. Two particularly plausible alternatives remain: either that bias is not domain-specific (and therefore could be subject to selection pressures not solely determined by the emerging cultural system); or it could be a bias weak enough not to have a strong impact on a single individual, but nevertheless be amplified by cultural transmission.

Biases That Shape Syntax

We have proposed that language has adapted to biases or constraints deriving from language learners and users: biases which may not be specific to language. But how far can these constraints be identified? To what extent can linguistic structure previously ascribed to an innate UG be identified as having a nonlinguistic basis? Clearly, establishing a complete answer to this question would require a vast program of research. Here we divide the constraints into four groups relating to thought, pragmatics, perceptuo-motor factors, and cognition (Christiansen and Chater 2008).

Constraints from thought. The structure of mental representation and reasoning must, we suggest, have a fundamental impact on the nature of language. The structure of human concepts and categorization must strongly influence lexical semantics; the infinite range of possible thoughts must drive the compositionality of natural language (as discussed above); the mental representation of time is likely to have influenced the linguistic systems of tense and aspect; and so on. While the Whorfian hypothesis that language influences thought remains controversial, there can be little doubt that thought profoundly influences language.

Pragmatic constraints. Similarly, language is likely to be substantially shaped by the pragmatic constraints involved in linguistic communication. Pragmatic processes may, indeed, be crucial in understanding many aspects of linguistic structure, as well as the processes of language change.

Levinson (2000) notes that "discourse" and syntactic anaphora have interesting parallels, which provide the starting point for a detailed theory of anaphora and binding. Levinson argues how initially pragmatic constraints may, over time, become "fossilized" in syntax, leading to some of the complex syntactic patterns described by binding theory. Thus, one of the paradigm cases for arbitrary UG constraints may derive, at least in part, from pragmatics.

Perceptuo-motor factors. The motor and perceptual machinery underpinning language seems, moreover, inevitably to influence language structure. The seriality of vocal output, most obviously, forces a sequential construction of messages. A perceptual system with a limited capacity for storing sensory input forces a code which can be interpreted incrementally (rather than the many practical codes in communication engineering, where information is stored in large blocks). The noisiness and variability (across contexts and speakers) of vocal or signed signals may, moreover, force a "digital" communication system, with a small number of basic units (i.e., phonetic features or phonemes). These discrete units, in turn, appear closely related to the vocal apparatus and to "natural" perceptual boundaries.

Cognitive mechanisms of learning and processing. Another source of constraints derives from the nature of cognitive architecture, including learning, processing, and memory. In particular, language processing involves extracting regularities from highly complex sequential input, pointing to an obvious connection between sequential learning and language: both involve the extraction and further processing of discrete elements occurring in complex temporal sequences. It is therefore not surprising that sequential learning tasks have become an important experimental paradigm for studying language acquisition and processing (sometimes under the guise of "artificial grammar/language learning" or "statistical learning"; for reviews, see Gómez and Gerken 2000).

Syntax Shaped by Sequential Learning

If language has evolved to fit human sequential learning mechanisms, then constraints on the learning and processing of sequential structure should be reflected in the universal properties of human language. Importantly, many of the cognitive constraints that have shaped the evolution of language would still be at play in our current language ability. Thus, the study of how artificial sequential material is learned may reveal selectional pressures operating on the evolution of natural languages. We summarize a series of modeling and experimental results that indicate how constraints on sequential learning may have given rise to certain word-order universals relating to head-ordering, as well as interactions between case and word-order flexibility.

Assuming that language acquisition and processing share mechanisms with sequential learning in other domains, then breakdown of language would be expected to be associated with impaired sequential learning. This prediction is particularly interesting, because breakdown in sequential learning does generally not co-occur with cognitive impairments. This prediction has been tested using an artificial grammar learning task involving agrammatic aphasics who typically have damage in or around Broca's area and have severe problems with the hierarchical structure of sentences (Christiansen and Ellefson 2002). Although both aphasics and normal controls, matched for age, socio-economic status, and abstract reasoning abilities, were able to complete a training task successfully involving same-different judgments on symbol strings, only the control group could correctly determine which of a set of novel strings was generated by the same rules as the training strings.

We would predict that basic word-order universals might arise from constraints on sequential learning, if sequential learning and language share common mechanisms. To pursue this hypothesis, let us begin with the *heads* of phrases: the word that determines the properties and meaning of the phrase as a whole (such as the noun *boy* in the noun phrase *the boy with the bicycle*). Across the world's languages, there is a statistical tendency toward a basic format in which the head of a phrase consistently is placed in the same position—either first or last—across different types of phrase. English is considered to be a head-first language, meaning that the head is most frequently placed first in a phrase, as when the verb is placed before the object noun phrase in a transitive verb phrase such as *eat curry*. A head-last language, such as Hindi, typically uses the opposite order, and hence the equivalent of *curry eat*. Likewise, head-first languages tend to have prepositions before the noun phrase in prepositional phrases (such as *with a fork*), whereas head-last languages tend to have postpositions following the noun phrase in postpositional phrases (such as *a fork with*). In the traditional UG framework, head-order consistency has been explained by innate language-specific constraints on the phrase structure of languages.

A very different picture emerges if we hypothesize that word order has evolved to fit human sequential learning mechanisms. Christiansen and Devlin (1997) trained simple recurrent networks[9] (Elman 1990; SRN) on corpora generated by 32 different grammars that differed in head-order consistency (i.e., inconsistent grammars would mix head-first and head-last phrases). The networks were trained to predict the next lexical category in a sentence. Although these networks had no built-in linguistic biases, their predictions were sensitive to the amount of head-order consistency found in the grammars, such that

[9] It is sometimes objected that these kinds of networks lack biological plausibility because they are typically trained using back-propagation. However, recent advances in neural computation undermine this objection by demonstrating that the kind of networks employed in the simulations reported here can be trained with similar results using reinforcement learning (Grüning 2007), which is a neurobiologically plausible learning algorithm.

there was a strong correlation between the degree of head-order consistency in a grammar and how successfully the networks learned the language: the more inconsistent the grammar, the harder it is to learn (Figure 15.5). Christiansen and Devlin further analyzed frequency data on the world's natural languages concerning the specific syntactic constructions used in the simulations. They found that languages incorporating patterns that the networks found hard to learn tended to be less frequent.

Incorporating systems of case marking, Lupyan and Christiansen (2002) were able to relate learnability in the networks with attested frequency of different orders of subjects (S), verbs (V), and objects (O), across the world's languages. Subject-first languages, which make up the majority of language types (SOV: 51% and SVO: 23%), were easily learned by the networks. Object-first languages, on the other hand, were not well learned, and have very low frequency in the world's languages (OVS: 0.75% and OSV: 0.25%). Using rule-based language induction, Kirby (1999) arrived at a similar account of typological universals.

Lupyan and Christiansen (2002) also modeled data from a study by Slobin and Bever (1982) showing differences in performance across English, Italian, Turkish, and Serbo-Croatian when children acted out reversible transitive sentences, such as *the horse kicked the cow*, using familiar toy animals. Like the children, the networks initially showed the best performance in Turkish, with English and Italian quickly catching up, and with Serbo-Croatian lagging behind. The close match between network performance across training and that of children across age is illustrated in Figure 15.6. Because of their consistent use of case and word order, respectively, Turkish and English were more easily

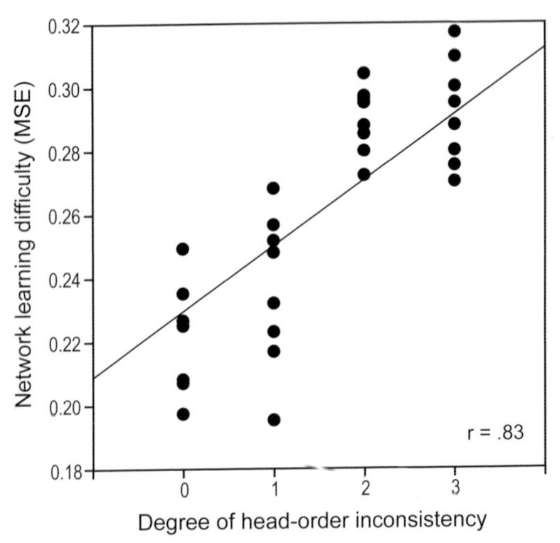

Figure 15.5 Relating the degree of head-order inconsistency and ease-of-learning in a connectionist network. Higher degrees of head-order inconsistency result in increased learning difficulty. Adapted from Christiansen and Devlin (1997).

learned than Italian and, in particular, the highly inconsistent Serbo-Croatian language. With repeated exposure, the networks learning Serbo-Croatian eventually caught up, as do the children learning this language.

To determine whether these sequential learning biases would result in the emergence of consistent head-ordering across successive generations of learners, Reali and Christiansen (2009) trained SRNs to map words onto grammatical roles. Prior to the introduction of language, the SRNs were first allowed to evolve "biologically" to improve their ability to perform a sequential learning task. Specifically, the initial weights from the best learner at each generation were chosen as the basis for the next, with copies of the parent's weights mutated slightly. After 500 generations, the SRNs had evolved a considerably better ability to deal with sequential structure. A language with no word-order constraints was introduced into the simulation. Crucially, both language and networks were allowed to change while the networks at the same time also had to maintain the same level of performance on the sequential learning task as obtained after initial evolution of sequential learning biases (on the assumption that this skill would still have been crucial for hominid survival after the emergence of language). Over generations, a consistent head-ordering emerged due to linguistic adaptation rather than biological adaptations (of initial weights). Indeed, the pressure toward maintaining a high level of sequential learning performance prevented the SRNs from adapting biologically to language.

If sequential learning is a fundamental human skill, as explored in these simulations, it should be possible to uncover the source of some of the universal constraints on language by studying human performance on sequential

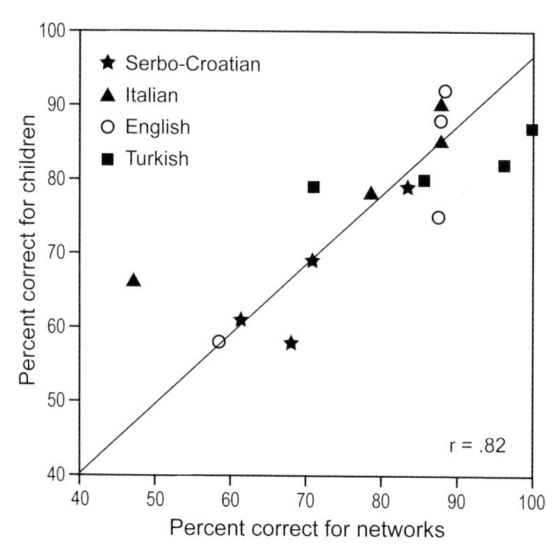

Figure 15.6 Using network performance as a function of training to predict the improvements in children's performance with increasing age in Turkish, English, Italian, and Serbo-Croatian. Network results from Lupyan and Christiansen (2002); child data from Slobin and Bever (1982).

learning tasks. In a sequential learning experiment (Christiansen and Ellefson 2002), human participants learned sequences generated by either a consistent or inconsistent grammar from Christiansen and Devlin (1997). When tested on novel sequences, the participants trained on the grammar with a consistent head-ordering were significantly better at distinguishing grammatical from ungrammatical items compared to participants trained on the inconsistently head-ordered grammar. Together, these simulations and human experiments suggest that sequential learning constraints may provide an alternative explanation of head-order consistency without UG. Specifically, constraints on basic word order may derive from nonlinguistic constraints on the learning and processing of complex sequential structure. Grammatical constructions with highly inconsistent head-ordering may simply be too hard to learn and therefore tend to disappear.

It is possible, moreover, that human sequential learning abilities are a crucial preadaptation to language. Conway and Christiansen (2001) reviewed evidence on sequential learning in nonhuman primates and concluded that although the performance of nonhuman primates on learning fixed sequences and certain types of statistical structure is similar to that of humans, the former has problems dealing with the kind of hierarchical sequential structure characteristic of human languages. This sequential learning may help explain why only humans have complex linguistic abilities.

Summary and Conclusion

The fundamental explanatory goal of linguistics is to answer the question: Why are languages the way they are and not some other way? We firmly believe that the only viable approach to this question is an evolutionary one: to answer why languages have the structure that they do, we need to ask how they came to be that way. However, language is not the result of a single adaptive system, and approaches that look to biological adaptation as their sole explanatory mechanism lead us to the wrong conclusions. In particular, we have highlighted the importance of taking into account the interactions between individual learning biases and cultural evolution in order to understand the sources of linguistic structure.

The problem is that the interactions between culture, biology, and individual learning are very complex, perhaps uniquely so when it comes to human language. The solution is to explore theories about their interactions by building, in miniature, models that take seriously the notion that population-level phenomena like languages must emerge from the lower-level interactions between individual learners. We have briefly summarized here a spectrum of different modeling approaches, from mathematical models, through computational simulations, to novel experimental frameworks. All the results we have so far come together to demonstrate that languages adapt culturally under influence from limitations on human learning and processing. Furthermore, this kind of cultural adaptation may reduce the influence of biological adaptation. We are left with the conclusion that many key features of language, such as syntactic structure, may be adaptations by language to the problem of being passed-on through generations of language learners.

Acknowledgments

The authors are listed in reverse alphabetical order and have contributed equally to this paper. SK was supported by ESRC Grant Number RES-062-23-1537 and by AHRC Grant Number AH/F017677/1. MHC was supported by a Charles A. Ryskamp Fellowship from the American Council of Learned Societies and by the Santa Fe Institute; NC was supported by a Major Research Fellowship from the Leverhulme Trust and by ESRC Grant Number RES-000-22-2768.

16

Cognition and Social Dynamics Play a Major Role in the Formation of Grammar

Luc Steels

Abstract

Modeling is an essential tool in all sciences and has contributed to the study of the origins and evolution of human languages. It can help us understand what kinds of mechanisms are necessary and sufficient for the origins and evolution of language. Through mathematical investigations and computational simulations, it is possible to examine whether certain basic assumptions of a theory are viable or not. This chapter examines models that explore the role of cognition and social interaction in the formation of grammar. After a brief survey of ongoing experiments, the kinds of structures and processes that have been shown to be effective will be discussed. It is argued that neither the grammatical structure of human languages nor the conceptual inventories expressed in language need to be strongly innate, and hence the role of biology is primarily to provide the powerful cognitive machinery which constitutes the foundation of human intelligence in general.

Introduction

There is a wide consensus that language rests on three aspects: biology, cognition, and culture. Biology provides the neuronal and physiological hardware; cognition is concerned with information representation, processing, and learning supported by this hardware; and culture is the shared system that emerges out of the linguistic activities of individuals and persists over time. At the present, there is no consensus on which of these three aspects should be taken as the *dominant* force in the theories of the origins and evolution of language.

- *Biology*: Some researchers emphasize the role of genetic evolution through natural selection (Pinker 2003; Stromswold 2001; Pinker and Jackendoff 2005; Bickerton 1984). They hypothesize that the language faculty is a highly specialized, genetically coded organ in the brain, which evolved through genetic evolution by natural selection. Accordingly, language is acquired primarily by setting parameters in the innate schemata provided by Universal Grammar (Lightfoot 1991; Thornton Wexler 1999), rather than through inductive learning or problem solving. This nativist approach to language origins has been explored with concrete operational models using the frameworks of genetic algorithms (Cangelosi and Parisi 1998; Szathmáry 2007) and evolutionary game theory (Nowak and Komarova 2001).

- *Culture*: Some researchers propose that cultural evolution is the primary force. They hypothesize that human neurobiology provides both general learning mechanisms and language-specific biases that speed up learning to overcome the "poverty of the stimulus" bottleneck (Elman 1996; Kirby and Hurford 2002; Boyd and Richerson 2005). Recent computational and mathematical models have indeed shown that if inductive learning is chained through a sequence of vertical transmissions from teacher to learner, language structures and conceptualizations gradually appear that reflect the inductive bias of the language learners (Brighton and Kirby 2001; Griffiths 2005; Nakamura 2003; Briscoe 2000).
- *Cognition*: Others emphasize social interaction patterns and cognition (Tomasello 1999; Steels 2005a). They argue that human neurobiology supports a large battery of cognitive mechanisms usable in a wide range of domains, and that language exploits most of these cognitive mechanisms to the fullest. Language production and interpretation are hence viewed as problem-solving processes, alternating between routine problem solving and the creative invention of new conceptualizations or language forms to handle novel situations or make production or interpretation become more efficient (Langacker 2008; Hopper 1987). These innovations become entrenched through well-documented grammaticalization processes (Heine 1997; Bybee 1994; Traugott and Heine 1990). Language learning is also viewed as a problem-solving process that engages all cognitive resources of the learner, including the ability to form new categories, to guess meaning through inference, to apply analogies and metaphors, to hypothesize the function of novel syntactic structures, to generate and test out new phonetic features, etc.

Here, I examine models built to explore the latter viewpoint. It needs to be emphasized that socio-cognitive models do not assume that anything is innate, on the contrary they assume a wide arsenal of tools for collectively building a symbolic communication system, but they argue that these tools are neither uniquely used for language nor specialized genetically.

Socio-cognitive models are necessarily quite complex as they require modeling as well as the operationalization of many aspects of cognition, including perception, memory, conceptualization, planning and plan recognition, metaphor, analogy, joint attention, and perspective reversal. Without adequate precise information-processing models of the processes that go into language production, language understanding, and language learning, we cannot begin to understand whether these processes are unique to language or shared by a wide spectrum of cognitive tasks (e.g., navigation, plan recognition, tool design, or social organization). We cannot know what can be learned and what needs to be provided innately as "principles and parameters" or as innate biases in intergenerational cultural evolution.

If we look at navigation, for example, we can clearly see that similar problem-solving and learning skills are needed, compared to language: Individuals must become sensitive to significant features of the environment; a sequence of actions must be planned and executed; the actions to navigate from one place to another often have a hierarchical structure; after finding one path it is often stored for later use so that navigation becomes more routine and more complex paths can be found by combining subpaths. There is also a social dimension to navigation: Individuals are able to recognize the navigation paths being executed by others or complete and correct a partial path once taken. They can solve a navigation problem by analogy with known solutions or explain to another person how to get somewhere, even with gestures or drawings. Spatial navigation provides a good example, because it is also a skill that develops progressively and requires constant adaptation, as the environment is changing and new origins and destination points or new cities are explored.

Of course, in the end, the three forces combine to play a role; they interact and impinge upon each other. However, for the purpose of scientifically advancing the field, it is useful to focus on one force to see what kind of explanations can be generated.

Language Games

The key ingredients of socio-cognitive models are: a *population* of individuals, modeled as *agents*, a *world* which acts as source of meaning for the agents, and a particular type of situated interaction between the agents, called a *language game*. An agent has a set of memory structures (e.g., for storing a lexicon) as well as a set of procedures for carrying out all the information processing that needs to go into language production, comprehension, and learning (e.g., procedures for parsing a sentence to extract a syntactic structure, or for guessing the meaning of an unknown word and storing it in the lexicon). In addition to procedures used for producing or interpreting utterances with "ready-made" linguistic solutions, agents have diagnostic and repair strategies to handle cases for which they have no solutions yet, and alignment strategies to coordinate their inventories based on the outcome of the language game (Pickering and Garrod 2005).

Agents are *autonomous* in the sense that there is no telepathy (one agent cannot inspect or change the state of another agent) and no central control. Thus, there is never any form of direct meaning transfer nor is there is prior "innate" language or "innate" conceptual system given by design, because the purpose of the models is precisely to show how these may form and propagate in a population. Agents are *embodied* when the memory and procedures are incorporated into a physical body with sensors and effectors and interaction takes place within the real world. In *simulated* language games, agents, populations, and the world are modeled abstractly and tested in software simulations. In *grounded* language games, the world is the real physical world, the agents are embodied, and the language games involve physical interactions such as pointing gestures or executing actions. Grounded language games can only be tested through robotic experiments (see Figure 16.1).

The Naming Game

One of the first and by now most widely studied language games is the *naming game*. It was first introduced and studied with computer simulations in 1995 (Steels 1995) and grounded experiments followed soon thereafter (Steels and Vogt 1997). The naming game has proven to be a good "E. coli" for studying the formation of linguistic conventions and has acquired a similarly prominent role in the study of "semiotic dynamics" as the Prisoner's Dilemma occupies in the social sciences (Loreto and Steels 2007). In the naming game, the speaker tries to draw the attention of the hearer to an object in the world by naming a

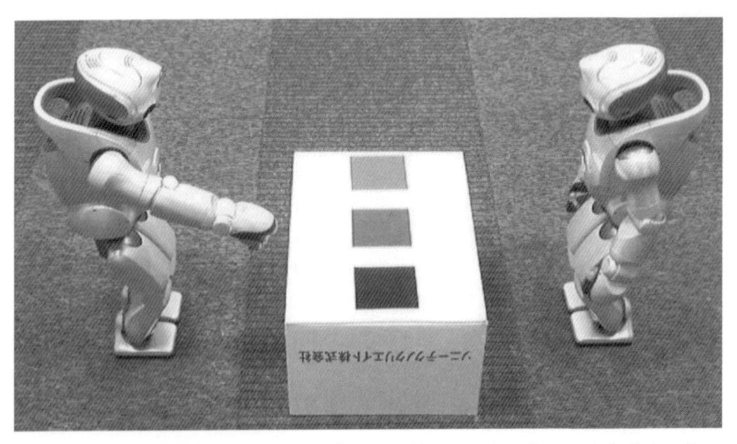

Figure 16.1 Investigations in the socio-cognitive mechanisms underlying language evolution are partially carried out through robotic experiments. Autonomous robots are set up to play language games that require making distinctions (e.g., hue distinctions) and expressing them via words or grammatical constructions. The experiment depicted here involves the formation of color categories and color names in a population of humanoid robots (Steels and Belpaeme 2005).

category that distinguishes the object from other objects in the context. The scenario is as follows:

1. A speaker and hearer (randomly drawn from the population) are situated within a shared context. Any agent can be both speaker and hearer.
2. The speaker selects one object in the context, further called the topic.
3. The speaker conceives of a category, which discriminates the topic from the other objects in the context.
4. The speaker looks up the name of the category in his own lexicon and transmits it to the hearer.
5. The hearer looks up the name in his lexicon and retrieves the category.
6. The hearer applies the category to the shared context to identify the topic and points to that object.
7. If the hearer's topic is the same as the speaker's, then the speaker signals success.
8. Otherwise the speaker signals failure and points to his original choice of topic.
9. The speaker and hearer update their internal memories based on the outcome of the game.

Speakers and hearers need a bidirectional associative memory (Kosko 1998) to store their lexicons, a way to categorize for discrimination, for example with discrimination trees or prototypes (Steels 1996), mechanisms to follow interaction scripts, as well as mechanisms for updating their inventories after a game, similar to those commonly used in reinforcement learning. Updating may imply that new words are added to the lexicon, the score of an association between a word and a category is increased or decreased, a new category is created, the prototype for a category shifts, etc. Note that both the categorical repertoire and the lexicon emerge simultaneously.

Computer simulations have now shown abundantly that well-chosen alignment operations carried out by each agent will give rise to global coherence (see Figure 16.2). Many solutions are possible (Oliphant 1997; Wagner et al. 2003), ranging from a lateral inhibition dynamics which progressively increases the score of winning associations and pushes down their competitors (Steels 1995) to a more discrete dynamics that eliminates all competitors as soon as one association wins (Baronchelli 2006b), or a replicator dynamics in which there is a selectionist struggle between populations of lexical rules (Steels and Szathmáry 2008). In-depth mathematical investigations of the naming game exist that show under what conditions convergence occurs (Lenaerts 2005; De Vylder and Tuyls 2006). We understand many other aspects as well: (a) how naming game behavior scales with respect to its main parameters (the size of the population and the size of the category inventory that needs to be expressed) (see Figure 16.3, after Baronchelli et al. 2008); (b) what influence the agents' network topology exerts, for example, in spatially or socially distributed naming games (Steels 1999; Lu 2008; Dall'Asta 2006b); (c) what

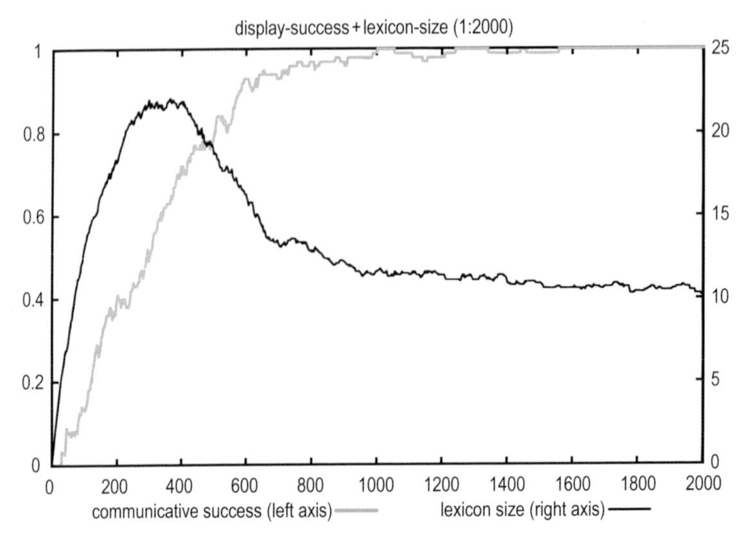

Figure 16.2 Example of semiotic dynamics generated by language games in a population of 10 agents self-organizing a lexicon to express 10 meanings. The X-axis shows the number of games played, each time only involving two agents; the Y-axis indicates communicative success as well as global size of the inventory. The lexicon overshoots initially because agents invent words for certain meanings not knowing that there are already other words for them in the population. In a second phase, agents align their inventories and reach an optimal lexicon of 10 words. We see that agents quickly reach 100% success.

influence stochasticity or population change has on the stability and evolution of the language (Steels and Kaplan 1998; Ke 2002).

Progress has also been made on the grounded naming game through actual robotic experiments, even though this poses immense additional challenges. Now the total semiotic cycle needs to be modeled and operationalized, including perception, joint attention, and the physical behavior required for setting up a shared context or for providing additional feedback to repair a failed game (Steels 2003). Several experiments have now shown this to be entirely feasible; they provide a solid foundation for moving towards more complex language. It is therefore possible to say that today we understand quite well how a symbolic communication system may form. Note that in all these experiments, there is no role for the genetic coding of the communication system or selection based on the biological fitness of individuals. There is also no critical role for cultural evolution based on biased iterated learning. Cultural transmission happens automatically as a side effect in these models. New agents can be made to enter the population and others can be made to leave, introducing a kind of population flux. However, no new mechanisms need to be introduced to handle vertical transmission because incoming agents pick up the conventions of their peers quickly based on the diagnostic and repair strategies shown to be needed for the emergence of vocabularies in the first place.

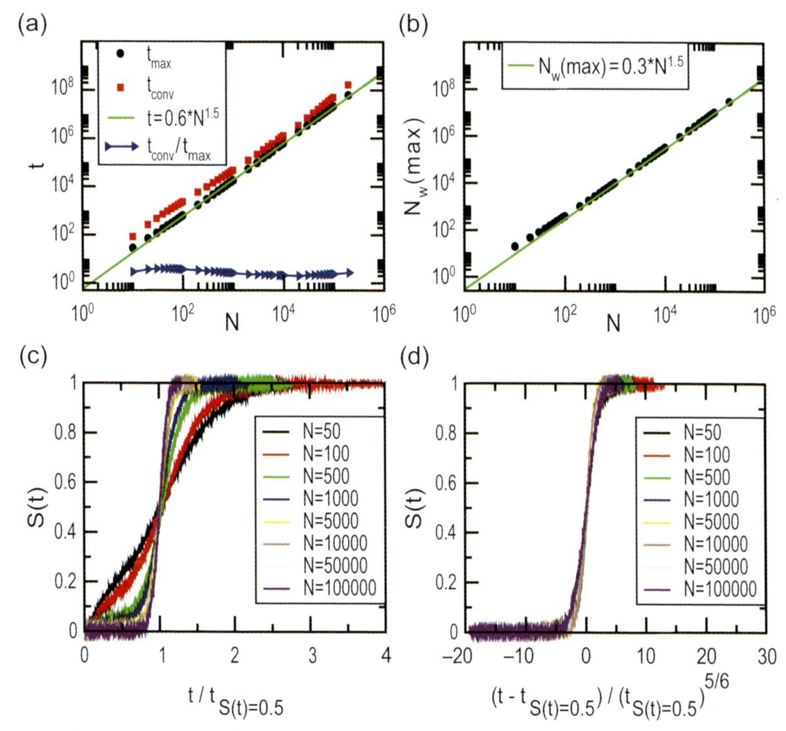

Figure 16.3 Log–log plots showing the power law behavior of the naming game (after Baronchelli et al. (2008). N, the size of the population, is shown on the X-axis and t_{max}, the time when the number of words is maximum, and t_{conv}, the time when the population reaches convergence, is shown on the Y-axis. Both show power law behavior with exponent 1.5. The bottom graphs show the success curves which are S-shaped. The disorder-order transition is fastest for large N.

The Guessing Game

In-depth naming game experiments have been carried out now for naming individual objects, i.e., objects which are sufficiently unique that they are considered individuals (Steels, Loetzsch, and Spranger, submitted), colors (Steels and Belpaeme 2005; Puglisi et al. 2008; Komarova 2007), spatial categories (Steels and Loetzsch 2008), or actions (Steels and Spranger 2008). Depending on the domain, we talk about the object naming game, the color naming game, the spatial naming game, or the action naming game. The naming game makes, however, one crucial assumption; namely that the semantic domain of categorization is restricted and implicitly known to the agent. For the color naming game, agents are basically saying *Point to the object whose color is red.* The situation changes drastically if this framework is taken away, and results in the so-called guessing game (Steels and Kaplan 2002). The script for the guessing game is similar to that of the naming game, but now agents are free to use any semantic dimension or combination of dimensions. For example, in the talking heads experiment (Steels and Kaplan 2002)—the first experiment in grounded guessing games—color, size, position, or shape of objects in the world were all possible dimensions for finding distinctive categories or combinations of categories.

The guessing game introduces two additional challenges. First, agents are confronted with Quine's "Gavagai" problem (Quine 1960), causing increased uncertainty for guessing the meaning of unknown words. This creates more complex semiotic dynamics and requires the use of heuristics to manage the combinatorial explosions that inevitably occur. Second, there is an opportunity for compositional language to emerge when combinations of categories need to be expressed (Van Looveren 1999). A compositional language means that several lexical items are used to cover the combination of categories rather than a single holistic coding. It can be shown that a compositional hierarchical language will arise when agents maximize the reuse of existing words (de Beule 2008), showing that overcoming the cultural transmission bottleneck (Brighton and Kirby 2001) is not the only way in which compositional coding may become selected over holistic coding.

The guessing game has now been studied intensely as well from the viewpoint of computational, mathematical, and robotic experiments. The biggest challenge has been to set up a world and an interaction pattern between the agents so that certain types of meanings become relevant to identify the heuristics that aid in category formation and the coordination of categories across the agents, and to set up the collective and individual search processes by which agents can establish agreements about the meaning of words. Figure 16.4, adapted from Wellens et al. (2008), shows the results of some recent guessing game experiments using real-world visual input captured by humanoid robots. The meaning of words is now a cluster of possible features, each of which has a score that can independently change. Many phenomena observed in human word meaning, such as the metaphorical extension of the use of a word or the gradual shifting of word meaning over time, appear as emergent phenomena in these experiments.

Grammatical Language Games

The naming game and the guessing game explore how an inventory of categories can form and become expressed in an emerging lexicon. Such communication systems do not yet require grammar. Grammar becomes necessary when we move up one level in semantic power to include predicates and arguments. This is almost always necessary in description games, for example, where the speaker describes to the hearer the scene before both of them, and the game is a success if the hearer agrees that the description given by the speaker fits with the current scene. For instance, a sentence like *Jill pushes the block to Jack*

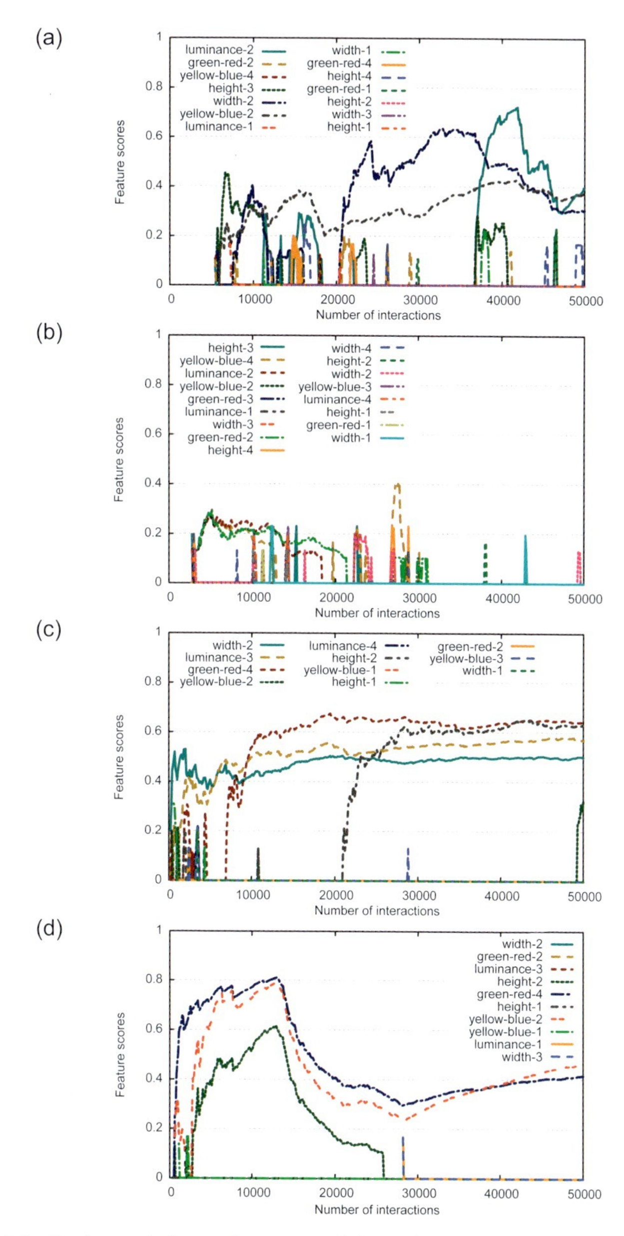

Figure 16.4 Each graph shows the scores of the various associations between a word and perceptually grounded categories for different sensory channels (width, height, luminance, green-red, yellow-blue) in the lexicon of an agent playing guessing games with other members of a population of 25 agents (adapted from Wellens et al. 2008). We see how clusters of senses get associated with a word and there is constant evolution as word meanings flexibly adapt to environmental needs and conventions emerging in the population.

involves reference to a number of objects and events (*Jill*, *Jack*, *the block*, *push*), which would already be handled by naming and guessing games, but to make it clear that it is Jill and not Jack who is pushing the block and that the block is pushed to Jack and not Jack to the block, additional information must be expressed beyond the meaning of individual words. Grammar is also necessary if semantic power is increased further, up to the level of second-order semantics, in the sense that some predicates operate over other predicates instead of individuals (Partee 2003). For example, an adverb, such as *very* in *very big*, modifies the meaning of the adjective. It is not a predicate over objects. When second-order semantics is introduced, the grammar not only has to signal what belongs to what but also what semantic function is intended. For example, the predicate *green* is used in *green is a color*, *the green ball*, *a greenish blue*, *she has a green finger*, *a greener tree*, but with different semantic functions signaled by different syntactic contexts.

The first significant experiments with grammatical language games were carried out by Batali (1998) who constructed a simulation where agents were endowed with the capacity to construct sequential patterns that cover a combination of predicates disambiguating the predicate–argument relations between them and ways to combine patterns by an appropriate mapping of the variables. Batali used an exemplar-based learning technique for retrieving form-meaning associations, which most closely match the meaning to be expressed (in production) or the form (in parsing). Exemplar-based or memory-based learning is a general learning technique that has been used successfully in a wide variety of problem domains. Agents align patterns because they tend to use the ones which occur most frequently in the population.

Other experiments in grounded grammatical languages, specifically for the formation of case grammars, have been reported (Steels 2004; Van Trijp 2008). These experiments start from a sophisticated vision system capable of interpreting, categorizing, and conceptualizing event structures (Steels and Baillie 2003; Siskind 2001). They include diagnostics to detect the need for introducing grammar, for example, because certain predicate–argument relations need to be expressed explicitly to avoid ambiguity or misunderstanding (Steels 2005b) or because combinatorial search needs to be avoided in parsing (Steels and Wellens 2006). The repair strategies possibly introduce or expand new syntactic or semantic categories as well as new constructions to express additional information or tighten the applicability of constructions with additional constraints. The same alignment strategies are used as in the naming game but are now at the level of the grammar. For example, if a grammatical construction could be applied and has led to a successful interaction, then the score of this construction and the different syntactic and semantic categories that were used is increased and competitors decreased (Steels et al. 2007).

An example of results from grammatical language experiments (taken from Van Trijp 2008) is shown in Figure 16.5. Here, agents play description games about events they have seen. Agents try to maximize communicative success

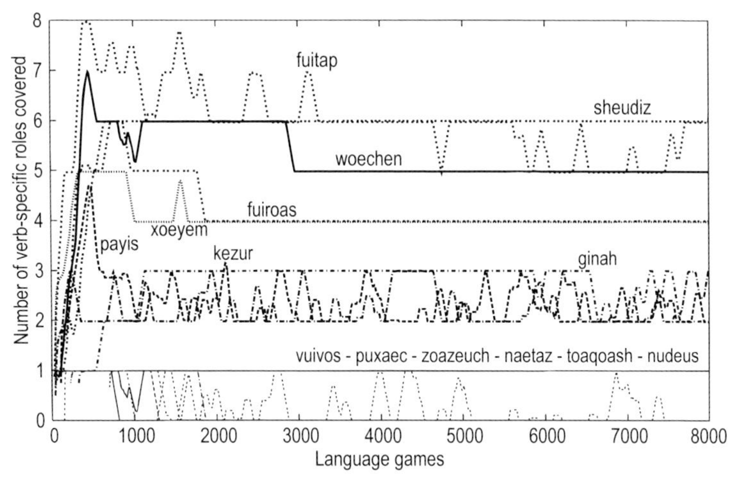

Figure 16.5 Results from an experiment in which visually embodied agents play description games about scenes perceived through their cameras. They describe events for which the roles of participants need to be made explicit to avoid ambiguity or misunderstanding. Agents begin with ad hoc markers for specific roles (e.g., the pusher of a push-event) but then generalize to more abstract semantic roles. The graph shows how many verb-specific participant roles each case marker covers. For example, *fuitap* covers 8 roles after 600 games, but is in conflict with other markers and in the end covers 6 roles. The graph shows that there is a continuum between more specific and more generalized markers as the grammar unfolds.

while minimizing cognitive effort; this is only possible if they express explicitly the roles of participants. Agents are set up to start with a kind of pidgin English, in the sense that their lexicon already contains English words for the basic concepts in their ontology. Thus, we can follow more easily what is going on, much the same way that we can see the emergence of grammar quite clearly in creole formation (Mufwene 2001). Diagnostics detect possible ambiguities and repair strategies introduce markers of participant roles. The markers are random combinations of syllables. Alignment progressively coordinates both the language categories (semantic and syntactic) and the grammatical constructions. Initially, the markers are entirely ad hoc: there is a marker, e.g., for the pusher of a push-event, but agents always try to reuse existing material as much as possible. This progressively pushes the grammar towards a higher level of abstraction with markers now expressing semantic roles similar to "agent" or "beneficiary," and the grammar now has a layer of syntactic categories that resemble cases like "nominative" or "accusative."

Some sentences that appear in the example grammar experiment from Figure 16.5 are as follows:

(1) *jack* *-fuitap* *jill* *-kezu* *walk-to*
jack sem-role-3 jill sem-role-44 walk-to
"Jack walks to Jill."

(2) *house* *-payis* *move-inside* *boy* *-fuitap*
house-1 sem-role-10 move-inside boy sem-role-3
"The boy moves inside house-1."

(3) *touch* *jill* *-fuitap* *house* *-ginah*
touch jill sem-role-3 house-1 sem-role-29
"Jill touches house-1."

Note that there is no word order in this artificial language except between the word for an object (like *jack*) and its marker (like *-fuitap*). *-fuitap* has been generalized over several specific participant roles to cover a more abstract role (sem-role-3).

Mathematical, computational, and robotic modeling of grammatical language games is still in an early stage, but the first solid results are undeniable. Today, experiments are ongoing to examine how tense–aspect–mood–modality systems can arise, how there can be an emergent system of determiners, how grammars for second-order semantics can form, and how long-term dependencies or anaphora may emerge. We can expect concrete results in the very near future, as complex technologies needed for the experiments become progressively available and are mastered by an increasing number of researchers, and as our understanding of the cognitive processes required for the formation and sustenance of grammatical languages increases.

Fluid Construction Grammar

Although achieving situated embodied language games requires the operationalization of many cognitive capabilities, for the remainder of this chapter I will focus solely on the structures and processes that are directly relevant for grammar. First, however, some initial remarks.

It is important to realize that information processing is always strongly dependent on the choice of representations and the structure of the procedures used to achieve certain tasks. It is easy, for example, to look up the telephone number in a telephone directory, when the name is known, but it is practically impossible to find the name if only the telephone number is known. Likewise, the representation of the grammar and nature of the representations used during linguistic processing will determine whether a particular process will be feasible or not. Thus, generative grammar may make it possible to enumerate all the syntactic structures that may occur in a language, but to use the same grammatical representation for parsing and production is not really possible because a lot of relevant knowledge is simply not expressed or only implicitly expressed. The same is true for learnability. The syntactic structure of human languages is often claimed to be unlearnable given the poverty of the stimulus. However, if we assume that learners are intelligent and use their cognitive capacities to their full extent, then the nature of the learning problem changes entirely. Learners can reconstruct and infer meaning beyond what they understand directly and formulate strong hypotheses on the structure of the grammar. Moreover, dramatic progress in machine learning over the last decade has shown that much more can be learned than was previously thought possible.

Second, it has *not* turned out to be possible to take existing formalisms (such as minimalist syntax) or existing parsing and production procedures (such as push-down automata or chart parsing) "off the shelf" and apply them to implement the grammatical processing needed in language games where the grammar is evolving. Coping with grammars that are constantly undergoing change and are not yet crystallized requires great flexibility and fluidity in the application of a linguistic inventory. Agents are not trying to decide whether a sentence is grammatical but must extract as much structure as possible so that they can at least partially reconstruct its meaning. Because there is unavoidable variation in the population, each language user must keep track of many variations in his inventory and progressively decide on which one is dominant.

One of the formalisms developed for supporting experiments in grammatical language games is fluid construction grammar (FCG) (Steels 2004; Steels et al. 2005), which belongs to a class of recent formalisms that attempts to operationalize the basic principles of construction grammar (Bergen and Chang 2004). FCG uses a similar architecture as is commonly found in rule-based inference systems, which is widely used for modeling problem solving in a large variety of domains (Newell 1990). Without going into too much detail, I will outline the basics of FCG as they constitute a theory of what the neurobiology of language must provide.

Feature Structures

In the course of interpretation and production, all the information being accumulated about the utterance must be collected in a data structure. In early syntactic formalisms, such as phrase structure grammar, this data structure takes the form of a parse tree which implicitly represents the hierarchical structure and the ordering of constituents. This solution is not viable in the present context, partly because a lot more must be represented than phrase structure, and partly because the implicit coding of hierarchy and ordering is not flexible enough. An alternative approach, pioneered in unification-based grammars such as lexical–functional grammar (Kaplan and Bresnan 1982) or HPSG (Sag 2006), is to code all information explicitly with attributes, values and relations, and represent it as feature structures.

In the FCG formalism employed in the grammatical language games discussed earlier, a feature structure decomposes into a semantic feature structure and a syntactic feature structure which are coupled together. They each consist of units roughly corresponding to lexical items. The same names are used for units in both so that rules can operate simultaneously over syntax and semantics. The semantic structure contains information on the communicative goal that is to be reached and the conceptualization (meaning) that is going to be conveyed explicitly to achieve the goal. Goals are decomposed into subcomponents and meanings are split up and distributed over different lexical items. The semantic structure contains further information on the information structure being imposed (topic/comment), the re-categorization of the meaning in terms of language-specific semantic categories (such as semantic roles), etc. The syntactic structure contains information on the syntactic units, their hierarchical relations, their syntactic functions, parts of speech, feature values for number, gender, tense, aspect, etc., as well as form constraints such as word order, agreement, intonation or stress patterns. Even for a simple sentence, a complete feature structure can take up a few pages.

Rather than representing words or phrases in their order of appearance, feature structures use predicates like "meets" or "precedes." And rather than representing hierarchical structure implicitly, it is done with predicates like "syn-subunits" or "sem-subunits." This explicit representation of hierarchy and ordering has a number of important advantages. First, the inventory of syntactic tools available in the language can be expanded by extending the set of possible predicates with which constraints on form are defined, rather than by changing the formalism. For example, if intonation becomes a carrier of meaning in the grammar (e.g., rising intonation for questions, falling intonation for commands), then it suffices to introduce a predicate intonation contour, and the grammar can make use of it. Of course intonation contour must be detected in the input and reproducible in the output, but recent research in feature generation (Pachet 2007) and in the origins of sound systems (De Boer 2000) shows that this is entirely feasible. Because of this flexibility it is no longer necessary for a language to use word order as the main vehicle for syntax. If the word order in a particular construction becomes more restricted, then this can simply be represented by adding an additional word-order constraint to the construction involved and the remaining parts of the utterance can still have free word order. There is also much greater flexibility in processing. It is entirely possible to parse sentences whose word order does not strictly adhere to that which is expected by the standard grammar. Such a variation in the language would break typical phrase structure parsers but can be handled now easily as violations of specific constraints. Even free word order or discontinuous phrases can be handled easily, and decisions on hierarchical structure can be postponed until enough evidence is available during parsing or production. Finally, no intricate movement operators are needed, because whatever information had to be expressed by moving a constituent in the tree can simply be represented explicitly.

Because feature structures are "declarative" representations of all the information that is required during language production or language comprehension, they are no longer so specialized compared to the kinds of representations which would be needed in other problem solving domains. For example, it would be entirely feasible to build a problem solver for spatial navigation that uses feature structures. It would then contain predicates for describing aspects of the environment, the hierarchical structure, and ordering of steps in the plan, possible solutions to subgoals.

Unify and Merge

Phrase structure formalisms represent the lexicon and grammar in terms of derivational rules, which give possible expansions of a tree, starting from an initial non-terminal category like S (sentence). This approach is not a viable solution in the present context partly because it is not flexible enough and partly because communication requires mapping meaning to form and form to meaning, rather than generating the set of possible syntactic structures in a language. An alternative approach, also pioneered in unification-based grammars, is to view linguistic rules (lexical entries, syntactic and semantic categorization rules, grammatical constructions) as constraints that license the inferential steps by which a feature structure is expanded until it contains enough information to allow interpretation (in parsing) or rendering in speech (in production). Some constraints can be violated or new constraints can be abducted on the fly by repair strategies to allow maximum flexibility (Steels 2004). Moreover all rules have a score which reflects the confidence of the agent in how far that rule has the highest probability to have success in the game, typically because it has most frequently been part of successful games in the past.

In FCG, the linguistic rules are represented the same way as feature structures with a semantic and a syntactic pole. The poles now have variables for some of their units or features. A rule is applied in a two-step process and the process is entirely analogous for parsing and production. In language production the semantic pole is matched against the semantic part of the current feature structure and if they can be unified, (i.e., if a binding of the variables can be found such that the semantic pole becomes a subset of the current feature structure), then the syntactic pole is merged with the current feature structure after instantiating its variables. Conversely in language parsing, the syntactic pole of a rule is matched against the syntactic part of the current feature structure and if they can be unified, the semantic pole is merged with the current feature structure after instantiating its variables (Steels and de Beule 2006; see Figure 6 in VanTrijp 2008).

Starting from an initial feature structure representing the goal of communication and the meaning in the case of language production or the incoming utterance in the case of language comprehension, consecutive rule applications expand the feature structure until enough information is available to render the sentence (in production) or apply the meaning to the present context (in parsing). It is unavoidable that there is a search process because there is usually more than one way to express the same meaning or more than one meaning for the same word or the same syntactic pattern. Speakers and hearers must be able to cope with variation and unstable grammars, so they must often entertain additional hypotheses, and they must try to apply rules even if their conditions match only partially. Every rule has a score reflecting how strongly the agent believes this rule to be part of the common grammar, and heuristic search can be used guided by the score of the rules involved in the search path so far and on how well these rules are matching with the growing feature structure. After a language game, the score of all rules that were used is updated using the same strategies as already pioneered for the naming game and the guessing game.

Once again, it is important to stress that the architecture of this system is entirely similar to rule-based problem-solving systems which have been explored for hundreds of different domains, particularly in the context of knowledge systems (Russell and Norvig 2003).

Hierarchy

As in all grammar formalisms, there must be a way to handle hierarchical structure. This is, at first sight, not so easy within a unification-based feature structure approach, particularly if the bi-directionality of rule application has to be maintained. The solution adopted in FCG is an operator (called the J-operator) (de Beule and Steels 2005), which triggers on parts of an existing feature structure, creates a new unit, adds the triggers to this unit, and possibly adds additional features. Units governed by this J-operator are ignored during the unify phase but they are enacted in the merge phase.

Thus to represent an English-like "caused-motion construction," as in *He swept the dust off the floor* (Goldberg 1995), the semantic (left) pole could contain a semantic frame with units and constraints for the event (it has to be a caused motion), the agent (it has to be animate), the moving object (which has to be moveable), and the path of motion (which has to be a surface). The syntactic (right) pole could contain word-order constraints expressed with the "meets" predicate (true if two units directly follow each other), agreement constraints (the subject's and the verb's number and person have to be equal), constraints on syntactic categories (the event has to be expressed with a a verb), and a syntactic frame with a role for subject, direct object and oblique.

In FCG style, a rule for this construction is represented in the expression below, of the form "left-pole <=> right-pole"' (see also Figure 16.6). It translates into the internal data structures that support the Unify and Merge operations during parsing or production. Variables are denoted with a question mark. Units for the subject and object are governed by a J-operator on the syntactic side.

(4) Construction: caused-motion

```
?top-unit
  sem-subunits: {?unit-a ?unit-b ?unit-c ?unit-d}
?unit-a
  sem-frame:    {sem-role-agent(?unit-b,?obj-x)
                 sem-role-moveable(?unit-c,?obj-y)
                 sem-role-path(?unit-d,?obj-z)}
  sem-cat: caused-motion
```

```
?unit-b
  referent: ?obj-x
  sem-cat: animate-object
?unit-c
  referent: ?obj-y
  sem-cat: moveable-object
?unit-d
  referent: ?obj-z
  sem-cat: surface
(J ?unit-e ?top-unit)
  sem-cat: description
<=>
?top-unit
  syn-subunits: {?unit-a ?unit-b ?unit-c ?unit-d}
  form   {meets(?unit-b,?unit-a), meets(?unit-a,
 ?unit-c)
          meets(?unit-c,?unit-d)}
?unit-a
  syn-cat:      {part-of-speech(verb),
                 agreement(?unit-a, ?unit-b, number),
                 agreement(?unit-a, ?unit-b, person)}
  syn-frame:    syn-role-subject(?unit-b),
                syn-role-object(?unit-c),
                syn-role-oblique(?unit-d)}
?unit-d
  syn-role: syn-role-oblique
?unit-b
  syn-role: syn-role-subject
?unit-c
  syn-role: syn-role-object
(J ?unit-e ?top-unit)
  syn-cat: sentence
```

Note the explicit representation of the semantic hierarchy (with sem-subunits) and syntactic hierarchy (with syn-subunits) and of the word-order constraints. The construction establishes the correct co-reference relations between the referents of the pending constituents (denoted with ?obj-x, ?obj-y, ?obj-z) and the participant roles (agent, moveable object, path) of the caused-motion event. ?unit-e is constructed through the J-operator both on the syntactic side and on the semantic side. It grabs together all the subunits that triggered the application of this construction and pulls them together as a new subunit from the top-unit. Figure 16.7 shows an example of application with the following bindings from unification: ?unit-a . unit-1, ?unit-b . unit-2, ?unit-c . unit-3, and ?unit-d . unit-4.

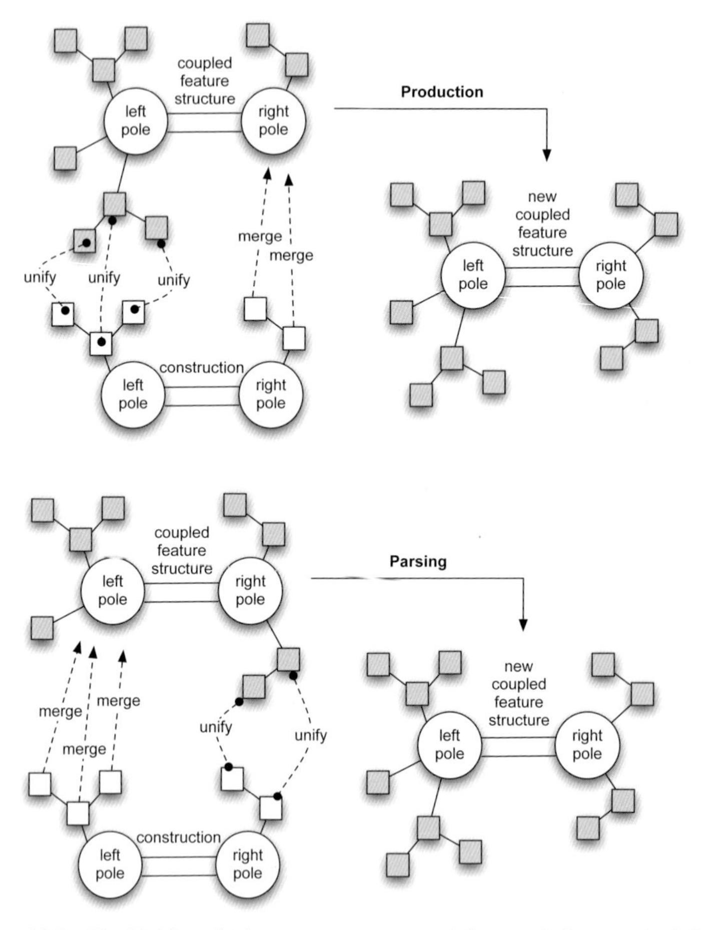

Figure 16.6 The Unify and Merge operators expand the coupled semantic (left pole) and syntactic (right pole) feature structures. Rules are constraints that are always applicable in both directions. They are chained until a solution is found, guided by heuristic search based on the score of the rules, how well they are matching, how much they cover, etc.

Grammar Networks

The final ingredient that turned out to be necessary for getting grammatical language games off the ground is a network superimposed on lexical entries and rules representing grammatical constructions. Sentences such as the following:

(5) (a) Jack sweeps the floor.
(b) Jack sweeps dust off the floor.
(c) Jack sweeps dust.
(d) The floor is swept by Jack.

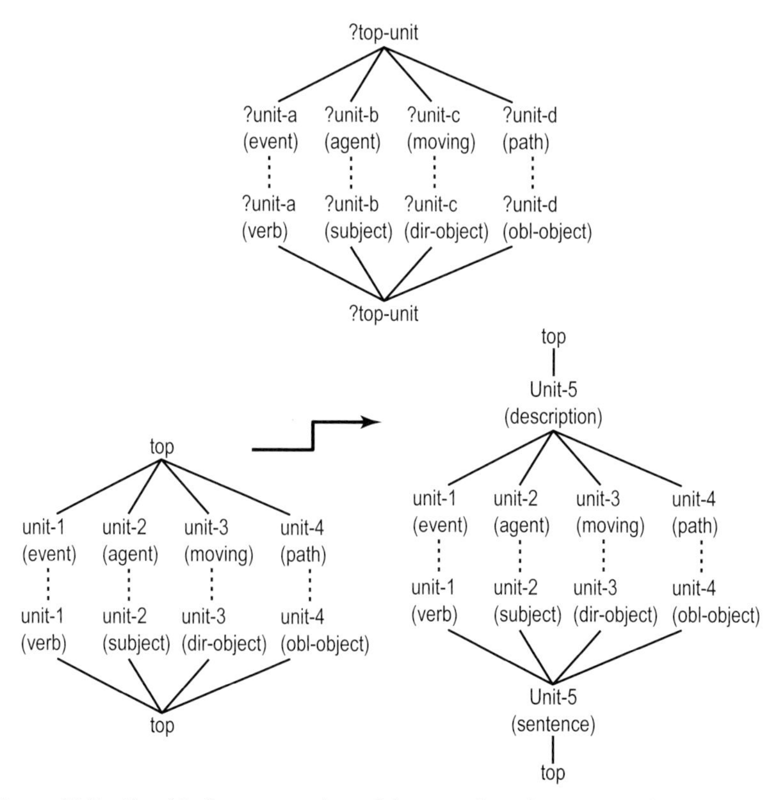

Figure 16.7 Graphical representation of the caused-motion construction (left) as defined in the text. It links a semantic frame for an event with an agent, moveable object and path, to a syntactic frame with subject, object and oblique. Application of this rule in parsing or production (shown on the right) combines all subunits under a new unit and adds information or modifies the feature structure.

show that the same participant roles are not always expressed in the same way. Variations occur depending on which participant roles are made explicit or to express additional aspects of meaning, such as the information structure (Lambrecht 1994). Thus, lexical items should be seen as introducing the possible participant roles. These roles are potentially realized with different abstract semantic roles or combinations of roles, and out of this potential, a grammatical construction then selects a particular constellation (Goldberg 1995). All of this can be handled by "rule networks" that link the different rules (lexical items and constructions) in the inventories of the agents (Figure 16.8). As the grammar evolves, this network is constructed and extended, and the scores of the links are updated based on success in the game (Van Trijp 2008).

The various mechanisms discussed above are all components that can be used to have an operational grammar system that is adequate with respect to the kinds of phenomena observed in human languages. A large part, not further

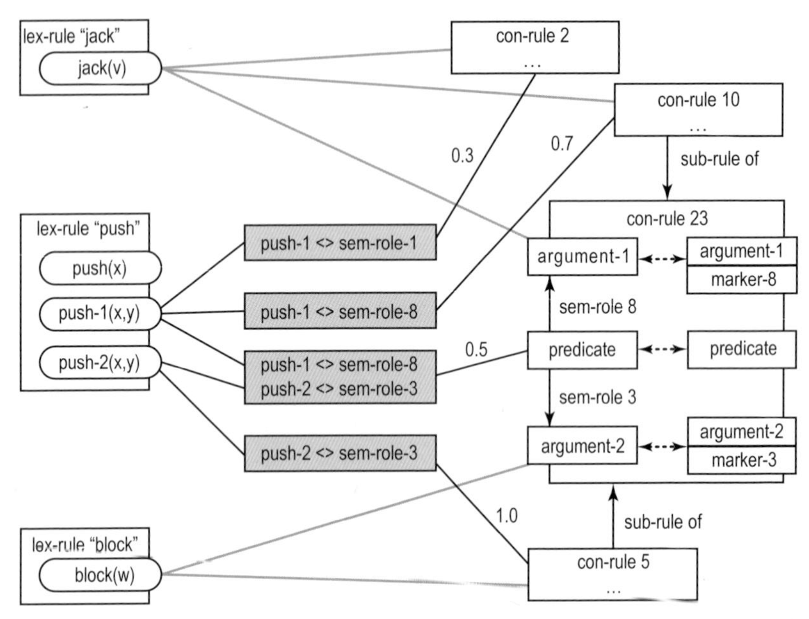

Figure 16.8 Example of a rule network linking potential semantic roles suggested by lexical entries to constructions and their surface realizations. All these links are dynamically updated based on repair and alignment strategies.

discussed due to space limitations, concerns the various repair and alignment strategies that are needed to construct these rules. These strategies make use of meta-rules that specify how one particular area of grammar will be dealt with. They form a template for the construction of rules. Much more research is needed to understand the processes by which strategies for new areas of grammar originate.

This very brief overview of technical issues in the representation and processing of grammatical structures is not exhaustive but should give some idea of the complexities involved. We see that many issues that have been discussed in linguistics must be addressed. The main novelty here is that the grammar representation and grammatical processing must cope with constant change. All aspects of the grammar, therefore, need to have scores that are updated based on the outcome of language games, and the application of the grammar needs to be very flexible, allowing the parsing as well as production of ungrammatical, incomplete or inadequate sentences which then get repaired by further interactions.

Implications for Biological Foundations

Let us now turn to a discussion of what can be learned from these models and experiments, particularly with respect to the questions of what language-specific constraints on neurobiology have to be genetically provided, what aspects of language can evolve in intergenerational cultural evolution, and what can be achieved by cognition and social interaction. Three aspects need to be considered:

1. The inventory for a specific language: the building blocks for constructing meanings (e.g., the distinction between red and green), the syntactic and semantic categories (e.g., the distinction between nouns and verbs or agent and beneficiary), the syntactic tools (e.g., word order, intonation, tones, stress, morphology), and the possible grammatical constructions (e.g., the Passive Construction, the di-transitive construction).
2. The highly complex mechanisms needed for parsing and producing language, given a particular inventory (e.g., the machinery for detecting the hierarchical structure in an utterance according to the grammar of a specific language.
3. The invention and learning mechanisms to expand the inventory of a particular language spoken in a particular community (e.g., mechanisms for coercing an existing word into a new usage or for guessing and thus acquiring the meaning of an unknown word).

Most linguists of all schools (structuralist as well as cognitive) tend to assume that the mechanisms needed for language are highly specific and hence that at least Pts. (2) and (3) must be genetically determined. Much has been made of recursion (e.g., Fitch and Hauser 2004), even though the ability to make and execute recursive plans or recognize recursive structures is surely not restricted to language; neither does recursion require particular architectural "breakthroughs" when constraint-based grammar formalisms, such as FCG, are used. In contrast, the architecture of the language systems used by agents in language games studied so far is not substantially different from architectures developed for modeling problem solvers in other domains. The creation of a search space, the representation of intermediate states (using data structures like feature structures), the use of rules to license possible steps in the inference process, the need for heuristics to guide the search, the use of meta-level diagnostics and repair strategies, reinforcement learning and credit assignment, chunking to speed up processing, and many of the other techniques, which have proven their worth in grammatical language game experiments, have all been developed and operationalized earlier in the context of problem solvers for other domains (Newell 1990). So the models and experiments reported here generate doubt about the idea that language-specific structures and processes need to be genetically coded.

Many linguists also assume that the structures used in a particular language are too complex to be induced from impoverished data, and hence that language learners must come to the task with rich prior innate knowledge of possible language structures (e.g., of what inventory of syntactic and semantic categories might be expected or what kinds of constructions might occur) or at least with very strong biases in their learning algorithms that will then show up through intergenerational cultural evolution. If that is the case then there is indeed an important role for genetics in Pt. (1) as well. Language must then have evolved primarily through genetic evolution and natural selection, with an important additional role for cultural intergenerational transmission.

The experiments reported here, however, argue for the opposite point of view. It appears entirely possible that language-like communication systems get established using the sophisticated human problem-solving and social coordination capacities that are also observed in other domains of human intelligence. The empirical data gathered by language typologists and historical linguists show that there is huge variety, which is difficult to capture in a fixed "universal" inventory (Haspelmath 2007; Talmy 2000), and that language systems appear to change all the time (Labov 1994), even in terms of what linguistic categories they employ or what syntactic tools they use (Brenier 2006; Vankemenade 1987). Socio-cognitive models suggest how new concepts, new syntactic and semantic categories, new lexical and grammatical materials, new constructions, or new syntactic tools can arise through generic cognitive processes and how they can propagate in a population based on alignment. Data of the actual evolution of human languages can be used as a way of empirically validating which cognitive mechanisms are being used (Heine 1997). Experiments, such as the one in the formation of case grammars reported here, are beginning to demonstrate this very concretely. Remarkably, no role for language-specific genetics or intergenerational cultural transmission was needed to see the formation of a case grammar, including the system of semantic roles and syntactic cases and their markings.

The socio-cognitive approach to language does not imply that there is a complete tabula rasa; to the contrary, it assumes a rich battery of cognitive mechanisms (problem-solving skills, representational structures, inference, learning mechanisms, social interaction patterns, etc.). It also does not imply that a single generic learning mechanism (e.g., recurrent neural networks) is able to do the job, rather it implies that a rich collection of learning techniques are necessary, including the induction of regularity from data, the constructive formulation and subsequent testing of hypotheses, the use of analogy, reinforcement learning, feature generation, memory-based learning, etc. In addition, the socio-cognitive approach does not imply that there is no role at all for genetics. Obviously, human neurobiology must support the enormous complex and demanding information processing that is necessary for this task. The "biological hardware" that supports the mental faculties and activities implicated in language must have undergone positive selection. What we are talking about here, however, are very gross features of brain architecture and functioning, such as increased growth in available memory or flexible "cabling" to allow information flow from any part of the brain to any other part, rather than a highly specific coding of linguistic information or highly specialized mechanisms for handling recursion.

The human language faculty is often conceived as a genetically determined modular organ. The models discussed here suggest instead that the human language faculty should be conceived as a dynamic configuration of brain mechanisms, which grows and adapts, like an organism (Szathmáry 2001), recruiting available cognitive/neural resources for optimally achieving the task of communication (i.e., for maximizing expressive power and communicative success while minimizing cognitive effort in terms of processing and memory). The mechanisms are not specific for language, but need to be recruited and configured dynamically in the service of language by each individual anew (Steels 2007).

One example of recruitment concerns egocentric perspective transformation (computing what the world looks like from another viewpoint). This activity is normally carried out in the parietal-temporal-occipital junction (Zacks 1999) and used for a wide variety of nonlinguistic tasks, such as prediction of the behavior of others or navigation (Iachini and Logie 2003). All human languages have ways to change and mark perspective (as in "your left" versus "my left"), which is only possible if speaker and hearer can conceptualize the scene from the listener's perspective—if they have recruited egocentric perspective transformation into their language system (Steels and Loetzsch 2008). Another example of a universal feature of human languages is that the emotional state of the speaker can be expressed by modulating the speech signal. For example, in the case of anger, the speaker may increase rhythm and volume, use a higher pitch, a more agitated intonation pattern, etc. This requires that the neural subsystems involved in emotion (such as the amygdala) are somehow linked into the language system so that information on emotional states can influence speech production and that information from speech recognition can flow towards the brain areas involved with emotion.

Conclusions

This chapter has examined the role of cognition and embodied, situated social interactions in the origins and evolution of language. A very brief survey was presented of various models and experiments that have been developed to explore this role. They show that we are beginning to understand better what cognitive processes and interaction patterns may give rise to communication systems that are increasingly closer to human natural languages. Although many more issues need to be explored, this line of work already suggests quite clearly that there might be a much smaller role for genetics and intergenerational cultural transmission than previously assumed, and that a population of individuals with cognitive capacities similar to those found in humans can invent and coordinate a communication system rather quickly if the need arises.

A number of conclusions can be drawn with respect to the biological prerequisites and origins of language. The first neurobiological prerequisite for language is that massive amounts of processing and storage capacities must be available to sustain the enormously complex information processing required for language. Language requires capacities comparable to about a million present-day computers linked in dense high speed networks and reaching peta-flop performance. Moreover, these capacities are embedded in a body that has millions of sensors and actuators with extremely fine-grained rapid control. The second prerequisite is that neural pathways must be able to grow across many different areas of the brain so that information can flow quickly to all areas that are participating in language. The third prerequisite is remarkable plasticity. Different processing areas and capacities must be able to be recruited into the language faculty. Instead of highly specialized brain areas and strong genetic coding of language, socio-cognitive models suggest just the opposite. They demonstrate that flexibility, plasticity, and adaptive reconfiguration of a broad repertoire of powerful but generic cognitive mechanisms must have been crucial to get language off the ground, once human populations were able to engage in the kind of cooperative social interactions that form the ultimate basis of verbal communication.

Acknowledgment

Background research for this paper was carried out at the Sony Computer Science Laboratory in Paris and at the AI Laboratory of the Free University of Brussels (VUB). Research was partly sponsored by the EU FET ECAgents project, and by the FP7 ALEAR Project.

17

What Can Formal or Computational Models Tell Us about How (Much) Language Shaped the Brain?

Ted Briscoe

Abstract

I review arguments for and against language-specific learning bias or constraint on human languages in the light of recent claims that innate constraints are not necessary to account for human language acquisition and that putative linguistic universals could be the result of convergent linguistic evolution. I argue that grammar-specific bias is essential for any psychologically feasible and precise account of language acquisition and that, recent simulation work notwithstanding, genetic assimilation remains the most evolutionarily plausible mechanism for its emergence in the linguistic environment of adaptation.

Introduction

In this chapter, I expand on and update the arguments presented earlier (Briscoe 2003) on the evolutionary emergence and maintenance of an innate language acquisition device (LAD). By a LAD, I mean nothing more or less than a learning mechanism which incorporates some language-specific inductive learning bias in favor of some proper subset of the space of possible grammars.[1] The existence of an innate LAD has remained controversial, and it is certainly the case that many arguments that have been proposed in its favor are questionable

[1] The term, LAD, is taken from Chomsky (1965a). In more recent work, it has been dropped in favor of Universal Grammar (e.g., Chomsky 1981), reflecting the increasing focus on constraints on the space of learnable grammars. Here I stick to the older term, as I believe that it is only possible to evaluate empirically claims about UG when they are embedded within a precise account of the acquisition of grammar.

or wrong (e.g., Pullum and Scholz 2002; Sampson 1989, 1999; Lappin and Shieber 2007). However, I will still argue that all adequate extant models of language acquisition do presuppose a LAD in the sense noted above. These arguments put the onus on nonnativists to demonstrate an adequate, detailed and precise account of the acquisition of grammar which does not rely on a LAD.

Chomsky has consistently downplayed the role of evolution in the emergence of an LAD, emphasized the discontinuities between human language and animal communication systems, and speculated that the LAD arose as a result of a macromutation or saltationist jump, even in his most recent work (e.g., Hauser et al. 2002). Pinker and Bloom (1990) developed an account of the gradual evolutionary emergence of the LAD via genetic assimilation (or in their terms, the Baldwin effect). More recently, Briscoe (1997), Deacon (1997), and others have argued that languages themselves are adaptive systems and that the universal constraints on grammar that underpin much argumentation for the LAD can be explained as a consequence of convergent evolution under similar linguistic selection pressure. I will argue, however, that this important insight does not undermine the existence of the LAD, though it certainly undermines arguments for the LAD based solely on the existence of linguistic universals.

Genetic assimilation is a neo-Darwininan mechanism (e.g., Waddington 1942) by which organisms can appear to inherit acquired characteristics though, in fact, it is changes in their behavior or more generally their environment (e.g., niche construction) which create novel selection pressures and thus cause information to be assimilated into the genome.[2] Genetic assimilation of grammatical information exemplified in the environment of adaptation of the LAD potentially would facilitate more rapid and robust acquisition of grammar by first language learners. Thus, if mastery of language increases fitness, we might expect natural selection to improve language learning. I have argued (e.g., Briscoe 2005) for a coevolutionary account incorporating this process in which natural languages are treated as complex adaptive systems undergoing often conflicting selection pressures, only some of which emanate from the LAD or indeed more general cognitive mechanisms, and where the LAD itself evolved via genetic assimilation in response to (proto)languages in the environment of adaptation.

In terms of Kirby et al. (this volume, Figure 15.2), the question we are considering is whether it is appropriate to add a further arrow to their diagram depicting the interaction of natural selection and linguistic selection which "closes the loop" between natural selection for cognitive machinery and the linguistic environment created via cultural transmission, as illustrated in Figure 17.1. Thus I am not arguing that cultural evolution and linguistic

[2] I have reviewed the evidence for genetic assimilation in areas other than language evolution elsewhere (e.g., Briscoe 2003). For a recent more extensive review and discussion, see Pigliucci et al. (2006).

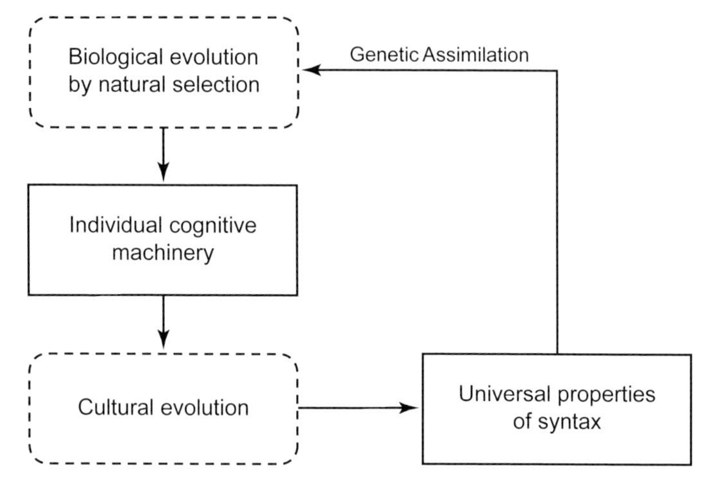

Figure 17.1 The emergent linguistic environment potentially creates new (natural) selection pressures on our cognitive machinery

selection have not had a profound effect on the nature of natural languages, nor that many if not most linguistic universals have emerged as a consequence of cultural or linguistic selection for more learnable or otherwise cognitively or socially advantageous linguistic forms or constructions. The only question I will consider is whether genetic assimilation of language-specific grammatical information into the LAD is also plausible given the coevolutionary scenario entailed by Figure 17.1. In particular, I discuss the extent to which formal or computational models could and do contribute to an answer to this question.

I do not intend to revisit all the arguments for and against genetic assimilation, reviewed in Briscoe (2003), or to review the various models discussed there again. Instead, here I focus on some recent arguments, sometimes supported by models and simulations against genetic assimilation of linguistic information. Before addressing these arguments, I will define grammatical acquisition and present a Bayesian account of the task.

Grammatical Acquisition

In Briscoe (2003), I discuss five desiderata that adequate accounts of grammatical acquisition during first language learning must satisfy:

1. Coverage of attested grammatical constructions.
2. Realistic input to the learner consisting of a finite, positive, but partly noisy sample from the target language.
3. Realistic contextual enrichment of this sample with only partial, noisy representations of the form-meaning mapping.
4. Selectivity in which a consistent grammar is acquired and random noise is rejected.

5. Accuracy in which the acquired grammar captures the form-meaning mappings of the target grammar; that is, learners do not "hallucinate" or invent grammatical properties regardless of the input, though they do (over)generalize and, in this sense, "go beyond the data."

If accuracy is defined in terms of formal learnability from realistic, finite, positive but noisy sentence-meaning pairs over a hypothesis space with adequate coverage, even when drawn from a single stationary target grammar, then inductive bias in the acquisition model is essential.[3]

The term inductive bias is utilized in learning theory to characterize both hard constraints on the hypothesis space considered by a learner, usually imposed by a restricted representation language for hypotheses, and soft constraints which create preferences within the hypothesis space, usually encoded in terms of cost metric or prior probability distribution on hypotheses (e.g., Mitchell 1997, p. 39f). Bayesian probabilistic learning theory is a general domain-independent formulation of learning (for an introduction, see Mitchell 1997, p. 154f) that relies on statistical inference and can thus cope with noise. Bayes's theorem provides a general formula and justification for the integration of prior bias with experience:

$$P(H|D) = \frac{P(H)P(H|D)}{P(D)}. \tag{17.1}$$

We compute the posterior probability of a hypothesis, H, given some data, D, by multiplying the prior probability of the hypothesis by its likelihood given the data, and normalize to obtain a probability by dividing the result by the overall probability of the data. We typically choose the hypothesis with the highest posterior probability. If we do not need to know this exact probability we can skip the normalization step and simply choose the highest value hypothesis after multiplying prior and likelihood:

$$H = \arg\max P(H)P(D|H). \tag{17.2}$$

The most general formulation of learning in this framework (Kolmogorov Complexity) posits a learner able to learn any generalization with a domain-independent bias (the so-called "universal prior") in favor of the smallest, most compressed hypothesis (e.g., Li and Vitanyi 1997). However, nobody has demonstrated that this general formulation could, even in principle, result in a learning algorithm capable of accurately acquiring a specific grammar of a human language from realistic input. However, there have been many demonstrations that grammars from more restrictive though infinite hypothesis

[3] See also Lappin and Shieber (2007) and Nowak et al. (2002) for related discussions of learning theory, where similar conclusions are drawn.

spaces, such as the class of context-free grammars (CFGs), can be acquired given a general bias in favor of the smallest or most probable hypothesis (e.g., Horning 1969). When such a general bias is applied to a domain-specific and restrictive representation, it will create bias in favor of certain form-meaning mappings. This is where domain-specific inductive bias appears to be unavoidable if the desideratum of learning accuracy is to be met. Thus, this is the basis upon which a LAD is unavoidable in any adequate account of grammatical acquisition. To relate this back to the equations above, if the space of possible hypotheses, H, is that of unrestricted rewrite rules or Turing machines, then we might argue reasonably that we have a domain-independent inductive bias. On the other hand, if this space is defined as the (infinite) class of context-free or indexed grammars, which cannot express some types of possible dependencies within sequential strings and thus some possible mappings between meaning and form, then we are positing a LAD, possibly with additional soft bias deriving from the prior.

Gold's (1967) original negative "in the limit" learnability results are founded on the intuition that any amount of finite, positive data from a target grammar in a class containing grammars capable of generating an infinite set of sentences is always compatible with a hypothesized grammar generating all and only the data seen so far and also with any one of a potentially infinite set of other grammars from the candidate class which generate some superset of the learning sample. Notwithstanding more recent developments in learnability theory and machine learning (e.g., Nowak et al. 2002; Lappin and Shieber 2007), this basic point still holds. A prior distribution or cost metric encoding a preference, for example, for smaller, more compressed grammars will, in general, select a single grammar which predicts the grammaticality of a specific superset of the learning sample. The exact form of the representation language in which candidate grammars are couched and/or the addition of factors other than just size to the prior distribution or cost metric will determine which of the potentially infinitely many grammars generating a superset of the learning sample is selected by the learner.

Consider a potential class of languages consisting of clauses constructed from a verb (V), a subject (S), and an object (O), where S and O are always realized as single (pro)nouns (N) or as noun phrases (NP) consisting of a noun and a (relative) clause; the S and O labels are a shorthand for the mapping from forms to meanings (in this instance, just predicate–argument structure). By stipulation, there is one root clause per sentence and all relative clauses (RCs) modify the immediately preceding or following N. Potentially, grammatical sentences in this class of languages can consist of any infinite sequence of Ss, Vs, and/or Os, where we will use subscripts to indicate which S or O is an argument of which V, when there is more than one V in a sentence. Thus, any clausal ordering of S, O, and V is possible, as well as any arrangement of root and relative clauses:

(1) (a) $S_i V_i O_i S_j V_j O_j$ (e.g., cats like dogs$_i$ who$_i$ like cats)
(b) $S_i V_i O_i S_j V_j O_j$ (e.g., who$_i$ like dogs cats$_i$ like cats)
(c) $S_i V_j O_j S_j V_i V_k O_k S_k O_i$ (e.g., cats$_i$ like dogs who$_i$ like eat mice who$_j$ cats$_j$).

These examples illustrate that post-and prenominal relative clauses with clause-initial and clause-final relative pronouns are all potentially grammatical sequences. A learner over CFGs with preterminals N and V will be capable, in principle, of acquiring any target grammar in this space. Suppose that the learner prefers, a priori, the smallest grammar compatible with the input, defined as the grammar with the least number of nonterminals and the least number of rules with the least number of daughters (where each nonterminal and rule costs one and each daughter of each rule costs one). Then a learner exposed to a sample of unembedded SVO sequences and (1a) might learn the grammar:[4]

(2) (a) Sent $\rightarrow$ NP^S V NP^O
(b) NP $\rightarrow$ NP Sent
(c) NP $\rightarrow$ N.

This grammar has a cost of 2 for nonterminals, 3 for rules, and 6 for daughters (making 11), and predicts the grammaticality of postnominal subject-modifying relative clauses and of center-embedded and right-branching sequences of relative clauses. (Given this cost metric, the learner could equally well learn a nonrecursive variant of (2b) with N substituted for NP as leftmost daughter.) Without the preference for smaller grammars, as defined above, a learner might have acquired the less predictive grammar:

(3) (a) Sent $\rightarrow$ N^S V N^O
(b) Sent $\rightarrow$ N^S_i V_i N^O_i N^S_j V_j N^O_j.

This grammar has a cost of 1 for nonterminals, 2 for rules, and 10 for daughters (making 13), and it does not predict the grammaticality of subject-modifying relative or multiply-embedded relative clauses. Moreover, a cost metric which assigned a cost of 2 to each rule would also select (3) in preference to (2).[5]

4 Once again, I use superscripted S and O and subscripted indices to show the mapping to predicate–argument structure and leave implicit that required to characterize the predicate–argument structure of sentences containing relative pronouns. The details of how this mapping is actually realized formally are not important to the argument, but either a rule-to-rule semantics based on the typed lambda calculus or a unification-based analogue would suffice.

5 This point is not new, of course. Chomsky (1965a, p. 38) recognized the need for an evaluation measure based on simplicity to choose between grammars during language acquisition, and others criticized the arbitrariness of such measures. The Kolmogorov Complexity (e.g., Li and Vitanyi 1997) and the related Minimum Description Length (MDL) principle (e.g., Rissanen 1989) provide a less arbitrary metric based on the cost of compressing a hypothesis. The MDL principle can be, and has been, applied to grammatical acquisition (e.g., Ristad and Rissanen 1994), but once again coupled with restricted hypothesis representation languages. These complexities are ignored here to keep the example simple as they do not alter the fundamental point about the domain specificity of cost metrics or prior distributions defined over restricted hypothesis representation languages.

If the input also includes (1b), containing a prenominal subject-modifying relative clause, then a learner utilizing grammar (2) might acquire a further right-recursive rule analogous to (2b), predicting complementary distribution of pre-and postmodifying relative clauses. A learner utilizing (3) might acquire a further rule analogous to (3b), predicting only subject-modifying prenominal relative clauses.

Example (1c) provides evidence for a root SVO language containing postnominal VOS relative clauses. A learner with no cost metric might well acquire a grammar with a rule analogous to (3b) with 9 daughters predicting this and only this exact sequence. A learner with the cost metric exposed to SVO unembedded sequences and (1c) would acquire grammar (4) with a total cost of 16:

(4) (a) Sent → NP^S V NP^O
(b) RC → V NP^O NP^S
(c) NP → NP RC
(d) NP → N.

Thus, this learning model predicts that mixed root and embedded constituent orders is a dispreferred or more marked option that will only be adopted when the learner is forced to do so by positive evidence.

By contrast, if the learner represents the class of context-free languages in ID/LP notation instead of standard CFG, acquiring immediate dominance (ID) rules independently of linear precedence (LP) rules (e.g., Gazdar et al. 1985) but utilizing a similar cost metric which also assigns a cost of one to each LP rule, then the preference ordering on specific ID/LP grammars predicts that order-free variants of the above grammars with no LP rules will be preferred and that the inclusion of examples like (1b) or (1c) in the input will not alter the learner's hypothesis. Thus, by changing the hypothesis representation language but keeping the cost metric the same, we create inductive bias in favor of different grammars which generalize in different ways from the evidence. Similarly, by keeping the representation language the same but modifying the cost metric, we can also create differing inductive biases.

The Bayesian learning framework also provides a general and natural way to understand and model how stronger grammar-specific inductive biases might have come to be integrated with the LAD, in terms of the evolution of increasingly more accurate prior distributions over the hypothesis space with an ever better "fit" with languages in the environment of adaptation. Cosmides and Tooby (1996), Geisler and Diehl (2003), and Staddon (1988) argue in detail that Bayesian learning theory is an appropriate framework for modeling learning in animals and humans, and that evolution can be understood within this framework as a mechanism for optimizing priors to "fit" the environment and thus increase fitness. Thus, it provides a framework for making precise the effects of genetic assimilation, as will be detailed later. Cost metrics applied to such restricted hypothesis representation languages entail that learners will "go beyond the evidence" in different ways and, thus, will have different specifically linguistic inductive biases (i.e., different mappings between form and meaning). However, learners without cost metrics, or equivalently prior distributions, cannot acquire target grammars accurately, as Gold's (1967) and Horning's (1969) work demonstrated.

All extant models which learn form-meaning mappings assume a LAD because they utilize prior distributions or cost metrics defined over restricted hypothesis representation languages selected to facilitate encoding of grammars for human languages. The onus is on nonnativists to develop a precise account of grammatical acquisition which meets the above desiderata and does not utilize a LAD in this sense. Work utilizing simple recurrent neural networks (RNNs) or other forms of statistical classification, purporting to address issues of grammar learning, is largely irrelevant as such models can at most learn to classify segments of the input and/or predict the class of the next unit of input. They do not learn a form-meaning mapping that requires the ability to construct a relational encoding using two-place predicates over constituents or lexical heads, such as "subject-of," "object-of," and so forth

Independently of these logical and theoretical arguments, there is psycholinguistic evidence that human language learners are biased in linguistically specific ways. There are learning stages in which overgeneralization of regular morphology is common, tense is assigned to auxiliaries and main verbs in subject–auxiliary inverted constructions, and so forth. While the exact interpretation of such phenomena is a matter of complex analysis within a theoretical framework, psycholinguists most often describe them as linguistically specific biases. For instance, Wanner and Gleitman (1982, p. 12f) argue that children are predisposed to learn lexical compositional systems in which "atomic" elements of meaning, such as negation, are mapped to individual words. This leads to transient production errors where languages, for example, mark negation morphologically.

In summary, the onus is on nonnativists to define an effective grammar learning procedure which meets the desiderata outlined in the opening paragraph of this section. Until this is done, we must continue to assume that grammar learning requires at least a weak inductive bias able to choose between different form-meaning mapping rule sets (grammars) which predict the grammaticality of different supersets of the learning data.

Linguistic Evolution

Linguistic evolution proceeds via cultural transmission (primarily, first language acquisition) at a faster rate than biological evolution. The populations involved are generally smaller (speech communities, rather than entire species), and language acquisition is a more flexible and efficient method of information transfer than genetic mutation. Clearly, vocabulary learning and, at least, peripheral grammatical development are ongoing processes that last beyond childhood, so that linguistic inheritance is less delineated or constrained than the biological mechanisms of genetic evolution. Several consequences emerge from the evolutionary account of languages as adaptive systems which must be taken into consideration by any plausible account of grammar learning. First, several researchers have considered what type of language acquisition procedure could not only underlie accurate learning of modern human languages but also predict the emergence of protolanguage(s) with un-decomposable form-meaning correspondences and the (subsequent) emergence of protolanguage(s) with decomposable (minimally grammatical) sentence-meaning correspondences (e.g., Oliphant 2002; Kirby 2002; Brighton 2002). They conclude that the language acquisition procedure must incorporate inductive bias resulting in generalization, and consequent regularization of the input, in order for repeated rounds of cultural transmission of language to regularize random variations into consistent and coherent communication systems.[6] Second, the account of languages as adaptive systems entails that linguistic universals no longer constitute strong evidence for a LAD. Deacon (1997), Briscoe (1997), and others make the point that universals may equally be the result of convergent evolution in different languages as a consequence of similar evolutionary pathways and linguistic selection pressures.

Zuidema (2003) has argued, following Deacon (2007), that if languages have evolved to be learnable this undermines the learnability arguments of Nowak et al. (2002), which hold that for speech communities to evolve, the probability of children being able to learn a target grammar must be higher than a "coherence threshold," below which no single communal grammar, and thus language, can be maintained. Zuidema presents a simulation of an iterated learning model in which early generations of learners do not acquire the target language, but a compression-based prior bias for small CFGs leads to the evolution of languages which can be acquired accurately by this learning procedure. Thus over generations, the population of learners evolves languages which meet the coherence threshold even though the starting conditions do not. To achieve this result, Zuidema must assume that the learners in his population come equipped with an invariant learning algorithm equivalent to that of Horning (1969), as a prior bias for small stochastic CFGs is equivalent to a compression-based learner of CFGs (e.g., Rissanen 1989). Thus, contrary to his claims, the model does not really address Gold's "in the limit" negative results, because of the assumed inductive bias for smaller grammars. The model does, however, show very elegantly how the fit between languages and prior bias is predicted to become very close in many if not all such models (e.g.,

---

[6] Newport (1999) reports the results of experiments on sign language acquisition from poor and inconsistent signers which clearly exhibit exactly this bias to impose regularity where there is variation unconditioned by social context or other factors.

Griffiths and Kalish 2007; Kirby et al. 2007).[7] The question I wish to address here is where might this grammar-specific bias have originated, given that it is evolutionarily implausible to assume that it simply emerged de novo before the emergence of (proto)language.

Genetic Assimilation

Although Pinker and Bloom (1990) and many others use the term Baldwin effect, I prefer Waddington's (1942, 1975) notion of genetic assimilation to describe the process by which changes in the behavior of a population (i.e., niche construction) can cause changes to the environment of adaptation and thus create novel selection pressures on that population. Unlike Baldwin, and others writing before the modern synthesis, Waddington was able to demonstrate experimentally with fruit flies that environmental changes and artificial selection for flies that responded in a phenotypically specific way to such changes results in canalization of the response in which phenotypic plasticity gave way to a genetically encoded invariant response in the evolved population, which no longer relied on the original environmental stimulus.[8]

Deacon (2003a and unpublished) argues that in addition to the unmasking of genes to novel selection pressure demonstrated by Waddington, niche construction may also mask selection for other genes. He gives the example of the loss of the ability in the primate lineage to synthesize ascorbic acid internally as a consequence of masking of selection for a gene which coded for a protein essential to this process after adoption of a diet containing fruit, and thus an external supply of ascorbic acid. Deacon characterizes the process of genetic assimilation as the unmasking of selection pressure on genes coding for cognitive neural mechanisms (e.g., the LAD) as a consequence of niche construction (e.g., the emergence of (proto)language). However, he argues that masking of selection on the genes coding for neural mechanisms and their consequent

[7] Kirby et al. (2007) argue, contra Griffiths and Kalish, that cultural transmission in the form of an "information bottleneck" (i.e., exposure to a finite positive sample of a language which doesn't completely determine the target grammar) can overcome prior bias for learners who select the most probable grammar rather than selecting a grammar with a bias determined by the posterior probability distribution over grammars. However, this result is questionable given the need for noise in linguistic production which essentially reincorporates the effect of posterior biased selection of a grammar in the original simulation.

[8] Longa (2006) argues, rather incoherently, that I resort to Waddington's mechanism of genetic assimilation to motivate my account of the emergence of a LAD via the Baldwin effect. He claims that I conflate the two processes and that somehow my arguments and simulation model rest on the parity of the two processes. In fact, I only refer to the Baldwin effect at all because of its widespread use by others to mean something like genetic assimilation where phenotypic plasticity is supplied by a within-lifetime learning mechanism. I am not particularly concerned with the premodern synthesis speculations of Baldwin and others or with the various (re)interpretations of these speculations, and the coevolutionary model and account of the emergence and maintenance of the LAD in no way rests on them.

"relaxation" is a more plausible explanation for our linguistic abilities because "highly distributed synergistic organization emerges from this type of process" and because "epigenetic parsimony" entails that the genes should only encode what cannot be offloaded to "self-organizing developmental processes" in interaction with the environment.

If Deacon is right (and the analogy with the development of more complex song in the Bengalese finch is certainly compelling), then masking in the linguistic niche means that we evolved into a "degenerated ape" rather than a more finely adapted one. However, even under this scenario, "stabilizing selection" for the suite of epigenetic responses to linguistic stimuli is still required for maintenance of our language learning abilities. So under this scenario, genetic assimilation still plays a role, but a reduced one in which the emergent complexity of language and its acquisition is more a consequence of serendipitous synergies among various less-constrained epigenetic developmental processes, rather than of active selection for a genetically encoded LAD. One possible problem with this account is that it relies on synergies whose probability may not turn out to be much higher than those required by saltationist accounts of the emergence of the LAD (Pinker and Bloom 1990). Whether one places the emphasis more on masking or unmasking, it is hard to see why this would impact on the issue of language specificity given the arguments above. Deacon endorses the simulation and modeling work of Yamauchi (e.g., 2001) apparently undermining the plausibility of genetic assimilation. However, in my own modeling work (Briscoe 2005), replicating Yamauchi's decorrelation of genotypic and phenotypic space within a coevolutionary model of the evolution of the LAD and of languages themselves, I showed that these results rest more on the simplifying assumptions of his model than on any substantive extension of Mayley's (1996) original work on decorrelation and genetic assimilation. I believe the distinction between an unmasking and a masking account reduces to one of causation in an evolutionary (pre)history to which we have only very indirect access. Either way, there is a critical role for genetic assimilation and, on balance, I believe current evolutionary theory suggests unmasking (i.e., genetic assimilation) would play the larger causative role in the development of novel traits.

Models and Simulations

The value of formal modeling and computational simulation of linguistic evolution and of associated cognitive neural evolution is that it can lend greater precision to argumentation concerning interactions between at least two complex and only partially understood domains. However, if a model supports a particular argument, this does not mean the argument is correct. Rather the required precision and detail needed to make a particular prediction exposes the assumptions, some perhaps implicit, whose plausibility can then be more directly evaluated.

For instance, Deacon (1997) argues, along with others, that genetic assimilation could not have been a significant factor in the development of the LAD because the speed of linguistic evolution so outpaces biological evolution that genes tracking grammatical regularities would not have time to go to fixation in the population before these changed and the associated selection pressure they entailed disappeared. This sounds plausible, but when tested by modeling and simulation it turns out to require an unstated assumption that the full range of grammatical possibilities available in the hypothesis space of grammars be manifest during the period of adaptation. No matter how much apparent linguistic change is manifested, if this only covers a proper subset of the hypothesis space of grammars, then there will be selection pressure for genes which constrain the hypothesis space to just this proper subset under the assumptions that this makes learning more robust and efficient and that mastery of language confers a fitness benefit. Of course, this does not prove that genetic assimilation of this kind occurred, but it does suggest Deacon's argument is flawed in this form.[9]

In Briscoe (2003, 2005) as well as in earlier work referenced there, I review, evaluate, and model a number of arguments and models both for and against genetic assimilation of grammatical information and argue that this remains a coherent and evolutionarily plausible account of the emergence and maintenance of the LAD. One theme that is often implicit but always present in this work is that designing a useful model and deriving results from it is a nontrivial business which, although apparently largely a mathematical and computational exercise, is in fact replete with complex judgments about the appropriate level of abstractness to adopt and what simplifying assumptions it is legitimate to make.

For instance, Christiansen and Reali (2006), summarized in Kirby et al. (this volume), revisit the relative speed of change argument, albeit without considering either Deacon's arguments or my own work, and present a series of simulations which they argue demonstrate that only functionally motivated features of language can become genetically encoded because of the rapidity of linguistic change compared to biological evolution. They take this as a refutation of Pinker and Bloom's (1990) claim that arbitrary features of language might become encoded in Universal Grammar (the LAD) to make language learning more robust. The model of learning is based on that of Hinton and Nowlan (1987), and language change is simulated by introducing a new language at each time step of the model. They do not measure the communicative

[9] I'd like to make clear at this point that I focus on Terry Deacon's work not because I think it is generally flawed but, on the contrary, because I find it very stimulating and often very convincing, and this provokes me to evaluate it carefully, even to the extent of building and modifying quite complex computational models.

success of the evolving learning agents after each time step and they do not investigate the proportion of the original hypothesis space explored during an average simulation run. The description of the simulation is not detailed enough to infer either; however, it seems likely that the former will be low, contrary to attested language change, and the latter high when change is not closely correlated with the genetic makeup of the population at the previous time step. It is, therefore, neither surprising nor particularly interesting that genetic assimilation does not occur under these conditions. Christiansen and Reali performed a second simulation run in which agents are selected on the basis of their communicative success and observed genetic assimilation, but they took this to mean that only functionally motivated traits can be assimilated. They appear to miss the point that arbitrary features of grammar, if assimilated, become functional in this sense if they make learning more efficient and thus increase communicative success—Pinker and Bloom's original point.

Similarly, Reali and Christiansen (2009) argue that there is evidence of considerable overlap in the cognitive mechanisms used in sequential learning and language learning. They evolve the initial weights of a population of simple RNNs to perform optimally on a sequential learning task. They then used these evolved simple RNNs to "learn" (i.e., predict string sequences of) languages in an iterated learning model, also allowing the simple RNN weights to evolve further, subject to the proviso that they maintained the same performance on the original sequential learning task. The result was that languages emerged with consistent head ordering, but the networks themselves did not evolve further. Reali and Christiansen interpret this to mean that a sequential learning mechanism exapted for language learning predicts that languages will evolve in typologically plausible ways without any specific linguistic biases being genetically assimilated. I think this is a potentially interesting claim and line of research which is unfortunately undermined by the use of simple RNNs which are incapable, in principle, of grammar learning and by the fact that the specific class of simple RNNs deployed may be unable to even reliably predict the sequences of many languages in the space explored given any possible weight settings. Rather than using the iterated learning paradigm with a population of learners, it would have been more informative to demonstrate learnability (predictability) of plausible and implausible word-order sequences by networks with various weight settings and then demonstrate that a network optimized for nonlinguistic sequential learning incorporates a bias against certain word-order sequences.

Conclusions

Modeling and simulation are potentially very valuable in such a complex domain of enquiry where the constraints on theory are weak given the available evidence. However, such modeling has a largely negative impact, mostly exposing the flaws and implicit implausible assumptions in arguments. Even to achieve this much, models must meet certain criteria before they become relevant. They must model the acquisition task realistically, track communicative success in many contexts, and make realistic assumptions about rates and types of language change.

To date, I believe that the evolutionarily most plausible account of the emergence and maintenance of the LAD is that a representation language evolved out of the compositional "language of thought" capable of mapping meaning to sequential or spatial realizations which disambiguate argument relations to predicates. Most likely the simplest mappings, requiring the least additional apparatus, embody substantive constraints on such mappings and thus are low or intermediate on the Chomsky hierarchy of language classes and associated automata. In this sense, the LAD already incorporated grammar-/language-specific bias. However, the linguistic niche created new selection pressures for robust and efficient language acquisition, and genetic assimilation provided the mechanism by which adaptations encoding ever more informative prior biases could evolve. These would most likely be weak biases rather than hard constraints, in the face of continuing linguistic evolution and then subsequent change within the space of modern human languages, and would asymptote at the point where, given such variation in the linguistic environment of adaptation, no further gains were possible or all relevant genetic variation had gone to fixation.

Acknowledgment

Thanks to Simon Kirby for modifying and allowing me to (re)use Figure 15.2 from Kirby et al. (this volume).

Overleaf (left to right, top to bottom):
Andrea Baronchelli, Herbert Jaeger, Tom Griffiths
Gerhard Jäger, Luc Steels, Group report session
Simon Kirby, Pete Richerson, Jochen Triesch
Natalia Komarova, Ted Briscoe, Morten Christiansen

Simon Kirby

18

What Can Mathematical, Computational, and Robotic Models Tell Us about the Origins of Syntax?

Herbert Jaeger, Luc Steels, Andrea Baronchelli, Ted Briscoe, Morten H. Christiansen, Thomas Griffiths, Gerhard Jäger, Simon Kirby, Natalia L. Komarova, Peter J. Richerson, and Jochen Triesch

Abstract

This working group staked out the landscape of mathematical and computational modeling approaches to language evolution. The authors attempt both to provide a survey of existing research in this field and to suggest promising novel directions for investigations. The focus was set on cultural evolution; biological evolution is covered in other chapters of this volume. The field is young, manifold, and highly productive, with contributions not only from theoretical linguists but also from theoretical physics, robotics, computational neuroscience, and machine learning. This chapter provides a systematic presentation and comparison of different modeling paradigms, which range from a rigorous mathematical analysis of macroscopic language transmission dynamics to computational (even robotic) studies of complex systems populated by learning agents. Original contributions of modeling research to the understanding of cultural language evolution are highlighted, shedding light on the magnification of learning bias through cultural transmission, restricting the space of possible grammars, coevolution of categories and names, and the emergence of linguistic ontologies.

Introduction

Mathematical and computational models play a crucial role in all sciences and can clearly be helpful to the study of language evolution as well. A model makes certain theoretical assumptions about evolutionary forces and linguistic representations and then shows what the consequences of these assumptions are on the outcome of language evolution. Of course, the model does not in itself prove that the assumptions are empirically valid but shows whether the assumptions are coherent, have the effect they are believed to have, and are in principle sufficient to generate the phenomena they are claimed to generate. Thus, models provide the opportunity to test different hypotheses about the ingredients that are necessary for languages with certain properties to emerge through processes of transmission and interaction between agents. In turn, they make predictions about the relationship between language acquisition, communicative interaction, and language change that can be assessed through experiments with human participants or robotic agents and through comparison with historical data.

In the field of language evolution, a wide range of models has already been explored, but this is only the beginning. The complexity of the models, the questions they address, and the techniques used to check the validity of current models vary widely. Consequently, the discussions in our group did not and could not be expected to lead to a unified and complete picture, partly because researchers have been looking at entirely different aspects of the enormously complex problem of language evolution and have been using very different methods. Instead we tried to sample the landscape of existing modeling efforts and the representations of grammar and grammatical processing that are used in them. We then surveyed arguments regarding why and how modeling can contribute to the overall language evolution research enterprise, and outlined future research, including possible collaboration with biologists and linguists.

To avoid a possible misunderstanding, we point out that the discussions in our group, and consequently the material in this chapter, focused primarily on the *cultural* evolution of language, to be distinguished from the *biological* evolution, which is in the focus of other contributions in this volume. Nonetheless, investigations of cultural language evolution have implications for research on biological evolution, because if it is found that certain traits of language can be naturally explained by the former, biological mechanisms are relieved from an explanatory load. Conversely, biologically evolved, generic, nonlinguistic information-processing capabilities (e.g., sequential processing mechanisms) yield the scaffolding for cultural evolution.

Another preparatory remark is that modeling efforts adopt approaches that are quite standard in other domains of complex systems science, but may be relatively new to linguists. For example, there is often an effort to seek simplified models to clearly pin down the assumptions and, in many cases, to make the models tractable from a mathematical point of view. Modelers typically focus on replicating statistical distributions of language phenomena rather than matching directly the particulars of a given human language. They will first consider communication systems that have only a rudimentary resemblance to language before increasing the complexity further, step by step. Alternatively, they will make assumptions about certain aspects of language interaction (such as joint attention or perception) to make simulations viable at all. Some models are not about language per se but address the preconditions for language, such as cooperation. It is therefore important to keep in mind that the modeling work discussed here is primarily concerned with investigating the consequences of hypotheses rather than trying to model in detail and in a realistic way the origins and evolution of human language.

Paradigms for Studying Language Evolution

Discussions in our group arose from the multifaceted experience of the participants with computer-based simulations of language dynamics, robotic experiments, and mathematical analysis. We are not aware of any generally accepted way of characterizing or classifying computational modeling approaches in the natural or social sciences. In the present context, we could nevertheless identify a number of different modeling paradigms that have grown up historically based on the shared interests of the researchers involved in exploring them. Each paradigm frames the process of language evolution in a particular way, focuses on some of the forces that might play a role, and then examines specific fundamental questions through concrete models and experiments. Within each paradigm we have seen the development of mathematical models, computational or robotic experiments, and psychological experiments with human subjects. Of course, the distinctions between paradigms made here are to some extent arbitrary and not always clear-cut. There are continuous dimensions linking these paradigms and hence considerable opportunities for cross fertilization. Moreover, we anticipate that additional modeling paradigms may spring up in the future to explore other aspects of the vast research domain of language evolution.

A first distinction that can be made is between agent-based models, which try to pin down the cognitive and social processes that could give rise to forms of language, and macroscopic models, which aggregate the behavior of a population and then formulate equations defining the evolution over time among these aggregate quantities. Another dimension for categorizing the models concerns the importance given to cultural transmission, cognition, or biology, which has given rise to iterated learning models, language games, and genetic evolution models. In Figure 18.1 we illustrate schematically two main dimensions on which the paradigms differ.

Agent-based Models

Agent-based models center on models of individual language users as members of populations. The agents are given certain cognitive capabilities (e.g., a particular learning strategy) and made to interact (e.g., in the simulation of a teacher–learner situation or a communicative interaction between two

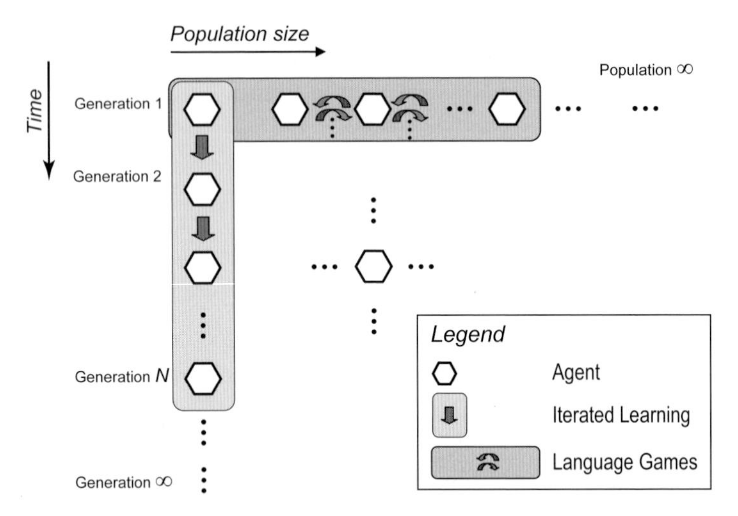

Figure 18.1 Schematic "coordinate system" comparing agent-based language evolution paradigms. The simplest models within the iterated learning paradigm focus on transmission across *generations* of agents in a singleton chain of teacher–learner dyades; language games focus on how language constructs emerge and evolve in interactions between agents. Numerous other paradigms can be seen as mixtures and ramifications of these two.

individuals). By simulating the effect of a large number of interactions, agent-based models can study under what conditions language systems with similar properties as human natural languages can appear. Agent models vary greatly in complexity, ranging from simple statistical "bag of words" language models to robots using complex grammatical and semantical representation formalisms to communicate with each other in a dynamical environment.

Three types of agent-based models have been developed: iterated learning models which focus on understanding the role of cultural transmission, language game models which emphasize the role of communication and cognition, and genetic models which explore the role of biological evolution.

Iterated Learning

The first paradigm, which has already been explored quite extensively, is known as the *iterated learning paradigm*. It focuses on understanding the relationship between properties of the individual and the resulting structure of language by embedding a model of an individual learner in a so-called "transmission chain" (also known as "diffusion chain"; for further details see Kirby et al. and Briscoe, both this volume; for a review of this approach to studying cultural evolution more generally, see Mesoudi 2007). In these models, the linguistic behavior of one individual becomes the learning experience of another individual, who in turn goes on to produce behavior that will be input for a third individual and so forth. The focus of this framework is on the contribution of learning in shaping the process of cultural transmission, with the goal of specifying precisely the relationship between constraints and biases provided by biology and the universal properties of linguistic structure. The idea is that a fundamental challenge for language is to be repeatedly transmitted between individuals over generations, and the transmission process is imperfect in important ways (e.g., learners have particular biases, they only see a subset of the language, there is noise in the world). The result is an adaptive system whereby language evolves culturally in such a way so as to give the appearance of being designed for transmission fidelity.

The main simplification in many (but not all) of the models of this "iterated learning" process is that the transmission chain consists of a single individual at each generation and involves only vertical transmission (i.e., transmission between generations). This simplification allows researchers to focus on the sole contribution of the learning bias plus the nature of the selection of training data (e.g., number of examples), although it leaves out many of the factors associated with horizontal transmission (e.g., selection of models to learn from, having shared communicative goals, and population structure). One avenue for future research is to explore the implications of other, more realistic models of populations, while maintaining the emphasis on the role of transmission in shaping language structure. For a recent review of general cultural evolution models, see McElreath and Henrich (2008).

Examples of iterated learning models are presented by Kirby et al. (this volume). An emphasis in many of these models has been the explanation of the emergence of compositional structure in language. Compositionality, along with recursion, is the fundamental feature of human syntax that gives us open-ended expressivity. It is also arguably absent in any other species, despite the prevalence of communication in nature. Accordingly, it is an important target for explanation. Using mathematical, computational, and experimental models, researchers have examined the conditions under which compositionality and the relationship between compositionality and frequency may emerge. Specifically, these models suggest that compositionality arises when there is a "bottleneck" on the cultural transmission of language; in other words, where learning data is sparse.

Language Games

The second class of models investigates the role of embodiment, communication, cognition, and social interaction in the formation of language. Instead of modeling only teacher–learner situations, as in iterated learning approach, it models the communicative interactions themselves in the form of language games. A language game is a situated embodied interaction between two individuals within a shared world that involves some form of symbolic communication. For example, the speaker asks for "a cup of coffee" and the hearer gives it to her. When speaker and hearer have shared conventions for solving a particular communicative problem, they use their existing inventory in a routine way. When this is not the case, the speaker requires the necessary cognitive capabilities to extend his inventory (e.g., expanding the meaning of a word or coercing an existing word into a new grammatical role), and the hearer requires the ability to infer meanings and functions of unknown items, thereby expanding his knowledge of the speaker's inventory.

In typical language game models, the individuals playing language games are always considered to be members of a population. They interact only in pairs without any centralized control or direct meaning transfer. There is unavoidable variation in the population because of different histories of interaction with the world and others; however, proper selectionist dynamics, implemented by choosing the right alignment and credit assignment strategies for each individual, causes certain variants to be preferred over others. Language game models often operate with a fixed population because they examine the thesis that language emerges and evolves by the invention, adoption, and alignment strategies of individuals in embodied communicative interactions. In addition, many experiments have been done in which a flow is organized in the population with members leaving or entering the population, thus demonstrating that the model handles cultural evolution as well.

By now there have been dozens of experiments in language games that explore how different aspects of language may arise (see Steels, this volume). The simplest and earliest game studied is the *naming game*, in which agents draw attention to individual objects in the world by using (proper) names (Steels 1995). Guessing games have been used to study the coevolution of perceptually grounded categories and words (Steels and Belpaeme 2005), flexible word meanings (Wellens et al. 2008), and the emergence of spatial language (Steels and Loetzsch 2008). Description games have been used in experiments on the emergence of grammar, in particular, case grammar (Van Trijp 2008).

Language games have been explored further from three angles: through mathematical analysis, particularly using the methods of statistical physics; through computational simulations and robotic experiments; and through experiments with human subjects as carried out, for example, by Galantucci (2005) and Pickering and Garrod (2004). Robotic experiments are particularly useful if one wants to study the question of how embodiment plays a role in language evolution. Data on actual language change, which comes from historical linguistics and socio-linguistics, are currently being used to constrain the repair and consolidation strategies of agents in grammatical language games; data from cognitive linguistics and cognitive semantics, in particular, are used to constrain the range of possible conceptualizations that could be the target of experiments. The theoretical tools developed in statistical physics and complex systems science have recently acquired a central role for the study of language games. The suite of methods developed in these fields has indeed allowed us to address quantitatively such issues as the scaling of relevant features of the models with the system size (e.g., convergence time or memory requirements; Baronchelli et al. 2006b, 2008), the impact of different underlying topology on global behaviors (e.g., homogeneous mixing [Baronchelli et al. 2008] vs. regular lattices [Baronchelli et al. 2006a] vs. complex networks [Dall'Asta et al. 2006a, b]), and the detailed study of convergence dynamics (Baronchelli et al. 2008). Thus, for example, it has been shown that complex networks are able to yield, at the same time, the fast convergence observed in unstructured populations and the finite memory requirements of low dimensional lattices (Dall'Asta et al. 2006a, b). Moreover, agents' architectures and interaction rules have been significantly simplified to allow thorough analysis, and this has enabled us to pinpoint the crucial ingredients responsible for the desired global coordination. The pursuit of simplicity, along with the novelty of the complex systems approach to this field, has thus far mostly limited the investigations to the study of the naming game and category game (in which the population ends up with a shared repertoire of categories) (Puglisi et al. 2008). Current research is trying to tackle higher-order problems, such as the emergence of compositionality (de Beule 2008). Experiments with human subjects show that humans can evolve communication systems, although some are better than others, mostly because of differences in social attitudes. Of course the greatest challenge is to scale these experiments up to the level of grammatical languages. Recent examples already showing the formation of case grammars, tense–aspect–mood systems, or determiner systems give reason for optimism (see, e.g., Van Trijp 2008).

Genetic Evolution

A third class of models explores the role of biology by modeling the genetic transmission of language. Agents are created based on a model of a genome that directly codes the lexicon or grammar of their language. Agents then engage in interactions that determine their fitness, and based on communicative success they have a higher chance of reproducing in the next generation. Due to random mutations and crossover, offspring have slightly different genomes, possibly giving higher communicative fitness which then leads to further propagation. These models use very similar techniques to those used in genetic algorithms, and they sprang up first in the context of artificial life (Cangelosi and Parisi 1998). Given that the explicit genetic coding of lexicon and grammar is highly implausible from a biological point of view, more recent models have considerably weakened this assumption and encode only strong biases and universal constraints on possible languages. This is particularly the case for the ENGA model (Szathmáry et al. 2007). ENGA is an ambitious framework that covers not only the genetics but also the neurodevelopmental processes in a biologically realistic way. Linguistic inventories are not coded genetically; they are acquired through a learning process. The ENGA model, therefore, attempts to cover the entire spectrum: from genetic to developmental and learning processes.

Aggregate Models

In addition to agent-based models, extensive research has constructed macroscopic models of language evolution and language dynamics.

Game Theoretic Models of Language Evolution

The main paradigm being explored here draws from the tradition of evolutionary game theory to focus on the role of imitation in cultural transmission. Imitation (or reuse) applies both to the adaptation of linguistic performance between adult speakers and the acquisition of language by infants. Imitation is framed as a form of replication. An evolutionary dynamics ensues in any population of replicating entities, provided the entities in the population vary in certain heritable characteristics, and replicative success is correlated with this variation. This is a crucial difference to the iterated learning paradigm, where every individual grammar participates equally in language replication. However, the game theoretic model—as a form of a selectionist model—assumes faithful replication, while replication under iterated learning may be imperfect. Under certain simplifying assumptions—like the postulation of an infinite population and a continuous time—such evolutionary dynamics can be described by a system of ordinary differential equations. In language evolution this dynamics is necessarily nonlinear because selection is frequency dependent. This can, for instance, be illustrated by the development of vocabulary: whether a candidate for a neologism catches on in a linguistic community (i.e., becomes replicated) depends on whether or not there already is another word for the same concept within this linguistic community. This indicates that the overall frequency distribution of words is a decisive factor for the fitness of each individual word. A similar point can be made for other linguistic units, ranging from phonemes to syntactic constructions.

Frequency-dependent selection can be modeled by means of replicator dynamics within the mathematical framework of evolutionary game theory (Maynard Smith 1982; Hofbauer and Sigmund 1998). In its simplest form, a model of a communication game consists of:

- a space of meanings and a space of forms,
- a space of production grammars (mappings from meanings to forms),
- a space of comprehension grammars (mappings from forms to meanings), and
- a utility function (i.e., a measure of success for a pairing of grammars, depending on the success of communication and complexity of the grammars involved).

Further parameters may be added, like a biased a priori probability distribution over meanings, or a confusion matrix for noisy transmissions of forms.

There are several off-the-shelf theorems from biomathematics regarding stability conditions for evolutionary games. Such theorems sometimes render it straightforward to identify the attractor states of the replicator dynamics without actually delving into the complexities of the underlying nonlinear differential equations.

The biomathematics literature contains a variety of results concerning the evolution of communication, where strategies ("grammars") are assumed to be innate and replication is interpreted in the biological sense (e.g., Wärneryd 1993; Trapa and Nowak 2000; Nowak and Krakauer 1999; Nowak et al. 1999; Jäger 2008a). These authors mainly consider biological evolution, and they assume that communicative success is correlated with biological fitness (i.e., the number of fertile offspring). However, their results are general enough that they can be extrapolated to cultural evolution. The background assumption here is that communicative success of a certain behavioral trait is positively correlated with its likelihood to be imitated (i.e., its cultural fitness). Possible applications of evolutionary game theory to the study of the cultural evolution of language (in the sense described above) have been investigated (Jäger 2007, 2008b; Jäger and van Rooij 2007).

Game theoretic research in language evolution has suggested a formal framework which is quite useful within this paradigm. Universal Grammar or a preexisting bias of grammar learning can be represented in the following abstract manner. Suppose we have a finite alphabet (i.e., a finite set of symbols) (Nowak et al. 2001; Komarova et al. 2001; Komarova and Nowak 2001, 2003; Nowak and Komarova 2001). A language is a probability distribution defined on a set of strings composed of the symbols of the alphabet. The allowed languages can be represented as probability distributions on a collection of (intersecting) sets. Then a learning mechanism is a way to "navigate" in this collection of sets. Pair-wise similarity among languages can be expressed as a matrix. The process of learning is thus a sequence of hypotheses of a learner in response to the input of a teacher (or teachers), which is a number of strings compatible with the teacher(s)' grammar. This framework allows one to use the machinery from mathematical learning theory and connect natural language evolution with insights from computer science/machine learning.

Summary

There are obvious relations, complementarities, and continuities between these approaches and paradigms. The game theoretic paradigm focuses on the selectionist dynamics of the language itself, whereas language game models use an agent-based approach, focusing on the cognitive mechanisms by which agents use, invent, and coordinate language so that the selectionist dynamics of language emerges. The iterated learning paradigm focuses on the role of bias and the vertical transmission bottleneck and tends, therefore, not to integrate the issue of communicative success, cognitive effort, or population dynamics into the models. The language game paradigm considers vertical transmission as an additional but not crucial effect on language evolution. Pursuing these different approaches provides us with the opportunity to explore how different factors (e.g., learning, communication, and population structure) influence the process of language evolution.

Linguistic Representations and Processes

Given that syntax was the focus of this Forum, it is relevant to examine what kind of representations for grammar are being used in language evolution models and what kind of syntactic operations and grammatical processes have been incorporated into these models. Researchers working on iterated learning and game theoretic approaches generally try to use existing "symbolic" formalisms or neural network models. Some have argued, however, that the requirements of evolvability put additional constraints on the nature of grammatical representations and processing, and this has led to some work on novel grammar formalisms which can cope with emergent grammar.

Symbolic Grammars

There are a variety of grammatical formalisms in the theoretical linguistics literature, some of which have been utilized in evolutionary models whereas others, such as Minimalism (Chomsky 1995), have not (possibly because they are less easily embedded in theories of processing). Examples of formalisms that have been deployed with minimal modification include optimality theory (Jäger 2004), extended categorial grammar (Briscoe 2000), and context-free grammars (Zuidema 2002). All such models require the embedding of the formalism into a theory of grammar learning and processing. Modelers have drawn on existing proposals from the literature, such as Bayesian parameter estimation, compression-based algorithms, or nonstatistical parameter-setting algorithms for implementing the learning mechanisms used in vertical transmission (Griffiths and Kalish 2007; Briscoe 2000).

Simple Recurrent Networks

Other language evolution models have avoided the explicit representation of hierarchical structures, syntactic and semantic categories, and grammatical rules, deploying distributed and subsymbolic representation. A popular alternative is simple recurrent networks (SRNs; Elman 1990). In SRNs, knowledge of language is learned from the presentation of multiple examples from which the networks learn to process syntactic structure. The general aim of such models is to capture observable language performance, rather than idealized linguistic competence (Christiansen 1992; Christiansen and Chater 1999). Much of this work has an emphasis on the integration of multiple sources of probabilistic information available in the input to the learner/speaker/hearer (e.g., from the perceptuo-motor system, cognition, socio-pragmatics, and thought as discussed by Kirby et al., this volume). Although much of this work tends to target small fragments of language for the purpose of close modeling of psycholinguistic results (e.g., Christiansen and Chater 1999; MacDonald and Christiansen 2002), some efforts have gone into scaling up models to deal with more realistic language samples, such as full-blown child-directed speech (Reali et al. 2003). In this framework, grammatical processing can be conceptualized as a trajectory through a high-dimensional state-space afforded by the hidden unit activations of the network (e.g., Elman 1990), potentially suggesting an alternative perspective on constituency and recursion in language (Christiansen and Chater 2003). These models do not include explicit grammar formalisms, but the behavior of the networks can in some cases be described in terms of such formalisms.

Formalisms Designed for Grammar Evolution

Some researchers have developed novel formalisms to be used specifically in language game experiments. This is particularly the case for *fluid construction grammar* (FCG). FCG (de Beule and Steels 2005) uses representational mechanisms already employed in several existing symbolic grammar formalisms like *head-driven phrase structure grammar* (HPSG; Sag et al. 2006) or lexical–functional grammar (Kaplan and Bresnan 1982) such as a feature/structure-based representation of intermediary structures during parsing and production, a constraint-based representation of linguistic rules so that they can be applied in a bidirectional fashion, and unification-style mechanisms for the application of these rules. FCG is in line with other construction grammar formalisms, such as embodied construction grammar (Bergen and Chang 2004), in the sense of supporting the explicit representation and processing of constructions, which is de-emphasized in Minimalism. However, FCG has various additional facilities to enable language evolution experiments:

1. Individual agents represent a multitude of hypotheses about the emerging language and are therefore able to handle variation in language use.
2. Rule application is flexible allowing the violation of constraints and robust parsing and production so that sentences can be understood even if they are not entirely grammatical (according to the preferred grammar of the agent).
3. The different variants compete within the individual when it has to make decisions about how to express something or interpret something

and, as an emergent effect, within the population for dominance in the emergent language.

4. Rather than coding systematicity in terms of more abstract rules, FCG maintains links between the rules, based on how the rules are formed through composition of other rules.

These links are then used for assigning credit or blame after a game, allowing the implementation of a multilevel selectionist dynamics. Because of these features, FCG exhibits dynamical systems properties seen in network representation systems of grammar which do not rely on symbolic structures, such as connectionist networks or recurrent neural networks (RNNs), while at the same time incorporating many ideas from decades of research into theoretical and computational linguistics.

Summary

The computer simulations carried out in the evolution of language research rest on a variety of formalisms to represent the inventory of the lexicon and grammar of the emerging language in the first place. In choosing a particular formalism, the researcher makes a commitment to what aspects of language are isolated for an inspection of their role in evolutionary dynamics, and what others are (implicitly) excluded.

How Can Modeling Inform the Study of Language Evolution?

Although mathematical and computational modeling of language evolution is still in its infancy, there are already quite a few results that show the power of the approach, and which may be of interest to biologists and linguistics. Computational modeling, like in any other field, enables two powerful avenues for accruing scientific insight:

1. *Formal analysis*. Computational models have to be rigorously formalized to make them operational on computers. When a simulation is running, all aspects of the simulation can be recorded, including the population aspects. The same is true in robotic experiments where all perceptual states, motor states, and the full details of all processes going into language production and understanding can be tracked; this is not possible with human subjects. This full access to relevant data makes the models amenable to mathematical analyses. Typical questions that can be answered by the analytical methods provided by nonlinear dynamics, game theory, and statistical physics concern asymptotic properties of evolutionary dynamics, the dependence of these dynamics on scaling parameters, or the prediction of sudden and dramatic changes (phase transitions).

2. *Simulation studies.* Carrying out simulations on a computer differs from carrying out real-life experiments in two crucial ways: (a) the simulated piece of reality is *completely* specified, and (b) one has *full control* over varying experimental conditions. There are risks and opportunities under these circumstances. An obvious pitfall is that the simulation may miss a crucial component of the real-life target system—this is the problem of abstraction. However, it should be noted that, in principle, the same problem is present in experimental designs involving human subjects: a particular experimental design may prevent real-life-relevant mechanisms from taking effect. The benefits added to empirical experiments (which remain indispensible) by simulation studies are, in our view, the following:

   - A systematic *exploration of large hypothesis spaces* is made possible due to the speed and low cost of simulations. This facilitates both the generation of new scientific hypotheses and the testing of existing ones.
   - Model simulations can give existence proofs for the efficiency of certain mechanisms to achieve a certain effect—always, of course, modulo the modeling assumptions.
   - In a related vein, model simulations can give nonuniqueness proofs if the same ultimate effect can be obtained by different mechanisms. Such demonstrations are helpful in precluding an early "contraction" to a single explanatory venue in theory development.
   - Simulations are replicable across different laboratories by sharing code.
   - Critiquing and improving simulation setups is transparent, because it is explicit how assumptions become operationalized in the designs.

If one is carefully conscious of the assumptions that go into a simulation model, research based on such models can decidedly "open up" the space of possible theories in a field, raising the awareness of alternative theories. To demonstrate this point, we present a number of examples that have arisen from ongoing work.

Magnification of Learning Bias through Cultural Transmission

Mathematical analyses of the iterated learning model provides some interesting insights into the relationship between the inductive biases of language learners—the factors that lead them to find it easier to learn one language than another, as might be the result of genetic constraints on language learning—and the kinds of languages that will be spoken in a community. As discussed by Kirby et al. (this volume) and Briscoe (this volume), one way to capture the inductive biases of learners is to assume that they identify a language from a set of utterances by applying Bayesian inference, with a "prior" distribution encoding which languages learners consider more probable before seeing any data. Languages with higher prior probability can be learned from less evidence, and the prior thus reflects the inductive biases of the learner. Analyses of iterated learning with Bayesian agents show that the relationship between the prior and the languages that are ultimately produced via cultural transmission can be complex (Griffiths and Kalish 2007; Kirby et al. 2007). Specifically, iterated learning can magnify weak inductive biases, with a slight difference in the prior probabilities of two languages resulting in a significant difference in the probability of those languages being produced via cultural transmission. These mathematical results suggest that strong genetically encoded constraints on learning may not be necessary to explain the structure of human languages, with cultural evolution taking on part of the role that might otherwise have been played by biological evolution.

Restricting the Space of Possible Grammars

It is tempting to reconstruct the notion of a linguistic universal as a property that every language with a grammar that can be cognitively represented and learned by humans (i.e., a language that conforms to "Universal Grammar" in the Chomskyan sense) shares. Evolutionary models indicate that there may be other sources of universals. Briefly put, a *possible language* must also be attainable under the evolutionary dynamics of language transmission.

In Jäger (2004), this basic idea is illustrated with a particular implementation. According to optimality theory, Universal Grammar defines a finite set of constraints, and each particular grammar is characterized by a linear ordering of these constraints. To account for certain strong typological tendencies, Aissen (2003) proposed restricting the space of possible grammars further by imposing certain subhierarchies of constraints that are never violated.

Following proposals by Boersma (1998), Jäger implements a stochastic learning algorithm for optimality theoretic grammars. However, unlike Boersma, Jäger assumes that language acquisition is bidirectional (i.e., the learner tries both to mimic the production behavior and the comprehension behavior of the teacher). It turned out that some constraint rankings are strictly not learnable at all. Among the remaining space of learnable grammars, some are more robustly learnable than others. After iterating the learning procedure a few dozen or hundred of times (where in each generation, the former learner becomes the teacher and produces utterances on the basis of his acquired grammar), only constraint rankings that conform to Aissen's prediction were observed.

The Coevolution of Categories and Names

One of the big debates in language studies concerns the question of how far perceptually grounded categories (e.g., such as colors) influence and are influenced by language that expresses these categories. From a Whorfian point of view there is a strong interaction, whereas those arguing for strong modularity argue that categories are innate or induced from empirical data and language just labels existing categories. Although color categorization and color naming does not relate directly to grammar, we include this theme here because it exemplifies the quality of insight that can be obtained from modeling studies, and because categorization and naming are prerequisites for grammatical language. Research on the use of language games for studying the coevolution of categories and names began with Steels and Belpaeme (2004); here, agent-based models of color naming and categorization were developed and systematically compared. This work showed that although a genetic evolution of color categories was possible, it not only took a long time, but also did not lead to a system that was adaptive, and it certainly did not lead to universal categories unless populations remained homogeneous. It also demonstrated that a purely learning-based approach did not lead to an explanation for trends in color categories nor to sufficient coherence in a population to explain how a successful communication was possible. More recently, this research has been extended in two directions.

The category game (Puglisi et al. 2008) is a language game that aims at describing how a population of agents can bootstrap a shared repertoire of linguistic categories out of pairwise interactions and without any central coordination. The main result is the emergence of a shared linguistic layer in which perceptual categories are grouped together into emerging linguistic categories to guarantee communicative success. Indeed, while perceptual categories are poorly aligned between individuals, the boundaries of the linguistic categories emerge as a self-organized property of the whole population and are therefore almost perfectly harmonized at a global level. Interestingly, the model reproduces a typical feature of natural languages: despite a very high resolution power and large population sizes, the number of linguistic categories is finite and small. Moreover, a population of individuals reacts to a given environment by refining the linguistic partitioning of the most stimulated regions, while nonuniform JNDs (e.g., the human JND function relative to hue perception) constrain to some extent the structure of the emergent ontology of linguistic categories.

Another simple framework has been designed to investigate the influence of various realistic features (linguistic, psychological, and physiological) on the shared color categorization (Komarova and Jameson 2008). As a result of a number of iterations of the appropriate game, a population of agents arrives at a shared categorization system, which possesses the following qualities: (a) the exemplar space is equipartitioned into a (predictable) number of distinct, deterministic color categories, (b) the size of color categories is uniquely defined by the pragmatic similarity parameter, and (c) the location of category boundaries possesses rotational symmetry. Empirical data of confusion spectra of abnormal color observers can be incorporated to generate specific color boundary predictions and to deduce how the color categorization of various populations is influenced by the population inhomogeneities (Komarova and Jameson 2008).

The Emergence of Linguistic Ontologies

The final example, which shows how modeling can lead to the opening up of new theoretical avenues and ideas in language evolution, comes from the domain of grammar. Grammar exploits syntactic devices (e.g., word order or morphology) to express additional aspects of meaning, such as discourse structure, thematic relations (predicate–argument structure), tense–aspect–mood, determination, scoping constraints on anaphora. In all current linguistic theories, the rules of grammar are expressed using an ontology of syntactic and semantic categories. These syntactic categories include, for example, parts of speech (e.g., noun, verb, adverb), types of constituents (e.g., noun phrase, relative clause), syntactic constraints (e.g., agreement, precedence), and syntactic features (e.g., nominative, masculine, neuter). The semantic categories include classifications of temporal aspects in terms of tense, aspect, or mood, semantic roles such as agent or beneficiary, categories used for conceptualizing discourse, like topic/comment, different shades of determination (e.g., definite/indefinite, count/mass), classifiers (as used in Bantu languages), deictic references both for use inside and outside discourse, epistemtic distinctions, and so on. A complex grammar undoubtedly requires hundreds of such categories. Thus, to understand the origins and evolution of language, we must know the origin of such a linguistic ontology.

There is a common (usually hidden) assumption among many theorists that linguistic ontology is universal and innate, but that does not explain how it originates. Typologists have argued that linguistic categories are to a large extent language dependent (Haspelmath 2007), and historical linguists have shown that categories change over time (Heine and Kuteva 2008). This suggests that linguistic categories may be similar to categories in other domains of cognition (e.g., the color categories), in the sense that they are culturally constructed and coordinated.

Recent language game experiments in the formation of a case grammar (Steels, this volume) have shown that the formation of linguistic ontologies is entirely possible. Concretely, semantic roles as needed in case grammar have been shown to arise when agents are trying to reuse by analogy semantic frames that have already been expressed in the emergent language. This reuse becomes licensed when particular predicate–argument relations are categorized in the same way as those already used in the existing semantic frames. Progressively, semantic roles get established and refined, partially driven by the semantic analogies that make sense in the real-world domain that generates the topics in the language game and partly by the conventions that are being enforced by the emergent language (Van Trijp 2008).

Summary

The examples discussed here illustrate some of the ways in which models of the cultural evolution of language can contribute to our understanding of its origins. By identifying what aspects of the properties of languages can be produced by cultural evolution alone, these models remove some of the explanatory burden from biological evolution, providing a more realistic target for research into the origins of language. In broad terms, these models illustrate how learning, communication, and population structure affect the languages that emerge from cultural evolution, providing potential explanations for two of the most important aspects of human languages: their consistent properties across communities—language universals—and the coherence of linguistic systems within communities. In iterated learning models, universals emerge as the result of learning biases or the goals of communication, and coherence is the result of the strength of these biases and the structure of the interactions with other individuals. In language game experiments, universal trends emerge due to constraints coming from embodiment, the cognitive mechanisms recruited for language, the challenge of communication, and the selectionist dynamics that emerges in populations of adaptive communicating agents. While there are still many questions to explore, these basic results help to illustrate the kinds of forces that influence the structure of human languages.

Suggestions for Future Research

Given that there is a broad variety of paradigms and modeling efforts, there are also many possible avenues for deepening current results or for exploring new avenues of research. Here we describe a number of suggestions without any claim to be exhaustive. In general, we can expect models to be developed that focus on quite different aspects of language evolution and that will be formulated at very different levels of abstraction. It will be important to establish the relationships between these models, such as identifying to what extent a simpler and more abstract model can be understood as an approximation to a more elaborate one.

Tighter Coupling between Models and Laboratory Experiments

It is important for future research to develop a tighter coupling between models and laboratory experiments. Currently, there are two ways in which conducting laboratory experiments in cultural evolution can complement the insights provided by mathematical and computational models. First, they provide a direct way of testing the predictions of these models, allowing us to ensure that the claims that we make about cultural evolution are actually borne out when these processes involve real people rather than abstract agents. For example,

Kalish et al. (2007) and Griffiths et al. (2008) have conducted direct tests of the key prediction that arises from models of iterated learning with Bayesian agents (i.e., structures that are easier to learn will be favored by the process of cultural transmission) by conducting laboratory experiments in which the structures transmitted by iterated learning were categories and functional relationships between variables for which previous research in cognitive psychology had established results on the difficulty of learning. Laboratory experiments, however, can also be valuable for a second reason: they provide a closer approximation to the true processes involved in language evolution. The models discussed earlier make assumptions both about how information is passed between agents and the learning mechanisms used by those agents. Conducting laboratory experiments in which information is passed between agents in the way described by a model, but where the agents are real human beings, removes one level of approximation from these models, allowing us to explore the plausibility of processes of cultural transmission as an account of why languages have the properties they do (Dowman et al. 2008). The experiment described by Kirby et al. (this volume) is of this kind, showing that iterated learning with human learners produces compositional structures. Further experiments that test models of language evolution and evaluate the impact of different forms of cultural transmission can help us develop models that provide a closer match to human behavior and assess the contributions of different kinds of evolutionary forces.

Tighter Coupling between Models and Data from Historical Linguistics

Much is known about the historical evolution of human languages over the past 5000 years. This research shows that there are recurrent patterns of grammaticalization and lexical change, and detailed case studies exist to explain, for example, how a language has developed determiners, a case system, or a tonal system (see, e.g., Heine and Kuteva 2008). It is therefore obvious that these results should constrain models of language evolution. Although it will never be possible to reconstruct the actual evolution of human languages, it might be possible to see similar grammaticalization phenomena as in human languages.

Modeling the Potential Role of Exaptation on Language Evolution

It is widely assumed that language in some form or another originated by piggybacking on a preexisting mechanism—exaptations—not dedicated to language. A possible avenue of language evolution modeling involves testing the possible effects for language evolution of particular hypothesized exaptations. For example, improved sequential learning of hierarchically organized structure in the human lineage has been proposed as a possible preadaptation for language (Christiansen and Chater 2008; Conway and Christiansen 2001), in part based on work in language acquisition (Gómez and Gerken 2000) and genetic data regarding the potential role of *FOXP2* in sequential learning (Hilliard and White, this volume). Reali and Christiansen (2009) have explored the implications of such assumptions by determining the effect of constraints derived from an earlier evolved mechanism for sequential learning on the interaction between biological and linguistic adaptation across generations of language learners. SRNs were initially allowed to evolve "biologically" to improve their sequential learning abilities, after which language was introduced into the population, comparing the relative contribution of biological and linguistic adaptation by allowing both networks and language to change over time. Reali and Christiansen's (2009) simulation results supported two main conclusions: First, over generations, a consistent head-ordering emerged due to linguistic adaptation. This is consistent with previous studies which suggest that some apparently arbitrary aspects of linguistic structure may arise from cognitive constraints on sequential learning. Second, when networks were selected to maintain a good level of performance on the sequential learning task, language learnability is significantly improved by linguistic adaptation but not by biological adaptation. Indeed, the pressure toward maintaining a high level of sequential learning performance prevented biological assimilation of linguistic-specific knowledge from occurring. Similarly, it may be possible to investigate the potential effects of other hypothesized exaptations on the relative contribution of cultural evolution and genetic assimilation to language evolution.

Along the same line, several language game experiments have examined how generic cognitive mechanisms could become recruited for language, pushed by the needs to solve specific problems in communication or in bootstrapping an efficient system (Steels 2007). For example, perspective and perspective reversal is often lexicalized in human languages to avoid ambiguity from which point of view a spatial relation should be interpreted.

Effects of Biased Unfaithful Copying

When empirical predictions are derived from dynamical models, the notion of an *equilibrium* is central. In the evolutionary context, we expect systems to spend most of their time in an *evolutionarily stable state*. The insights from historical linguistics, especially regarding *grammaticalization*, indicate that language never actually reaches such a stable state.[1] Rather, languages perpetually change in a partially predictable way. Complex morphology tends to be reduced over time and to disappear altogether eventually. An example is the loss of case distinctions from Latin (five cases) to French (no case distinctions). On the other hand, lexical morphemes are recruited to serve grammatical

[1] This statement might be too bold in its generality. Some aspects of language are certainly in equilibrium most of the time. A good example might be vowel systems.

functions. A recent example is the use of the item “going to” in contemporary English to express future. This recruitment usually concurs with phonological reduction, like the change from *going to* to *gonna*. Grammatical words tend to get further reduced to affixes; one example would be the regular German past tense morpheme *t* that is originally derived from the Germanic verb *tun* (in English: *do*).

The macroscopic consequence of these processes is that languages continually change their grammatical type, moving from synthetic to analytic, due to reduction of morphology, and back to synthetic, due to recruitment of lexical items for grammatical purposes and their subsequent reduction to affixes. The underlying microdynamics involves biased unfaithful copying (i.e., words and phrases are not imitated verbatim but phonetically reduced and semantically modified). The challenge for evolutionary models is to connect these two aspects in such a way that the directedness of language change is connected to empirical insight about unfaithful replication in language use. Deutscher (2005) has proposed a verbal model that resembles the socio-linguistic arguments of Labov (2001). Individuals often innovate new speech forms in an effort to find a more emphatic or colorful way of phrasing an idea or grammatical function. Conventional forms bore us while prose or speech stylists who play with the limits of convention attract attention. When prestigious people do this, the new speech form tends to spread. Sometimes the motivations for innovation are social; people seem to favor forms of speaking that differentiate them from social others. In other words, linguistic equilibria are weakly constrained in that communicating individuals must agree *sufficiently* on meanings for communication to be possible. Yet a speech community can easily cope with a modest rate of innovation driven by social and aesthetic forces. To our knowledge, these mechanisms have not been incorporated into formal models except in the special case of symbolic markers of group boundaries (McElreath et al. 2003).

Long-term Language Change Dynamics: A Mathematical Perspective

It appears that language modeling poses challenges for existing mathematical methods commonly used to describe emerging and dynamical real-life phenomenon. A ready example comes from language games. Language game solutions may vary with regards to their stability properties, depending on the type/purpose of the model use and on the exact question we address. In certain situations, interesting quasi-stable solutions are attained. One instantiation comes from modeling color categorization in people, where the shared population categorization solution cannot be described as a stable solution of a dynamical system, or a stationary probability distribution of a stochastic process. In the category game (Puglisi et al. 2008), even though the only absorbing state is the trivial one in which all the agents share the same unique word for all their perceptual categories, there are clear signatures of a saturation with time of metastable states with a finite and "small" number of linguistic categories. This observation suggests an analogy with glassy systems in physics (Mezard et al. 1987) and is also confirmed by quantitative observations. Thus, in this framework, interesting solutions would be long-lived (strongly, e.g., exponentially, dependent on the population size) pre-asymptotic states. In other models of color categorization, the shared population categorization solution appears dynamically stable on a certain timescale, but it may drift or cycle (while retaining global topological structure) on longer timescales, depending on the particular constraints. Mathematical properties of such solutions have not been investigated in detail but their understanding is important because conventional methods do not grasp the relevant properties of such solutions. The application of new mathematical technologies thus developed will be wide, as it has implications in the dynamics of populations of learners trying to achieve shared solutions on (possibly very complex) topological semantic spaces.

Selectively Neutral Mechanisms of Linguistic Evolution

A further direction for future research is to understand to what extent processes of selection are necessary to explain the properties of languages. In biology, selectively neutral processes such as mutation and genetic drift have been identified as playing a significant role in accounting for genetic variation (Kimura 1983). It remains to be seen whether linguistic variation is best analyzed as the result of selective pressures acting on the properties of languages, or the outcome of selectively neutral processes that are the cultural equivalents of mutation and drift. Answering this question requires developing a "neutral theory" for language evolution. In this case, the analog of mutation is the variation that is produced as a consequence of failed transmission of languages through the "learning bottleneck" produced by the fact that learners only observe a finite number of utterances. Iterated learning models thus provide a starting point for developing a neutral theory, and understanding which properties of languages can be produced by iterated learning and which properties cannot thus constitutes an interesting direction for future research.

Replicators

This discussion opens up the question of whether or not we should be thinking about cultural transmission/social interaction models in terms of competition among replications or, more excitingly, in terms of different levels or types of replication. It is tempting to propose (as outlined in Kirby 2006) that the emergence of syntax marks a change from one type of replicator (solitary replicators) to another (ensemble replicators), to use terms from Szathmáry (2000). However, this raises the issue of what exactly these replicators are, and whether their dynamics are best described in terms of selection at all.

It appears that the answers to these questions vary enormously depending on one's perspective on the best way to represent the knowledge being acquired/adapted by individuals and the mechanisms for acquiring that knowledge. For example, one view of language might propose that we internalize a set of constructions (e.g., Croft 2000) that have a fairly straightforward relationship with utterances. Accordingly, we might reasonably think of these constructions as replicators, with selection being driven by speakers choosing among constructions to use to produce an utterance. Alternatively, we could think of learners as providing selection pressure, with the constructions that produce the most evidence for their existence in the data available to the learner ending up being the most stable through the learning bottleneck. Here, we can imagine constructions competing for place in the learners' input.

Another view might be that a language is a hypothesis for which we select on the basis of evidence combined with an inductive bias. Where are the replicators here? Who is doing the selection? Given this latter perspective, the neutral model outlined earlier appears more appropriate.

Which of these perspectives is correct? It is possible that in fact they are compatible—that they are different ways of analyzing the same process, namely social/cultural *adaptation*. The challenge is in seeing how these analyses relate to one another and to the models that exist in the literature.

Incidentally, we need to be clear that when we refer to selection and replication here, we are not talking about selection of heritable genetic variation (although that is clearly relevant to language evolution and to models of language evolution). Nor are we referring to the natural selection of cultural variants, a mechanism by which fitter individuals are more likely to survive and pass on their cultural traits (although this, too, is likely to be important). Instead, we are talking about the kind of adaptation that occurs purely through the complex process of a repeated cycle of utterance creation, interpretation, and internalization that happens in language transmission—be it in an iterated learning model focusing on vertical transmission or a negotiation model focusing on social coordination.

Gene–Culture Coevolution

As we pointed out at the onset of this chapter, the current focus of language evolution modeling lies in cultural evolution. It is, however, clear that a complete picture must integrate cultural with biological (genetic) evolution. Formal modeling of gene–culture coevolution began in the mid-1970s (Cavalli-Sforza and Feldman 1981). Briscoe (2003, this volume) reviews models of gene–culture coevolution applied to language evolution. The basic idea is to use the population geneticists' recursion equation formalism for cultural as well as genetic evolution. The result is a system of linked dynamic equations that keep track of genes and culture as they change through time. In general, genes can influence culture via decision-making rules. An innate syntax might constrain the evolution of languages. The flow of causation will also generally work the other way. An element of a culturally transmitted protolanguage might exert selective pressure on the genes. If genetic variation exists in the innate supports for language, and if more efficient communication is favored, the variants that make the protolanguage more sophisticated will increase in the population. Since cultural evolution will tend to be faster than genetic evolution, cultural evolution will tend to be the driving partner in the coevolutionary circuit and genetic evolution the rate limiting step. This does not tell us anything about the division of labor between genes and culture at evolutionary equilibrium. That will depend upon many contingent costs and benefits of transmitting adaptations genetically versus culturally. Very broadly speaking, the genetic and cultural subsystems are both adaptive systems, and selection may be more or less indifferent as to how a given adaptation is transmitted.

Although a complete coverage of gene–cultural modeling remains as a task for the future, one question which has already been studied by gene–culture coevolutionists is whether and how the evolution of various human adaptations may facilitate or constrain the evolution of language (see Boyd and Richerson 1985, 1996, 2005). Language would seem to require a large measure of cooperation. Otherwise hearers could not trust speakers. The noncooperative case seems to exemplify the situation for most other species. Hence communication systems in other animals are rather limited. Even in humans, people who live in different societies may not be trustworthy sources of information. Hence the evolution of linguistic differences between human groups may be adapted to limit communications from untrustworthy others.

Advanced Recurrent Neural Network Models

There exist a number of RNN architectures designed to model complex linguistic (or visual) processing that are both computationally powerful and partly biologically plausible. These models have not yet been used as a basis for evolution of language studies. Due to their expressivity and the availability of advanced learning algorithms, they appear to be promising carrier formalisms for future evolutionary studies of grammatical processing.

The SHRUTI family of connectionist architectures (e.g., Shastri 1999) represents a long-standing research strand to explain fast-forward inferences in semantic text understanding. These models are very complex, hand-designed networks of semantic and syntactic processing nodes which communicate with each other by biologically motivated neural spike codes, enabling combinatorial binding of different representation nodes across the network.

In machine learning, a recent landmark paper (Hinton and Salakhutdinov 2006) has unleashed a flurry of research in deep belief networks (DBNs) and restricted Boltzmann machines (RBMs). With these models and novel learning algorithms, it has become feasible for the first time to train deep conceptual hierarchical representations from large-volume real-life data in an unsupervised way. While this field has been preoccupied thus far with visual learning tasks, the step toward speech/text input is imminent.

Another family of hierarchical RNN-based models for learning representations of very complex multiscale data is emerging. These models arise as hierarchical/multiscale extensions of *echo state networks* (Jaeger and Haas 2004) or *liquid state machines* (Maass et al. 2002), the two main exemplars of a new computational paradigm in neuroinformatics referred to as *reservoir computing* (Jaeger et al. 2005). They share a number of important characteristics with the neural models of speech recognition explored by Dominey (2005; Dominey et al. 2006). In this field, language and speech modeling is indeed an important target domain.

An important characteristic of all RNN models, which sets them apart from symbolic grammar formalisms, is that speech/language processing is construed as a fast, self-organizing dynamical system, which does not need search subroutines and does not build interim alternative interpretations. On the positive side, this leads to very fast processing (timescale of a few neuronal delays); on the negative side, if an interpretation trajectory goes astray, this has to be detected and separate repair mechanisms have to be invoked.

Creatures-based Modeling

Current simulation-based studies on language change concentrate primarily on cultural transmission dynamics. Neither brain structures nor genetic determinants for such brain structures are modeled. This makes simulation-based research blind to some of the questions that are raised in biological evolutionary accounts of the origins of language (in this volume, see chapters by Givón, Fedor et al., Hilliard and White). A potentially powerful avenue would be to simulate brain–body coevolution along the lines staked out in Artificial Life and Evolutionary Robotics (e.g., Sims 1994; Nolfi and Floreano 2000; Szathmáry 2007). In this research, artificial creatures are evolved in simulation or in physical robotic hardware. A creature has a body equipped with sensors and actuators, and is controlled by an artificial neural network that coevolves with the body. Research of this kind has resulted in the evolution of surprisingly complex and adaptive behavior repertoires driven by neurocontrollers of surprisingly small size. However, linguistic capabilities have so far largely fallen beyond the scope of this research (cf. Wischmann and Pasemann 2006). It appears a natural and fascinating endeavor, albeit computationally challenging, to implement simulation scenarios where body+brain creatures are evolved under selective pressures that favor efficient communication. In this way, simulation-based research might tell an almost complete (if duly simplified) story, connecting mechanisms of genetic coding of neural structures and the ensuing slow "biological" adaptations with the fast cultural transmission dynamics that are the hallmark of today's investigations.

Detailed Models of Language Learning During Development

Most models of language and syntax evolution treat an individual's learning of language during their "childhood" in a very simplistic fashion. However, the transmission of language from one generation to the next is clearly a central aspect of the evolution of language. Thus, more elaborate modeling of the acquisition of language during infancy and childhood is needed. Ideally, such models would take the embodied nature of language learning into account, capturing how the learner interacts with their physical and social environment. At the same time such models should be constrained by developmental psychology and developmental neuroscience, providing constraints regarding the underlying neural structures, representations, and learning mechanisms, as well as the nature of the language input to which infants are typically exposed. Thus far, such approaches have been mostly restricted to learning early precursors of language, such as gaze following (Triesch et al. 2006) or learning of word meanings (Yu et al. 2005), but the time seems ripe to extend such models toward the acquisition of grammatical structures.

Case Studies

Scientific fields often organize their research around key challenges that are accepted by a large group of researchers independently of the methods they are using. In technical fields, such as machine learning, robotics, or high performance computing, there are often well-defined challenges against which different research groups compete, often leading to very rapid progress (e.g., as seen in the Robocup). What would such key challenges look like in the case of language evolution? One possibility is to pick a certain domain which has been grammaticalized in many languages of the world, although often in different ways, and show what cognitive mechanisms and interaction patterns are needed to see the emergence of such a system in a population of agents. Another possibility is to develop evolutionary models that are also capable of capturing psycholinguistic data on actual human language behavior.

In terms of domain, consider the following example: Many languages have the means to express predicate–argument structure through a system of case grammar, either expressed morphologically or through word-order structure. The emergence of such a system requires not only the emergence of conventions but also the emergence of the semantic and syntactic categories that underly it. A lot of data is available from historical linguistics showing how such systems have arisen in human natural languages, often by the grammaticalization of verbs, and these data could be used to constrain potential models. Some attempts have already been made to explain the emergence of case grammar, from the viewpoints of each of the paradigms introduced earlier (Moy 2006; Jäger 2007; Van Trijp 2008), and these can act as a starting point for tackling this challenge. It is not difficult to imagine other aspects of grammar that could form the focus of well-defined challenges, and once these are addressed, more challenging ones could be attempted, such as clause structure with long-distance dependencies.

Conclusion

Mathematical and computational models of language evolution make it possible to examine the consequences of certain theoretical assumptions by mathematical deduction, large-scale computer simulation, or robotic experimentation. Several efforts are under way to apply this methodology to questions related to the problem of the origins and evolution of language. No single paradigm or methodology exists; instead, multiple paradigms explore different questions. At present, models focus primarily on the origins of lexicons, categories that can act as building blocks for conceptualization, and simple languages with few of the complex structuring principles found in human languages (cf. Briscoe, this volume). However, we are confident that the technological foundations and mathematical tools will become progressively more sophisticated and thus be able to tackle increasingly deeper and more intricate questions relating to the origins and evolution of syntax in language.

Glossary

ADV	Adverb
AF	Articulate fasciculus
AFP	Anterior forebrain pathway
AG	Angular gyrus
AMH	Anatomically modern humans
AP	Adjective phrase
ApoE4	Apolipoprotein E
BA	Brodmann's area
Baldwin effect	Proposed by James Mark Baldwin, the Baldwin effect, also known as Baldwinian evolution or ontogenic evolution, consists of a mechanism for specific selection for general learning ability. Selected offspring tend to have an increased capacity for learning new skills rather than being confined to genetically coded, relatively fixed abilities. Suppose a species is threatened by a new predator and there is a behavior that makes it more difficult for the predator to kill individuals of the species. Individuals who learn the behavior more quickly will obviously be at an advantage. As time goes on, the ability to learn the behavior will improve (by genetic selection), and at some point it will seem to be an instinct, thus the repertoire of genetically encoded abilities can ultimately increase. This ultimate phase of the Baldwin effect is related to the concept of genetic assimilation.
BG	Basal ganglia
Binding	Relationship between anaphors and their antecedents
Birdsong	Bird vocalization includes both bird calls and bird songs. The distinction between songs and calls is based upon inflection, length, and context. Songs are longer and more complex and are associated with courtship and mating, whereas calls tend to serve such functions as alarms or keeping members of a flock in contact.
Birdsong syntax	Temporal sequence in which discrete units of song (notes, syllables, phrases, motifs, bouts) are produced
BOLD signal	Blood-oxygen-level-dependent signal
CFG	Context-free grammar
CHiP	Chromatin immunoprecipitation
CNTNAP2	Contactin-associated protein-like 2 gene

Coevolution	In population biology, coevolution refers to the joint evolution of populations of two or more species interacting with each other, which is triggered by these interactions.
Control	Select the closest possible antecedent for the verb in question, subject to lexical exceptions that are determined by purely semantic considerations
CP	Complementizer phrase
CS	Central sulcus
CSF	Cerebrospinal fluid
CT	Computed tomography
CVP	Category violation point
CYCLE	Curtiss-Yamada comprehensive language evaluation
DA	Dopamine
DBN	Deep belief network
DLM	Dorsolateral thalamus
dobj	Direct object
DP	Determiner phrase
DSI	Diffusion spectrum MRI
DTI	Diffusion tensor imaging
DVD	Developmental verbal dyspraxia
DZ	Dizygotic
EEG	Electroencephalography
ELAN	Early left anterior negativity
Empty categories	Elements that are "understood" but not phonetically expressed
ENGA	Evolutionary neurogenetic algorithm
ENU	N-ethyl-N-nitrosourea
EOI	Extended optional infinitive theory
ERP	Event-related potential
Evolvability	Capacity of a population of organisms to respond to directional selection (e.g., sexual recombination)
Exaptation	Exaptation, co-option, and preadaptation are related terms referring to shifts in the function of a trait during evolution. For example, a trait can evolve because it served one particular function, but it may subsequently come to serve another, initially, in a mostly rudimentary form. Bird feathers are a classic example: initially they evolved to regulate temperature but were later adapted for flight.
F	Feature triggering one of the major syntactic operations
FCG	Fluid construction grammar
FDG	Fluorodeoxyglucose
FIN	Finite

FLS	Fasciculus longitudinalis superior
fMRI	Functional magnetic resonance imaging
FOC	Focus
fOP	Frontal operculum
FOXP2	Transcription factor identified in 2001 as the monogenetic locus of a mutation causing an inherited speech and language disorder. Human "FOXP2" is capitalized, mouse "Foxp2" is not, and "FoxP2" denotes the molecule in mixed groups of animals. Italics are used when referring to genetic material such as FoxP2 mRNA.
FSG	Finite state grammar
Genetic assimilation	Process by which the effect of an environmental condition, such as exposure to a substance, is used in conjunction with artificial selection or natural selection to create a strain of organisms with similar changes in phenotype that are encoded genetically. It also refers to cases when individuals initially have to learn some trait, followed by the appearance of organisms by selection that are "hardwired" for that trait (i.e., their nervous systems become tuned by genes rather than experience).
GPSG	Generalized phrase structure grammars
H	Functional head
Haplotype	Contraction of the term "haploid genotype." In genetics, a haplotype is a combination of alleles at multiple loci that are transmitted together on the same chromosome. Haplotype may refer to as few as two loci or to an entire chromosome depending on the number of recombination events that have occurred between a given set of loci. Haplotype is also a set of single nucleotide polymorphisms on a single chromatid that are statistically associated.
HPSG	Head-driven phrase structure grammar
HVC	High vocal center (a premotor-association cortex-like pallial region)
ID	Immediate dominance
IFC	Inferior frontal cortex
IFG	Inferior frontal gyrus
INT	Interrogative, a node label that triggers question forms
LAD	Language acquisition device
LAN	Left anterior negativity
LDD	Long-distance dependencies
LH	Left hemisphere
LIFC	Left inferior frontal cortex

LMAN	Lateral portion of the magnocellular nucleus of the anterior nidopallium
LP	Linear precedence
MDL	Minimum description length
MEG	Magnetoencephalography
Merge	Process that takes any two units (words, phrases, clauses…) and forms them into a single unit, subject to "feature-matching"
MFG	Middle frontal gyrus
MNI	Montreal Neurological Institute
MNS	Mirror neuron system
MOD	Modality
Move	The instruction that triggers movement (*vide infra*) in the course of a sentence derivation
Movement	Hypothesized process in syntax that generates a lexical item in one position in the original mental representation of a sentence but causes that item to appear in another position when the sentence is externalized.
MRI	Magnetic resonance imaging
MTG	Medial temporal gyrus
MZ	Monozygotic
N400 response	Negative deflection (topologically distributed over central-parietal sites on the scalp) elicited by unexpected linguistic stimuli peaking approximately 400 ms (300–500 ms) after the presentation of the stimulus
Niche construction	Creation (in whole or part) of an ecological niche by a particular species
NOM	Nominative case
NP	Noun phrase
O	Object
P&P approach	Principles and Parameter approach
P600 response	Positive deflection (appearing mostly over the posterior part of the center of the scalp) elicited by a violating word versus a correct control, peaking roughly around 600 ms, but more typically at 500–900 ms
PET	Positron emission tomography
PF	Phonological form
PHON	Sound (phonetic form)
pITG	Posterior inferior temporal gyrus
pMTG	Posterior middle temporal gyrus
PP	Prepositional phrase
PRO	Pronominal subject
PSG	Phrase structure grammar
Q	Question

RA	Robust nucleus of the arcopallium
RBM	Restricted Boltzmann machine
Recursion	Ability to insert one structure inside another of the same kind; any process that uses the output of one stage as the input to the next
RH	Right hemisphere
RNN	Recurrent neural network
ROI	Regions of interest
S	Subject
SAP	Sound Analysis Pro
SD	Standard deviation
SEM	Meaning (logical form)
SF	Sylvian fissure
Shruti model	Neurally motivated model of relational knowledge representation and rapid inference using temporal synchrony
SLF	Superior longitudinal fasciculus
SLI	Specific language impairment
SMG	Supramarginal gyrus
SNPs	Single nucleotide polymorphisms
Spell-out	Phonetic realization of the various expressions
SPM	Sentence-picture matching
SRN	Simple recurrent network
SRT	Serial-response time task
STG	Superior temporal gyrus
STM	Short-term memory
STS	Superior temporal sulcus
SVO	Subject, verb, object
Syntax	Process for progressively merging words into larger units, upon which algorithms are superimposed that determine the reference of items which might otherwise be misleading
T	Tense
TMA	Tense, modal, aspect
TMS	Transcranial magnetic stimulation
ToM	Theory of mind
TOP	Topic
TP	Tense phrase
UG	Universal Grammar
V	Verb
vlPFC	Ventrolateral prefrontal cortex
VLSM	Voxel-based lesion-symptom mapping
VP	Verb phrase
vPMC	Ventral premotor cortex

VSO	Verb, subject, object
WAB	Western aphasia battery
WGCNA	Weighted gene co-expression network analysis

Bibliography

Aboh, E. 2004. The Morphosyntax of Complement-Head Sequences. New York: Oxford Univ. Press.

Airey, D. C., A. I. Robbins, K. M. Enzinger, F. Wu, and C. E. Collins. 2005. Variation in the cortical map of C57BL/6J and DBA/2J inbred mice predicts strain identity. *BMC Neurosci.* **6**:18.

Aissen, J. 2003. Differential object marking: Iconicity vs. economy. *Natural Lang. Ling. Theory* **21(3)**:435–483.

Alarcón, M., B. S. Abrahams, J. L. Stone et al. 2008. Linkage, association, and gene-expression analyses identify CNTNAP2 as an autism-susceptibility gene. *Am. J. Hum. Gen.* **82**:150–159.

Amaral, D. G. 2000. The anatomical organization of the central nervous system. In: Principles of Neural Science, 4th edition, ed. E. R. Kandel, J. H Schwartz, and T. M. Jessell, pp. 317–336. New York: McGraw-Hill.

Amedi, A., N. Raz, P. Pianka, R. Malach, and E. Zohary. 2003. Early "visual" cortex activation correlates with superior verbal memory performance in the blind. *Nature Neurosci.* **6**:758–766.

Amunts, K., L. Jancke, H. Mohlberg, H. Steinmetz, and K. Zilles. 2000. Interhemispheric asymmetry of the human motor cortex related to handedness and gender. *Neuropsychologia* **38**:304–312.

Amunts, K., A. Schleicher, U. Burgel et al. 1999. Broca's region revisited: Cytoarchitecture and intersubject variability. *J. Comp. Neurol.* **412**:319–341.

Amunts, K., A. Schleicher, A. Ditterich, and K. Zilles. 2003. Broca's region: Cytoarchitectonic asymmetry and developmental changes. *J. Comp. Neurol.* **465**:72–90.

Amunts, K., and K. Zilles. 2001. Advances in cytoarchitectonic mapping of the human cerebral cortex. *Neuroimaging Clinics* **11**:151–169, vii.

——. 2006. A multimodal analysis of structure and function in Broca's region. In: Broca's Region, ed. Y. Grodzinsky and K. Amunts, pp. 17–31. Oxford: Oxford Univ. Press.

Andersen, R. 1979. Expanding Schumann's pidginization hypothesis. *Lang. Learn.* **29**:105–119.

Angluin, D. 1982. Inference of reversible languages. *J. Assn. Comp. Machinery* **29**:741–765.

Anwander, A., M. Tittgemeyer, D. Y. von Cramon, A. D. Friederici, and T. R. Knösche. 2007. Connectivity-based parcellation of Broca's area. *Cereb. Cortex* **17**:816–825.

Arbib, M. A. 2005. From monkey-like action recognition to human language: An evolutionary framework for neurolinguistics. *Behav. Brain Sci.* **28**:105–167.

——. 2008. From grasp to language: Embodied concepts and the challenge of abstraction. *J. Physiol. Paris* **102**:4–20.

Arbib, M. A., and M. Bota. 2003. Language evolution: Neural homologies and neuroinformatics. *Neural Netw.* **16**:1237–1260.

Argyropoulos, G. P. 2008. The subcortical foundations of grammaticalization. In: The Evolution of Language: Proc. 7th Intl. Conf. (EVOLANG7), ed. A. D. M. Smith, K. Smith, and R. Ferrer i Cancho, pp. 10–17. Barcelona: World Scientific Press.

Arnold, K., Y. Pohlner, and K. Zuberbühler. 2008. A forest monkey's alarm call series to predator models. *Behav. Ecol. Sociobiol.* **62**:549–559.

Arnold, K., and K. Zuberbühler. 2006. The alarm-calling system of adult male putty-nosed monkeys, *Cercopithecus nictintans martini*. *Anim. Behav.* **72**:643–653.

——. 2008. Meaningful call combinations in a nonhuman primate. *Curr. Biol.* **18(5)**:202–203.

Atkinson, R. C., and R. M. Shiffrin. 1968. Human memory: A proposed system and its control processes. In: The Psychology of Learning and Motivation, vol. 2, ed. K. W. Spence and T. Spence. New York: Academic Press.

Austin, J. 1962. How to Do Things with Words. Cambridge, MA: Harvard Univ. Press.

Avian Nomenclature Consortium. 2005. Avian brains and a new understanding of vertebrate brain evolution. *Nat. Rev. Neurosci.* **6(2)**:151–159.

Baddeley, A. D. 1986. Working Memory. Oxford: Clarendon Press.

Baddeley, A. D., S. Gathercole, and C. Papagno. 1998. The phonological loop as a language learning device. *Psychol. Rev.* **105**:158–173.

Badre, D., and A. D. Wagner. 2007. The left ventrolateral prefrontal cortex and the cognitive control of memory. *Neuropsychologia* **45**:2883–2901.

Bagic, A., and S. Sato. 2007. Principles of electroencephalography and magnetoencephalography. In: Functional Neuroimaging in Clinical Populations, ed. F. G. Hillary and S. M. Rao, pp. 71–96. New York: Guilford Press.

Bahlmann, J., R. Schubotz, and A. D. Friederici. 2008. Hierarchical sequencing engages Broca's area. *NeuroImage* **42**:525–534.

Baker, M. C. 2001. The Atoms of Language: The Mind's Hidden Rules of Grammar. New York: Basic Books.

Baldwin, J. M. 1896. A new factor in evolution. *Am. Nat.* **30**:441–451, 533–536.

Balsamo, L. M., B. Xu, and W. D. Gaillard. 2006. Language lateralization and the role of the fusiform gyrus in semantic processing in young children. *NeuroImage* **31**:1306–1314.

Bandettini, P. A. 2007. Principles of functional magnetic resonance imaging. In: Functional Neuroimaging in Clinical Populations, ed. F. G. Hillary and S. M. Rao, pp. 31–70. New York: Guilford Press.

Barabasi, A. L., and Z. N. Oltvai. 2004. Network biology: Understanding the cell's functional organization. *Nature Rev. Genet.* **5**:101–113.

Barham, L. S. 2002. Systematic pigment use in the Middle Pleistocene of south central Africa. *Curr. Anthro.* **31(1)**:181–190.

Barker, M., and T. Givón. 2002. On the pre-linguistic origins of language processing rates. In: The Evolution of Language out of Pre-Language, ed. T. Givón and B. F. Malle. Amsterdam: John Benjamins.

Barlow, H. B. 1995. The neuron in perception. In: The Cognitive Neurosciences, ed. M. S. Gazzaniga, pp. 415–434. Cambridge, MA: MIT Press.

Barnea-Goraly, N., V. Menon, M. Eckert et al. 2005. White matter development during childhood and adolescence: A cross-sectional diffusion tensor imaging study. *Cereb. Cortex* **15**:1848–1854.

Baronchelli, A., L. Dall'Asta, A. Barrat, and V. Loreto. 2006a. Topology induced coarsening in language games language. *Phys. Rev. E* **73**:015102(R).

Baronchelli, A., M. Felici, E. Caglioti, V. Loreto, and L. Steels. 2006b. Sharp transition towards shared vocabularies in multi-agent systems. *J. Stat. Mech.* P06014.

Baronchelli, A., V. Loreto, and L. Steels. 2008. In-depth analysis of the Naming Game dynamics: The homogeneous mixing case. *Intl. J. Mod. Phys.* **19(5)**:785–812.

Bartels, A., and S. Zeki. 2005. The chronoarchitecture of the cerebral cortex. *Phil. Trans. Roy. Soc. Lond. B* **360**:733–750.

Bartlett, C. W., J. F. Flax, M. W. Logue et al. 2002. A major susceptibility locus for specific language impairment is located on 13q21. *Am. J. Hum. Gen.* **71**:45–55.

Batali, J. 1998. Computational simulations of the emergence of grammar. In: Approaches to the Evolution of Language: Social and Cognitive Bases, ed. J. R. Hurford, M. Studdert-Kennedy, and C. Knight. Cambridge: Cambridge Univ. Press.

Bates, E. A. 1999. Plasticity, localization, and language development. In: The Changing Nervous System: Neurobehavioral Consequences of Early Brain Disorders, ed. S. H. Broman and J. M. Fletcher, pp. 214–253. New York: Oxford Univ. Press.

——. 2004. Commentary: Explaining and interpreting deficits in language development across clinical groups: Where do we go from here? *Brain Lang.* **88**:248–253.

Bates, E. A., L. Camioni, and V. Volterra. 1975. The acquisition of performatives prior to speech. *Merrill-Palmer Qtly.* 21.

Bates, E. A., and G. F. Carnevale. 1993. New directions in research on language development. *Devel. Rev.* **13**:436–470.

Bates, E., D. Thal, D. Trauner et al. 1997. From first words to grammar in children with focal brain injury. *Devel. Neuropsychol.* **13(3)**:447–476.

Bayer, J. 1984. COMP in Bavarian. *Ling. Rev.* **3**:209–274.

Belletti, A., ed. 2004. Structures and Beyond: The Cartography of Syntactic Structures, vol. 3. New York: Oxford Univ. Press.

Belton, E., C. H. Salmond, K. E. Watkins, F. Vargha-Khadem, and D. G. Gadian. 2003. Bilateral brain abnormalities associated with dominantly inherited verbal and orofacial dyspraxia. *Hum. Brain Map.* **18**:194–200.

Ben-Shachar, M., T. Hendler, I. Kahn, D. Ben-Bashat, and Y. Grodzinsky. 2003. The neural reality of syntactic transformations: Evidence from functional magnetic resonance imaging. *Psychol. Sci.* **14(5)**:433–440.

Ben-Shachar, M., D. Palti, and Y. Grodzinsky. 2004. Neural correlates of syntactic movement: Converging evidence from two fMRI experiments. *NeuroImage* **21**:1320–1336.

Ben Shalom, D., and D. Poeppel. 2008. Functional anatomic models of language: Assembling the pieces. *Neuroscientist* **14**:119–127.

Bergen, B., and N. Chang. 2004. Embodied construction grammar in simulation-based language understanding. In: Construction Grammar(s): Cognitive and Cross-Language Dimensions, ed. J.-O. Östman and M. Fried. Amsterdam: John Benjamins.

Berkowitz, A. L., and D. Ansari. 2008. Generation of novel motor sequences: The neural correlates of musical improvisation. *NeuroImage* **41**:535–543.

Berndt, R., C. Mitchum, A. Haendiges et al. 1996. Comprehension of reversible sentences in "agrammatism": A meta-analysis. *Cognition* **58**:289–308.

Besson, M., and F. Macar. 1987. An event-related potential analysis of incongruity in music and other non-linguistic contexts. *Psychophysiology* **24**:14–25.

Bever, T. 1971. The nature of cerebral dominance in speech behaviour of the child and adult. In: Language Acquisition: Models and Methods, ed. R. Huxley and E. Ingram. Oxford: Academic Press.

Bickerton, D. 1981. Roots of Language. Ann Arbor: Karoma.

——. 1984. The language bioprogram hypothesis. *Behav. Brain Sci.* **7**:17–388.

——. 1990. Language and Species. Chicago: Univ. Chicago Press.

——. 2008. Bastard Tongues. New York: Hill & Wang.

——. 2009. Adam's Tongue: How Humans Made Language, How Language Made Humans. New York: Hill & Wang.

Bickerton, D., and C. Odo. 1976. General phonology and pidgin syntax. Final Report on NSF Grant No. 39748, Univ. of Hawaii.

Binder, J. R., J. A. Frost, T. A. Hammeke et al. 1997. Human brain language areas identified by functional magnetic resonance imaging. *J. Neurosci.* **17**:353–362.

Binford, L. R. 1983. In Pursuit of the Past: Decoding the Archaeological Record. London: Thames and Hudson.

Bishop, D. V. M. 1988. Language development after focal brain damage. In: Language Development in Exceptional Circumstances, ed. D. V. M. Bishop and K. Mogford, pp. 203–219. Edinburgh: Churchill Livingstone.

——. 2002a. Putting language genes in perspective. *Trends Gen.* **18**:57–59.

——. 2002b. The role of genes in the etiology of specific language impairment. *J. Comm. Disorders* **35**:311–328.

——. 2005. DeFries–Fulker analysis of twin data with skewed distributions: Cautions and recommendations from a study of children's use of verb inflections. *Behav. Gen.* **35**:475–490.

Bishop, D. V. M., C. V. Adams, and C. F. Norbury. 2006. Distinct genetic influences on grammar and phonological short-term memory deficits: Evidence from 6-year-old twins. *Gen. Brain Behav.* **5**:158–169.

Bishop, D. V. M., S. J. Bishop, P. Bright et al. 1999. Different origin of auditory and phonological processing problems in children with language impairment: Evidence from a twin study. *J. Speech Lang. Hearing Res.* **42**:155–168.

Bishop, D. V. M., T. North, and C. Donlan. 1996. Nonword repetition as a behavioural marker for inherited language impairment: Evidence from a twin study. *J. Child Psychol. Psych.* **37**:391–403.

Bloom, L. 1973. One Word at a Time: The Use of Single Word Utterances before Syntax. The Hague: Mouton.

Bloom, P. 1994. Recent controversies in the study of language acquisition. In: Handbook of Psycholinguistics, ed. M. A. Gernsbacher, pp. 741–779. San Diego: Academic Press.

Bloomfield, L. 1933. Language. New York: Holt, Rinehart & Winston.

Blumstein, S. E., G. Byma, K. Kurowski et al. 1998. On-line processing of filler-gap constructions in aphasia. *Brain Lang.* **61**(2):149–169.

Blumstein, S. E., and W. Milberg. 1983. Automatic and controlled processing in speech/language deficits in aphasia. Symposium on Automatic Speech. Minneapolis: Academy of Aphasia.

Bodén, M., and J. Wiles. 2000. Context-free and context-sensitive dynamics in recurrent neural networks. *Connect. Sci.* **12**:197–210.

Boeckx, C. 2006. Linguistic Minimalism: Origins, concepts, methods, and aims. Oxford: Oxford Univ. Press.

Boersma, P. 1998. Functional phonology: Formalizing the interactions between articulatory and perceptual drives. Ph.D. thesis, Univ. of Amsterdam.

Bookheimer, S. 2002. Functional MRI of language: New approaches to understanding the cortical organization of semantic processing. *Ann. Rev. Neurosci.* **25**:151–188.

Booth, J. R., B. MacWhinney, and Y. Harasaki. 2000. Developmental differences in visual and auditory processing of complex sentences. *Child Dev.* **71(4)**:981–1003.

Bornkessel, I., and M. Schlesewsky. 2006. The extended argument dependency model: A neurocognitive approach to sentence comprehension across languages. *Psychol. Rev.* **113**:787–821.

Bornkessel, I., S. Zysset, A. D. Friederici, D. Y. von Cramon, and M. Schlesewsky. 2005. Who did what to whom? The neural basis of argument hierarchies during language comprehension. *NeuroImage* **26(1)**:221–233.

Borsley, R. D., and M. Tallerman. 1996. Phrases and soft mutation in Welsh. *J. Celtic Ling.* **5**:1–49.

Borsley, R. D., M. Tallerman, and D. Willis. 2007. The Syntax of Welsh. Cambridge: Cambridge Univ. Press.

Botha, R. 2006a. On the "Windows Approach" to language evolution. *Lang. Comm.* **26**:129–143.

——. 2006b. Pidgin languages as a putative window on language evolution. *Lang. Comm.* **26**:1–14.

——. 2007. On homesign systems as a potential window on language evolution. *Lang. Comm.* **27**:41–53.

——. 2008. Prehistoric shell beads as a window on language evolution. *Lang. Comm.* **28(3)**:197–212.

Bottjer, S. W., E. A. Miesner, and A. P. Arnold. 1984. Forebrain lesions disrupt development but not maintenance of song in passerine birds. *Science* **224**:901–903.

Boughman, J. W. 1998. Vocal learning by greater spear-nosed bats. *P. Roy. Soc. Lond. B* **V265**:227–233.

Bouzouggar, A., N. Barton, M. Vanhaeren et al. 2007. 82,000-year-old shell beads from North Africa and implications for the origins of modern human behavior. *PNAS* **104**:9964–9969.

Bowerman, M. 1973. Early Syntactic Development. Cambridge: Cambridge Univ. Press.

Boyd, R., and P. J. Richerson. 1985. Culture and the Evolutionary Process. Chicago: Univ. of Chicago Press.

——. 1996. Why culture is common but cultural evolution is rare. *Proc. Brit. Acad.* **88**:73–93.

——. 2005. The Origins and Evolution of Culture. Oxford: Oxford Univ. Press.

Braaten, R. F., M. Petzoldt, and A. Colbath. 2006. Song perception during the sensitive period of song learning in zebra finches (Taeniopygia guttata). *J. Comp. Psychol.* **120**:79–88.

Brainard, M. S., and A. J. Doupe. 2000. Interruption of a basal ganglia-forebrain circuit prevents plasticity of learned vocalizations. *Nature* **V404**:762–766.

Brauer, J., A. Anwander, and A. D. Friederici. 2008. Functional development and structural maturation of language areas in the human brain. *NeuroImage* **41**:S1.

Brauer, J., and A. D. Friederici. 2007. Functional neural networks of semantic and syntactic processes in the developing brain. *J. Cogn. Neurosci.* **19**:1609–1623.

Brenier, J., and L. Michaelis. 2006. Optimization via syntactic amalgam: Syntax-prosody mismatch and copula doubling. *Corpus Linguistics and Linguistic Theory* **1**:45–88.

Brenowitz, E. A., D. Margoliash, and K. W. Nordeen. 1997. An introduction to birdsong and the avian song system. *J. Neurobiol.* **33**:495–500.

Bresnan, J. W. 2001. Lexical-functional syntax. Oxford: Blackwell.

Brighton, H. 2002. Compositional syntax through cultural transmission. *Artif. Life* **8.1**:25–54.

Brighton, H., and S. Kirby. 2001. The survival of the smallest: Stability conditions for the cultural evolution of compositional language. In: Advances in Artificial Life, ed. J. Kelemen and P. Sosik, pp. 592–601. Berlin: Springer-Verlag.

Brighton, H., K. Smith, and S. Kirby. 2005. Language as an evolutionary system. *Physics Life Rev.* **2**:177–226.

Briscoe, E. J. 1997. Co-evolution of language and of the language acquisition device. Proc. 35th Assoc. for Comp. Ling., pp. 418–427. San Francisco: Morgan Kaufmann.

——. 2000. Grammatical acquisition: Inductive bias and coevolution of language and the language acquisition device. *Language* **76.2**:245–296.
——. 2003. Grammatical assimilation. In: Language Evolution: The States of the Art, ed. M. Christiansen and S. Kirby, pp. 293–316. Oxford: Oxford Univ. Press.
——. 2005. Coevolution of the language faculty and language(s) with decorrelated encodings. In: Language Origins: Perspectives on Evolution, ed. M. Tallerman, pp. 310–333. Oxford: Oxford Univ. Press.
Broca, P. 1861. Remarques sur le siège de la faculté de langage articulé: Suivies d'une observation d'aphemie. *Bull. Soc. Anat. Paris* **6**:330–357.
——. 1863. Localisation des fonctions cérébrales: Siège du langage articulé. *Bull. Soc. Anthropol. Paris* **4**:200–204.
——. 1865. Sur la siège de la faculté langage articule. *Bull. Soc. Anthropol.* **6**:377–396.
Brodmann, K. 1905. Beiträge zur histologischen Lokalisation der Grosshirnrinde. Dritte Mitteilung: Die Rindenfelder der niederen Affen. *J. Psychol. Neurol.* **4**:177–226.
——. 1909. Vergleichende Lokalisationslehre der Großhirnrinde. Leipzig: Barth.
Bruner, E., G. Manzi, and J. L. Arsuaga. 2003. Encephalization and allometric trajectories in the genus Homo: Evidence from the Neandertal and modern lineages. *PNAS* **100(26)**:15,335–15,340.
Bufill, E., and E. Carbonell. 2004. Are symbolic behaviour and neuroplasticity an example of gene-culture coevolution? (in Spanish) *Rev. Neurol.* **39**:48–55.
Bullock, S. 1998. A continuous evolutionary simulation model of the attainability of honest signalling equilibria. In: Artificial Life VI, ed. C. Adami, R. K. Belew, H. Kitano, and C. E. Taylor, pp. 339–348. Cambridge, MA: MIT Press (Bradford Books).
Burling, R. 1993. Primate calls, human language, and nonverbal communication. *Curr. Anthro.* **34**:25–53.
——. 2005. The Talking Ape: How Language Evolved. Oxford: Oxford Univ. Press.
Buzsáki, G., K. Kaila, and M. Raichle. 2007. Inhibition and brain work. *Neuron* **56**:771–783.
Bybee, J., R. Perkins, and W. Pagliuca. 1994. The Evolution of Grammar: Tense, Aspect, and Modality in the Languages of the World. Chicago: Univ. of Chicago Press.
Calvin, W. H. 1987. The brain as a Darwin machine. *Nature* **330**:33–34.
——. 1991. The Throwing Madonna. London: Bantam.
——. 1996. The Cerebral Code: Thinking a Thought in the Mosaics of the Mind. Cambridge, MA: MIT Press.
Calvin, W. H., and D. Bickerton. 2000. Lingua ex Machina: Reconciling Darwin and Chomsky with the Human Brain. Cambridge, MA: MIT Press.
Campbell, D. T. 1974. Evolutionary epistemology. In: The Philosophy of Karl R. Popper, ed. P. A. Schilpp, pp. 412–463. LaSalle, IL: Open Court.
Cangelosi, A., and D. Parisi. 1998. The emergence of a "language" in an evolving population of neural networks. *Connect. Sci.* **10(2)**:83–89.
Caplan, D. 1994. The cognitive neuroscience of syntactic processing. In: The Cognitive Neurosciences, ed. M. Gazzaniga, pp. 871–879. Cambridge, MA: MIT Press.
——. 1995. Issues arising in contemporary studies of disorders of syntactic processing in sentence comprehension in agrammatic patients. *Brain Lang.* **50**:325–338.
——. 2001a. The measurement of chance performance in aphasia, with specific reference to the comprehension of semantically reversible passive sentences: A note on issues raised by Caramazza, Capitani, Rey and Berndt (2000) and Drai, Grodzinsky and Zurif (2000). *Brain Lang.* **76**:193–201.

——. 2001b. Points regarding the functional neuroanatomy of syntactic processing: A response to Zurif. *Brain Lang.* **79**:329–332.

——. 2006a. Functional neuroimaging of syntactic processing: New claims and methodological issues. *Curr. Med. Imag. Rev.* **2**:443–451.

——. 2006b. Why is Broca's area involved in syntax? *Cortex* **42**:469–471.

——. 2007. Functional neuroimaging studies of syntactic processing in sentence comprehension: A critical selective review. *Lang. Ling. Compass* **1**:32–47.

Caplan, D., N. Alpert, and G. Waters. 1999. PET studies of syntactic processing with auditory sentence presentation. *NeuroImage* **9**:343–351.

Caplan, D., N. Alpert, G. Waters, and A. Olivieri. 2000a. Activation of Broca's area by syntactic processing under conditions of concurrent articulation. *Hum. Brain Map.* **9**:65–71.

Caplan, D., C. Baker, and F. Dehaut. 1985. Syntactic determinants of sentence comprehension in aphasia. *Cognition* **21**:117–175.

Caplan, D., E. Chen, and G. Waters. 2008. Task-dependent and task-independent neurovascular responses to syntactic processing. *Cortex* **44**:257–275.

Caplan, D., G. DeDe, and J. Michaud. 2006. Task-independent and task-specific deficits in aphasia comprehension. *Aphasiology* **9**:893–920.

Caplan, D., and N. Hildebrandt. 1988. Disorders of Syntactic Comprehension. Cambridge, MA: MIT Press (Bradford Books).

Caplan, D., N. Hildebrandt, and N. Makris. 1996. Location of lesions in stroke patients with deficits in syntactic processing in sentence comprehension. *Brain* **119**:933–949.

Caplan, D., S. Vijayan, G. Kuperberg et al. 2000b. Vascular responses to syntactic processing: Event-related fMRI study of relative clauses. *Hum. Brain Map.* **15**:26–38.

Caplan, D., and G. S. Waters. 1990. Short-term memory and language comprehension: A critical review of the neuropsychological literature. In: The Neuropsychology of Short-Term Memory, ed. T. Shallice and G. Vallar, pp. 337–389. Cambridge: Cambridge Univ. Press.

——. 1999. Verbal working memory capacity and language comprehension. *Behav. Brain Sci.* **22**:114–126.

Caplan, D., G. S. Waters, and G. DeDe. 2007a. Specialized verbal working memory for language comprehension. In: Variation in Working Memory, ed. A. Conway, C. Jarrold, M. Kane, A. Miyake, and J. Towse. Oxford: Oxford Univ. Press.

Caplan, D., G. Waters, and N. Hildebrandt. 1997. Syntactic determinants of sentence comprehension in aphasic patients in sentence-picture matching and enactment tasks. *J. Speech Lang. Hearing Res.* **40**:542–555.

Caplan, D., G. S. Waters, D. Kennedy et al. 2007b. A study of syntactic processing in aphasia. I: Psycholinguistic aspects. *Brain Lang.* **101**:103–150.

Caplan, D., G. S. Waters, D. Kennedy et al. 2007c. A study of syntactic processing in aphasia II: Neurological aspects. *Brain Lang.* **101**:151–177.

Caramazza, A. 2000. The organization of conceptual knowledge in the brain. In: The New Cognitive Neuroscience, ed. M. Gazzaniga. Cambridge, MA: MIT Press.

Caramazza, A., A. G. Basili, J. J. Koller, and R. S. Berndt. 1980. An investigation of repetition and language processing in a case of conduction aphasia. *Brain Lang.* **14**:235–271.

Caramazza, A., E. Capitani, A. Rey, and R. S. Berndt. 2001. Agrammatic Broca's aphasia is not associated with a single pattern of comprehension performance. *Brain Lang.* **76**:158–184.

Caramazza, A., and B. Z. Mahon. 2006. The organization of conceptual knowledge in the brain: The future's past and some future prediction. *Cogn. Neuropsychol.* **23**:13–38.

Caramazza, A., and E. R. Zurif. 1976. Dissociation of algorithmic and heuristic processes in language comprehension: Evidence from aphasia. *Brain Lang.* **3**:572–582.

Carlsson, P., and M. Mahlapuu. 2002. Forkhead transcription factors: Key players in development and metabolism. *Devel. Biol.* **250**:1–23.

Carroll, S. B. 2001. Chance and necessity: The evolution of morphological complexity and diversity. *Nature* **409**:1102–1109.

Carter, A. 1974. Communication in the sensory-motor period. Ph.D. thesis, Univ. of California, Berkeley.

Catani, M., D. K. Jones, R. Donato, and D. H. Ffytche. 2003. Occipito-temporal connections in the human brain. *Brain* **126**:2093–2107.

Cavalli-Sforza, L. L., and M. W. Feldman. 1981. Cultural Transmission and Evolution: A Quantitative Approach. Monographs in Population Biology, vol. 16. Princeton: Princeton Univ. Press.

Changeux, J.-P. 1983. L'Homme Neuronal. Paris: Librairie Arthème Fayard.

Chater, N., F. Reali, and M. H. Christiansen. 2009. Restrictions on biological adaptation in language evolution. *PNAS* **106**:1015–1020.

Chen, C.-H., and V. Honavar. 1999. A neural-network architecture for syntax analysis. *IEEE Trans. Neural Netw.* **10**:94–114.

Cheney, D. L., and R. M. Seyfarth. 1990. How Monkeys See the World. Chicago: Univ. of Chicago Press.

——. 2007. Baboon Metaphysics: The Evolution of a Social Mind. Chicago: Univ. of Chicago Press.

Chesi, C. 2005. Phases and cartography in linguistic computations. Ph.D. thesis, Univ. of Siena.

Chimpanzee Sequencing and Analysis Consortium. 2005. Initial sequence of the chimpanzee genome and comparison with the human genome. *Nature* **437**:69–87.

Chomsky, N. 1957. Syntactic Structures. The Hague: Mouton.

——. 1965a. Aspects of the Theory of Syntax. Cambridge, MA: MIT Press.

——. 1965b. Cartesian Linguistics. New York: Harper and Row.

——. 1968. Language and Mind. New York: Harcourt, Brace, Jovanovich.

——. 1975. Reflections on Language. New York: Pantheon Books.

——. 1981. Government and Binding. Dordrecht: Foris.

——. 1986. Knowledge of Language. New York: Praeger.

——. 1988. Language and Problems of Knowledge: The Managua Lectures. Cambridge, MA: MIT Press.

——. 1992. A minimalist program for linguistic theory. MIT Occasional Papers in Linguistics, No. 1. Cambridge, MA: MIT Press.

——. 1995. The Minimalist Program. Cambridge, MA: MIT Press.

——. 2000. Minimalist inquiries: The framework. In: Step by Step: Essays in Minimalist Syntax in Honor of Howard Lasnik, ed. R. Martin, D. Michaels, and J. Uriagereka. Cambridge, MA: MIT Press.

——. 2001. Derivation by phase. In: Ken Hale: A Life in Language, ed. M. Kenstowicz. Cambridge, MA: MIT Press.

——. 2004. Beyond explanatory adequacy. In: Structures and Beyond: The Cartography of Syntactic Structures (vol. 3), ed. A. Belletti, pp. 104–131. Oxford: Oxford Univ. Press.

——. 2006. On phases. In: Foundational Issues in Linguistic Theory, ed. R. Freidin, C. Otero, and M.-L. Zubizaretta. Cambridge, MA: MIT Press.

——. 2007. Approaching UG from below. In: Interfaces + Recursion = Language?, ed. U. Sauerland and H.-M. Gaertner, pp. 1–30. New York: Mouton de Gruyter.

Christiansen, M. H. 1992. The (non)necessity of recursion in natural language processing. In: Proc. 14th Annual Conf. of the Cognitive Science Society, pp. 665–670. Hillsdale, NJ: Lawrence Erlbaum.

Christiansen, M. H., and N. Chater. 1999. Toward a connectionist model of recursion in human linguistic performance. *Cogn. Sci.* **23**:157–205.

——. 2003. Constituency and recursion in language. In: The Handbook of Brain Theory and Neural Networks, ed. M. A. Arbib, 2nd ed., pp. 267–271. Cambridge, MA: MIT Press.

——. 2008. Language as shaped by the brain. *Behav. Brain Sci.* **31(5)**:489–509.

Christiansen, M. H., and J. T. Devlin. 1997. Recursive inconsistencies are hard to learn: A connectionist perspective on universal word order correlations. In: Proc. 19th Annual Conf. of the Cognitive Science Society, pp. 113–118. Mahwah, NJ: Lawrence Erlbaum.

Christiansen, M. H., and M. R. Ellefson. 2002. Linguistic adaptation without linguistic constraints: The role of sequential learning in language evolution. In: Transitions to Language, ed. A Wray, pp. 335–358. Oxford: Oxford Univ. Press.

Christiansen, M. H., L. Kelly, R. Shillcock, and K. Greenfield. 2008. Impaired artificial grammar learning in agrammatism. *Cogn. Psychol.* **58(2)**:250–271.

Christiansen, M. H., and S. Kirby, eds. 2003. Language Evolution. Oxford: Oxford Univ. Press.

Christiansen, M. H., F. Reali, and N. Chater. 2006. The Baldwin effect works for functional, but not arbitrary, features of language. In: The Evolution of Language: Proc. 6th Intl. Conf. (EVOLANG7), ed. A. Cangelosi, A. Smith, and K. Smith, pp. 27–34. London: World Scientific Publishing.

Cinque, G. 1999. Adverbs and Functional Heads: A Cross-linguistic Perspective. New York: Oxford Univ. Press.

——. 2002. The Structure of CP and DP: The Cartography of Syntactic Structures, vol. 1. New York: Oxford Univ. Press.

——. 2005. Deriving Greenberg's Universal 20 and its exceptions *Ling. Inquiry* **36**:315–332.

Claeys, K. G., P. Dupont, L. Cornette et al. 2004. Color discrimination involves ventral and dorsal stream visual areas. *Cereb. Cortex* **14**:803–822.

Clarke, E., U. H. Reichard, and K. Zuberbuhler. 2006. The syntax and meaning of wild gibbon songs. *PLoS ONE* **1**:73.

Collins, C. 2002. Eliminating labels. In: Derivation and Explanation in the Minimalist Program, ed. S. Epstein and T. D. Seely, pp. 42–64. Oxford: Blackwell.

Comrie, B. 2008. What is passive? In: Studies in Voice and Transitivity, ed. Z. Estrada-Fernández et al., pp. 1–18. Munich: Lincom.

Conard, N. 2003. Palaeolithic ivory sculptures from southwestern Germany and the origins of figurative art. *Nature* **426(6968)**:830–832.

Conard, N., M. Malina, S. Münzel, and F. Seeberge. 2004. Eine Mammutelfenbeinflöte aus dem Aurignacien des Geissenklösterle: Neue Belege für eine musikalische Tradition im Frühen Jungpaläolithikum auf der Schwäbischen Alb. *Archäol. Korr.* **34(4)**:447–462.

Constable, R. T., K. R. Pugh, E. Berroya et al. 2004. Sentence complexity and input modality effects in sentence comprehension: An fMRI study. *NeuroImage* **22**:11–21.

Conway, C. M., and M. H. Christiansen. 2001. Sequential learning in non-human primates. *Trends Cogn. Sci.* **5**:539–546.

Cooke, A., C. DeVita, W. Chen et al. 2001. Sentence comprehension in frontotemporal dementia subgroups: Evidence from neuroimaging. *Brain Lang.* **79**:171–173.

Cooke, A., E. B. Zurif, C. DeVita et al. 2002. Neural basis for sentence comprehension: Grammatical and short-term memory components. *Hum. Brain Map.* **15**:80–94.

Coolidge, F. L., and T. Wynn. 2004. A cognitive and neuropsychological perspective on the Châtelperronian. *J. Anthro. Res.* **60**:55–73.

Coop, G., K. Bullaughey, F. Luca, and M. Przeworski. 2008. The timing of selection at the human FOXP2 gene. *Mol. Biol. Evol.* **25**:1257–1259.

Corballis, M. C. 2002. From Hand to Mouth: The Origins of Language. Princeton: Princeton Univ. Press.

——. 2007a. On phrase structure and brain responses: A comment on Bahlmann, Gunter, and Friederici. *J. Cogn. Neurosci.* **19**:1581–1583.

——. 2007b. Recursion, language, and starlings. *Cogn. Sci.* **31**:697–704.

Cosmides, L., and J. Tooby. 1996. Are humans good intuitive statisticians after all? Rethinking some conclusions from the literature on judgement under uncertainty. *Cognition* **58**:1–73.

Coulson, S., J. W. King, and M. Kutas. 1998a. ERPs and domain specificity: Beating a straw horse. *Lang. Cogn. Proc.* **13**:653–672.

——. 1998b. Expect the unexpected: Event-related brain response to morphosyntactic violations. *Lang. Cogn. Proc.* **13**:21–58.

Crockford, C., and C. Boesch. 2003. Context-specific calls in wild chimpanzees, *Pan troglodytes verus*: Analysis of barks. *Anim. Behav.* **66**:115–125.

——. 2005. Call combinations in wild chimpanzees. *Behavior* **142**:397–421.

Croft, W. 2000. Explaining Language Change: An Evolutionary Approach. London: Longman.

——. 2001. Radical Construction Grammar: Syntactic Theory in Typological Perspective. Oxford: Oxford Univ. Press.

Crosson, B., H. Benefield, M. A. Cato et al. 2003. Left and right basal ganglia and frontal activity during language generation: Contributions to lexical, semantic, and phonological processes. *J. Intl. Neuropsychol. Soc.* **9**:1061–1077.

Crouch, J. E. 1978. Functional Human Anatomy, 3rd ed. Philadelphia: Lea and Fabiger.

Culicover, P. V., and R. Jackendoff. 2005. Simpler Syntax. Oxford: Oxford Univ. Press.

Cupples, L., and A. L. Inglis. 1993. When task demands induce "asyntactic" comprehension: A study of sentence interpretation in aphasia. *Cogn. Neuropsychol.* **10**:201–234.

Cynx, J., and U. Von Rad. 2001. Immediate and transitory effects of delayed auditory feedback on bird song production. *Anim. Behav.* **62**:305–312.

Dall'Asta, L., A. Baronchelli, A. Barrat, and V. Loreto. 2006a. Agreement dynamics on small-world networks. *Europhys. Lett.* **73**:969.

——. 2006b. Non-equilibrium dynamics of language games on complex networks. *Phys. Rev. E* **74**:036105.

Damasio, A. R., and H. Damasio. 1992. Brain and Language. *Sci. Am.* **267**:89–95.

Damasio, A. R., and N. Geschwind. 1984. The neural basis of language. *Ann. Rev. Neurosci.* **7**:127–147.

Dang, M. T., F. Yokoi, H. H. Yin et al. 2006. Disrupted motor learning and long-term synaptic plasticity in mice lacking NMDAR1 in the striatum. *PNAS* **103**:15,254–15,259.

Dapretto, M., and S. Y. Bookheimer. 1999. Form and content: Dissociating syntax and semantics in sentence comprehension. *Neuron* **24**:427–432.

Dawkins, R. 1971. Selective neurone death as a possible memory mechanism. *Nature* **229**:118–119.

——. 1982. The Extended Phenotype. Oxford: Oxford Univ. Press.

Deacon, T. 1997. The Symbolic Species: The Co-Evolution of Language and the Brain. New York: Norton.

——. 2003a. Multilevel selection in a complex adaptive system: The problem of language origins. In: Evolution and Learning: The Baldwin Effect Reconsidered, ed. B. Weber and D. Depew, pp. 81–106. Cambridge, MA: MIT Press.

——. 2003b. Universal grammar and semiotic constraints. In: Language Evolution, ed. M. Christiansen and S. Kirby, pp. 111–139. Oxford: Oxford Univ. Press.

——. 2004. Emerging from the tangled web of language evolution. *Sci. J. Kagaku* **74(7)**:882–886.

Dean, C., F. G. Scholl, J. Choih et al. 1999. Neurexin mediates the assembly of presynaptic terminals. *Nature Neurosci.* **6**:708–716.

de Beule, J. 2004. Creating temporal categories for an ontology of time. In: Proc. 16th Belgium-Netherlands AI Conf., ed. R. Verbrugge, N. Taatgen, and L. Schomaker, pp. 107–114. Groningen: BNAIC04.

——. 2008. The emergence of compositionality, hierarchy and recursion in peer-to-peer interactions. In: The Evolution of Language: Proc. 7th Intl. Conf. (EVOLANG7), ed. A. D. M. Smith, K. Smith, and R. Ferrer i Cancho, pp. 75–82. Barcelona: World Scientific Press.

de Beule, J., and L. Steels. 2005. Hierarchy in fluid construction grammar. In: Proc. of KI-2005, ed. U. Furbach, pp. 1–15. Lecture Notes in AI 3698. Berlin: Springer-Verlag.

De Boer, B. 2000. Self organization in vowel systems. *J. Phonetics* **28(4)**:441–465.

Decety, J., J. Grezes, N. Costes et al. 1997. Brain activity during observation of actions: Influence of action content and subject's strategy. *Brain* **120(10)**:1763–1777.

Dediu, D., and D. Ladd. 2007. Linguistic tone is related to the population frequency of the adaptive haplogroups of two brain size genes, Microcephalin and ASPM. *PNAS* **104(26)**:10,944–10,949.

DeFelipe, J., L. Alonso-Nanclares, and J. I. Arellano. 2002. Microstructure of the neocortex: Comparative aspects. *J. Neurocytol.* **31**:299–316.

DeFries, J. C., and D. W. Fulker. 1985. Multiple regression analysis of twin data. *Behav. Gen.* **15**:467–473.

Dehaene, S. 1997. The Number Sense: How the Mind Creates Mathematics. Oxford: Oxford Univ. Press.

Dehaene, S., and L. Cohen. 2007. Cultural recycling of cortical maps. *Neuron* **56(2)**:384–398.

DeLong, K. A., T. P. Urbach, and M. Kutas. 2005. Probabilistic word pre-activation during language comprehension inferred from electrical brain activity. *Nature Neurosci.* **8**:1117–1121.

Denenberg, V. H. 2005. Behavioral symmetry and reverse asymmetry in the chick and rat. *Behav. Brain Sci.* **28**:597.

d'Errico, F. 2003. The invisible frontier: A multiple species model for the origin of behavioral modernity. *Evol. Anthro.* **12**:182–202.

d'Errico, F., and C. Henshilwood. 2007. Additional evidence for bone technology in the southern African Middle Stone Age. *J. Hum. Evol.* **52(2)**:142–163.

d'Errico, F., C. Henshilwood, G. Lawson et al. 2003. The search for the origin of symbolism, music and language: A multidisciplinary endeavour. *J. World Prehist.* **17(1)**:1–70.

d'Errico, F., C. Henshilwood, M. Vanhaeren, and K. van Niekerk. 2005. *Nassarius kraussianus* shell beads from Blombos Cave: Evidence for symbolic behavior in the Middle Stone Age. *J. Hum. Evol.* **48**:3–24.

d'Errico, F., and G. Lawson. 2006. The sound paradox: How to assess the acoustic significance of archaeological evidence? In: Archaeoacoustics, ed. C. Scarre and G. Lawson, pp. 41–57. Cambridge: McDonald Institute Monographs.

d'Errico, F., and M. Vanhaeren. 2009. Earliest personal ornaments and their significance for the origin of language debate. In: The Cradle of Human Language, ed. R. Botha and C. Knight. Oxford: Oxford Univ. Press.

Desikan, R.S., F. Segonne, and B. Fischl et al. 2006. An automated labeling system for subdividing the human cerebral cortex on MRI scans into gyral based regions of interest. *NeuroImage* **31**:968–980.

Dessalles, J. 1998. Altruism, status, and the origin of relevance. In: Approaches to the Evolution of Language, ed. J. R. Hurford, M. Studdert-Kenedy, and C. Knight, pp. 130–147. Cambridge: Cambridge Univ. Press.

D'Eustachio, P., U. S. Rutishauser, and G. M. Edelman. 1977. Clonal selection and the ontogeny of the immune response. *Intl. Rev. Cytol.* **5**:1–60.

Deutsch, A., and S. Bentin. 2001. Syntactic and semantic factors in processing gender agreement in Hebrew: Evidence from ERPs and eye movements. *J. Mem. Lang.* **45**:200–224.

Deutscher, G. 2005. The Unfolding of Language: An Evolutionary Tour of Mankind's Greatest Invention. New York: Henry Holt.

Devlin, J. T. 2008. Current perspectives on imaging language. In: Neural Correlates of Thinking, ed. E. Kraft, B. Gulyas, and E. Poppel, pp. 123–139. Heidelberg: Springer.

Devlin, J. T., P. M. Matthews, and M. F. S. Rushworth. 2003. Semantic processing in the left prefrontal cortex: A combined functional magnetic resonance imaging and transcranial magnetic stimulation study. *J. Cogn. Neurosci.* **15**:71–84.

de Vries, M. H., P. Monaghan, S. Knecht, and P. Zwitserlood. 2008. Syntactic structure and artificial grammar learning: The learnability of embedded hierarchical structures. *Cognition* **107(2)**:763–774.

deWaal, F. 2001. The Ape and the Sushi Master. New Yokr: Basic Books.

deWaal, F., and F. Lanting. 1997. Bonobo: The Forgotten Ape. Berkeley: Univ. of California Press.

Dibble, H. L. 1989. The implications of stone tool types for the presence of language during the Lower and Middle Palaeolithic. In: The Human Revolution, ed. P. Mellars and C. Stringer, pp. 415–432. Edinburgh: Edinburgh Univ. Press.

Dick, F., E. Bates, and B. Wulfeck et al. 2001. Language deficits, localization, and grammar: Evidence for a distributive model of language breakdown in aphasic patients and neurologically intact individuals. *Psychol. Rev.* **108**:759–788.

Diessel, H. 2005. The Acquisition of Complex Sentences. Cambridge: Cambridge Univ. Press.

Dittmar, M., K. Abbot-Smith, E. Lieven, and M. Tomasello. 2008. German children's comprehension of word order and case marking in causative sentences. *Child Dev.* **79**:1152–1167.

Dominey, P. F. 2005. From sensorimotor sequence to grammatical construction: Evidence from simulation and neurophysiology. *Adap. Behav.* **13(4)**:347–361.
Dominey, P. F., M. Hoen, and T. Inui. 2006. A neurolinguistic model of grammatical construction processing. *J. Cogn. Neurosci.* **18(12)**:2088–2107.
Donald, M. 1991. Origins of the Modern Mind. Cambridge, MA: Harvard Univ. Press.
Donohue, M., and L. Brown. 1999. Ergativity: Some additions from Indonesia. *Australian J. Ling.* **19**:57–76.
Douglas, R. J., and K. A. Martin. 2004. Neuronal circuits of the neocortex. *Ann. Rev. Neurosci.* **27**:419–451.
Doupe, A. J., and P. K. Kuhl. 1999. Birdsong and human speech: Common themes and mechanisms. *Ann. Rev. Neurosci.* **22**:567–631.
Dowman, M., J. Xu, and T. L. Griffiths. 2008. A human model of colour term evolution. In: The Evolution of Language. Hackensack, NJ: World Scientific.
Drai, D., and Y. Grodzinksy. 1999. Comprehension regularity in Broca's aphasia? There's more of it than you ever imagined. *Brain Lang.* **70**:139-140.
Drenhaus, H., P. Beim Graben, D. Saddy, and S. Frisch. 2006. Diagnosis and repair of negative polarity constructions in the light of symbolic resonance analysis. *Brain Lang.* **96**:255–268.
Dronkers, N., D. Wilkin, R. Van Valin, B. Redfern, and J. Jaeger. 2004. Lesion analysis of the brain areas involved in language comprehension. *Cognition* **92**:145–177.
Dryer, M. 1991. SVO languages and the OV:VO typology. *J. Ling.* **27**:443–482.
——. 1992. The Greenbergian word order correlations. *Language* **68**:81–138.
——. 1997. Are grammatical relations universal? In: Essays on Language Function and Language Type: Dedicated to T. Givon, ed. J. Bybee, J. Haiman, and S. Thompson, pp. 115–143. Amsterdam: John Benjamins.
Dunbar, R. 1992. Co-evolution of cortex size, group size and language. *Behav. Brain Sci.* **16(4)**:681–735.
——. 1998. Theory of mind and the evolution of language. In: Approaches to the Evolution of Language, ed. J. R. Hurford, M. Studdert-Kenedy, and C. Knight, pp. 92–110. Cambridge: Cambridge Univ. Press.
Egnor, S. E., J. G. Wickelgren, and M. D. Hauser. 2007. Tracking silence: Adjusting vocal production to avoid acoustic interference. *J. Comp. Physiol. A* **193**:477–483.
Eickhoff, S. B., T. Paus, S. Caspers et al. 2007a. Assignment of functional activations to probabilistic cytoarchitectonic areas revisited. *NeuroImage* **36**:511–521.
Eickhoff, S. B., A. Schleicher, F. Scheperjans, N. Palomero-Gallagher, and K. Zilles. 2007b. Analysis of neurotransmitter receptor distribution patterns in the cerebral cortex. *NeuroImage* **34**:1317–1330.
Elman, J. L. 1990. Finding structure in time. *Cogn. Sci.* **14**:179–211.
Elman, J. L., E. A. Bates, M. Johnson et al. 1996. Rethinking Innateness: A Connectionist Perspective on Development. Cambridge, MA: MIT Press.
Embick, D., A. Marantz, Y. Miyashita, W. O'Neil, and K. L. Sakai. 2000. A syntactic specialization for Broca's area. *PNAS* **97**:6150–6154.
Emonds, J. E. 1970. Root and structure preserving transformations. Ph.D. thesis, MIT.
——. 1978. The verbal complex V'–V in French. *Ling. Inquiry* **9**:151–175.
——. 1980. Word order in generative grammar. *J. Ling. Res.* **1**:33–54.
Enard, W., M. Przeworski, S. E. Fisher et al. 2002. Molecular evolution of FOXP2, a gene involved in speech and language. *Nature* **418**:869–872.
Ericsson, A., and W. Kintsch. 1995. Long term working memory. *Psychol. Rev.* **102(2)**:211–245.

Esser, K. H., C. J. Condon, N. Suga, and J. S. Kanwal. 1997. Syntax processing by auditory cortical neurons in the FM-FM area of the mustached bat *Pteronotus parnellii*. *PNAS* **94**:14,019–14,024.

Everett, D. 2005. Cultural constraints on Pirahã grammar. *Curr. Anthro.* **46**:621–646.

Falcaro, M., A. Pickles, D. F. Newbury et al. 2008. Genetic and phenotypic effects of phonological short-term memory and grammatical morphology in specific language impairment. *Gen. Brain Behav.* **7**:393–402.

Falk, D. 2004. Prelinguistic evolution in early hominins: Whence motherese? *Behav. Brain Sci.* **27**:491–503.

Felsenfeld, S. 2002. Finding susceptibility genes for developmental disorders of speech: The long and winding road. *J. Comm. Disorders* **35**:329–345.

Ferland, R. J., T. J. Cherry, P. O. Preware, E. E. Morrisey, and C. A. Walsh. 2003. Characterization of Foxp2 and Foxp1 mRNA and protein in the developing and mature brain. *J. Comp. Neurol.* **460**:266–279.

Fernald, R. D., and S. A. White. 2000. Social control of brains: From behavior to genes. In: The New Cognitive Neuroscience, ed. M. Gazzaniga. Cambridge, MA: MIT Press.

Fernández-Duque, D. 2009. Cognitive and neural underpinnings of syntactic complexity. In: The Genesis of Syntactic Complexity, ed. T. Givon and M. Shibatani. Amsterdam: John Benjamins, in press.

Fernando, C., K. K. Karishma, and E. Szathmáry. 2008. Copying and evolution of neuronal topology. *PLoS ONE* **3(11)**:e3775.

Ferreira, F., and N. D. Patson. 2007. The "good enough" approach to language comprehension. *Lang. Ling. Compass* **1**:71–83.

Fiebach, C. J., M. Schlesewsky, G. Lohmann, D. Y. von Cramon, and A. D. Friederici. 2005. Revisiting the role of Broca's area in sentence processing: Syntactic integration versus syntactic working memory. *Hum. Brain Map.* **24**:79–91.

Fiete, I. R., R. H. Hahnloser, M. S. Fee, and H. S. Seung. 2004. Temporal sparseness of the premotor drive is important for rapid learning in a neural network model of birdsong. *J. Neurophys.* **92**:2274–2282.

Fiez, J. A. 1997. Phonology, semantics, and the role of the left inferior pre-frontal cortex. *Hum. Brain Map.* **5(2)**:79–83.

Finkel, L. H., and G. M. Edelman. 1985. Interaction of synaptic modification rules within populations of neurons. *PNAS* **82**:1291–1295.

Fischer, J. 2003. Developmental modifications in the vocal behavior of non-human primates. In: Primate Audition: Ethology and Neurobiology, ed. A. A. Ghazanfar, pp. 109–125. CRC Press: Boca Raton, Florida.

———. 2008. Transmission of acquired information in nonhuman primates. In: Encyclopedia of Learning and Memory, ed. D. Sweatt, R. Menzel, H. Eichenbaum, and H. Roediger, pp. 299–313. Oxford: Elsevier.

Fisher, S. E. 2005. Dissection of molecular mechanisms underlying speech and language disorders. *Appl. Psycholing.* **26**:111–128.

———. 2006. Tangled webs: Tracing the connections between genes and cognition. *Cognition* **101**:270–297.

Fisher, S. E., C. S. Lai, and A. P. Monaco. 2003. Deciphering the genetic basis of speech and language disorders. *Ann. Rev. Neurosci.* **26**:57–80.

Fisher, S. E., and G. F. Marcus. 2006. The eloquent ape: Genes, brains and the evolution of language. *Nature Rev. Genet.* **7**:9–20.

Fitch, W. T. 1997. Vocal tract length and formant frequency dispersion correlate with body size in rhesus macaques. *J. Acoustic Soc. Am.***102**:1213–1222.

Fitch, W. T., and M. D. Hauser. 2004. Computational constraints on syntactic processing in a nonhuman primate. *Science* **303**:377–380.
Flint, J. 1999. The genetic basis of cognition. *Brain* **122**:2015–2031.
Fodor, J. A. 1983. The Modularity of Mind. Cambridge, MA: Bradford Books.
Fox, D. 2000. Economy and Semantic Interpretation. Cambridge, MA: MIT Press.
Fox, D., and Y. Grodzinsky. 1998. Children's passive: A view from the by-phrase. *Ling. Inquiry* **29**:311–332.
Frazier, L. 1987a. Sentence processing. In: Attention and Performance, ed. M. Coltheart, pp. 559–586. Hillsdale: Erlbaum.
——. 1987b. Structure in auditory word recognition. *Cognition* **25**:157–187.
Freudenthal, D., J. M. Pine, J. Aguado-Orea, and F. Gobet. 2007. Modeling the developmental patterning of finiteness marking in English, Dutch, German, and Spanish using MOSAIC. *Cogn. Sci.* **31**:311–341.
Friederici, A. D. 1995. The time course of syntactic activation during language processing: A model based on neuropsychological and neurophysiological data. *Brain Lang.* **50**:259–281.
——. 2000. The neuronal dynamics of auditory language comprehension. In: Image, Language, Brain: Papers from the First Mind Articulation Project Symposium, ed. A. Marantz, Y. Miyashita, and W. O'Neil. Cambridge, MA: MIT Press.
——. 2002. Towards a neural basis of auditory sentence processing. *Trends Cogn. Sci.* **6**:78–84.
——. 2004a. The neural basis of syntactic processes. In: Cognitive Neurosciences III, ed. M. S. Gazzaniga, pp. 789–801. Cambridge, MA: MIT Press.
——. 2004b. Processing local transitions versus long-distance syntactic hierarchies. *Trends Cogn. Sci.* **8**:245–247.
——. 2006. Broca'a area and the ventral premotor cortex in language: Functional differentiation and specificity. *Cortex* **42**:472–475.
Friederici, A. D., and K. Alter. 2004. Lateralization of auditory language functions: A dynamic dual pathway model. *Brain Lang.* **89**:267–276.
Friederici, A. D., J. Bahlmann, S. Heim, R. I. Scubotz, and A. Anwander. 2006a. The brain differentiates human and non-human grammar: Functional localization and structural connectivity. *PNAS* **103(7)**:2458–2463.
Friederici, A. D., C. J. Fiebach, M. Schlesewsky, I. Bornkessel, and D. Y. von Cramon. 2006b. Processing linguistic complexity and grammaticality in the left frontal cortex. *Cereb. Cortex* **16(12)**:1709–1717.
Friederici, A. D., and S. Frisch. 2000. Verb-argument structure processing: The role of verb-specific and argument-specific information. *J. Mem. Lang.* **43**:476–507.
Friederici, A. D., A. Hahne, and A. Mecklinger. 1996. The temporal structure of syntactic parsing: Early and late event-related brain potential effects. *J. Exp. Psychol. Learn. Mem. Cogn.* **22**:1219–1248.
Friederici, A. D., A. Hahne, and D. Saddy. 2002. Distinct neurophysiological patterns reflecting aspects of syntactic complexity and syntactic repair. *J. Psycholing. Res.* **31**:45–63.
Friederici, A. D., A. Hahne, and D. Y. von Cramon. 1998. First-pass versus second-pass parsing processes in a Wernick's and a Broca's aphasic: Electrophysiological evidence for a double dissociation. *Brain Lang.* **62**:311–341.
Friederici, A. D., and S. A. Kotz. 2003. The brain basis of syntactic processes: Functional imaging and lesion studies. *NeuroImage* **20**:S8–S17.

Friederici, A. D., E. Pfeifer, and A. Hahne. 1993. Event-related brain potentials during natural speech processing: Effects of semantic, morphological and syntactic violations. *Cogn. Brain Res.* **1**:183–192.

Friederici, A. D., S.-A. Rüschemeyer, A. Hahne, and C. J. Fiebach. 2003. The role of left inferior frontal and superior temporal cortex in sentence comprehension: Localizing syntactic and semantic processes. *Cereb. Cortex* **13**:170–177.

Friederici, A. D., D. Y. von Cramon, and S. A. Kotz. 1999. Language related brain potentials in patients with cortical and subcortical left hemisphere lesions. *Brain* **122**:1033–1047.

Friederici, A. D., Y. Wang, C. S. Herrmann, B. Maess, and U. Oertel. 2000. Localization of early syntactic processes in frontal and temporal cortical areas: A magnetoencephalographic study. *Hum. Brain Map.* **11**:1–11.

Friedrich, M., and A. D. Friederici. 2004. N400-like semantic incongruity effect in 19-month-olds: Processing known words in picture contexts. *J. Cogn. Neurosci.* **16**:1465–1477.

——. 2005. Semantic sentence processing reflected in the event-related potentials of one- and two-year-old children. *Neuroreport* **16**:1801–1804.

Frisch, S., A. Hahne, and A. D. Friederici. 2004. Word category and verb-argument structure information in the dynamics of parsing. *Cognition* **91**:191–219.

Friston, K. J., J. T. Ashburner, S. J. Kiebel, T. E. Nichols, and W. D. Penny, eds. 2007. Statistical Parametric Mapping. Amsterdam: Academic Press.

Fujita, E., Y. Tanabe, A. Shiota et al. 2008. Ultrasonic vocalization impairment of FOXP2 (R552H) knockin mice related to speech-language disorder and abnormality of Purkinje cells. *PNAS* **105**:3117–3122.

Funabiki, Y., and M. Konishi. 2003. Long memory in song learning by zebra finches. *J. Neurosci.* **23**:6928–6935.

Fuster, J. M. 1995. Temporal processing. *Ann. NY Acad. Sci.***769**:173–181.

Gaillard, W. D., L. Hertz–Pannier, S. H. Mott et al. 2000. Functional anatomy of cognitive development: fMRI of verbal fluency in children and adults. *Neurology* **54**:180–185.

Gaillard, W. D., M. Pugliese, C. B. Grandin et al. 2001. Cortical localization of reading in normal children: An fMRI language study. *Neurology* **57**:47–54.

Gainotti, G. 2007. Face familiarity feelings, the right temporal lobe and the possible underlying neural mechanisms. *Brain Res. Rev.* **56**:214–235.

Galantucci, B. 2005. An experimental study of the emergence of human communication systems. *Cogn. Sci.* **29**:737–767.

Galuske, R.A.W., W. Schlote, H. Bratzke, and W. Singer. 2000. Interhemispheric asymmetries of the modular structure in human temporal cortex. *Science* **289**:1946–1949.

Gandour, J., and R. Dardarananda. 1983. Identification of tonal contrasts in Thai aphasic patients. *Brain Lang.* **18**:98–114.

Gandour, J., D. Wong, L. Hsieh et al. 2002. A crosslinguistic PET study of tone perception. *J. Cogn. Neurosci.* **12**:207–222.

Gannon, P. J., R. L. Holloway, D. C. Broadfield, and A. R. Braun. 1998. Asymmetry of chimpanzee planum temporale: Humanlike pattern of Wernickes brain language area homolog. *Science* **279**:220–222.

Gargalovic, P. S., M. Imura, B. Zhang et al. 2006. Identification of inflammatory gene modules based on variations of human endothelial cell responses to oxidized lipids. *PNAS* **103**:12,741–12,746.

Garrido, M. I., J. M. Kilner, S. J. Kiebel, and K. J. Friston. 2007. Evoked brain responses are generated by feedback loops. *PNAS* **104**:20,961–20,966.

Gathercole, S. E., and A. D. Baddeley. 1990. Phonological memory deficits in language disordered children: Is there a causal connection? *J. Mem. Lang.* **29**:336–360.

——. 1993. Working Memory and Language. Hillsdale, NJ: Erlbaum.

Gazdar, G., E. Klein, G. Pullum, and I. Sag. 1985. Generalized Phrase Structure Grammar. Oxford: Blackwell.

Geary, D. C. 2005. The Origin of Mind: The Evolution of Brain, Cognition and General Intelligence. Washington, DC: Am. Psych. Association.

Geisler, W., and R. Diehl. 2003. A Bayesian approach to the evolution of perceptual and cognitive systems. *Cogn. Sci.* **27**:379–402.

Geissman, T. 2000. Gibbon songs and human music from an evolutionary perspective. In: The Origins of Music, ed. L. Nils, B. M. Wallin and S. Brown, pp. 103–123. Cambridge, MA: MIT Press.

Gelfand, J. R., and S. Y. Bookheimer. 2003. Dissociating neural mechanisms of temporal sequencing and processing phonemes. *Neuron* **38**:831–842.

Gentilucci, M., F. Benuzzi, M. Gangitano, and S. Grimaldi. 2001. Grasp with hand and mouth: A kinematic study on healthy subjects. *J. Neurophys.* **86**:1685–1699.

Gentilucci, M., and M. C. Corballis. 2006. From manual gesture to speech: A gradual transition. *Neurosci. Biobehav. Rev.* **30**:949–960.

Gentner, T. Q., K. M. Fenn, D. Margoliash, and H. C. Nusbaum. 2006. Recursive syntactic pattern learning by songbirds. *Nature* **440**:1204–1207.

Gentner, T. Q., and S. H. Hulse. 1998. Perceptual mechanisms for individual vocal recognition in European starlings, Sturnus vulgaris. *Anim. Behav.* **56**:579–594.

Gentner, T. Q., S. H. Hulse, G. E. Bentley, and G. F. Ball. 2000. Individual vocal recognition and the effect of partial lesions to HVc on discrimination, learning, and categorization of conspecific song in adult songbirds. *J. Neurobiol.* **42**:117–133.

George, I., B. Vernier, J. P. Richard, M. Hausberger, and H. Cousillas. 2004. Hemispheric specialization in the primary auditory area of awake and anesthetized starlings (*Sturnus vulgaris*). *Behav. Neurosci.* **118**:597–610.

Gerdes, L. U., C. Gerdes, P. S. Hansen, I. C. Klausen, and O. Faergman. 1996. Are men carrying the apolipoprotein ε4- or ε2 allele less fertile than ε3ε3 genotypes? *Hum. Genet.* **98**:239–242.

Gernsbacher, M. A. 1990. Language Comprehension as Structure Building. Hillsdale, NJ: Erlbaum.

Geschwind, N. 1970. The organization of language and the brain. *Science* **170**.

——. 1979. Specialization of the human brain. *Sci. Am.* **241**:180–199.

——. 1985. Mechanisms of change after brain lesions. *Ann. NY Acad. Sci.* **457**:1–11.

Geschwind, N., and W. Levitsky. 1968. Human brain: Left-right asymmetries in temporal speech region. *Science* **161**:186–187.

Gibson, E. 1998. Linguistic complexity: Locality of syntactic dependencies. *Cognition* **68**:1–76.

Gibson, E., and K. Wexler. 1994. Triggers. *Ling. Inquiry* **25**:407–454.

Giedd, J. N., J. Blumenthal, N. O. Jeffries et al. 1999. Brain development during childhood and adolescence: A longitudinal MRI study. *Nat. Neurosci.* **2**:861–863.

Gilbert, S. L., W. B. Dobyns, and B. T. Lahn. 2005. Genetic links between brain development and brain evolution. *Nature Rev. Genet.* **6**:581–590.

Giles, H. 1984. The dynamics of speech accommodation. *Int. J. Sociol. Lang.* **46**:1–155.

Giorgio, A., K. E. Watkins, G. Douaud et al. 2008. Changes in white matter microstructure during adolescence. *NeuroImage* **39**:52–61.

Gisiger, T., M. Kerszberg, and J.-P. Changeux. 2005. Acquisition and performance of delayed-response tasks: A neural network model. *Cereb. Cortex* **15**:489–506.

Givón, T. 1971. Historical syntax and synchronic morphology: An archaeologist's field trip. *Chicago Ling. Soc.* **7**:394–415.

——. 1979. On Understanding Grammar. New York: Academic Press.

——. 1989. Mind, Code and Context: Essays in Pragmatics. Hillsdale, NJ: Erlbaum.

——. 1990. Natural language learning and organized language teaching. In: Variability in Second Language Acquisition, ed. H. Burmeister and P. Rounds. Eugene: Univ. of Oregon.

——. 1995. Functionalism and Grammar. Amsterdam: John Benjamins.

——. 2001. Syntax. Amsterdam: John Benjamins.

——. 2002. Bio-linguistics. Amsterdam: John Benjamins.

——. 2005. Context as Other Minds: The Pragmatics of Cognition and Communication. Amsterdam: John Benjamins.

——. 2008. On the relational properties of passive clauses: A diachronic perspective. In: Studies in Voice and Transitivity, ed. Z. Estrada-Fernández et al. Munich: Lincom.

Glasser, M. F., and J. K. Rilling. 2008. DTI tractography of the human brain's language pathways. *Cereb. Cortex* **18**:2471–2482.

Gold, E. 1967. Language identification in the limit. *Information and Control* **10**:447–474.

Goldberg, A. E. 1995. Constructions: A Construction Grammar Approach to Argument Structure. Chicago: Univ. of Chicago Press.

——. 2006. Constructions at Work: The Nature of Generalization in Language. Oxford Univ. Press.

Goldberg, T. E., and D. R. Weinberger. 2004. Genes and the parsing of cognitive processes. *Trends Cogn. Sci.* **8**:325–335.

Goldin-Meadow, S., and C. Mylander. 1998. Spontaneous sign systems created by deaf children in two cultures. *Nature* **391(6664)**:279–281.

Goldin-Meadow, S., W. C. So, A. Ozyurek et al. 2008. The natural order of events: How speakers of different languages represent events nonverbally. *PNAS* **105**:9163–9168.

Golestani, N., F. X. Alario, S. Meriaux et al. 2006. Syntax production in bilinguals. *Neuropsychologia* **44**:1029–1040.

Gómez, R. L., and L. A. Gerken. 2000. Infant artificial language learning and language acquisition. *Trends Cogn. Sci.* **4**:178–186.

Goodale, M. A. 2000. Perception and action in the human visual system. In: The New Cognitive Neuroscience, ed. M. Gazzaniga. Cambridge, MA: MIT Press.

Gopnik, M. 1990. Feature-blind grammar and dysphasia. *Nature* **344**:715.

——. Familial language impairment: More English evidence. *Folia Phoniatr. Logop.* **51**:5–19.

Gould, S. J. 1977. Ontogeny and Phylogeny. Cambridge, MA: Harvard Univ. Press.

Grainger, J., A. Rey, and S. Dufau. 2008. Letter perception: From pixels to pandemonium. *Trends Cogn. Sci.* **12**:381–387.

Greenberg, J. 1963. Universals of Language. Cambridge, MA: MIT Press.

——. 1974. The relation of frequency to semantic feature in a case language (Russian). *Working Papers on Language Universals* **16**:21–46.

——. 1976. Language Universal, With Special Reference to Feature Hierarchies. The Hague: Mouton.
——. 1978. Diachrony, synchrony and language universals. In: Universals of Human Language, ed. J. Greenberg, C. Ferguson, and E. Moravcsik. Stanford: Stanford Univ. Press.
——. 1979. Rethinking linguistic diachrony. *Language* **55**:275–290.
Greenfield, P. M. 1991. Language, tools and brain: The ontogeny and phylogeny of hierarchically organized sequential behaviour. *Behav. Brain Sci.* **14**:531–595.
Grice, H. P. 1975. Logic and conversation. In: Speech Acts, ed. P. Cole and J. Morgan. New York: Academic Press.
Griffiths, T. L., B. R. Christian, and M. L. Kalish. 2008. Using category structures to test iterated learning as a method for revealing inductive biases. *Cogn. Sci.* **32**:68–107.
Griffiths, T. L., and M. Kalish. 2005. A Bayesian view of language evolution by iterated learning. In: Proc. 27th Annual Conf. of the Cognitive Science Society, ed. B. G. Bara, L. Barsalou, and M. Bucciarelli. Mahwah, NJ: Lawrence Erlbaum.
——. 2007. Language evolution by iterated learning with Bayesian agents. *Cogn. Sci.* **31**:441–480.
Griswold, C. K. 2006. Pleiotropic mutation, modularity, and evolvability. *Evol. Dev.* **8**:81–93.
Grodzinsky, Y. 2000. The neurology of syntax: Language use without Broca'a area. *Behav. Brain Sci.* **23**:47–117.
Grodzinsky, Y., and K. Amunts, eds. 2006. Broca's Region. Oxford: Oxford Univ. Press.
Grodzinsky, Y., and A. D. Friederici. 2006. Neuroimaging of syntax and syntactic processing. *Curr. Op. Neurobiol.* **16**:240–246.
Grodzinsky, Y., and A. Santi. 2008. The battle for Broca's region. *Trends Cog. Sci.* **12**:474-480.
Grodzinsky, Y., K. Wexler, Y.-C. Chien, S. Marakovitz, and J. Solomon. 1993. The breakdown of binding relations. *Brain Lang.* **45**:396–422.
Gross, C. G. 1992. Representation of visual stimuli in inferior temporal cortex. *Phil. Trans. Roy. Soc. Lond. B* **335**:3–10.
Groszer, M., D. A. Keays, R. M. Deacon et al. 2008. Impaired synaptic plasticity and motor learning in mice with a point mutation implicated in human speech deficits. *Curr. Biol.* **18(5)**:354–362.
Grüning, A. 2007. Elman backpropagation as reinforcement for Simple Recurrent Networks. *Neural Comp.* **19**:3108–3131.
Gulyás, B. 2001. The dynamics of cortical macronetworks in the human brain. *Brain Res. Bull.* **53**:251–253.
Gulyás, B., C. A. Heywood, D. B. Popplewell, A. Cowey, and P. E. Roland. 1994. Visual form discrimination from color or motion cues: Functional anatomy by positron emission tomography. *PNAS* **91**:9965–9969.
Gulyás, B., and P. E. Roland. 1991. Cortical fields participating in form and color discrimination in the human brain. *Neuroreport* **2**:585–588.
——. 1994a. Binocular disparity detection in human visual cortex: Functional anatomy by positron emission tomography. *PNAS* **91**:1239–1243.
——. 1994b. Processing and analysis of form, color and binocular disparity in the human brain: Functional anatomy by positron emission tomography. *Eur. J. Neurosci.* **6**:1811–1828.

——. 1995. Visual cortical fields participating in spatial frequency and orientation discrimination: Functional anatomy by positron emission tomography. *Hum. Brain Map.* **3**:133–152.

Gulyás, B., P. E. Roland, A. Cowey, C. A. Heywood, and D. Popplewell. 1998. Visual form discrimination from texture cues: A PET study. *Hum. Brain Map.* **6**:115–127.

Gulyás, B., and N. Sjöholm. 2007. Principles of positron emission tomography. In: Functional Neuroimaging in Clinical Populations, ed. F. G. Hillary and J. DeLuca, pp. 3–30. New York: Guilford Press.

Gunter, T. C., A. D. Friederici, and H. Schriefers. 2000. Syntactic gender and semantic expectancy: ERPs reveal early autonomy and late interaction. *J. Cogn. Neurosci.* **12**:556–568.

Haarmann, H. J., M. A. Just, and P. A. Carpenter. 1997. Aphasic sentence comprehension as a resource deficit: A computational approach. *Brain Lang.* **59**:76–120.

Haarmann, H. J., and H. H. Kolk. 1991. Syntactic priming in Broca's aphasics: Evidence for slow activation. *Aphasiology* **5**:247–263.

——. 1994. On-line sensitivity to subject-verb agreement violations in Broca's aphasics: The role of syntactic complexity and time. *Brain Lang.* **46**:493–516.

Habib, R., L. Nyberg, and E. Tulving. 2003. Hemispheric asymmetries of memory: The HERA model revisited. *Trends Cogn. Sci.* **7**:241–245.

Haegeman, L. 1996. Introduction to Government and Binding Theory. Oxford: Blackwell.

Haesler, S., C. Rochefort, B. Georgi et al. 2007. Incomplete and inaccurate vocal imitation after knockdown of FOXP2 in songbird basal ganglia nucleus Area X. *PLoS Biol.* **5(12)**:e321.

Haesler, S., K. Wada, A. Nshdejan et al. 2004. FoxP2 expression in avian vocal learners and non-learners. *J. Neurosci.* **24**:3164–3175.

Hagoort, P. 2003. How the brain solves the binding problem for language: A neurocomputational model of syntactic processing. *NeuroImage* **20**:S18–S29.

——. 2005. On Broca, brain and binding: A new framework. *Trends Cogn. Sci.* **9**:416–423.

Hagoort, P., C. Brown, and J. Groothusen. 1993. The syntactic positive shift (SPS) as an ERP measure of syntactic processing. *Lang. Cogn. Proc.* **8**:439–483.

Hagoort, P., C. Brown, and L. Osterhout. 1999a. The neurocognition of syntactic processing. In: Neurocognition of Language, eds. C. Brown and P. Hagoort, pp. 273–316. Oxford: Oxford Univ. Press.

Hagoort, P., L. Hald, M. Bastiaansen, and K. M. Petersson. 2004. Integration of word meaning and world knowledge in language comprehension. *Science* **304**:438–441.

Hagoort, P., P. Indefrey, C. Brown et al. 1999b. The neural circuitry involved in the reading of German words and pseudowords: A PET study. *J. Cogn. Neurosci.* **11**:383–398.

Hagoort, P., M. Wassenaar, and C. M. Brown. 2003. Syntax-related ERP-effects in Dutch. *Cogn. Brain Res.* **16**:38–50.

Hahne, A., and A. D. Friederici. 1999. Electrophysiological evidence for two steps in syntactic analysis: Early automatic and late controlled processes. *J. Cogn. Neurosci.* **11**:194–205.

Hahne, A., and J. D. Jescheniak. 2001. What's left if the Jabberwock gets the semantics? An ERP investigation into semantic and syntactic processes during auditory sentence comprehension. *Cogn. Brain Res.* **11**:199–212.

Hahnloser, R. H., A. A. Kozhevnikov, and M. S. Fee. 2002. An ultra-sparse code underlies the generation of neural sequences in a songbird. *Nature* **419**:65–70.

Hakuta, K., E. Bialystok, and E. Wiley. 2003. Critical evidence: A test of the critical-period hypothesis for second-language acquisition. *Psychol. Sci.* **14**:31–38.

Haller, S., E. W. Radue, M. Erb, W. Grodd, and T. Kircher. 2005. Overt sentence production in event-related fMRI. *Neuropsychologia* **43**:807–814.

Hammerschmidt, K., and J. Fischer. 2008. Constraints in primate vocal production. In: The Evolution of Communicative Creativity: From Fixed Signals to Contextual Flexibility, ed. U. Griebel and K. Oller, pp. 93–119. Cambridge, MA: MIT Press.

Hamzei, F., M. Rijntjes, C. Dettmers et al. 2003. The human action recognition system and its relationship to Broca's area: An fMRI study. *NeuroImage* **19**:637–644.

Hara, E., L. Kubikova, N. A. Hessler, and E. D. Jarvis. 2007. Role of the midbrain dopaminergic system in modulation of vocal brain activation by social context. *Eur. J. Neurosci.* **25**:3406–3416.

Harris, D., and S. Bullock. 2002. Enhancing game theory with coevolutionary simulation models of honest signalling. In: Congress on Evolutionary Computation, ed. D. Fogel, pp. 1594–1599. Washington, DC: IEEE Press.

Harris, T., K. Wexler, and P. Holcomb. 2000. An ERP investigation of binding and coreference. *Brain Lang.* **75(3)**:313–346.

Haspelmath, M. 1990. The grammaticalization of passive morphology. *Stud. Lang.* **14**:25–72.

——. 2007. Pre-established categories don't exist: Consequences for language description and typology. *Ling. Typol.* **11(1)**:119–132.

Hauk, O., I. Johnsrude, and F. Pulvermüller. 2004. Somatotopic representation of action words in human motor and premotor cortex. *Neuron* **41**(2):301–307.

Hauser, M. D., N. Chomsky, and W. T. Fitch. 2002. The faculty of language: What is it, who has it, and how did it evolve? *Science* **298**:1569–1579.

Hawkins, J. A. 1990. A parsing theory of word order universals. *Ling. Inquiry* **21**:223–262.

——. 1994. A Performance Theory of Order and Constituency. Cambridge: Cambridge Univ. Press.

——. 2004. Efficiency and Complexity in Grammars. Oxford: Oxford Univ. Press.

Haxby, J. V., M. I. Gobbini, M. L. Furey et al. 2001. Distributed and overlapping representations of faces and objects in ventral temporal cortex. *Science* **293**:2425–2430.

Heim, S., B. Opitz, and A. D. Friederici. 2003. Distributed cortical networks for syntax processing: Broca's area as the common denominator. *Brain Lang.* **85**:402–408.

Heine, B. 1997. Cognitive Foundations of Grammar. Oxford: Oxford Univ. Press.

Heine, B., and T. Kuteva. 2002. World Lexicon of Grammaticalization. Cambridge: Cambridge Univ. Press.

——. 2007. The Genesis of Grammar. Oxford: Oxford Univ. Press.

——. 2008. The Genesis of Grammar: A Reconstruction. Oxford: Oxford Univ. Press.

Helekar, S. A., G. G. Espino, A. Botas, and D, B. Rosenfield. 2003. Development and adult phase plasticity of syllable repetitions in the birdsong of captive zebra finches (Taeniopygia guttata). *Behav. Neurosci.* **117**:939–951.

Henrich, J., and R. McElreath. 2003. The evolution of cultural evolution. *Evol. Anthro.* **12**:123–135.

Henrich, N., and J. Henrich. 2007. Why Humans Cooperate: A Cultural and Evolutionary Explanation. Oxford: Oxford Univ. Press.

Henshilwood, C., F. d'Errico, M. Vanhaeren et al. 2004. Middle Stone Age shell beads from South Africa. *Science* **304**:404.

Henshilwood, C., F. d'Errico, R. Yates et al. 2002. Emergence of modern human behavior: Middle Stone Age engravings from South Africa. *Science* **295**:1278–1280.

Henshilwood, C., and B. Dubreuil. 2009. Reading the artefacts: Gleaning language skills from the Middle Stone Age in southern Africa. In: The Cradle of Language, Volume 2: African perspectives, pp. 41–61. Oxford: Oxford Univ. Press.

Henshilwood, C., and C. W. Marean. 2003. The origin of modern human behavior critique of the models and their test implications. *Curr. Anthro.* **44(5)**:627–651.

Hessler, N. A. and A. J. Doupe. 1999. Social context modulates singing in the songbird forebrain. *Nature Neurosci.* **2**:209–211.

Hewes, G. 1973. Primate communication and the gestural origin of language. *Curr. Anthro.* **14**:5–25.

Hickok, G., B. Buchsbaum, C. Humphries, and T. Muftuler. 2003. Auditorymotor interaction revealed by fMRI: Speech, music, and working memory in area Spt. *J. Cogn. Neurosci.* **15**:673–682.

Hickok, G., and D. Poeppel. 2007. Opinion: The cortical organization of speech processing. *Nature Rev. Neurosci.* **8**:393–402.

Hildebrandt, N., D. Caplan, and K. Evans. 1987. The man left without a trace: A case study of aphasic processing of empty categories. *Cogn. Neuropsychol.* **4(3)**:257–302.

Hill, R. S., and C. A. Walsh. 2005. Molecular insights into human brain evolution. *Nature* **437**:64–67.

Hinton, G. E., and S. J. Nowlan. 1987. How learning can guide evolution. *Complex Systems* **1**:495–502.

Hinton, G. E., and R. R. Salakhutdinov. 2006. Reducing the dimensionality of data with neural networks. *Science* **313**:504–507.

Hjelmfelt, A., E. D. Weinberger, and J. Ross. 1992. Chemical implementation of finite-state machines. *PNAS* **89**:383–387.

Hoekstra, T., and J. Kooij. 1988. The innateness hypothesis. In: Explaining Language Universals, ed. J. A. Hawkins. Oxford: Blackwell.

Hoen, M., and P. F. Dominey. 2000. ERP analysis of cognitive sequencing: A left anterior negativity related to structural transformation processing. *Neuroreport* **11**:3187–3191.

Hofbauer, J., and K. Sigmund. 1998. Evolutionary Games and Population Dynamics. Cambridge: Cambridge Univ. Press.

Holland, J. H. 1992. Adaptation in Natural and Artificial Systems. Cambridge, MA: MIT Press.

Holland, S. K., E. Plante, A. Weber Byars et al. 2001. Normal fMRI brain activation patterns in children performing a verb generation task. *NeuroImage* **14**:837–843.

Holland, S. K., J. Vannest, M. Mecoli et al. 2007. Functional MRI of language lateralization during development in children. *Intl. J. Audiol.* **46**:533–551.

Homae, F., H. Watanabe, T. Nakano, K. Asakawa, and G. Taga. 2006. The right hemisphere of sleeping infant perceives sentential prosody. *Neurosci. Res.* **54**:276–280.

Hopcroft, J. E., and J. D. Ullman. 1979. Introduction to Automata Theory, Languages, and Computation. Reading, MA: Addison-Wesley.

Hopkins, W. D., and C. Cantalupo. 2008. Theoretical speculations on the evolutionary origins of hemispheric specialization. *Curr. Dir. Psychol. Sci.* **17**:233–237.

Hopper, P. 1987. Emergent grammar. *Berkeley Ling. Conf.* **13**:139–157.

Horning, J. 1969. A study of grammatical inference. Ph.D. thesis, Computer Science Dept., Stanford University.

Hornstein, N., and D. Lightfoot, eds. 1981. Explanation in Linguistics: The Logical Problem of Language Acquisition. London: Longman.

Horvath, S., B. Zhang, M. Carlson et al. 2006. Analysis of oncogenic signaling networks in glioblastoma identifies ASPM as a molecular target. *PNAS* **103**:17,402–17,407.

Hovers, E., S. Ilani, O. Bar-Yosef, and B. Vandermeersch. 2003. An early case of color symbolism ochre use by modern humans in Qafzeh Cave. *Curr. Anthro.* **44**:491–511.

Hultsch, H. 1991. Early experience can modify singing styles: Evidence from experiments with nightingales, luscinia-megarhynchos. *Anim. Behav.* **42(6)**:883–889.

Hultsch, H., and D. Todt. 2004. Learning to sing. In: Nature's Music: The Science of Birdsong, ed. P. Marler and H. Slabbekorn, pp. 80–107. Amsterdam: Elsevier.

Humphreys, G. W., and J. M. Riddoch, eds. 1987. Visual Object Processing: A Cognitive Neuropsychological Approach. London: Erlbaum.

Hurd, P. L. 1995. Communication in discrete action-response games. *J. Theor. Biol.* **174**:217–222.

Hurford, J. R. 1990. Nativist and functional explanations in language acquisition. In: Logical Issues in Language Acquisition, ed. I. M. Roca, pp. 85–136. Dordrecht: Foris.

——. 2000. Social transmission favors linguistic generalization. In: The Evolutionary Emergence of Language: Social Function and the Origins of Linguistic Form, ed. C. Knight, M. Studdert-Kennedy, and J. Hurford, pp. 324–352. Cambridge: Cambridge Univ. Press.

——. 2007. The Origins of Meaning. Oxford: Oxford Univ. Press.

——. 2009. The Origins of Grammar. Oxford: Oxford Univ. Press.

Hurst, J. A., M. Baraitser, E. Auger, F. Graham, and S. Norell. 1990. An extended family with a dominantly inherited speech disorder. *Devel. Med. Child Neurol.* **32**:352–355.

Huttenlocher, P. R. 2002. Neural Plasticity. Cambridge, MA: Harward Univ. Press

Iachini, T., and R. Logie. 2003. The role of perspective in locating position in a real-world, unfamiliar environment. *Appl. Cognit. Psychol.* **17**:904.

Iacoboni, M. and M. Dapretto. 2006. The mirror neuron system and the consequences of its dysfunction. *Nature Rev. Neurosci.* **7**:942–951.

Indefrey, P. 2004. Hirnaktivierungen bei syntaktischer Sprachverarbeitung: Eine Meta-Analyse. In: Neurokognition der Sprache, eds. H. M. Müller and G. Rickheit, pp. 31–50. Tübingen: Stauffenburg Verlag.

Indefrey, P., and A. Cutler. 2004. Prelexical and lexical processing in listening. In: The Cognitive Neurosciences III, ed. M. S. Gazzaniga, 3rd ed., pp. 759–774. Cambridge, MA: MIT Press.

Indefrey, P., and W. J. M. Levelt. 2004. The spatial and temporal signatures of word production components. *Cognition* **92**:101–144.

Jackendoff, R. 2002. Foundations of Language: Brain, Meaning, Grammar, Evolution. Oxford: Oxford Univ. Press.

Jaeger, H., and H. Haas. 2004. Harnessing nonlinearity: Predicting chaotic systems and saving energy in wireless communication. *Science* **304**:78–80.

Jaeger, H., W. Maass, and J. Principe. 2007. Special issue on echo state networks and liquid state machines. *Neural Netw.* **20(3)**:287–289.

Jäger, G. 2004. Learning constraint subhierarchies: The bidirectional gradual learning algorithm. In: Optimality Theory and Pragmatics, ed. R. Blutner and H. Zeevat, pp. 251–287. New York: Palgrave MacMillan.

——. 2007. Evolutionary game theory and typology: A case study. *Language* **83(1)**:74–109.

——. 2008a. Evolutionary stability conditions for signaling games with costly signals. *J. Theor. Biol.* **253(1)**:131–141.

——. 2008b. The evolution of convex categories. *Ling. Phil.* **30(5)**:551–564.

Jäger, G., and R. van Rooij. 2007. Language structure: Psychological and social constraints. *Synthese* **159(1)**:99–130.

Jamain, S., H. Quach, C. Betancur et al. 2003. Mutations of the X–linked genes encoding neuroligins NLGN3 and NLGN4 are associated with autism. *Nature Genet.* **34**:27–29.

James, W. 1890. The Principles of Psychology. Dover Publications.

Janik, V. M., L. S. Sayigh, and R. S. Wells. 2006. Signature whistle shape conveys identity information to bottlenose dolphins. *PNAS* **103**:8293–8297.

Jarvis, E. D. 2004. Learned birdsong and the neurobiology of human language. *Ann. NY Acad. Sci.* **1016**:749–777.

Jarvis, E. D., C. Scharff, M. R. Grossman, J. A. Ramos, and F. Nottebohm. 1998. For whom the bird sings: Context-dependent gene expression. *Neuron* **21**:775–788.

Jerne, N. K. 1985. The generative grammar of the immune system. *Science* **229**:1057–1059.

Jin, D. Z., F. M. Ramazanoglu, and H. S. Seung. 2007. Intrinsic bursting enhances the robustness of a neural network model of sequence generation by avian brain area HVC. *J. Comput. Neurosci.* **23**:283–299.

Joshi, A. K., and Y. Schabes. 1997. Tree-adjoining grammars. In: Handbook of Formal Languages and Automata, ed. A. Salomma and G. Rosenberg, vol. 3, pp. 69–124. Heidelberg: Springer-Verlag.

Julien, M. 2002. Syntactic Heads and Word Formation. New York: Oxford Univ. Press.

Just, M. A., and P. A. Carpenter. 1992. A capacity theory of comprehension: Individual differences in working memory. *Psychol. Rev.* **99(1)**:122–149.

Just, M. A., P. A. Carpenter, T. A. Keller, W. F. Eddy, and K. R. Thulborn. 1996. Brain activation modulated by sentence comprehension. *Science* **274**:114–116.

Kaan, E. 2007. Event-related potentials and language processing: A brief overview. *Lang. Ling. Compass* **1**:571–591.

Kaan, E., A. Dallas, and C. M. Barkley. 2006. Processing bare quantifiers in discourse. *Brain Res.* **1146**:199–209.

Kaan, E., A. Harris, E. Gibson, and P. Holcomb. 2000. The P600 as an index of syntactic integration difficulty. *Lang. Cogn. Proc.* **15**:159–201.

Kaan, E., and T. Y. Swaab. 2002. The neural circuitry of syntactic comprehension. *Trends Cogn. Sci.* **6**:350–356.

——. 2003. Repair, revision and complexity in syntactic analysis: An electrophysiological investigation. *J. Cogn. Neurosci.* **15**:98–110.

Kakishita, Y., K. Sasahara, T. Nishino, M. Takahashi, and K. Okanoya. 2007. Pattern extraction improves automata-based syntax analysis in songbirds. ACAL2007. *Lecture Notes in Artificial Intelligence* **828**:321–333.

Kalish, M. L., T. L. Griffiths, and S. Lewandowsky. 2007. Iterated learning: Intergenerational knowledge transmission reveals inductive biases. *Psychon. Bull. Rev.* **14**:288–294.

Kandel, E. R., J. H. Schwartz, and T. M. Jessell. 2000. Principles of Neural Science, 4th ed. New York: McGraw-Hill.

Kang-Kwong, L., H. L. Liu, Y. Y. Wai, Y. L. Wan, and L. H. Tan. 2002. Functional anatomy of syntactic and semantic processing in language comprehension. *Hum. Brain Map.* **16**:133–145.

Kao, M. H., A. J. Doupe, and M. S. Brainard. 2005. Contributions of an avian basal ganglia-forebrain circuit to real-time modulation of song. *Nature* **433**:638–643.

Kaplan, R., and J. Bresnan. 1982. LFG: A formal system for grammatical representation. In: The Mental Representation of Grammatical Relations, ed. J. Bresnan, pp. 173–281. Cambridge, MA: MIT Press.

Karmiloff-Smith, A. 2006. The tortuous route from genes to behavior: A neuroconstructivist approach. *Cogn. Affect. Behav. Neurosci.* **6**:9–17.

Kayne, R. S. 1984. Connectedness and Binary Branching. Dordrecht: Foris Publications.

——. 1994. The Antisymmetry of Syntax. Cambridge, MA: MIT Press.

——. 2000. Parameters and Universals. New York: Oxford Univ. Press.

Ke, J., J. Minett, C.-P. Au, and W. Wang. 2002. Self-organization and selection in the emergence of vocabulary. *Complexity* **7(3)**:41–54.

Keenan, E. L., and B. Comrie. 1977. Noun phrase accessibility and universal grammar. *Ling. Inquiry* **8**:63–99.

Keller, M. C., and G. Miller. 2006. Resolving the paradox of common, harmful, heritable mental disorders: Which evolutionary genetic models work best? *Behav. Brain Sci.* **29**:385–452.

Kempler, D., S. Curtiss, E. Metter, C. Jackson, and W. Hanson. 1991. Grammatical comprehension, aphasic syndromes and neuroimaging. *J. Neuroling.* **6**:301–318.

Kéri, S., and B. Gulyás. 2003. Four facets of a single brain: Behavior, cerebral blood flow/metabolism, neuronal activity and neurotransmitter dynamics. *Neuroreport* **14**:1097–1106.

Kimura, M. 1983. The Neutral Theory of Molecular Evolution. Cambridge: Cambridge Univ. Press.

King, J., and M. A. Just. 1991. Individual difference in syntactic processing: The role of working memory. *J. Mem. Lang.* **30**:580–602.

King, J., and M. Kutas. 1995. Who did what when? Using word- and clause-related ERPs to monitor working memory usage in reading. *J. Cogn. Neurosci.* **7**:378–397.

Kintsch, W. 1992. How readers construct situation models for stories: The role of syntactic cues and causal inference. In: Essays in Honor of William K. Estes, ed. A. F. Healy, S. Kosslyn and R. M. Shiffrin. Hillsdale, NJ: Erlbaum.

——. 1994. The psychology of discourse processing. In: Language Comprehension as Structure Building, ed. M. A. Gernsbacher. Hillsdale, NJ: Erlbaum.

Kirby, S. 1999. Function, Selection and Innateness: The Emergence of Language Universals. Oxford: Oxford Univ. Press.

——. 2001. Spontaneous evolution of linguistic structure: An iterated learning model of the emergence of regularity and irregularity. *IEEE Trans. Evol. Comp.* **5(2)**:102–110.

——. 2002. Learning, bottlenecks and the evolution of recursive syntax. In: Language Acquisition and Linguistic Evolution: Formal and Computational Approaches, ed. E. Briscoe, pp. 173–204. Cambridge: Cambridge Univ. Press.

——. 2006. Major transitions in the evolution of language. In: The Evolution of Language, ed. A. Cangelosi, A. D. M. Smith, and K. Smith. Singapore: World Scientific Press.

Kirby, S., H. Cornish, and K. Smith. 2008. Cumulative cultural evolution in the laboratory: An experimental approach to the origins of structure in human language. *PNAS* **105(31)**:10,681–10,686.

Kirby, S., M. Dowman, and T. L. Griffiths. 2007. Innateness and culture in the evolution of language. *PNAS* **104(12)**:5241–5245.

Kirby, S., and J. Hurford. 1997. Learning, culture and evolution in the origin of linguistic constraints. In: ECAL97, ed. P. Husbands and I. Harvey, pp. 493–502. Cambridge, MA: MIT Press.

——. 2002. The emergence of linguistic structure: An overview of the iterated learning model. In: Simulating the Evolution of Language, ed. A. Cangelosi and D. Parisi, pp. 121–148. London: Springer-Verlag.

Kirby, S., K. Smith, and H. Brighton. 2004. From UG to universals: Linguistic adaptation through iterated learning. *Stud. Lang.* **28(3)**:587–607.

Klein, R. G. 2000. Archeology and the evolution of human behavior. *Evol. Anthro.* **9**:17–36.

Kluender, R., and M. Kutas. 1993. Bridging the gap: Evidence from ERPs on the processing of unbounded dependencies. *J. Cogn. Neurosci.* **5**:196–214.

Knight, C. 1998. Ritual/speech coevolution: A solution to the problem of deception. In: Approaches to the Evolution of Language, ed. J. R. Hurford, M. Studdert-Kenedy, and C. Knight, pp. 68–91. Cambridge: Cambridge Univ. Press.

——. 2007. Language co-evolved with the rule of law. *Mind and Society* **7(1)**:109–128.

Koelsch, S., T. C. Gunter, M. Wittforth, and D. Sammler. 2005. Interaction between Syntax Processing in Language and in Music: An ERP Study. *J. Cogn. Neurosci.* **17**:1565–1577.

Koenig, P., E. E. Smith, G. Glosser et al. 2003. The neural basis for novel semantic categorization. *Neuroimage* **24**:369–383.

Köhler, W. 1925. The Mentality of Apes. New York: Harcourt Brace.

Komarova, N. L., and K. A. Jameson. 2008. Population heterogeneity and color stimulus heterogeneity in agent-based color categorization. *J. Theor. Biol.* **253(4)**:680–700.

Komarova, N. L., K. A. Jameson, and L. Narens. 2007. Evolutionary models of color categorization based on discrimination. *J. Math. Psychol.* **51**:359–382.

Komarova, N. L., P. Niyogi, and M. A. Nowak. 2001. Evolutionary dynamics of grammar acquisition. *J. Theor. Biol.* **209(1)**:43–59.

Komarova, N. L., and M. A. Nowak. 2001. Population dynamics of grammar acquisition. In: Simulating the Evolution of Language, ed. A. Cangelosi and D. Parisi, pp. 149–164. London: Springer-Verlag.

——. 2003. Language, learning and evolution. In: Language Evolution: The States of the Art, ed. S. Kirby and M. H. Christiansen, pp. 317–337. Oxford: Oxford Univ. Press.

Konishi, M., and F. Nottebohm. 1969. Experimental studies in the ontogeny of avian vocalizations. In: Bird Vocalizations, ed. R. A. Hinde, pp. 29–48. London: Cambridge Univ. Press.

Kosko, B. 1988. Bidirectional associative memories. *IEEE Trans. Syst. Man Cybern.* **18**:49–60.

Kovács, G., B. Gulyás, and P. E. Roland. 1998. Processing of 2D and 3D shapes in the visual association cortex. *NeuroImage* **7**:S335.

Kozhevnikov, A. A., and M. S. Fee. 2007. Singing-related activity of identified HVC neurons in the zebra finch. *J. Neurophys.* **97**:4271–4283.

Krause, J., C. Lalueza-Fox, L. Orlando et al. 2007. The derived FOXP2 variant of modern humans was shared with Neandertals. *Curr. Biol.* **17**:1908–1912.

Krubitzer, L., and J. Kaas. 2005. The evolution of the neocortex in mammals: How is phenotypic diversity generated? *Curr. Op. Neurobiol.* **15**:444–453.

Kuperberg, G. 2007. Neural mechanisms of language comprehension: Challenges to syntax. *Brain Res.* **1146**:23–49.

Kutas, M., and S. A. Hillyard. 1980. Reading senseless sentences: Brain potentials reflect semantic incongruity. *Science* **207**:203–206.

Kutas, M., C. K. Van Petten, and R. Kluender. 2006. Psycholinguistics electrified II (1994–2005). In: Handbook of Psycholinguistics, eds. M. A. Gernsbacher and M. Traxler, 2nd ed., pp. 659–724. New York: Elsevier.

Labov, W. 2001. Principles of Linguistic Change (vol. 1: Internal Factors, 1994; vol. 2: Social Factors, 2001). Oxford: Blackwell Publ.

Lachlan, R. F., and P. J. B. Slater. 1999. The maintenance of vocal learning by gene-culture interaction: The cultural trap hypothesis. *P. Roy. Soc. Lond. B* **266**:701–706.

Lachmann, M., S. Számadó, and C. T. Bergstrom. 2001. Cost and constraints in animals and in human language. *PNAS* **28**:13,189–13,194.

Lai, C. S. L., S. E. Fisher, J. A. Hurst, F. Vargha-Khadem, and A. P. Monaco. 2001. A forkhead-domain gene is mutated in a severe speech and language disorder. *Nature* **413**:519–523.

Lai, C. S. L., D. Gerelli, A. P. Monaco, S. E. Fisher, and A. J. Copp. 2003. FOXP2 expression during brain development coincides with adult sites of pathology in a severe speech and language disorder. *Brain* **126(11)**:2455–2462.

Laland, K. N., F. J. Odling-Smee, and N. W. Feldman. 1999. Niche construction: Biological evolution and cultural change. *Behav. Brain Sci.* **23**:131–175.

Lambrecht, K. 1994. Information Structure and Sentence Form: Topic, Focus, and the Mental Representation of Discourse Referents. Cambridge: Cambridge Univ. Press.

Land, M. F., and R. D. Fernald. 1992. The evolution of eyes. *Ann. Rev. Neurosci.* **15**:1–29.

Langacker, R. 2008. Cognitive Grammar: A Basic Introduction. Oxford: Oxford Univ. Press.

Lappin, S., and S. M. Shieber. 2007. Machine learning theory and practice as a source of insight into universal grammar. *J. Ling.* **43**:1–34.

Larsen, T., and W. Norman. 1979. Correlates of ergativity in Mayan grammar. In: Ergativity, ed. F. Plank, pp. 347–370. New York: Academic Press.

Lashley, K. S. 1950. In search of the engram. *Symp. Soc. Exp. Biol.* 454–482.

Lau, E., C. Stroud, S. Plesch, and C. Phillips. 2006. The role of structural prediction in rapid syntactic analysis. *Brain Lang.* **98**:74–88.

Lebel, C., L. Walker, A. Leemans, L. Phillips, and C. Beaulieu. 2008. Microstructural maturation of the human brain from childhood to adulthood. *NeuroImage* **40**:1044–1055.

Lee, B. C. P., K. Kuppusamy, R. Grueneich et al. 1999. Hemispheric language dominance in children demonstrated by functional magnetic resonance imaging. *J. Child Neurol.* **14**:5.

Lelekov, T., P. F. Dominey, and L. Garcia-Larrea. 2000. Dissociable ERP profiles for processing rules vs. instances in a cognitive sequencing task. *Neuroreport* **11**:1129–1132.

Lenaerts, T., B. Jansen, K. Tuyls, and B. de Vylder. 2005. The evolutionary language game: An orthogonal approach. *J. Theor. Biol.* **235**:566–582.

Lennie, P. 2003. The cost of cortical computation. *Curr. Biol.* **13**:493–497.

Levi, O., A. L. Jongen-Relo, J. Feldon, A. D. Roses, and D. M. Michaelson. 2003. ApoE4 impairs hippocampal plasticity isoform-specifically and blocks the environmental stimulation of synaptogenesis and memory. *Neurobiol. Disease* **13**:273–282.

Levinson, S. 2000. Presumptive Meaning. Cambridge, MA: MIT Press.

Li, G., J. Wang, S. J. Rossiter, G. Jones, and S. Zhang. 2007. Accelerated FOXP2 evolution in echolocating bats. *PLoS ONE* **2(9)**:e900.

Li, M., and P. Vitanyi. 1997. An Introduction to Kolmogorov Complexity and its Applications. Berlin: Springer-Verlag.

Lieberman, P. 2002. On the nature and evolution of the neural bases of human language. *Am. J. Phys. Anthro.* **35**:36–62.

——. 2006. Towards an Evolutionary Biology of Language. Cambridge, MA: Harvard Univ. Press.

——. 2007. The evolution of human speech. Its anatomical and neural bases. *Curr. Anthro.* **48**:39–66.

Liegeois, F., T. Baldeweg, A. Connelly et al. 2003. Language fMRI abnormalities associated with FOXP2 gene mutation. *Nature Neurosci.* **6**:1230–1237.

Liegeois, F., A. Connelly, J. H. Cross et al. 2004. Language reorganization in children with early-onset lesions of the left hemisphere: An fMRI study. *Brain* **127**:1229–1236.

Lightfoot, D. 1991. How to Set Parameters: Arguments from Language Change. Cambridge, MA: MIT Press (Bradford Books).

——. 2000. The spandrels of the linguistic genotype. In: The Evolutionary Emergence of Language: Social Function and the Origins of Linguistic Form, ed. C. Knight, M. Studdert-Kennedy, and J. R. Hurford, pp. 231–247. Cambridge: Cambridge Univ. Press.

Lohmann, H., B. Dräger, S. Müller-Ehrenberg, M. Deppe, and S. Knecht. 2005. Language lateralization in young children assessed by functional transcranial Doppler sonography. *NeuroImage* **24**:780–790.

Longa, V. 2006. A misconception about the Baldwin effect. *Folia Linguistica* **40**:305–318.

Loreto, V., and L. Steels. 2007. Emergence of language. *Nature Physics* **3(11)**:758–760.

Lu, M. M., S. Li, H. Yang, and E. E. Morrisey. 2002. Foxp4: A novel member of the Foxp subfamily of winged-helix genes co-expressed with Foxp1 and Foxp2 in pulmonary and gut tissues. *Gene Expr. Patterns* **2**:223–228.

Lu, Q., G. Korniss, and B. K. Szymanski. 2008. Naming games in two-dimensional and small-world-connected random geometric networks. *Phys. Rev. E* **77(1)**:016111.

Luck, S. J. 2005. An introduction to the event-related potential technique. Cambridge, MA: MIT Press.

Lueck, C. J., S. Zeki, K. J. Friston et al. 1989. The colour centre in the cerebral cortex of man. *Nature* **340**:386–389.

Lupyan, G., and M. H. Christiansen. 2002. Case, word order, and language learnability: Insights from connectionist modeling. In: Proc. 24th Annual Conf. of the Cognitive Science Society, pp. 596–601. Mahwah, NJ: Lawrence Erlbaum.

Maass, W., T. Natschläger, and H. Markram. 2002. Real-time computing without stable states: A new framework for neural computation based on perturbations. *Neural Comp.* **14(11)**:2531–2560.

Macdermot, K. D., E. Bonora, N. Sykes et al. 2005. Identification of FOXP2 truncation as a novel cause of developmental speech and language deficits. *Am. J. Hum. Gen.* **76**:1074–1080.

MacDonald, M. C., and M. H. Christiansen. 2002. Reassessing working memory: A comment on Just & Carpenter (1992) and Waters & Caplan (1996). *Psychol. Rev.* **109**:35–54.

MacDonald, M. C., N. J. Pearlmutter, and M. S. Seidenberg. 1994. Lexical nature of syntactic ambiguity resolution. *Psychol. Rev.* **101**:676–703.

MacSweeney, M., C. M. Capek, R. Campbell, and B. Wolf. 2008. The signing brain: The neurobiology of sign language. *Trends Cog. Sci.* **12**:432–440.

MacWhinney, B. 2002. The gradual emergence of language. In: The Evolution of Language out of Pre-Language, ed. T. Givón and B. F. Malle. Amsterdam: John Benjamins.

MacWhinney, B., C. Pléh, and E. Bates. 1985. The development of sentence interpretation in Hungarian. *Cogn. Psychol.* **17**:178–209.

MacWhinney, B., and C. Snow. 1985. The Child Language Data Exchange System. *J. Child Lang.* **12**:271–295.

Makuuchi, M., J. Bahlmann, A. Anwander, and A. D. Friederici. 2009. Segregating the core computation of human language from working memory. *PNAS*, in press.

Marcus, G. F. 1998. Rethinking eliminative connectionism. *Cogn. Psychol.* **37**:243–282.

Marcus, G. F., and S. E. Fisher. 2003. FOXP2 in focus: What can genes tell us about speech and language. *Trends Cogn. Sci.* **7**:257–262.

Marcus, G. F., and H. Rabagliati. 2006. What developmental disorders can tell us about the nature and origins of language. *Nature Neurosci.* **9**:1226–1229.

Marean, C. W., M. Bar-Matthews, J. Bernatchez et al. 2007. Early human use of marine resources and pigment in South Africa during the Middle Pleistocene. *Nature* **449**:905–908.

Margoliash, D., and E. S. Fortune. 1992. Temporal and harmonic combination-sensitive neurons in the zebra finch's HVc. *J. Neurosci.* **12**:4309–4326.

Markman, E. M. 1992. Constraints on word learning: Speculations about their nature, origins, and domain specificity. In: Modularity and Constraints in Language and Cognition, ed. M. Gunnar and M. Maratsos, pp. 59–101. Hillsdale, NJ: Erlbaum.

Marler, P. 1972. A comparative approach to vocal learning: Song development in white-crowned sparrows. In: Biological Boundaries of Learning, ed. M.E.P. Seligman and J.L. Hager, pp. 336–376. New York: Appleton Century Crafts.

——. 1977. Recognition of Complex Acoustic Signals, ed. T. H. Bullock. Verlag Chemie: Weingein/Bergstr.

Marshack, A. 1981. On Paleolithic ochre and the early uses of color and symbol. *Curr. Anthro.* **22**:188–191.

Marslen-Wilson, W. D., and L. K. Tyler. 1980. The temporal structure of spoken language understanding. *Cognition* **8**:1–71.

Martin, A., and L. L. Chao. 2001. Semantic memory and the brain: Structure and processes. *Curr. Op. Neurobiol.* **11**:194–201.

Martin, A., C. L. Wiggs, L. G. Ungerleider, and J. V. Huxby. 1996. Semantic memory and the brain: Structure and processes. *Nature* **379**:649–652.

Martínez, I., J. L. Arsuaga, R. Quam et al. 2008. Human hyoid bones from the middle Pleistocene site of the Sima de los Huesos (Sierra de Atapuerca, Spain). *J. Hum. Evol.* **54(1)**:118–124.

Matelli, M., G. Luppino, and G. Rizzolatti. 1985. Patterns of cytochrome-oxidase activity in the frontal agranular cortex of the macaque monkey. *Behav. Brain Res.* **18**:125–136.

Matsen, F. A., and M. A. Nowak. 2004. Win-stay, lose-shift in language learning from peers. *PNAS* **101(52)**:18,053–18,057.

Mayley, G. 1996. Landscapes, learning costs and genetic assimilation. In: Evolution, Learning and Instinct: 100 Years of the Baldwin Effect, ed. P. Turney, D. Whitley, and R. Anderson. Cambridge, MA: MIT Press.

Maynard Smith, J. 1982. Evolution and the Theory of Games. Cambridge: Cambridge Univ. Press.

——. 1991. Honest signalling: The Philip Sidney game. *Anim. Behav.* **42**:1034–1035.

——. 1998. Evolutionary Genetics. Oxford: Oxford Univ. Press.

Maynard Smith, J., and E. Szathmáry. 1995. The Major Transitions in Evolution. Oxford: Freeman.

——. 1999. The Origins of Life. Oxford: Oxford Univ. Press.

Mayr, E. 1969. Population, Species and Evolution, Cambridge, MA: Harvard Univ. Press.

——. 1976. Evolution and the Diversity of Life, Cambridge, MA: Harvard Univ. Press.

Mazza, P. P. A., F. Martini, B. Sala et al. 2006. A new Palaeolithic discovery: Tar-hafted stone tools in a European Mid-Pleistocene bone-bearing bed. *J. Archaeol. Sci.* **33**:1310–1318.

Mazziotta, J., A. Toga, and A. Evans et al. 2001. A probabilistic atlas and reference system for the human brain. *Phil. Trans: Biol. Sci.* **356(1412)**:1293–1322.

McBrearty, S., A. S. Brooks. 2000. The revolution that wasn't: A new interpretation of the origin of modern human behavior. *J. Hum. Evol.* **39**:453–563.

McDonald, J. L. 2008. Grammaticality judgments in children: The role of age, working memory and phonological ability. *J. Child Lang.* **35**:247–268.

McElreath, R., R. Boyd, and P. J. Richerson. 2003. Shared norms and the evolution of ethnic markers. *Curr. Anthro.* **44**:122–129.

McElreath, R., and J. Henrich. 2008. Modeling cultural evolution. In: Oxford Handbook of Evolutionary Psychology, ed. R. Dunbar and L. Barrett, pp. 571–585. Oxford: Oxford Univ Press.

McGeoch, P. D., D. Brang, and V. S. Ramachandran. 2007. Apraxia, metaphor and mirror neurons. *Med. Hypotheses* **69**:1165–1168.

McKinnon, R., and L. Osterhout. 1996. Constraints on movement phenomena in sentence processing: Evidence from event-related brain potentials. *Lang. Cogn. Proc.* **11**:495–523.

Mecklinger, A., H. Schriefers, K. Steinhauer, and A. D. Friederici. 1995. Processing relative clauses varying on syntactic and semantic dimensions: An analysis with event-related potentials. *Mem. Cogn.* **23**:477–494.

Menn, L., and L. Obler, eds. 1990. Agrammatic Aphasia, vol. 1. Amsterdam: John Benjamins.

Mesoudi, A. 2007. Using the methods of experimental social psychology to study cultural evolution. *J. Soc. Evol. Cul. Psychol.* **1(2)**:35–58.

Mesulam, M.-M. 1990. Large-scale neurocognitive networks and distributed processing for attention, language, and memory. *Ann. Neurol.* **28**:597–613.

——. 1998. Form sensation to cognition. *Brain* **121**:1013–1052.

——. 2000. Principles of Behavioral and Cognitive Neurology. New York: Oxford Univ. Press.

——. 2002. The human frontal lobes: Transcending the default mode through contingent encoding. In: Principles of Frontal Lobe Function, ed. D. T. Stuss and R. T. Knight, pp. 8–31. Oxford: Oxford Univ. Press.

Meyer, M., K. Alter, and A. D. Friederici. 2003. Functional MR imaging exposes differential brain responses to syntax and prosody during auditory sentence comprehension. *J. Neuroling.* **16**:277–300.

Meyer, M., K. Alter, A. D. Friederici, G. Lohmann, and D. Y. von Cramon. 2002. fMRI reveals brain regions mediating slow prosodic modulations in spoken sentences. *Hum. Brain Map.* **17**:73–88.

Meyer, M., K. Steinhauer, K. Alter, A. D. Friederici, and D. Y. von Cramon. 2004. Brain activity varies with modulation of dynamic pitch variance in sentence melody. *Brain Lang.* **89**:277–289.

Mezard, M., G. Parisi, and M. A. Virasoro. 1987. Spin Glass Theory and Beyond. Teaneck, NJ: World Scientific Press.

Miller, E. K. 2000. The prefrontal cortex and cognitive control. *Nature Rev. Neurosci.* **1**:59–65.

Miller, G. 2001. The Mating Mind. New York: Anchor Books.

Miller, J. E., E. Spiteri, M. C. Condro et al. 2008. Birdsong decreases protein levels of FoxP2, a molecule required for human speech. *J. Neurophys.* **100(4)**:2015–2025.

Minett, J. W., and W. S.Y. Wang. 2005. Language Acquisition, Change and Emergence: Essays in Evolutionary Linguistics. Hong Kong: City Univ. of Hong Kong Press.

Mitchell, T. 1997. Machine Learning. New York: McGraw Hill.

Miyake, A. K., P. Carpenter, and M. Just. 1994. A capacity approach to syntactic comprehension disorders: Making normal adults perform like brain-damaged patients. *Cogn. Neuropsychol.* **11**:671–717.

Mohr, J. P., M. S. Pessin, and S. Finkelstein et al. 1978. Broca aphasia: Pathologic and clinical. *Neurology* **28**:311–324.

Molnar-Szakacs, I., J. Kaplan, P. M. Greenfield, and M. Iacoboni. 2006. Observing complex action sequences: The role of the fronto-parietal mirror neuron system. *Neuroimage* **33**:923–935.

Momo, K., H. Sakai, and K. L. Sakai. 2008. Syntax in a native language still continues to develop in adults: Honorification judgment in Japanese. *Brain Lang.* **107**:81–89.

Morgan, C. L. 1896. On modification and variation. *Science* **4**:733–740.

Mormann, F., S. Kornblith, R. Quian Quiroga et al. 2008. Latency and selectivity of single neurons indicate hierarchical processing in the human medial temporal lobe. *J. Neurosci.* **28(36)**:8865–8872.

Moro, A., M. Tettamanti, D. Perani et al. 2001. Syntax and the Brain: Disentangling grammar by selective anomalies. *NeuroImage* **13**:110–118.

Moy, J. 2006. Word order and case in models of simulated language evolution. Ph.D. thesis, Univ. of York.

Mufwene, S. 2001. The Ecology of Language Evolution. Cambridge: Cambridge Univ. Press.

Müller, R.-A., R. D. Rothermel, M. E. Behen et al. 1999. Language organization in patients with early and late left-hemisphere lesion: A PET study. *Neuropsychologia* **37**:545–557.

Mummery, C. J., K. Patterson, J. R. Hodges, and R. J. S. Wise. 1996. Generating 'tiger' as an animal name or a word beginning with T: Differences in brain activation. *P. Roy. Soc. Lond. B* **263**:989–995.

Musso, M., C. Weiller, S. Kiebel et al. 1999. Training-induced brain plasticity in aphasia. *Brain* **122**:1781–1790.

Naeser, M. A., and R. W. Hayward. 1978. Lesion localization in aphasia with cranial computed tomography and the Boston diagnostic aphasia examination. *Neurology* **28**:545–551.

Naeser, M. A., R. W. Hayward, S. Laughlin, and L. M. Zatz. 1979. Quantitative CT scans studies in aphasia. I. Infarct size and CT numbers. *Brain Lang.* **12**:140–164.

Nakamura, M., T. Hashimoto, and S. Tojo. 2003. The language dynamics equations of population-based transition: A scenario for Creolization. In: IC-AI'03. CSREA Press.

Neville, H. 1995. Developmental specificity in neurocognitive development in humans. In: The New Cognitive Neuroscience, ed. M. Gazzaniga. Cambridge, MA: MIT Press.

Neville, H., and D. Bavelier. 1998. Neural organization and plasticity of language. *Curr. Op. Neurobiol.* **8**:254–258.

Neville, H., D. L. Mills, and D. S. Lawson. 1992. Fractionating language: Different neural systems with different sensitive periods. *Cereb. Cortex* **2**:244–258.

Neville, H., J. L. Nicol, A. Barss, K. I. Forster, and M. F. Garrett. 1991. Syntactically based sentence processing classes: Evidence from event-related brain potentials. *J. Cogn. Neurosci.* **3**:151–165.

Newbury, D. F., and A. P. Monaco. 2008. The application of molecular genetics to the study of language impairments. In: Understanding Developmental Language Disorders, ed. C. F. Norbury, J. B. Tomblin, and D. V. M. Bishop, pp. 79–91. Hove: Psychology Press.

Newell, A. 1990. Unified Theories of Cognition. Cambridge, MA: Harvard Univ. Press.

Newmeyer, F. J. 1991. Functional explanation in linguistics and the origins of language. *Lang. Comm.* **11**:3–28.

Newport, E. 1999. Reduced input in the acquisition of signed languages: Contributions to the study of creolization. In: Language Creation and Language Change: Creolization, Diachrony, and Development, ed. M. DeGraff. Cambridge MA: MIT Press.

Nissenbaum, J. 2000. Investigations of covert phrasal movement. Ph.D. thesis, MIT.

Noble, J. 2000. Cooperation, competition and the evolution of prelinguistic communication. In: The Evolutionary Emergence of Language: Social Function and the Origins of Linguistic Form, ed. C. Knight, M. Studdert-Kenedy and J. R. Hurford, pp. 40–61. Cambridge: Cambridge Univ. Press.

Nobre, A. C., and K. Plunkett. 1997. The neural system of language: Structure and development. *Curr. Op. Neurobiol.* **7**:262–268.

Nolfi, S., and D. Floreano. 2000. Evolutionary Robotics: The Biology, Intelligence, and Technology of Self-Organizing Machines. Cambridge, MA: MIT Press.

——. 2002. Synthesis of autonomous robots through evolution. *Trends Cogn. Sci.* **6**:31–37.

Norbury, C. F., D. V. M. Bishop, and J. Briscoe. 2002. Does impaired grammatical comprehension provide evidence for an innate grammar module? *Appl. Psycholing.* **23**:247–268.

Nordeen, K. W., and E. J. Nordeen. 1992. Auditory feedback is necessary for the maintenance of stereotyped song in adult zebra finches. *Behav. Neural Biol.* **57**:58–66.

Nottebohm, F., and A. P. Arnold. 1976. Sexual dimorphism in vocal control areas of the songbird brain. *Science* **194**:211–213.

Nottebohm, F., D. B. Kelley, and J. A. Paton. 1982. Connections of vocal control nuclei in the canary telencephalon. *J. Comp. Neurol.* **207**:344–357.

Novick, J. M., J. C. Trueswell, and S. L. Thompson-Schill. 2005. Cognitive control and parsing: Reexamining the role of Broca's area in sentence comprehension. *Cogn. Affect. Behav. Neurosci.* **5**:263–281.

Nowak, M. A., and N. L. Komarova. 2001. Toward an evolutionary theory of language. *Trends Cogn. Sci.* **5(7)**:288–295.

Nowak, M. A., N. Komarova, and P. Niyogi. 2001. Evolution of universal grammar. *Science* **291**:114–118.

——. 2002. Computational and evolutionary aspects of language. *Nature* **417**:611–617.

Nowak, M. A., and D. C. Krakauer. 1999. The evolution of language. *PNAS* **96(14)**:8028–8033.

Nowak, M. A., D. C. Krakauer, and A. Dress. 1999. An error limit for the evolution of language. *P. Roy. Soc. Lond. B* **266**:2131–2136.

Núñez-Peña, M. I., and M. L. Honrubia-Serrano. 2004. P600 related to rule violation in an arithmetic task. *Cogn. Brain Res.* **18**:130–141.

Oberecker, R., and A. D. Friederici. 2006. Syntactic event-related potential components in 24-month-olds' sentence comprehension. *Neuroreport* **17**:1017–1021.

Oberecker, R., M. Friedrich, and A. D. Friederici. 2005. Neural correlates of syntactic processing in two-year-olds. *J. Cogn. Neurosci.* **17**:1667–1678.

O'Connell-Rodwell, C. E. 2007. Keeping an "ear" to the ground: Seismic communication in elephants. *Physiology* **22**:287–294.

O'Connor, V. M., O. Shamotienko, E. Grishin, and H. Betz. 1993. On the structure of the synaptosecretosome: Evidence for a neurexin/synaptotagmin/syntaxin/Ca^{2+} channel complex. *FEBS Lett.* **326**:255–260.

Odling-Smee, F. J., K. N. Laland, and N. W. Feldman. 2003. Niche Construction: The Neglected Process in Evolution. Princeton: Princeton Univ. Press.

Okanoya, K. 2004a. The Bengalese finch: A window on the behavioral neurobiology of birdsong syntax. *Ann. NY Acad. Sci.* **1016**:724–735.

——. 2004b. Song syntax in Bengalese finches: Proximate and ultimate analyses. *Adv. Study Behav.* **34**:297–346.

Okuhata, S., and N. Saito. 1987. Synaptic connections of a forebrain nucleus involved with vocal learning in zebra finches. *Brain Res. Bull.* **18**:35–44.

Oldham, M .C., S. Horvath, and D. H. Geschwind. 2006. Conservation and evolution of gene coexpression networks in human and chimpanzee brains. *PNAS* **103**:17,973–17,978.

Oliphant, M. 1997. Communication as altruistic behavior. Ph.D. thesis, Univ. of California, San Diego.

——. 2002. Learned systems of arbitrary reference: The foundation of human linguistic uniqueness. In: Language Acquisition and Linguistic Evolution: Formal and Computational Approaches, ed. E. Briscoe, pp. 23–52. Cambridge: Cambridge Univ. Press.

Opitz, B., and A. D. Friederici. 2004. Brain correlates of language learning: The neuronal dissociation of rule-based versus similarity-based learning. *J. Neurosci.* **24**:8436–8440.

O'Reilly, R. C. 2006. Biologically based computational models of high-level cognition. *Science* **314**:91–94.

Orr, H. A. 2000. Adaptation and the cost of complexity. *Evolution* **54**:13–20.

Osborn, H. F. 1896. A mode of evolution requiring neither natural selection nor the inheritance of acquired characteristics. *NY Acad. Sci. Trans.* **15**:141–142, 148.

Osterhout, L., and P. Hagoort. 1999. A superficial resemblance does not necessarily mean you are part of the family: Counterarguments to Coulson, King and Kutas (1998) in the P600/SPS-P300 debate. *Lang. Cogn. Proc.* **14**:1–14.

Osterhout, L., and P. J. Holcomb. 1992. Event-related brain potentials elicited by syntactic anomaly. *J. Mem. Lang.* **31**:785–806.

——. 1993. Event-related potentials and syntactic anomaly: Evidence of anomaly detection during the perception of continuous speech. *Lang. Cogn. Proc.* **8**:413–438.

Osterhout, L., P. J. Holcomb, and D. A. Swinney. 1994. Brain potentials elicited by garden-path sentences: Evidence of the application of verb information during parsing. *J. Exp. Psychol. Learn. Mem. Cogn.* **20**:786–803.

Osterhout, L., and L. A. Mobley. 1995. Event-related brain potentials elicited by failure to agree. *J. Mem. Lang.,* **34**:739–773.

Osterhout, L., and J. Nicol. 1999. On the distinctiveness, independence, and time course of the brain responses to syntactic and semantic anomalies. *Lang. Cogn. Proc.* **14**:283–317.

Owen, A. M., C. E. Stern, R. B. Look et al. 1998. Functional organization of spatial and nonspatial working memory processing within the human lateral frontal cortex. *PNAS* **95**:7721–7726.

Oyakawa, C., H. Koda, and H. Sugiura. 2007. Acoustic features contributing to the individuality of wild agile gibbon (Hylobates agilis agilis) songs. *Am. J. Primatol.* **69**:777–790.

Pachet, F., and P. Roy. 2007. Exploring billions of audio features. In: Content-Based Multimedia Indexing, pp. 227–235. CBMI 07.

Pandya, D. N., E. H. Yeterian, S. Fleminger, and S. B. Dunnett. 1996. Comparison of prefrontal architecture and connections. *Phil. Trans. Roy. Soc. Lond. B* **351**:1423–1432.

Partee, B. H. 2003. Compositionality in Formal Semantics: Selected Papers of Barbara Partee. Oxford: Blackwell Publishers.

Passingham, R. E. 2002. The frontal cortex: Does size matter? *Nature Neurosci.* **5**:190–192.

——. 2008. What is Special about the Human Brain? Oxford: Oxford Univ. Press.

Patel, A. D. 2003. Language, music, syntax and the brain. *Nature Neurosci.* **6**:674–681.

Patel, A. D., E. Gibson, J. Ratner, M. Besson, and P. J. Holcomb. 1998. Processing syntactic relations in language and music: An event-related potential study. *J. Cogn. Neurosci.* **10**:717–733.

Patel, A. D., J. Iversen, M. Wassenaar, and P. Hagoort. 2008. Musical syntactic processing in agrammatic Broca's aphasia. *Aphasiology* **22**:776–789.

Paul, H. 1890. Principles of the History of Language (trans. by H. A. Strong). London: Swan, Sonnenschein & Co.

Paus, T., A. Zijdenbos, K. Worsley et al. 1999. Structural maturation of neural pathways in children and adolescents: *In vivo* study. *Science* **283**:1908–1911.

Payne, K. 2000. The progressively changing songs of humpback whales: A window on the creative process in a wild animal. In: The Origins of Music, ed. L. Nils, B. M. Wallin, and S. Brown, pp. 135–150. Cambridge, MA: MIT Press.

Payne, R. S., and S. McVay. 1971. Songs of humpback whales. *Science* **173**:585–597.

Peck, K. K., C. E. Wierenga, A. B. Moore et al. 2004. Comparison of baseline conditions to investigate syntactic production using functional magnetic resonance imaging. *NeuroImage* **23**:104–110.

Penn, D. C., and D. J. Povinelli. 2007. Causal cognition in human and nonhuman animals: A comparative, critical review. *Ann. Rev. Psychol.* **58**:97–118.

Pepperberg, I. M. 1999. The Alex Studies: Cognitive and Communicative Abilities of Grey Parrots. Cambridge, MA: Harvard Univ. Press.

Perrett, D. L., H. M. Harries, R. Bevan et al. 1989. Framework of analysis of the neural representation of animate objects and actions. *J. Exp. Biol.* **146**.

Perruchet, P., and A. Rey. 2005. Does the mastery of center-embedded linguistic structures distinguish humans from nonhuman primates? *Psychon. Bull. Rev.* **12**:307–313.

Petersen, S. E., P. T. Fox, M. I. Posner, M. Mintun, and M. E. Raichle. 1988. Positron emission tomographic studies of the cortical anatomy of single-word processing. *Nature* **331**:585–589.

Petri, H. L., and M. Mishkin. 1994. Behaviorism, cognitivism and the new psychology of memory. *Am. Sci.* **82**:30–37.

Petrides, M., G. V. Cadoret, and S. Mackey. 2005. Orofacial somatomotor responses in the macaque monkey homologue of Broca's area. *Nature* **435**:1235–1238.

Petrides, M., and D. N. Pandya. 1994. Comparative cytoarchitectonic analysis of the human and the macaque frontal cortex. In: Handbook of Neuropsychology, ed. F. Boller, and J. Grafman, vol. 9, pp. 17–58. Amsterdam: Elsevier.

——. 1999. Dorsolateral prefrontal cortex: Comparative cytoarchitectonic analysis in the human and the macaque brain and corticocortical connection patterns. *Eur. J. Neurosci.* **11**:1011–1036.

——. 2002a. Association pathways of the prefrontal cortex and functional observations. In: Principles of Frontal Lobe Function, ed. D. T. Stuss and R. T. Knight, pp. 31–51. Oxford: Oxford Univ. Press.

——. 2002b. Comparative cytoarchitectonic analysis of the human and the macaque ventrolateral prefrontal cortex and corticocortical connection patterns in the monkey. *Eur. J. Neurosci.* **16**:291–310.

Pettitt, P. 2002. The Neanderthal dead: Exploring mortuary variability in Middle Palaeolithic Eurasia. *Before Farming* **1(4)**:1–19.

Phillips, C. 1997. Order and structure. Ph.D thesis, MIT.

——. 2003. Linear order and constituency. *Ling. Inquiry* **34**:37–90.

Pica, P., C. Lemer, V. Izard, and S. Dehaene. 2004. Exact and approximate arithmetics in an Amazonian indigene group. *Science* **306**:499–503.

Pickering, M. J., and S. Garrod. 2004. Toward a mechanistic psychology of dialogue. *Behav. Brain Sci.* **27(2)**:169–226.

——. 2005. Alignment as the basis of successful communication. *Res. Lang. Comp.* **4**:203–228.

Pigliucci, M., C. J. Murren, and C. D. Schlichting. 2006. Phenotypic plasticity and evolution by genetic assimilation. *J. Exp. Biol.* **209**:2362–2367.

Pine, J. M., and E. V. M. Lieven. 1997. Slot and frame patterns and the development of the determiner category. *Appl. Psycholing.* **18**:123–138.

Pinker, S. 1994. The Language Instinct. New York: Harper Collins.

——. 2003. Language as an adaptation to the cognitive niche. In: Language Evolution: States of the Art, ed. M. Christiansen and S. Kirby, pp. 16–37. New York: Oxford Univ. Press.

Pinker, S., and P. Bloom. 1990. Natural language and natural selection. *Behav. Brain Sci.* **13**:707–786.

Pinker, S., and R. Jackendoff. 2005. The faculty of language: What's special about it? *Cognition* **95(2)**:201–236.

Platzack, C. 1987. The Scandinavian languages and the null subject parameter. *Natural Lang. Ling. Theory* **5**:377–401.

Pléh, C. 1982. Sentence interpretation strategies and dichotic asymmetries in Hungarian children between 3 and 6 years. In: Psychophysiology 1980, ed. R. Sinz and M. R. Rosenzweig, pp. 443–448. Amsterdam: Fischer–Elsevier.

——. 1989. The development of sentence interpretation in Hungarian. In: The Cross-linguistic Study of Sentence Processing, ed. B. MacWhinney and E. Bates, pp. 158–184. Cambridge: Cambridge Univ. Press.

——. 1990. Word order and morphonological factors in the development of sentence interpretation in Hungarian. *Linguistics* **28**:1449–1469.

Plomin, R., and P. S. Dale. 2000. Genetics and early language development: A UK study of twins. In: Speech and Language Impairments in Children: Causes, Characteristics, Intervention and Outcome, ed. D. V. M. Bishop and L. B. Leonard, pp. 35–51. Hove: Psychology Press.

Pollard, K. S., S. A. Salama, N. Lambert et al. 2006. An RNA gene expressed during cortical development evolved rapidly in humans. *Nature* **443(7108)**:149–150.

Pollock, J.-Y. 1989. Verb movement universal grammar and the structure of IP. *Ling. Inquiry* **20**:365–424.

Poole, J. H., P. L. Tyack, A. S. Stoeger-Horwath, and S. Watwood. 2005. Animal behaviour: Elephants are capable of vocal learning. *Nature* **434**:455–456.

Posner, M. I., and J. Fan. 2008. Attention as an organ system. In: Neurobiology of Perception and Communication: From Synapses to Society, eds. J. R. Pomerantz and M. C. Crair, pp. 31–61. Cambridge: Cambridge Univ. Press.

Posner, M. I., and A. Pavese. 1998. Anatomy of word and sentence. *PNAS* **95(3)**:899–905.

Power, C. 1998. Old wives' tales: The gossip hypothesis and the reliability of cheap signals. In: Approaches to the Evolution of Language, ed. J. R. Hurford, M. Studdert-Kenedy, and C. Knight, pp. 111–129. Cambridge: Cambridge Univ. Press.

Prather, J. F., S. Peters, S. Nowicki, and R. Mooney. 2008. Precise auditory-vocal mirroring in neurons for learned vocal communication. *Nature* **451**:305–310.

Premack, D. 2004. Is language the key to human intelligence? *Science* **303**:318–320.

Premack. D., and G. Woodruff. 1978. Does the chimpanzee have a theory of mind? *Behav. Brain Sci.* **4**:515–526.

Preuss, T. M. 2000. Taking the measure of diversity: Comparative alternatives to the model-animal paradigm in cortical neuroscience. *Brain Behav. Evol.* **55**:287–299.

Price, C. J. 2000. The anatomy of language: Contributions from functional neuroimaging. *J. Anatomy* **197**:335–359.

Price, C. J., C. J. Moore, G. W. Humphreys, and R. J. S. Wise. 1997. Segregating semantic from phonological processes during reading. *J. Cogn. Neurosci.* **9(6)**:727–733.

Price, C. J., C. J. Mummery, C. J. Moore, R. S. Frakowiak, and K. J. Friston. 1999. Delineating necessary and sufficient neural systems with functional imaging studies of neuropsychological patients. *J. Cogn. Neurosci.* **11**:371–382.

Puglisi, A., A. Baronchelli, and V. Loreto. 2008. Cultural route to the emergence of linguistic categories. *PNAS* **105**:7936.

Pujol, J., C. Soriano-Mas, H. Ortiz et al. 2006. Myelination of language-related areas in the developing brain. *Neurology* **66**:339–343.

Pullum, G., and B. Scholz. 2002. Empirical assessment of stimulus poverty arguments. *Ling. Rev.* **19(1-2)**:1–50.

Pulvermüller, F. 2002. A brain perspective on brain mechanisms: From discrete neural assemblies to serial order. *Prog. Neurobiol.* **67**:85–111.

——. 2003. The Neuroscience of Language. Cambridge: Cambridge Univ. Press.

Pulvermüller, F., and R. Assadollahi. 2007. Grammar or serial order? Discrete combinatorial brain mechanisms reflected by the syntactic Mismatch Negativity. *J. Neurosci.* **19**:971–980.

Pulvermüller, F., and Y. Shtyrov. 2003. Automatic processing of grammar in the human brain as revealed by the Mismatch Negativity. *NeuroImage* **20**:159–172.

Pulvermüller, F., Y. Shtyrov, and R. P. Carlyon. 2008. Syntax as reflex: Neurophysiological evidence for early automaticity of grammatical processing. *Brain Lang.* **104(3)**:244–253.

Quian Quiroga, R., R. Mukamel, E. A. Isham, R. Malach, and I. Fried. 2008. Human single-neuron responses at the threshold of conscious recognition. *PNAS* **105(9)**:3599–3604.

Quine, W. 1960. Word and Object. Cambridge, MA: MIT Press.

Raber, J., D. Wong, G. Yu et al. 2000. Apolipoprotein E and cognitive performance. *Nature* **404**:352–354.

Rademacher, J., A. M. Galaburda, D. N. Kennedy, P. A. Filipek, and V. S. Caviness Jr. 1992. Human cerebral cortex: Localization, parcellation, and morphometry with magnetic resonance imaging. *J. Cogn. Neurosci.* **4**:352–374.

Raichle, M. E. 1998. The neural correlates of consciousness: An analysis of cognitive skill learning. *Phil. Trans. Roy. Soc. Lond. B* **29**:1889–1901.

Raichle, M. E., J. A. Fietz, T. O. Videen et al. 1994. Practice-related changes in human brain functional anatomy during non-motor learning. *Cereb. Cortex* **4**:8–26.

Rapoport, S. I. 1990. How did the human brain evolve? A proposal based on new evidence from *in vivo* brain imaging during attention and ideation. *Brain Res. Bull.* **50**:149–165.

Reali, F., and M. H. Christiansen. 2009. Sequential learning and the interaction between biological and linguistic adaptation in language evolution. *Interaction Studies* **10**:5–30.

Reali, F., M. H. Christiansen, and P. Monaghan. 2003. Phonological and distributional cues in syntax acquisition: Scaling up the connectionist approach to multiple-cue integration. In: Proc. 25th Annual Conf. of the Cognitive Science Society, pp. 970–975. Mahwah, NJ: Lawrence Erlbaum.

Recanzone, G. H., and M. M. Merzenich. 1988. Intracortical microstimulation in somatosensory cortex in adult rats and owl monkeys results in a large expansion of the cortical zone of representation of a specific cortical receptive field. *Soc. Neurosci. Abstr.* **14**:223.

Reinhart, T. 2006. Interface Strategies. Cambridge, MA: MIT Press.

Renfrew, C. 1996. The sapient behavior paradox: How to test for potential? In: Modelling Early Human Mind, ed. P. Mellars and K. Gibson, pp. 11–14. Cambridge: McDonald Institute Research Monographs.

Rice, M. L. 2000. Grammatical symptoms of specific language impairment. In: Speech and Language Impairments in Children: Causes, Characteristics, Intervention and Outcome, ed. D. V. M. Bishop and L. B. Leonard, pp. 17–34. Hove: Psychology Press.

Rice, M. L., K. Wexler, and S. M. Redmond. 1999. Grammaticality judgments of an extended optional infinitive grammar: Evidence from English-speaking children with specific language impairment. *J. Speech Lang. Hearing Res.* **42**:943–961.

Rilling, J. K., M. F. Glasser, T. M. Preuss et al. 2008. The evolution of the arcuate fasciculus revealed with comparative DTI. *Nature Neurosci.* **11**:426–428.

Rissanen, J. 1989. Stochastic Complexity in Statistical Inquiry. Singapore: World Scientific.

Ristau, C. A., and D. Robbins. 1982. Language in the great apes: A critical review. *Adv. Study Behav.* **12**:141–255.

Rizzi, L. 1982. Issues in Italian Syntax. Dordrecht: Foris Publications.

——. 1990. Relativized Minimality. Cambridge, MA: MIT Press.

——. 1997. The fine structure of the left periphery. In: Elements of Grammar, ed. L. Haegeman, pp. 281–337. Dordrecht: Kluwer.

——. 2004a. Locality and left periphery. In: Structures and Beyond. The Cartography of Syntactic Structures, ed. A. Belletti, pp. 223–252. Oxford: Oxford Univ. Press.

——. 2004b. The Structure of CP and IP: The Cartography of Syntactic Structures, vol. 3. Oxford: Oxford Univ. Press.

Rizzolatti, G., and M. A. Arbib. 1998. Language within our grasp. *Trends Cogn. Sci.* **21**:188–194.

Roberts, I. 1993. Verbs and Diachronic Syntax. Dordrecht: Kluwer.

——. 2005. Principles and Parameters in a VSO Language: A Case Study in Welsh. New York: Oxford Univ. Press.

Rodriguez, J., J. Wiles, and J. L. Elman. 1999. A recurrent neural network that learns to count. *Connect. Sci.* **11**:5–40.

Roeder, B., O. Stock, H. Neville, S. Bien, and F. Rösler. 2002. Brain activation modulated by the comprehension of normal and pseudo-word sentences of different processing demands: A functional magnetic resonance imaging study. *NeuroImage* **15**:1003–1014.

Rogers, L. J. 2008. Development and function of lateralization in the avian brain. *Brain Res. Bull.* **76**:235–244.

Rohrbacher, B. 1999. Morphology-Driven Syntax: A Theory of V to I Raising and Pro-Drop. Amsterdam: Benjamins.

Roland, P. E. 1993. Brain Activation. New York: Wiley-Liss.

Roland, P. E., S. Geyer, K. Amunts et al. 1997. Cytoarchitectural maps of the human brain in standard anatomical space. *Hum. Brain Map.* **5**:222–227.

Rouveret, A. 1994. Syntaxe du gallois: Principes généraux et typologie. Paris: CNRS Éditions.

Ruchkin, D. S., R. Johnson, H. Canoune, and W. Ritter. 1990. Short-term memory storage and retention: An event-related brain potential study. *Electro. Clin. Neurophys.***76**:419–439.

Rüschemeyer, S. A., C. J. Fiebach, V. Kempe, and A. D. Friederici. 2005. Processing lexical semantic and syntactic information in first and second language: fMRI evidence from German and Russian. *Hum. Brain Map.* **25**:266–286.

Russell, S. J., and P. Norvig. 2003. Artificial Intelligence: A Modern Approach. Upper Saddle River, NJ: Prentice Hall.

Rutherford, S. L., and S. Lindquist. 1998. Hsp90 as a capacitor for morphological evolution. *Nature* **396**:336–342.

Sag, I., T. Wasow, and E. Bender. 2006. Syntactic Theory. A Formal Introduction. Chicago: Univ. of Chicago Press.

Sakata, J. T., and M. S. Brainard. 2006. Real-time contributions of auditory feedback to avian vocal motor control. *J. Neurosci.* **26**:9619–9628.

Sakata, J. T., C. M. Hampton, and M. S. Brainard. 2008. Social modulation of sequence and syllable variability in adult birdsong. *J. Neurophys.* **99**:1700–1711.

Sampson, G. 1989. Language acquisition: Growth or learning? *Phil. Papers* **XVIII.3**:203–240.

——. 1999. Educating Eve: The Language Instinct Debate. London: Continuum Intl.

Sanides, F. 1962. Die Architektonik des menschlichen Stirnhirns. Berlin: Springer-Verlag.
Santi, A., and Y. Grodzinsky. 2007a. Taxing working memory with syntax: Bihemispheric modulations. *Hum. Brain Map.* **28**:1089–1097.
——. 2007b. Working memory and syntax interact in Broca's area. *NeuroImage* **37**:8–17.
Sasaki, A., T. D. Sotnikova, R. R. Gainetdinov, and E. D. Jarvis. 2006. Social context-dependent singing-regulated dopamine. *J. Neurosci.* **26**:9010–9014.
Saussure, F. 1916/1985. Cours de linguistique générale. Paris: Payot.
Savage-Rumbaugh, E. S., J. Murphy, R. A. Sevcik et al. 1993. Language comprehension in ape and child. *Monogr. Soc. Res. Child Devel.* **58**:1–242.
Scharff, C., and S. Haesler. 2005. An evolutionary perspective on FOXP2: Strictly for the birds? *Curr. Op. Neurobiol.* **15**:694–703.
Scharff, C., and F. Nottebohm. 1991. A comparative study of the behavioral deficits following lesions of various parts of the zebra finch song system: Implications for vocal learning. *J. Neurosci.* **11**:2896–2913.
Schlesewsky, M., and I. Bornkessel. 2004. On incremental interpretation: Degrees of meaning accessed during sentence comprehension. *Lingua* **114**:1213–1234.
Schmahmann, J. D., D. N. Pandya, R. Wang et al. 2007. Associate fibre pathways of the brain: Parallel observations from diffusion spectrum imaging and autoradiography. *Brain* **130(3)**:602–605.
Schmitt, A. M., J. Shi, A. M. Wolf et al. 2006. Wnt-Ryk signalling mediates medial-lateral retinotectal topographic mapping. *Nature* **439**:31–37.
Schneider, W., and J. M. Chien. 2003. Controlled and automatic processing: Behavior, theory, and biological mechanisms. *Cogn. Sci.* **27**:525–559.
Schubotz, R. I., and D. Y. von Cramon. 2002. A blueprint for target motion: fMRI reveals perceived sequential complexity to modulate premotor cortex. *NeuroImage* **16**:920–935.
Schumann, J. 1978. The Pidginization Process: A Model for Second Language Acquisition. Rowley, MA: Newbury House.
Scollon, R. 1976. Conversations with a One-Year Old Child. Honolulu: Univ. of Hawaii Press.
Searle, J. 1970. Speech Acts. Cambridge: Cambridge Univ. Press.
Senghas, A., S. Kita, and A. Özyürek. 2004. Children creating properties of language: Evidence from an emerging sign language in Nicaragua. *Science* **305**:1779–1782.
Sereno, M. 2005. Language origins without the semantic urge. *Cogn. Sci. Online* **3**:1–12.
Seyfarth, R. M., and D. L. Cheney. 2003. Signalers and receivers in animal communication. *Ann. Rev. Psychol.* **54**:145–173.
Seyfarth, R. M., D. L. Cheney, and P. Marler. 1980. Monkey responses to three different alarm calls: Evidence of predator classification and semantic communication. *Science* **210**:801–803.
Shastri, L. 1999. Advances in Shruti: A neurally motivated model of relational knowledge representation and rapid inference using temporal synchrony. *Artif. Intell.* **11**:79–108.
Shu, W., J. Y. Cho, Y. Jiang et al. 2005. Altered ultrasonic vocalization in mice with a disruption in the FOXP2 gene. *PNAS* **102**:9643–9648.
Sims, K. 1994. Evolving virtual creatures. Siggraph '94 Proc. *Computer Graphics* **1994**:15–22.

Singer, W. 1999. Neuronal synchrony: A versatile code for the definition of relations? *Neuron* **24**:49–65, 111–125.

Siskind, J. M. 2001. Grounding the Lexical Semantics of Verbs in Visual Perception using Force Dynamics and Event Logic. *J. AI Res.* **15**:31–90.

SLI Consortium. 2002. A genomewide scan identifies two novel loci involved in specific language impairment. *Am. J. Hum. Gen.* **70**:384–398.

——. 2004. Highly significant linkage to SLI1 locus in an expanded sample of individuals affected by Specific Language Impairment (SLI). *Am. J. Hum. Gen.* **94**:1225–1238.

Slobin, D. 2002. Language evolution, acquisition and diachrony: Probing the parallels. In: The Evolution of Language out of Pre-Language, ed. T. Givón and B. F. Malle. Amsterdam: John Benjamins.

Slobin, D., and T. G. Bever. 1982. Children use canonical sentence schemas: A cross-linguistic study of word order and inflections. *Cognition* **12**:229–265.

Smith, K., and S. Kirby. 2008. Cultural evolution: Implications for understanding the human language faculty and its evolution. *Phil. Trans. Roy. Soc. Lond. B* **363**:3591–3603.

Soha, J. A., and P. Marler. 2001. Vocal syntax development in the white-crowned sparrow (*Zonotrichia leucophrys*). *J. Comp. Psychol.* **115**:172–180.

Sohrabji, F., E. J. Nordeen, and K. W. Nordeen. 1990. Selective impairment of song learning following lesions of a forebrain nucleus in the juvenile zebra finch. *Behav. Neural Biol.* **53**:51–63.

Soltoggio, A., P. Dürr, C. Mattiussi, and D. Floreano. 2007. Evolving neuromodulatory topologies for reinforcement-like problems. In: Congress on Evolutionary Computation, pp. 2471–2478. IEEE Press.

Soma, M., M. Hiraiwa-Hasegawa, and K. Okanoya. 2009. Early ontogenetic effects on song quality in the Bengalese finches (*Lonchura striata var. Domestica*): Laying order, sibling competition, and song syntax. *Behav. Ecol. Sociobiol.* **63**:363–370.

Soressi, M., and F. d'Errico. 2007. Pigments, gravures, parures: Les comportements symboliques controversés des Néandertaliens. In: Les Néandertaliens. Biologie et cultures, ed. B. Vandermeersch and B. Maureille, pp. 297–309. Paris: Éditions du CTHS.

Sowell, E. R., B. S. Peterson, P. M. Thompson et al. 2003. Mapping cortical change across the human life span. *Nature Neurosci.* **6**:309–315.

Spiteri, E., G. Konopka, G. Coppola et al. 2007. Identification of the transcriptional targets of FOXP2, a gene linked to speech and language, in developing human brain. *Am. J. Hum. Gen.* **81**:1144–1157.

Spitzer, M. 1999. The Mind within the Net: Models of Learning, Thinking and Acting, Cambridge, MA: MIT Press.

Squire, L. R. 1987. Memory and Brain. Oxford: Oxford Univ. Press.

Staddon, J. 1988. Learning as inference. In: Evolution and Learning, ed. R. Bolles and M. Beecher. Hillside, NJ: Lawrence Erlbaum.

Staudt, M., I. Krageloh-Mann, and W. Grodd. 2000. Normal myelination of the child brain on MRI: A meta-analysis. *Fortsch. Geb. Rontgen. Neuen Bildgeb. Verfahr.* **172**:802–811.

Steedman, M. 2000. The Syntactic Process. Cambridge, MA: MIT Press.

Steels, L. 1995. A self-organizing spatial vocabulary. *Artif. Life* **2(3)**:319–332.

——. 1996. Perceptually grounded meaning creation. In: Proc. 2nd Intl. Conf. on Multi-Agent Systems, ICMAS-2, ed. M. Tokoro, pp. 338–344. Menlo Park: AAAI Press.

——. 2000. Language as a complex adaptive system. Lecture Notes in Computer Science. Parallel Problem Solving from Nature, PPSN-VI, ed. M. Schoenauer et al. Berlin: Springer-Verlag.

——. 2003. Evolving grounded communication for robots. *Trends Cogn. Sci.* **7(7)**:308–312.

——. 2004. Constructivist development of grounded construction grammars. In: Proc. 42nd Annual Meeting on Association for Computational Linguistics, ed. D. Scott, W. Daelemans, and M. Walker, pp. 9–19. Morristown: Association for Computational Linguistics.

——. 2005a. The emergence and evolution of linguistic structure: From lexical to grammatical communication systems. *Connect. Sci.* **17(3-4)**:213–230.

——. 2005b. What triggers the emergence of grammar? In: Proc. AISB 2005 Conf., ed. A. Cangelosi and C. L. Nehaniv. Hatfield, UK: Univ. of Herfordshire.

——. 2007. The recruitment theory of language origins. In: Emergence of Language and Communication, ed. C. Lyon, C. L. Nehaniv, and A. Cangelosi, pp. 129–151. Berlin: Springer-Verlag.

Steels, L., and J.-C. Baillie. 2003. Shared grounding of event descriptions by autonomous robots. *Robotics & Autonomous Systems* **43(2-3)**:163–173.

Steels, L., and T. Belpaeme. 2005. Coordinating perceptually grounded categories through language. A case study for colour. *Behav. Brain Sci.* **24(6)**:469–489.

Steels, L., and J. de Beule. 2006. Unify and Merge in fluid construction grammar. In: Symbol Grounding and Beyond, ed. P. Vogt et al., pp. 197–223. Berlin: Springer-Verlag.

Steels, L., J. de Beule, and N. Neubauer. 2005. Linking in Fluid Construction Grammar. *Trans. Belgian Royal Soc. Sci. Arts* October edition.

Steels, L., and F. Kaplan. 1998. Spontaneous Lexicon Change. Proc. COLING-ACL 1998, pp. 1243–1249. Montreal: ACL.

Steels, L., F. Kaplan, A. McIntyre, and J. Van Looveren. 2002. Crucial factors in the origins of word-meaning. In: The Transition to Language, ed. A. Wray. Oxford: Oxford Univ. Press.

Steels, L., and M. Loetzsch. 2008. Perspective alignment in spatial language. In: Spatial Language in Dialogue, ed. K. Coventry, J. Bateman, and T. Tenbrink. Oxford: Oxford Univ. Press.

Steels, L., and A. McIntyre. 1999. Spatially distributed naming games. In: Advances in Complex Systems, vol. 1, nr. 4, pp. 301–323. Paris: Hermes Science Publications.

Steels, L., and M. Spranger. 2008. The robot in the mirror. *Connect. Sci.* **20(4)**:337–358.

Steels, L., and E. Szathmáry. 2008. The replicator dynamics of language processing. In: The Evolution of Language: Proc. 7th Intl. Conf. (EVOLANG7), ed. A. D. M. Smith, K. Smith, and R. Ferrer i Cancho, pp. 503–504. Barcelona: World Scientific Press.

Steels, L., R. Van Trijp, and P. Wellens. 2007. Multi-level selection in the emergence of language systematicity. In: ECAL0, ed. F. Almeida e Costa et al., pp. 425–434. Berlin: Springer-Verlag.

Steels, L., and P. Vogt. 1997. Grounding adaptive language games in robotic agents. In: Proc. 4th European Conf. on Artificial Life, ed. I. Harvey and P. Husbands, pp. 474–482. Cambridge, MA: MIT Press.

Steels, L., and P. Wellens. 2006. How grammar emerges to dampen combinatorial search in parsing. In: Symbol Grounding and Beyond, ed. P. Vogt et al., pp. 76–88. Berlin: Springer-Verlag.

Steklis, H. B., S. R. Harnard, and J. Lancaster, eds. 1976. Origins and evolution of language and speech. *Ann. NY Acad. Sci.* **280**.

Stephan, D. A. 2008. Unraveling autism. *Am. J. Hum. Gen.* **82**:7–9.

Stoeger-Horwath, A. S., S. Stoeger, H. M. Schwammer, and H. Kratochvil. 2007. Call repertoire of infant African elephants: First insights into the early vocal ontogeny. *J. Acoustic Soc. Am.* **121**:3922–3931.

Stout, D., and T. Chaminade. 2007. The evolutionary neuroscience of tool making. *Neuropsychologia* **45(5)**:1091–1100.

Stout, D., N. Toth, K. Schick, and T. Chaminade. 2008. Neural correlates of Early Stone Age toolmaking: Technology, language and cognition in human evolution. *Phil. Trans. Roy. Soc. Lond. B* **363**:1939–1949.

Stout, D., N. Toth, K. Schick, J. Stout, and G. Hutchins. 2000. Stone tool-making and brain activation: Position Emission Tomography (PET) studies. *J. Archaeol. Sci.* **27(12)**:1215–1223.

Stowe, L. A., M. Haverkort, and F. Zwarts. 2005. Rethinking the neurological basis of language. *Lingua* **115**:997–1042.

Stromswold, K. 2001. The heritability of language: A review and meta-analysis of twin, adoption, and linkage studies. *Language* **77**:647–723.

Stromswold, K., D. Caplan, N. Alpert, and S. Rauch. 1996. Localization of syntactic comprehension by positron emission tomography. *Brain Lang.* **52**:452–473.

Sun, G. Z., C. L. Giles, H. H. Chen, and Y. C. Lee. 1998. The neural network pushdown automaton: Architecture, dynamics and training. In: Adaptive Processing of Sequences and Data Structures, ed. C. L. Giles and M. Gori, pp. 296–345. Berlin: Springer-Verlag.

Sur, M., and C. A. Learney. 2001. Development and plasticity of cortical areas and networks. *Nature Rev. Neurosci.* **2**:251–262.

Suzuki, R., J. R. Buck, and P. L. Tyack. 2006. Information entropy of humpback whale songs. *J. Acoustic Soc. Am.* **119**:1849–1866.

Swinney, D. A. 1979. Lexical access during sentence comprehension: (Re)consideration of context effects. *J. Verbal Learn. Verbal Behav.* **18**:645–659.

Swinney, D., and E. Zurif. 1995. Syntactic processing in aphasia. *Brain Lang.* **50**:225–239.

Számadó, S. 1999. The validity of the handicap principle in discrete action-response games. *J. Theor. Biol.* **198**:593–602.

Szathmáry, E. 2000. The evolution of replicators. *P. Roy. Soc. Lond. B* **355(1403)**:1669–1676.

——. 2001. Origin of the human language faculty: The language amoeba hypothesis. In: New Essays on the Origin of Language, ed. J. Trabant and S. Ward, pp. 41–51. New York: Mouton/de Gruyter.

——. 2003. Cultural processes: The latest major transition in evolution. In: Encyclopedia of Cognitive Science, ed. L. Nadel. London: Macmillan.

——. 2007. Towards an understanding of language origins. In: Codes of Life, ed. M. Barbieri, pp. 283–313. Springer-Verlag.

——. 2008. Towards an understanding of language origins. In. Codes of Life, ed. M. Barbieri, pp. 283–313. New York: Springer-Verlag.

Szathmáry, E., and S. Számadó. 2008a. Being human. Language: A social history of words. *Nature* **456**:40–41.

——. 2008b. A social history of words. *Nature* **456**:2–3.

Szathmáry, E., Z. Szatmáry, P. Ittzés et al. 2007. *In silico* evolutionary developmental neurobiology and the origin of natural language. In: Emergence of Communication and Language, ed. C. Lyon, C. Nehaniv, and A. Cangelosi, pp. 151–187. London: Springer-Verlag.

Taglialatela, J. P., J. L. Russell, J. A. Schaeffer, and W. D. Hopkins. 2008. Communicative signaling activates Broca's homolog in chimpanzees. *Curr. Biol.* **18**:343–348.

Takahashi, K., F. C. Liu, K. Hirokawa, and H. Takahashi. 2003. Expression of Foxp2, a gene involved in speech and language, in the developing and adult striatum. *J. Neurosci. Res.* **73**:61–72.

Takahashi, T. 1989. Neuromechanisms of Sound Localization in the Barn Owl. Inst. of Cognitive and Decision Sciences, Univ. of Oregon.

Tallal, P. 2000. Experimental studies of language learning impairments: From research to remediation. In: Speech and Language Impairments in Children: Causes, Characteristics, Intervention and Outcome, ed. D. V. M. Bishop and L. B. Leonard, pp. 131–155. Hove: Psychology Press.

Tallerman, M. 2006. The syntax of Welsh "direct object mutation" revisited. *Lingua* **116**:1750–1776.

Talmy, L. 2000. Towards a Cognitive Semantics. Cambridge, MA: MIT Press.

Tanenhaus, M. K., M. J. Spivey-Knowlton, K. M. Eberhard, and J. C. Sedivy. 1995. Integration of visual and linguistic information in spoken language comprehension. *Science* **268**:1632–1634.

Tattersall, I., and J. H. Schwartz. 2001. Extinct Humans. Boulder: Westview Press.

Tchernichovski, O., P. P. Mitra, T. Lints, and F. Nottebohm. 2001. Dynamics of the vocal imitation process: How a zebra finch learns its song. *Science* **V291**:2564–2569.

Tchernichovski, O., F. Nottebohm, C. E. Ho, B. Pesaran, and P. P. Mitra. 2000. A procedure for an automated measurement of song similarity. *Anim. Behav.* **59**:1167–1176.

Teramitsu, I., L. C. Kudo, S. E. London, D. H. Geschwind, and S. A. White. 2004. Parallel FOXP1 and FOXP2 expression in songbird and human brain predicts functional interaction. *J. Neurosci.* **24**:3152–3163.

Teramitsu, I., and S. A. White. 2006. FoxP2 regulation during undirected singing in adult songbirds. *J. Neurosci.* **26**:7390–7394.

——. 2008. Motor learning: The FoxP2 puzzle piece. *Curr. Biol.* **18**:R335–337.

Terrace, H. S., L. A. Petitto, R. J. Sanders, and T. G. Bever. 1979. Can an ape create a sentence? *Science* **200**:891–902.

Teter, B., P. Xu, J. R. Gilbert et al. 2002. Defective neuronal sprouting by human apolipoprotein E4 is a gain-of-negative function. *J. Neurosci. Res.* **68**:331–336.

Tettamanti, M., H. Alkadhi, A. Moro et al. 2002. Neural correlates for the acquisition of natural language syntax. *NeuroImage* **17**:700–709.

Tettamanti, M., and D. Weniger. 2006. Broca's area: A supramodal hierarchical processor? *Cortex* **42**:491–494.

Thoenissen, D., K. Zilles, and I. Toni. 2002. Differential involvement of parietal and precentral regions in movement preparation and motor intention. *J. Neurosci.* **22**:9024–9034.

Thompson-Schill, S. L., M. D'Esposito, and E. P. Kan. 1999. Effects of repetition and competition on activity in left prefrontal cortex during word generation. *Neuron* **23**:513–522.

Thornton, R., and K. Wexler. 1999. Principle B, VP Ellipsis, and Interpretation in Child Grammar. Cambridge MA: MIT Press.

Thorpe, S., D. Fize, and C. Marlot. 1996. Speed of processing in the human visual system. *Nature* **381**:520–522.

Toga, A. W., P. M. Thompson, and E. R. Sowell. 2006. Mapping brain maturation. *Trends Neurosci.* **29**:148–159.

Toma, D. T., K. P. White, J. Hirsch, and R. J. Greenspan. 2002. Identification of genes involved in *Drosophila melanogaster* geotaxis: A complex behavioral trait. *Nature Genet.* **31**:349–353.

Tomaiuolo, F., J. D. MacDonald, Z. Caramanos et al. 1999. Morphology, morphometry, and probability mapping of the pars opercularis of the inferior frontal gyrus: An *in vivo* MRI analysis. *Eur. J. Neurosci.* **11**:3033–3046.

Tomasello, M. 1999. The Cultural Origins of Human Cognition. Cambridge, MA: Harvard Univ. Press.

——. 2000. Acquiring syntax is not what you think. In: Speech and Language Impairments in Children: Causes, Characteristics, Intervention and Outcome, ed. D. V. M. Bishop and L. B. Leonard, pp. 1–15. Hove: Psychology Press.

Tomasello, M., and J. Call. 1997. Primate Cognition. Oxford: Oxford Univ. Press.

Tomasello, M., and M. Carpenter. 2007. Shared intentionality. *Devel. Sci.* **10**:121–125.

Tomasello, M., M. Carpenter, J. Call, T. Behne, and H. Moll. 2005. Understanding and sharing intentions: The origins of cultural cognition. *Behav. Brain Sci.* **28**:675–735.

Tomblin, J. B., and J. Pandich. 1999. Lessons from children with specific language impairment. *Trends Cogn. Sci.* **3**:283–285.

Tramo, M. J., K. Baynes, and B. T. Volpe. 1988. Impaired syntactic comprehension and production in Broca's aphasia: CT lesion localization and recovery patterns. *Neurology* **38**:95–98.

Trapa, P. E., and M. A. Nowak. 2000. Nash equilibria for an evolutionary language game. *J. Math. Biol.* **41**:172–188.

Traugott, C., and B. Heine, eds. 1990. Approaches to Grammaticalization (2 vols.). Amsterdam: John Benjamins.

Treisman, A. M., and B. DeSchepper. 1996. Object tokens, attention and visual memory. In: Attention and Performance XVI: Information Integration in Perception and Communication, ed. T. Inui and J. McClelland. Cambridge, MA: MIT Press.

Triesch, J., C. Teuscher, G. Deak, and E. Carlson. 2006. Gaze following: Why (not) learn it? *Devel. Sci.* **9(2)**:125–147.

Troyer, T. W., and A. J. Doupe. 2000. An associational model of birdsong sensorimotor learning II. Temporal hierarchies and the learning of song sequence. *J. Neurophys.* **V84**:1224–1239.

Tschauner, H. 1996. Middle-range theory, behavioral archaeology, and post-empiricist philosophy of science in archaeology. *J. Archaeol. Meth. Theory* **3**:3–30.

Tucker, D. M. 1991. Developing emotions and cortical networks. In: Developmental Neuroscience, Minnesota Symposium on Child Psychology, ed. M. Gunnar and C. Nelson, vol. 24. Hillsdale, NJ: Erlbaum.

Uddén, J., V. Folia, C. Forkstam et al. 2008. The inferior frontal cortex in artificial syntax processing: An rTMS study. *Brain Res.* **1224C**:69–78.

Ullman, M. T. 2004. Contributions of memory circuits to language: the declarative/procedural model. *Cognition* **92**:231–270.

Ullman, M. T., and E. I. Pierpont. 2005. Specific language impairment is not specific to language: The procedural deficit hypothesis. *Cortex* **41**:399–433.

Ulualp, S. O., B. B. Biswal, F. Z. Yetkin, and T. M. Kidder. 1998. Functional magnetic resonance imaging of auditory cortex in children. *Laryngoscope* **108**:1782–1786.

Ungerleider, L. A., and M. Mishkin. 1982. Two cortical visual systems. In: Analysis of Visual Behavior, ed. D. G. Ingle, M. A. Goodale, and R. J. Q. Mansfield. Cambridge, MA: MIT Press.

Uylings, H. B. M., L. I. Malofeeva, I. N. Bogolepova, K. Amunts, and K. Zilles. 1999. Broca's language area from a neuroanatomical and developmental perspective. In: Neurocognition of Language, ed. C. M. Brown and P. Hagoort, pp. 319–336. Oxford: Oxford Univ. Press.

Van Berkum, J. J. A., C. M. Brown, P. Zwitserlood, V. Kooijman, and P. Hagoort. 2005. Anticipating upcoming words in discourse: Evidence from ERPs and reading times. *J. Exp. Psychol. Learn. Mem. Cogn.* **31**:443–467.

Van den Brink, D., and P. Hagoort. 2004. The influence of semantic and syntactic context constraints on lexical selection and integration in spoken-word comprehension as revealed by ERPs. *J. Cogn. Neurosci.* **16**:1068–1084.

Van der Lely, H. J. K. 2005. Domain-specific cognitive systems: Insight from Grammatical-SLI. *Trends Cogn. Sci.* **9**:53–59.

Van der Lely, H. J. K., S. Rosen, and A. McClelland. 1998. Evidence for a grammar-specific deficit in children. *Curr. Biol.* **8**:1253–1258.

Vaneechoutte, M., and J. R. Skoyles. 1998. The memetic origin of language: Modern humans as musical primates. *J. Memetics Trans.* **2(2)**.

Vanhaeren, M., and F. d'Errico. 2006. Aurignacian ethno-linguistic geography of Europe revealed by personal ornaments. *J. Archaeol. Sci.* **33**:1105–1128.

Vanhaeren, M., F. d'Errico, C. Stringer et al. 2006. Middle Paleolithic shell beads in Israel and Algeria. *Science* **312**:1785–1788.

Van Hoek, K. 1997. Anaphora and Conceptual Structure. Chicago: Univ. Chicago Press.

Vanier, M., and D. Caplan. 1989. CT-Scan correlates of agrammatism. In: Agramatic Aphasia, ed. L. Menn and L. K. Obler. New York: J. Benjamin.

Van Kemenade, A. 1987. Syntactic Case and Morphological Case in the History of English. Dordrecht: Forist Publ.

Van Looveren, J. 1999. Multiple word naming games. In: Proc. 11th Belgium-Netherlands AI Conf. Maastricht: Universiteit Maastricht.

Van Petten, C. K., and B. J. Luka. 2006. Neural localization of semantic context effects in electromagnetic and hemodynamic studies. *Brain Lang.* **97**:279–293.

Van Trijp, R. 2008. The emergence of semantic roles in fluid construction grammar. In: The Evolution of Language: Proc. 7th Intl. Conf. (EVOLANG7), ed. A. D. M. Smith, K. Smith, and R. Ferrer i Cancho, pp. 346–353. Barcelona: World Scientific Press.

Vargha-Khadem, F., L. J. Carr, E. Isaacs et al. 1997. Onset of speech after left hemispherectomy in a nine-year-old boy. *Brain* **120**:159–182.

Vargha-Khadem, F., D. G. Gadian, A. Copp, and M. Mishkin. 2005. FOXP2 and the neuroanatomy of speech and language. *Nature Rev. Neurosci.* **6**:131–138.

Vargha-Khadem, F., K. E. Watkins, C. J. Price et al. 1998. Neural basis of an inherited speech and language disorder. *PNAS* **95**:12,695–12,700.

Vernes, S. C., D. F. Newbury, B. S. Abrahams et al. 2008. A functional genetic link between distinct developmental language disorders. *NEJM* **359**:2337–2345.

Vernes, S. C., J. Nicod, F. M. Elahi et al. 2006. Functional genetic analysis of mutations implicated in a human speech and language disorder. *Hum. Mol. Genet.* **15**:3154–3167.

Vernes, S. C., E. Spiteri, J. Nicod et al. 2007. High-throughput analysis of promoter occupancy reveals direct neural targets of FOXP2, a gene mutated in speech and language disorders. *Am. J. Hum. Gen.* **81**:1232–1250.

Vidnyánszky, Z., B. Gulyás, and P. E. Roland. 2000. Visual exploration of form and position with identical stimuli: Functional anatomy with PET. *Hum. Brain Map.* **11**:104–116.

Vigneau, M., V. Beaucousin, P. Y. Hervé et al. 2006. Meta-analyzing left hemisphere language areas: Phonology, semantics, and sentence processing. *NeuroImage* **30**:1414–1432.

Vikner, S. 1997. V to I and inflection for person in all tenses. In: The New Comparative Syntax, ed. L. Haegeman, pp. 189–213. London: Longman.

Villa, P., and F. d'Errico. 2001. Bone and ivory points in the Lower and Middle Paleolithic of Europe. *J. Hum. Evol.* **41**:69–112.

Vogt, C., and O. Vogt. 1919. Allgemeinere Ergebnisse unserer Hirnforschung. Vierte Mitteilung. Die physiologische Bedeutung der architektonischen Rindenfelderung auf Grund neuer Rindenreizungen. *J. Psychol. Neurol.* **25**:399–462.

Von Economo, C., and G. N. Koskinas. 1925. Die Cytoarchitektonik der Hirnrinde des erwachsenen Menschen. Berlin: Springer-Verlag.

Vosse, T., and G. A. M. Kempen. 2000. Syntactic structure assembly in human parsing: A computational model based on competitive inhibition and lexicalist grammar. *Cognition* **75**:105–143.

Vu, E. T., M. E. Mazurek, and Y. C. Kuo. 1994. Identification of a forebrain motor programming network for the learned song of zebra finches. *J. Neurosci.* **14**:6924–6934.

Vylder, B. D., and K. Tuyls. 2006. How to reach linguistic consensus: A proof of convergence for the Naming Game. *J. Theor. Biol.* **242(4)**:818–831.

Waddington, C. 1942. Canalization of development and the inheritance of acquired characters. *Nature* **150**:563–565.

——. 1975. The Evolution of an Evolutionist. Edinburgh: Edinburgh Univ. Press.

Wagner, K., J. A. Reggia, J. Uriagereka, and G. S. Wilkinson. 2003. Progress in the simulation of emergent communication and language. *Adapt. Behav.* **11(1)**:37–69.

Wanner, E., and L. Gleitman. 1982. Introduction. In: Language Acquisition: The State of the Art, ed. E. Wanner and L. Gleitman, pp. 3–48. Cambridge MA: MIT Press.

Wärneryd, K. 1993. Cheap talk, coordination and evolutionary stability. *Games Econ. Behav.* **5**:532–546.

Wartenburger, I., J. Steinbrink, S. Telkemeyer et al. 2007. The processing of prosody: Evidence of interhemispheric specialization at the age of four. *NeuroImage* **34**:416–425.

Washburn, S. L., and C. Lancaster. 1968. The evolution of hunting. In: Man the Hunter, ed. R. B. Lee and I. DeVore, pp. 293–303. Chicago: Aldine.

Waters, G., D. Caplan, N. Alpert, and L. Stanczak. 2003. Individual differences in rCBF correlates of syntactic processing in sentence comprehension: Effects of working memory and speed of processing. *NeuroImage* **19**:101–112.

Watkins, K. E., N. F. Dronkers, and F. Vargha-Khadem. 2002. Behavioural analysis of an inherited speech and language disorder: Comparison with acquired aphasia. *Brain* **125**:452–464.

Webb, D. M., and J. Zhang. 2005. FoxP2 in song-learning birds and vocal-learning mammals. *J. Hered.* **96**:212–216.

Wellens, P., M. Loetzsch, and L. Steels. 2008. Flexible word meaning in embodied agents. *Connect. Sci.* **20(2)**:173–191.

Wernicke, C. 1874. Der aphasische Symptomenkomplex. Breslau: Cohen und Weigert.

White, S. A., S. E. Fisher, D. H. Geschwind, C. Scharff, and T. E. Holy. 2006. Singing mice, songbirds, and more: Models for FOXP2 function and dysfunction in human speech and language. *J. Neurosci.* **26**:10,376–10,379.

Wild, J. M. 1993. Descending projections of the songbird nucleus robustus archistriatalis. *J. Comp. Neurol.* **338**:225–241.

Williams, H., and N. Mehta. 1999. Changes in adult zebra finch song require a forebrain nucleus that is not necessary for song production. *J. Neurobiol.* **39**:14–28.

Wilson, R. A., and F. C. Keil, eds. 1999. The MIT Encyclopedia of the Cognitive Sciences. Cambridge, MA: MIT Press.

Wischmann, S., and F. Pasemann. 2006. The emergence of communication by evolving dynamical systems. In: From Animals to Animats 9: Proc. 9th Intl. Conf. on Simulation of Adaptive Behaviour, ed. S. Nolfi et al., pp. 777–788. Heidelberg: Springer-Verlag.

Wolpert, L. 2003. Causal belief and the origins of technology. *Phil. Trans. Roy. Soc. A* **361**:1709–1719.

Woods, G. 2005. Is size everything? Controlling human brain size. Wellcome Trust News, issue 1, October.

Woods, G., J. Bond, and W. Enard. 2005. Autosomal recessive primary microcephaly (MCPH): A review of clinical, molecular, and evolutionary findings. *Am. J. Hum. Genet.* **76**:717–728.

Woolley, S. M., and E. W. Rubel. 1997. Bengalese finches Lonchura Striata domestica depend upon auditory feedback for the maintenance of adult song. *J. Neurosci.* **17(16)**:6380–6390.

Wyles, J. S., J. G. Kunkel, and A. C. Wilson. 1983. Birds, behavior, and anatomical evolution. *PNAS* **80**:4394–4397.

Yamauchi, H. 2001. The difficulty of the Baldwinian account of linguistic innateness. In: Advances in Artificial Life, ed. J. Keleman and P. Sosik. Heidelberg: Springer-Verlag.

Ye, Z., Y. J. Luo, A. D. Friederici, and X. L. Zhou. 2006. Semantic and syntactic processing in Chinese sentence comprehension: Evidence from event-related potentials. *Brain Res.* **1071**:186–196.

Yellen, J. E., A. S. Brooks, E. Cornelissen, J. Mehlman, and K. Stewart. 1995. A Middle Stone Age worked bone industry from Katanda, Upper Semliki Valley, Zaire. *Science* **268**:553–556.

Yu, A. C., and D. Margoliash. 1996. Temporal hierarchical control of singing in birds. *Science* **273**:1871–1875.

Yu, C., D. Ballard, and R. Aslin. 2005. The role of embodied intention in early lexical acquisition. *Cogn. Sci.* **29(6)**:961–1005.

Zacks, J., B. Rypma, J. D. E. Gabrieli, B. Tversky, and G. H. Glover. 1999. Imagined transformations of bodies: An fMRI investigation. *Neuropsychologia* **37**:1029–1040.

Zatorre, R. J., and P. Belin. 2001. Spectral and temporal processing in human auditory cortex. *Cereb. Cortex* **11**:946–953.

Zatorre, R. J., P. Belin, and V. B. Penhune. 2002. Structure and function of auditory cortex: Music and speech. *Trends Cogn. Sci.* **6**:37–46.

Zeki, S., J. D. Watson, C. J. Lueck et al. 1991. A direct demonstration of functional specialization in human visual cortex. *J. Neurosci.* **11**:641–649.

Zhang, B., and S. Horvath. 2005. A general framework for weighted gene co-expression network analysis. *Stat. Appl. Genet. Mol. Biol.* **4**:Article17.

Zhang, J., D. M. Webb, and O. Podlaha. 2002. Accelerated evolution and origins of human-specific features: FOXP2 as an example. *Genetics* **162**:1825–1835.

Zilhão, J. 2006. Neandertals and moderns mixed, and it matters. *Evol. Anthro.* **15**:183–195.

Zilles, K., and N. Palomero-Gallagher. 2001. Cyto-, myelo-, and receptor architectonics of the human parietal cortex. *NeuroImage* **14**:8–20.

Zilles, K., N. Palomero-Gallagher, C. Grefkes et al. 2002. Architectonics of the human cerebral cortex and transmitter receptor fingerprints: Reconciling functional neuroanatomy and neurochemistry. *Eur. Neuropsychopharm.* **12**:587–599.

Zilles, K., N. Palomero-Gallagher, and A. Schleicher. 2004. Transmitter receptors and functional anatomy of the cerebral cortex. *J. Anatomy* **205**:417–432.

Zilles, K., M. S. Qü, H. Schröder, and A. Schleicher. 1991. Neurotransmitter receptors and cortical architecture. *J. Hirnforsch.* **32**:343–356.

Zuberbühler, K. 2002. A syntactic rule in forest monkey communication. *Anim. Behav.* **63**:293–299.

Zuidema, W. 2002. Language adaptation helps language acquisition: A computational model study. In: From Animals to Animats 7: Proc. 7th Intl. Conf. on Simulation of Adaptive Behavior, ed. B. Hallam, D. Floreano, J. Hallam, G. Hayes, and J.-A. Meyer. Cambridge, MA: MIT Press.

——. 2003. How the poverty of the stimulus solves the poverty of the stimulus. In: Advances in Neural Information Processing Systems 15, ed. S. Becker, S. Thrun, and K. Obermayer. Cambridge, MA: MIT Press.

Zurif, E., D. Swinney, P. Prather, J. Solomon, and C. Bushell. 1993. An on-line analysis of syntactic processing in Broca's and Wernicke's Aphasia. *Brain Lang.* **45**:448–464.

Subject Index